DISTANCE AND MIDPOINT FORMULAS

The **distance** between $P(x_1, y_1)$ and $Q(x_2, y_2)$ is
$$PQ = \sqrt{(x_2 - x_1)^2 + (y_2 - y_1)^2}$$
and the coordinates of the **midpoint** of line segment $\overline{PQ}$ are
$$\left(\frac{x_1 + x_2}{2}, \frac{y_1 + y_2}{2}\right)$$

EQUATION OF A CIRCLE

The equation of a circle with center (h, k) and radius r is
$$(x - h)^2 + (y - k)^2 = r^2$$

LINES

1. The **slope** of a line through $P(x_1, y_1)$ and $Q(x_2, y_2)$ is
$$m = \frac{y_2 - y_1}{x_2 - x_1}$$

2. The **slope-intercept form** of a line with slope m and y-intercept b is
$$y = mx + b$$

3. The **point-slope form** of a line through $P(x_1, y_1)$ with slope m is
$$y - y_1 = m(x - x_1)$$

VERTEX FORMULA

The graph of $f(x) = ax^2 + bx + c$ is a *parabola*. The coordinates of its **vertex** are
$$\left(-\frac{b}{2a}, f\left(-\frac{b}{2a}\right)\right)$$

LOGARITHMS

1. $y = \log_b x$ is equivalent to $b^y = x$
2. $\log_b 1 = 0$
3. $\log_b b = 1$
4. $\log_b b^x = x$
5. $b^{\log_b x} = x$
6. $\log_b xy = \log_b x + \log_b y$
7. $\log_b \frac{x}{y} = \log_b x - \log_b y$
8. $\log_b x^n = n \log_b x$
9. $\log_b x = \frac{\log_a x}{\log_a b}$

GRAPHS O

1. **Constant**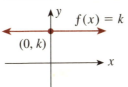

2. **Identity Function**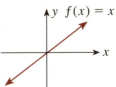

3. **Absolute Value Function** 4. **Squaring Function**

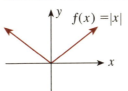 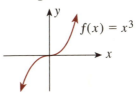

3. **Absolute Value Function**
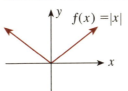

4. **Squaring Function**
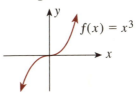

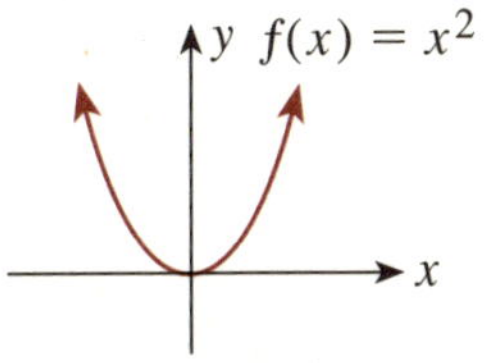

5. **Cubing Function**
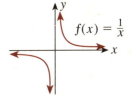

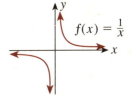

6. **Reciprocal Function**

7. **Square Root Function**
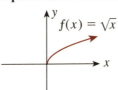

8. **Cube Root Function**
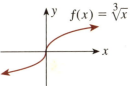

9. **Exponential Function**
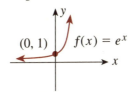

10. **Logarithmic Function**
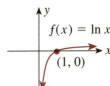

BINOMIAL THEOREM

$$(x + y)^n = \sum_{i=0}^{n} \binom{n}{i} x^{n-i} y^i$$

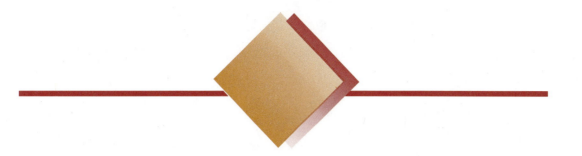

College Trigonometry

College Trigonometry

David E. Stevens
Wentworth Institute of Technology

WEST PUBLISHING COMPANY
Minneapolis/St. Paul ◆ New York
Los Angeles ◆ San Francisco

◆ PRODUCTION CREDITS

Copyediting: Katherine Townes/Tech*arts*
Design: Tech*arts*
Composition: The Clarinda Company
Interior artwork: Scientific Illustrators
Cover photograph: Graeme Outerbridge

◆ WEST'S COMMITMENT TO THE ENVIRONMENT

In 1906, West Publishing Company began recycling materials left over from the production of books. This began a tradition of efficient and responsible use of resources. Today, up to 95 percent of our legal books and 70 percent of our college and school texts are printed on recycled, acid-free stock. West also recycles nearly 22 million pounds of scrap paper annually—the equivalent of 181,717 trees. Since the 1960s, West has devised ways to capture and recycle waste inks, solvents, oils, and vapors created in the printing process. We also recycle plastics of all kinds, wood, glass, corrugated cardboard, and batteries, and have eliminated the use of styrofoam book packaging. We at West are proud of the longevity and the scope of our commitment to the environment.

◆ PRODUCTION, PRINTING AND BINDING BY WEST PUBLISHING COMPANY.

◆ PHOTO CREDITS FOLLOW THE INDEX.

◆ COPYRIGHT © 1994   BY WEST PUBLISHING COMAPNY
610 Opperman Drive
P.O. Box 64526
St. Paul, MN 55164-0526

◆ ALL RIGHTS RESERVED
PRINTED IN THE UNITED STATES OF AMERICA
01 00 99 98 97 96 95 94   8 7 6 5 4 3 2 1 0

◆ LIBRARY OF CONGRESS CATALOGING-IN-PUBLICATION DATA
STEVENS, DAVID E.
   College trigonometry / David E. Stevens.
      p.   cm.
   Includes index.
   ISBN 0-314-01214-1
   1. Trigonometry.   I. Title.
QA 531.S784 1994
516.24--dc20

93-51494
CIP

 TEXT IS PRINTED ON 10% POST CONSUMER RECYCLED PAPER

CONTENTS

Preface ix

1 An Introduction to Functions and Graphs — 1

- **1.1** Real Numbers and Interval Notation — 2
- **1.2** Working in the Cartesian Plane — 10
- **1.3** Functions — 23
- **1.4** Techniques of Graphing Functions — 35
- **1.5** Composite and Inverse Functions — 47
- **1.6** Applied Functions and Variation — 58

  Chapter 1 Review — 68
  - ◆ Questions for Group Discussion — 68
  - ◆ Review Exercises — 69

2 Trigonometric Functions — 73

- **2.1** Angles and Their Measures — 74
- **2.2** Trigonometric Functions of Angles — 85
- **2.3** Evaluating Trigonometric Functions of Angles and Real Numbers — 99
- **2.4** Properties and Graphs of the Sine and Cosine Functions — 114
- **2.5** Properties and Graphs of the Other Trigonometric Functions — 129
- **2.6** The Inverse Trigonometric Functions — 145

  Chapter 2 Review — 159
  - ◆ Questions for Group Discussion — 159
  - ◆ Review Exercises — 160

3 Applications of the Trigonometric Functions — 165

- **3.1** Applications Involving Right Triangles and Harmonic Motion — 166
- **3.2** Law of Sines — 178
- **3.3** Law of Cosines — 189
- **3.4** Vectors — 201

  Chapter 3 Review — 216
  - ◆ Questions for Group Discussion — 216
  - ◆ Review Exercises — 217

CUMULATIVE REVIEW EXERCISES FOR CHAPTERS 1, 2, AND 3 — 223

4 Analytic Trigonometry — 227

4.1	Algebraic Manipulations of Trigonometric Expressions	228
4.2	Trigonometric Equations	238
4.3	Sum and Difference Formulas	251
4.4	Multiple-Angle Formulas	264
4.5	Product-to-Sum Formulas and Sum-to-Product Formulas	275
	Chapter 4 Review	285
	◆ Questions for Group Discussion	285
	◆ Review Exercises	286

5 Complex Numbers — 289

5.1	The Complex Number System	290
5.2	Trigonometric Form of Complex Numbers	298
5.3	Powers and Roots of Complex Numbers	308
	Chapter 5 Review	318
	◆ Questions for Group Discussion	318
	◆ Review Exercises	319

6 Conic Sections — 321

6.1	The Parabola	322
6.2	The Ellipse	334
6.3	The Hyperbola and the Common Definition of the Conic Sections	346
6.4	Rotation of Axes	361
6.5	Polar Equations	373
6.6	Parametric Equations	389
	Chapter 6 Review	400
	◆ Questions for Group Discussion	400
	◆ Review Exercises	402

7 Exponential and Logarithmic Functions — 405

7.1	Properties and Graphs of Exponential Functions	406
7.2	Logarithmic Functions	417
7.3	Properties of Logarithms	429
7.4	Graphs of Logarithmic Functions	438
7.5	Logarithmic and Exponential Equations	446
	Chapter 7 Review	455
	◆ Questions for Group Discussion	455
	◆ Review Exercises	456

CUMULATIVE REVIEW EXERCISES FOR CHAPTERS 4, 5, 6, AND 7	**459**
APPENDIX A: SIGNIFICANT DIGITS	**A1**
APPENDIX B: ASYMPTOTES	**A4**
APPENDIX C: TABLES	**A6**
SOLUTIONS TO PROBLEMS AND ANSWERS TO ODD-NUMBERED EXERCISES	**A17**
INDEX	**I1**

PREFACE

Intent

A course in college trigonometry must meet the needs of students with diverse mathematical backgrounds and goals. Science and engineering students enroll in this course before beginning a traditional three-semester calculus sequence. Business and social science students enroll in this course before taking courses in finite mathematics, statistics, or introductory calculus. Other students enroll in this course because it is a prerequisite for elementary courses in astronomy, biology, physics, chemistry, computer science, architecture, or geology. In writing this text, I have tried to be sensitive to the needs of *all* these students by preparing them for more advanced courses and illustrating through real-life applied problems that knowledge of college trigonometry is fundamental to many disciplines.

Approach

For success in a college trigonometry course, it is essential that students become active rather than passive readers of the text. Therefore, I have written this text using an *interactive approach.* Each key mathematical concept is supported with a step-by-step text example with margin annotations and explanatory notes, and each text example is followed by a practice problem for the student to work out. The practice problem may ask the reader to check the preceding text example, work the preceding text example using an alternative approach, extend the preceding text example by asking for additional information, or try an entirely different problem that has similar mathematical steps. In effect, the practice problems require the student to become involved with the mathematics and constitute a built-in workbook for the student. *A complete detailed solution of each practice problem is given in the back of the text.*

Features

Written in a warm and user-friendly style, this is a traditional college trigonometry text with a contemporary flair in the sense that it starts to address the concerns of writing across the curriculum, group learning, critical thinking, and the use of modern technology in the math classroom. The following features distinguish this text from the existing texts in the market.

Applied Problems: To arouse student interest, each chapter opens with an applied problem and a related photograph. The solution to the applied problem occurs within the chapter after the necessary mathematics has been developed. Other applied problems from the fields of science, engineering, and business are introduced at every reasonable opportunity. They are clearly visible throughout the text and occur in separate subsections at the end of most sections. They are a highlight of this text and tend to show how college trigonometry relates to real-life situations.

Exercise Sets: The heart of any math textbook is its end-of-section exercise sets. It is here that students are given an opportunity to practice the mathematics that has been developed. The exercise sets in this text are broken into three parts: *Basic Skills*, *Critical Thinking*, and *Calculator Activities.*

Basic Skills: These exercises are routine in nature and tend to mimic the text examples that are worked-out in each section.

Critical Thinking: These exercises require the student to "think critically" and transcend the routine application of the basic skills to the next level of difficulty. Exercises in this group may require the student to draw upon skills developed in earlier chapters.

Calculator Activities: These exercises require the use of a calculator to solve basic-skill- and critical-thinking-type problems. Some of the exercises in this group ask the student to use a *graphing calculator* to solve problems in which standard algebraic methods do not apply. Problems that require the use of a graphing calculator are identified by the logo 📱.

Some of the exercise sets also contain problems that are *calculus-related.* Designed for students who are taking a college trigonometry course as a prerequisite to calculus, these exercises illustrate the algebraic support that is needed for simplifying derivatives, solving optimization problems, finding the area bounded by two or more curves, and so on. In calculus, the ratio of the change in the variable y to the change in the variable x is designated by $\Delta y/\Delta x$. In this text, we use the logo $\frac{\Delta y}{\Delta x}$ to identify the problems in the exercise sets that are calculus-related.

Chapter Reviews: To help students prepare for chapter exams, each chapter in this text concludes with an extensive chapter review. The chapter reviews are broken into two parts: *Questions for Group Discussion* and *Review Exercises.*

Questions for Group Discussion: In keeping with the interactive approach, these questions allow students to state in their own words what they have learned in the chapter. Since many of these questions have open-ended answers, they are ideally suited for class or group discussions and are extremely valuable to those who believe in cooperative or collaborative learning.

Review Exercises: These exercises reinforce the ideas that are discussed in the chapter and allow the instructor to indicate to the student the types of problems that may appear on a chapter test.

Cumulative Reviews: To help students pull together ideas from several chapters, cumulative review exercises are strategically placed after Chapters 3 and 7. The problems in these exercises are ungraded as far as difficulty and presented in a random order. Some problems are basic and similar to those already studied, while others are more challenging and require some creative thinking.

Pedagogy: Every effort has been made to make this a text from which students can learn and succeed. The following pedagogical features attest to this fact: *Caution notes*, flagged by the symbol ⚠, help eliminate misconceptions and bad mathematical habits by pointing out the most common errors that students make.

Introductory comments at the start of many sections introduce vocabulary and inform the reader of the purpose of the section.

Boxed definitions, formulas, laws, and properties state key mathematical ideas and provide the reader with quick and easy access to this information.

Step-by-step procedural boxes indicate the sequence of steps that a student can follow in order to simplify trigonometric expressions, solve certain types of equations and inequalities, sketch the graph of various functions, find the inverse of a function, and so on.

Development

Chapter 1 provides an extensive review of the basic algebraic topics that are needed in the study of trigonometry such as interval notation, the coordinate plane, graphing techniques, and the functional concept. Section 1.4 lists eight basic functions and their graphs (constant, identity, absolute value, squaring, cubing, reciprocal, square root, and cube root functions) and then applies the vertical and horizontal shift rules, the x- and y-axis reflection rules, and the vertical stretch and shrink rules to sketch the graphs of several other related functions. These eight basic functions and their graphs are then used to discuss composition of functions, inverse functions, applied functions, and variation. The graphing calculator is introduced in this chapter and some of the exercises suggest using this tool to verify results or to explore new ideas.

Chapters 2, 3, and 4 represent the core of a college trigonometry course. *Chapter 2* introduces the trigonometric functions using both angle domains and real number domains. After an introductory section on angles and their measure, Section 2.2 defines the six trigonometric ratios for any angle θ and develops the fundamental trigonometric identities from these definitions. This section also introduces the trigonometric ratios of right triangles and the cofunction relationships. Section 2.3 evaluates the trigonometric functions of angles and uses the unit circle to show that the trigonometric functions of a real number may be found by considering the real number as the radian measure of its corresponding central angle. The graphs of the sine and cosine functions (Section 2.4) are obtained by observing the changes in the x-coordinate and y-coordinate of a point $P(x,y)$ on the unit circle. These graphs are then used to help list the important properties of the sine and cosine functions and to help develop the graphs of the other trigonometric functions. The restricted sine, restricted cosine, and restricted tangent functions are defined in Section 2.6 and the inverse of these functions are then discussed.

Chapter 3 discusses several applications of the trigonometric functions. Section 3.1 shows how the inverse trigonometric functions, in conjunction with a calculator, may be used to help solve a right triangle. This section also includes a variety of applied problems involving angles of elevation, angles of depression, and simple harmonic motion. The method of solving an oblique triangle by using the law of sines and law of cosines is developed in the next two sections of this chapter. Area formulas for oblique triangles, including Hero's formula, are also developed and applied to several problems. Section 3.4 defines a vector, introduces the vector operations of addition, subtraction, and scalar multiplication, and applies vectors to problems involving forces and displacements.

Chapter 4 discusses analytic trigonometry—a branch of mathematics in which algebraic procedures are applied to trigonometry. Section 4.1 suggests a general scheme for verifying trigonometric identities and Section 4.2 states a general procedure for solving trigonometric equations. Some of the exercises in these sections ask the student to use a graphing calculator to verify a trigonometric identity or to solve a trigonometric equation. Several important trigonometric formulas are developed in the other sections of this chapter. They include the sum and difference formulas, multiple-angle formulas, product-to-sum formulas, and sum-to-product formulas. The chapter concludes with a summary of all the important trigonometric identities and formulas discussed in Chapters 2 and 4.

Chapters 5, 6, and 7 include several other topics of interest in a college trigonometry course. *Chapter 5* discusses operations with complex numbers. The rules for adding, subtracting, multiplying, and dividing complex numbers in standard form are discussed in the first section. Section 5.2 defines the trigonometric form of a complex number and develops the rules for multiplying and dividing complex numbers in trigonometric form. DeMoivre's theorem is developed in Section 5.3 and used to find powers and roots of complex numbers. The chapter concludes by finding all the roots of a polynomial equation of degree n and showing that these roots, when plotted in the complex plane, represent the vertices of a regular n-sided polygon whenever $n \geq 3$.

Chapter 6 discusses the conic sections using Cartesian, polar, and parametric equations. The first three sections of this chapter state the geometric properties of each conic section and the distinguishing characteristics of their equations. Section 6.4 uses the rotation formulas and the discriminant to help sketch the graph of general quadratic equations in two unknowns in which $B \neq 0$. The polar coordinate system is discussed in Section 6.5 and the common definition of the conic sections is used to develop the polar equations for the parabola, ellipse, and hyperbola. Parametric equations of the conic sections are developed in Section 6.6 and the parametric mode on a graphing calculator is used to generate the graph of some polar equations.

Chapter 7 discusses the properties of real exponents, defines the exponential function, and develops the logarithmic function as the inverse of the exponential function. By letting the number of compounding periods in the compound interest formula increase without bound, the reader is shown how the number e develops in a real-life situation. The properties of logarithms are used to help graph functions containing logarithmic expressions (Section 7.4) and also to help solve exponential and logarithmic equations (Section 7.5). In Section 7.5, the graphing calculator is used to help solve some exponential equations that are not solvable by ordinary algebraic methods.

Supplements

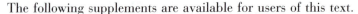

The following supplements are available for users of this text.

1. *Instructor's Solution Manual* by Eleanor Canter—includes complete worked-out solutions to all the even-numbered exercises.
2. *Student's Solution Manual* by Eleanor Canter—provides worked-out solutions for the odd-numbered exercises from the text.
3. *Instructor's Manual with Test Bank* by Cheryl Roberts—includes sample syllabi, suggested course schedule, chapter outlines with references to videos, homework assignments, chapter tests, and a test bank of multiple choice questions and open-ended problems.
4. *Graphing Calculator Lab Manual*

5. *WESTEST* 3.0—computer generated testing programs include algorithmically-generated questions and are available to qualified adopters. Macintosh and IBM-compatible versions are available.
6. *Video series*—"In simplest Terms" produced by Anneberg/CPB Collection. The videos are referenced in the Instructor's Manual accompanying *College Trigonometry*.
7. *Mathens Tutorial Software*—generates problems of varying degrees of difficulty and guides students through step-by-step solutions.

Acknowledgments

The chapters in this text were class-tested with over 1000 students at Wentworth Institute of Technology. I would like to thank these students and their professors Donald Filan, Michael John, Marcia Kemen, Anita Penta, and Charlene Solomon for their assistance throughout the development of this project. Special thanks go to my friend and colleague Eleanor Canter for her work in checking the answers and writing complete worked-out solutions to the more than 3000 exercises in this text. I also express my sincere thanks to the following reviewers. Their ideas were extremely helpful in shaping this text into its present form.

Haya Adner,
Queensborough Community College

Daniel D. Anderson,
University of Iowa

Thomas A. Atchison,
Stephen F. Austin State University

Jerald T. Ball,
Las Positas College

Kathleen Bavelas,
Manchester Community College

Louise M. Boyd,
Livingston University

Susan N. Boyer,
University of Maryland-Baltimore County

James R. Brasel,
Phillips County Community College

John E. Bruha,
University of Northern Iowa

Kathleen B. Burk,
Pensacola Junior College

John W. Coburn,
St. Louis Community College at Florissant Valley

Sally Copeland,
Johnson County Community College

Judith Covington,
University of Southwest Louisiana

Deborah A. Crocker,
Miami University

Terry Czerwinski,
University of Illinois-Chicago

Bettyann Daley,
University of Delaware

R. G. Dean,
Stephen F. Austin State University

Ryness A. Doherty,
Community College of Denver

Michael W. Ecker,
The Pennsylvania State University

Kathryn A. Engebus,
New Mexico State University

Paula C. Gnepp,
Cleveland State University

Stuart Goff,
Keene State College

Sarita Gupta,
Northern Illinois University

Louis Hoelzle,
Bucks County Community College

Ed Huffman,
Southwest Missouri State University

Nancy Hyde,
Broward Community College

Sylvia M. Kennedy,
Broome Community College

Margaret Kothmann,
University of Wisconsin-Stout

Keith Kuchar,
Northern Illinois University

Jeuel LaTorre,
Clemson University

James W. Lea, Jr.,
Middle Tennessee State University

Michael E. Mays,
West Virginia University

Myrna L. Mitchell,
Pima Community College

Maurice L. Monahan,
South Dakota State University

Jeri A. Nichols,
Bowling Green State University

Jean M. Prendergast,
Bridgewater State College

William Radulovich,
Florida Community College

James A. Reed, Sr.,
University of Hartford

Betty Rehfuss,
North Dakota State University

Cheryl V. Roberts,
Northern Virginia Community College

Stephen B. Rodi,
Austin Community College

Thomas Roe,
South Dakota State University

William B. Rundberg,
College of San Mateo

Ned W. Schillow,
Lehigh County Community College

George W. Schultz,
St. Petersburg Junior College

Richard D. Semmler,
Northern Virginia Community College

Lynne B. Small,
University of San Diego

C. Donald Smith,
Louisiana State University

John Spellman,
Southwest Texas State University

Charles Stone,
DeKalb College

Michael D. Taylor,
University of Southern Florida

Beverly Weatherwax,
Southwest Missouri State University

William H. White,
University of South Carolina-Spartanburg

Bruce Williamson,
University of Wisconsin-River Falls

Jim Wooland,
Florida State University

Marvin Zeman,
Southern Illinois University at Carbondale

The production of a textbook is a team effort between the editorial staff and the author. My editor, Ron Pullins, always offered the support and guidance that I needed to complete this project. Denise Bayko organized our reviewers' comments into a format that revealed where extra work was needed. Kathi Townes copyedited the manuscript and prodded me to provide additional information that would benefit the reader, and Tamborah Moore kept the project moving despite some unusual circumstances. I would like to thank each of you for the encouragement and enthusiasm you provided in the preparation of this book. I can't imagine working with a better team.

D.E. Stevens
Boston, Massachusetts, 1993

College Trigonometry

CHAPTER  1

An Introduction To Functions and Graphs

The real estate tax T on a property varies directly as its assessed value V.
(a) Express T as a function of V if $T = \$2800$ when $V = \$112{,}000$.
(b) State the domain of the function defined in part (a), then sketch its graph.

(For the solution, see Example 3 in Section 1.6.)

1.1   **Real Numbers and Interval Notation**
1.2   **Working in the Cartesian Plane**
1.3   **Functions**
1.4   **Techniques of Graphing Functions**
1.5   **Composite and Inverse Functions**
1.6   **Applied Functions and Variation**

1.1 Real Numbers and Interval Notation

◆ **Introductory Comments**

A chef requires 3 eggs and $\frac{2}{3}$ cup of sugar for a cake recipe. A meteorologist reports that the temperature is $-4\,°C$ and the barometric pressure is 29.35 inches. A student determines that the side of a right triangle is $\sqrt{5}$ units and the area of a circle is π square units. Each of the numbers

$$3,\quad \frac{2}{3},\quad -4,\quad 29.35,\quad \sqrt{5},\quad \text{and}\quad \pi$$

is an element of the **set of real numbers.*** These are the type of numbers that we work with every day. The set of real numbers has five important subsets.

1. **Natural numbers**
   or
   Positive integers: $\{1, 2, 3, 4, \ldots\}$
2. **Whole numbers:** $\{0, 1, 2, 3, \ldots\}$
3. **Integers:** $\{\ldots, -3, -2, -1, 0, 1, 2, 3, \ldots\}$
4. **Rational numbers:** {All real numbers of the form $\frac{p}{q}$, where p and q are integers and $q \neq 0$.}
   or
   {All decimal numbers that either terminate or repeat the same block of digits.}
   Examples: $\frac{3}{5},\ \frac{-8}{3},\ \frac{16}{1},\ \frac{0}{4},\ 0.75,\ -5.343434\ldots$
5. **Irrational numbers:** {All real numbers that are not rational.}
   or
   {All decimal numbers that neither terminate nor repeat the same block of digits.}
   Examples: $\sqrt{2},\ -\sqrt{3},\ \pi,\ \sqrt[3]{6},\ 3.050050005\ldots$

A geometric interpretation of the real numbers can be shown on the *real number line,* as illustrated in Figure 1.1. For each point on this line there corresponds exactly one real number, and for each real number there corresponds exactly one point on this line. This type of relationship is called a **one-to-one correspondence.**

The real number associated with a point on the real number line is called the

---

*The concept of a set is often used in mathematics. A *set* is a collection of objects, and these objects are called the *elements* of the set. A *subset* of a given set is formed by selecting particular elements of the set. Braces { } are used to enclose the elements of sets and subsets.

FIGURE 1.1
On the real number line there exists a *one-to-one correspondence* between the set of real numbers and the set of points on the line.

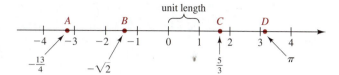

coordinate of the point. Referring to Figure 1.1, $-\frac{13}{4}$ is the coordinate of point A, $-\sqrt{2}$ is the coordinate of point B, $\frac{5}{3}$ is the coordinate of point C, and π is the coordinate of point D.

In this section, we discuss various sets of real numbers and the notation that is used to describe these sets. We begin by reviewing the inequality symbols that we use to compare two distinct real numbers.

◆ Inequality Symbols

The real number line gives us a convenient way to compare two distinct real numbers a and b. If a is to the *right* of b on the real number line, then **a is greater than b,** and we write

$$a > b$$

Referring to Figure 1.1, since π is to the right of $\frac{5}{3}$, we have $\pi > \frac{5}{3}$.

If a is to the left of b on the real number line, then **a is less than b,** and we write

$$a < b$$

Referring to Figure 1.1, since $\frac{5}{3}$ is to the left of π, we have $\frac{5}{3} < \pi$. In general, for real numbers a and b,

$$a < b \quad \text{if and only if} \quad b > a$$

Note: The phrase *if and only if* occurs frequently in mathematics. In the preceding statement, it implies two statements:

1. If $a < b$, then $b > a$ and, conversely,
2. If $b > a$, then $a < b$.

The symbols $>$ and $<$ are called **inequality symbols,** and expressions such as $a > b$ and $a < b$ are called **inequalities.** Two other inequality symbols used frequently are

$\leq$ read "less than or equal to"   and $\geq$ read "greater than or equal to."

Inequalities can be used to indicate if a number is *positive, negative, nonnegative,* or *nonpositive,* as shown in Table 1.1

Table 1.1
Some inequalities and their meanings

Inequality	Meaning
$a > 0$ or $0 < a$	a is positive
$a < 0$ or $0 > a$	a is negative
$a \geq 0$ or $0 \leq a$	a is nonnegative
$a \leq 0$ or $0 \geq a$	a is nonpositive

FIGURE 1.2
Three distinct real numbers on the real number line with $a < b < c$.

Figure 1.2 shows three distinct real numbers a, b, and c on a real number line. To indicate that b is between a and c on this line, we can write either

$$a < b < c \quad \text{or} \quad c > b > a$$

Each of these expressions is a **double inequality.** When using expressions like $a < b < c$ and $c > b > a$, be sure all the inequality symbols point in the same direction. For example, the expression $a < c > b$ is completely meaningless.

EXAMPLE 1 Rewrite each statement using inequality symbols.

(a) a is at most 6.   **(b)** b is at least -2.

(c) c is nonnegative and less than 10.

SOLUTION

(a) a is at most 6 is written as $a \leq 6$

(b) b is at least -2 is written as $b \geq -2$

(c) c is nonnegative and less than 10 is written as $0 \leq c$ and $c < 10$ or, more compactly, $0 \leq c < 10$. ◆

PROBLEM 1 Repeat Example 1 for each statement.

(a) a is not more than 8.   **(b)** b is negative and at least -4. ◆

◆ **Distance between Two Points on the Real Number Line**

The distance between zero and a number a on the real number line, without regard to direction, is called the *absolute value* of a and is denoted $|a|$. Because distance is independent of direction and is always nonnegative, the absolute value of any real number is also nonnegative. That is, $|a| \geq 0$. A more formal definition of **absolute value** follows.

Absolute Value

For any real number a,

$$|a| = \begin{cases} a & \text{if } a \geq 0 \\ -a & \text{if } a < 0. \end{cases}$$

To find $|4|$ from this definition, we use $|a| = a$, since $4 \geq 0$. Thus,

$$|4| = 4.$$

To find $|-4|$ from this definition, we use $|a| = -a$, since $-4 < 0$. Thus,

$$|-4| = -(-4) = 4.$$

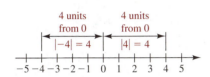

FIGURE 1.3
The absolute value of 4, denoted $|4|$, and the absolute value of -4, denoted $|-4|$, are both equal to 4.

Figure 1.3 illustrates $|4|$ and $|-4|$.

If on the real number line, points A and B have coordinates a and b, respectively, then the **distance** between points A and B, denoted AB, is defined as follows.

Distance between Two Points on the Real Number Line

For any points A and B with coordinates a and b, respectively,

$$AB = |a - b|.$$

For example, if the coordinate of point A is -8 and the coordinate of point B is 5, then

$$AB = |-8 - 5| = |-13| = 13 \text{ units},$$

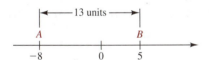

FIGURE 1.4
The distance between -8 and 5 is 13 units.

as shown in Figure 1.4

Note also that $BA = |5 - (-8)| = |13| = 13$ units. In general, for any points A and B, the distance is the same in both directions, that is, $AB = BA$.

EXAMPLE 2 For the given coordinates of points A and B, find AB.

(a) The coordinate of point A is $-\dfrac{13}{4}$, and the coordinate of point B is $\dfrac{5}{3}$.

(b) The coordinate of point A is 2, and the coordinate of point B is $\sqrt{3}$.

SOLUTION

(a) $AB = \left|\dfrac{-13}{4} - \dfrac{5}{3}\right| = \left|\dfrac{-39 - 20}{12}\right| = \left|\dfrac{-59}{12}\right| = \dfrac{59}{12}.$

(b) Because $\sqrt{3} \approx 1.73$, $2 - \sqrt{3} > 0$. Hence, we have

$$AB = |2 - \sqrt{3}| = 2 - \sqrt{3}.$$

PROBLEM 2 For Examples 2(a) and 2(b), verify that $BA = AB$.

◆ Interval Notation

Any unbroken portion of the real number line is called an **interval.** We may describe the set of all real numbers in an interval in either *set-builder notation* or *interval notation*. For example, consider the interval that is designated by the graph in Figure 1.5. The closed dot indicates that -3 is part of the interval, whereas the open circle indicates that 5 is *not* part of the interval. Hence, the graph in the figure represents the set of all real numbers x such that x is greater than or equal to -3 but less than 5. We may describe the set of real numbers in this interval by writing

FIGURE 1.5
Graph of all real numbers x such that x is greater than or equal to -3 but less than 5.

Set-builder notation: $\{x \mid -3 \leq x < 5\}$

or

Interval notation: $[-3, 5)$.

When using interval notation, a bracket, [or], indicates that an endpoint is included in the interval and a parenthesis, (or), indicates that the endpoint is excluded from the interval. An interval that includes both its endpoints is called a **closed interval,** and an interval that excludes both its endpoints is an **open interval.** An interval that contains one of its endpoints but not the other, such as $[-3, 5)$ or $(-3, 5]$, is called a **half-open interval.**

When using interval notation, be sure the smaller of the two numbers is written first. To express the interval in Figure 1.5 as

$(5, -3]$ is **WRONG!**

Remember, we always record the numbers in the order they appear as we read the number line from left to right.

In interval notation, the symbol ∞, read *infinity,* indicates that an interval has no right-hand boundary and the symbol $-\infty$, *negative infinity,* indicates that an interval has no left-hand boundary. We refer to an interval with no right-hand boundary, or no left-hand boundary, or neither, as an **unbounded interval.** The symbols ∞ and $-\infty$ are *not* real numbers and, therefore, cannot be included in an interval. Thus, for unbounded intervals, $-\infty$ is always preceded by a parenthesis, and ∞ is always followed by a parenthesis.

In summary, we list nine possible types of intervals. In each case, a and b are fixed yet unspecified real numbers (constants) and x is a variable. A closed dot on a graph indicates that the point is included as part of the interval, whereas an open circle indicates that the point is not part of the interval.

Graph	Set-builder notation	Interval notation
Unbounded intervals:		
→x at a (open)	$\{x \mid x > a\}$	(a, ∞)
→x at a (closed)	$\{x \mid x \geq a\}$	$[a, \infty)$
←x at a (open)	$\{x \mid x < a\}$	$(-\infty, a)$
←x at a (closed)	$\{x \mid x \leq a\}$	$(-\infty, a]$
←→x	$\{x \mid x \text{ is real}\}$	$(-\infty, \infty)$
Open interval:		
a ○——○ b	$\{x \mid a < x < b\}$	(a, b)
Closed interval:		
a ●——● b	$\{x \mid a \leq x \leq b\}$	$[a, b]$
Half-open intervals:		
a ○——● b	$\{x \mid a < x \leq b\}$	$(a, b]$
a ●——○ b	$\{x \mid a \leq x < b\}$	$[a, b)$

Note: For consistency, throughout the remainder of this text we shall use interval notation, instead of set-builder notation, to describe the set of all real numbers in an interval.

EXAMPLE 3 Use interval notation to describe the set of all real numbers designated by each graph.

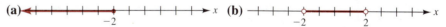

SOLUTION

(a) This graph shows all real numbers x such that x is less than or equal to -2. We may describe this set of real numbers by writing

$$\text{Interval notation:} \quad (-\infty, -2].$$

(b) This graph shows all real numbers x that are greater than -2 but less than 2. We may describe this set of real numbers by writing

$$\text{Interval notation:} \quad (-2, 2). \qquad \blacklozenge$$

PROBLEM 3 Draw a graph for each set of numbers:

(a) $(3, \infty)$ (b) $[4, 9)$ ◆

The **union** of two sets A and B is denoted $A \cup B$, which we read as "A union B". This operation denotes the set of all elements that are members of set A, or members of set B, or members of both sets A and B. In the next example, we use the union operation in conjunction with interval notation to describe the set of all real numbers designated by two or more intervals.

EXAMPLE 4   Use interval notation to describe the set of all real numbers designated by each graph.

(a)

(b)

SOLUTION

(a) This graph shows two intervals that represent all real numbers x such that x is less than or equal to -3 *or* greater than or equal to 3. We may describe the set of real numbers in these two intervals by writing

Interval notation: $(-\infty, -3] \cup [3, \infty)$.

(b) This graph shows three intervals that represent all real numbers x except -1 and 7. We may describe the set of real numbers in these three intervals by writing

Interval notation: $(-\infty, -1) \cup (-1, 7) \cup (7, \infty)$.   ◆

PROBLEM 4   Draw a graph for each set of numbers:

(a) $[-2, -1) \cup (-1, 2]$ (b) $(-\infty, -3] \cup [3, 5) \cup (5, \infty)$ ◆

Exercises 1.1

   Basic Skills

In Exercises 1–10, rewrite each statement using inequality symbols.

1. x is negative.
2. y is positive.
3. a is at most 7.
4. b is at least -9.
5. p is greater than 2 and less than or equal to 10.
6. q is less than or equal to 4 and greater than -1.
7. c is positive and less than 8.
8. d is negative and more than -5.
9. t is nonpositive and at least -2.
10. k is nonnegative and at most 4.

In Exercises 11–20, the coordinates of points A and B are given. Find AB, the distance from A to B.

11. The coordinate of point A is -9 and the coordinate of point B is 6.
12. The coordinate of point A is 13 and the coordinate of point B is -2.
13. The coordinate of point A is -1.9 and the coordinate of point B is -3.8.
14. The coordinate of point A is 0.16 and the coordinate of point B is 0.09.
15. The coordinate of point A is $\frac{7}{8}$ and the coordinate of point B is $-\frac{13}{8}$.
16. The coordinate of point A is $-\frac{3}{4}$ and the coordinate of point B is $-\frac{9}{10}$.
17. The coordinate of point A is $-3\frac{1}{2}$ and the coordinate of point B is $5\frac{2}{3}$.
18. The coordinate of point A is 19 and the coordinate of point B is $6\frac{5}{6}$.
19. The coordinate of point A is $\sqrt{2}$ and the coordinate of point B is π.
20. The coordinate of point A is $-\sqrt{2}$ and the coordinate of point B is $\sqrt{3}$.

In Exercises 21–40, use interval notation to describe the set of all real numbers designated by each graph.

21. 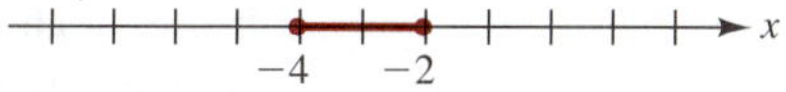

22. 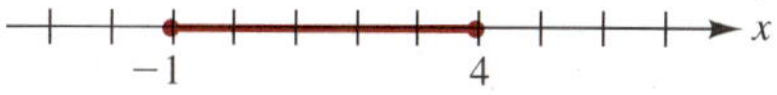

23. 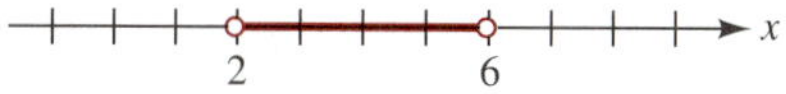

24. 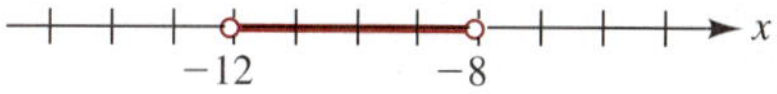

25. 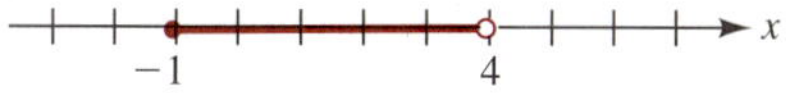

26. 

27. 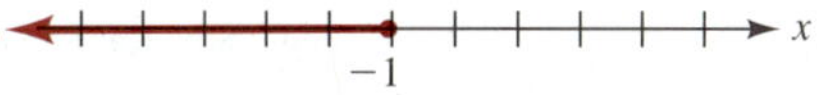

28. 

29. 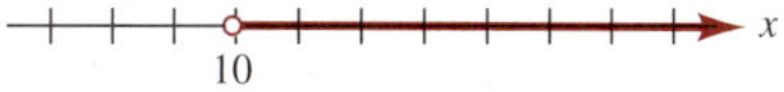

30. 

31. 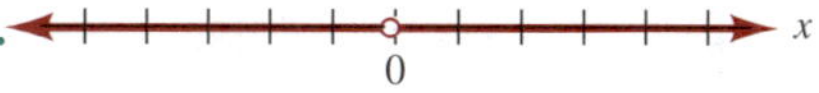

32. 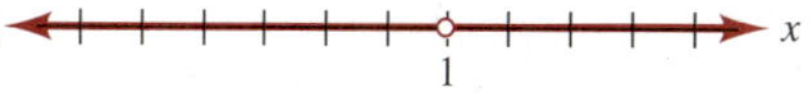

33. 

34. 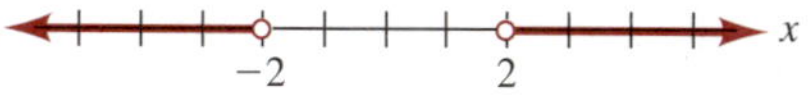

35. 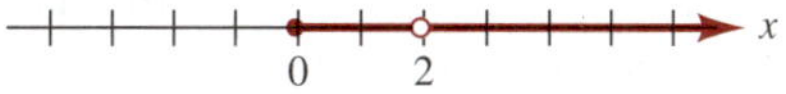

36. 

37. 

38. 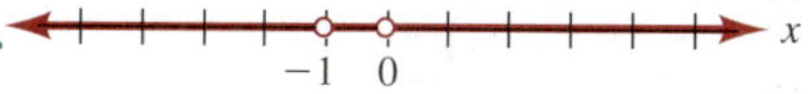

39. 

40. 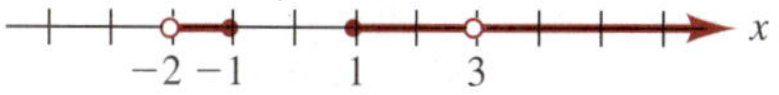

In Exercises 41–50, draw a graph of each set of numbers.

41. $(-3, 0]$

42. $[-6, -3]$

43. $(-\infty, \infty)$

44. $[2, \infty)$

45. $[0, 1) \cup (3, \infty)$

46. $(-\infty, 1) \cup [2, 5]$

47. $(-\infty, 7) \cup (7, \infty)$

48. $[-1, 0) \cup (0, 2]$

49. $(-\infty, 0] \cup [1, 2) \cup (2, \infty)$

50. $(0, 1) \cup (1, 3) \cup (3, \infty)$

Critical Thinking

In Exercises 51–56, rewrite each expression so that it does not contain absolute value bars.

51. $|\pi - x|$, if $x \geq \pi$

52. $|\pi - x|$ if $x < \pi$

53. $|x - 3| + |x - 4|$ if $3 < x < 4$

54. $|x - 3| - |x - 4|$ if $x < 3$

55. $|x| < 5$

56. $|y| > 3$

In Exercises 57–60, rewrite each statement using absolute value bars and inequality symbols.

57. The distance between a and 7 is at least 3 units.

58. c is less than 6 units from 0.

59. d is closer to 1 than to 0.

60. b is farther from -2 than from 5.

61. What meaning (if any) can be assigned to the interval notation $[a, a]$? (a, a)?

62. Explain what is *wrong* with each of the following interval notations.

(a) $(0, -6)$ (b) $[-2, -\infty)$
(c) $[-\infty, 5)$ (d) $[1, \infty]$

63. Use interval notation to describe the set of all real numbers that are less than their reciprocals.

64. Use interval notation to describe the set of all real numbers that are less than or equal to their squares.

Calculator Activities

65. Use your calculator to help list the following real numbers in order from smallest to largest:

$$3.145 \quad \pi \quad \frac{22}{7} \quad 3.2 \quad \sqrt{10} \quad \frac{157}{50}$$

66. Use your calculator to help list the following real numbers in order from largest to smallest:

$$\frac{7}{5} \quad \sqrt{2} \quad 1.414 \quad \frac{71}{50} \quad \frac{8\pi}{17} \quad 1.5$$

In Exercises 67–70, the coordinates of points A and B are given. Find AB, the distance from A to B.

67. The coordinate of point A is $-18{,}912{,}988$ and the coordinate of point B is $2{,}398{,}975$.

68. The coordinate of point A is $-52{,}988.06$ and the coordinate of point B is $-12{,}398.974$.

69. The coordinate of point A is 2.451×10^{12} and the coordinate of point B is 3.675×10^{13}.

70. The coordinate of point A is 1.397×10^{-11} and the coordinate of point B is -2.231×10^{-10}.

1.2 Working in the Cartesian Plane

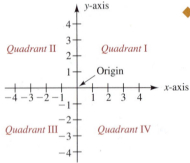

FIGURE 1.6
The coordinate plane, or Cartesian plane.

◆ Introductory Comments

Recall from Section 1.1 that for each point on the real number line, there corresponds a real number called its *coordinate*. In this section, we extend this idea by assigning to each point in a plane a pair of real numbers. To do this, we construct horizontal and vertical real number lines that intersect at the zero points of the two lines (see Figure 1.6). The two lines are called *coordinate axes* and the *plane* in which they lie is called the **coordinate plane** or **Cartesian plane**, after the French mathematician René Descartes (1596–1650). The point where the axes intersect is called the **origin**. The horizontal number line has its positive direction to the right and is usually called the **x-axis**. The vertical number line has its positive direction upward and is usually called the **y-axis**. The coordinate axes divide the plane into four regions, or **quadrants**, which are labeled with Roman numerals, as shown in Figure 1.6.

◆ Plotting Points

We can now assign to each point P in the coordinate plane a unique pair of numbers, called its **rectangular coordinates** or **Cartesian coordinates.** By drawing horizontal and vertical lines through P, we find that the vertical line intersects the

SECTION 1.2 Working in the Cartesian Plane

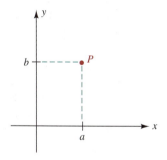

FIGURE 1.7
The coordinates of point P are (a, b).

x-axis at some point with coordinate a, and the horizontal line intersects the y-axis at some point with coordinate b, as shown in Figure 1.7. Thus, we assign the pair (a, b) to the point P. In the pair (a, b), the first number a is called the **x-coordinate** or **abscissa** or P, and the second number b is called the **y-coordinate** or **ordinate** of P. Since the x-coordinate is always written first, we refer to (a, b) as an **ordered pair** of numbers.

Conversely, we can locate any ordered pair (a, b) in the coordinate plane by constructing a vertical line through a on the x-axis and a horizontal line through b on the y-axis. The intersection of these lines determines a unique point P, which we designate as $P(a, b)$. Thus, for each point in the coordinate plane there corresponds a unique ordered pair of real numbers, and for each ordered pair of real numbers there is a unique point in the coordinate plane. Hence, we have a *one-to-one correspondence* between pairs of real numbers and points in a coordinate plane.

EXAMPLE 1 Plot the points $A(2, 3)$, $B(-3, 2)$, $C(-4, -3)$, and $D(\frac{3}{2}, -4)$ on a coordinate plane. Specify the quadrant in which each point lies.

SOLUTION The point $A(2, 3)$ is in quadrant I, $B(-3, 2)$ is in quadrant II, $C(-4, -3)$ is in quadrant III, and $D(\frac{3}{2}, -4)$ is in quadrant IV. Figure 1.8 shows the points on a coordinate plane. ◆

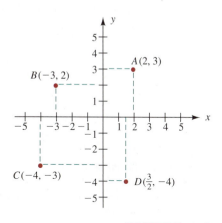

FIGURE 1.8
Plotting points associated with ordered pairs.

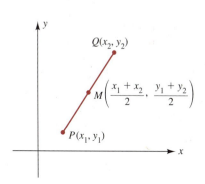

FIGURE 1.9
The midpoint M of $\overline{PQ}$.

PROBLEM 1 Plot the points $A(-3, 1)$, $B(0, -4)$, $C(4, 0)$, and $D(1\frac{1}{4}, 2)$ on a coordinate plane. ◆

◆ The Midpoint Formula and Distance Formula

Given two points $P(x_1, y_1)$ and $Q(x_2, y_2)$ in the coordinate plane (see Figure 1.9), we can find the coordinates of the midpoint M of $\overline{PQ}$ (read "line segment PQ") by finding the average value of the x-coordinates and the average value of the

Midpoint Formula

The coordinates of the **midpoint** M of a line segment joining the points $P(x_1, y_1)$ and $Q(x_2, y_2)$ are

$$\left(\frac{x_1 + x_2}{2}, \frac{y_1 + y_2}{2} \right)$$

By convention, we designate the *length* of $\overline{PQ}$ by writing PQ. To determine PQ, we begin by constructing a right triangle as shown in Figure 1.10. The length of the horizontal leg of the triangle is $|x_2 - x_1|$ and the length of the vertical leg is $|y_2 - y_1|$. By the Pythagorean theorem (see the inside front cover), we can find the length of $\overline{PQ}$:

$$(PQ)^2 = |x_2 - x_1|^2 + |y_2 - y_1|^2$$
$$PQ = \sqrt{|x_2 - x_1|^2 + |y_2 - y_1|^2}$$
$$PQ = \sqrt{(x_2 - x_1)^2 + (y_2 - y_1)^2}$$

We refer to this last equation as the **distance formula.**

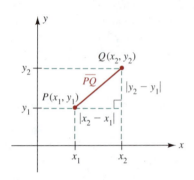

FIGURE 1.10
A formula for the length of $\overline{PQ}$ (denoted PQ) may be found by using the Pythagorean theorem.

Distance Formula

The **distance** between two points $P(x_1, y_1)$ and $Q(x_2, y_2)$ in the coordinate plane is given by

$$PQ = \sqrt{(x_2 - x_1)^2 + (y_2 - y_1)^2}$$

Note: When we use the distance formula, it does not matter which point is called (x_1, y_1) and which is called (x_2, y_2). This is because $(x_2 - x_1)$ and $(x_1 - x_2)$ are negatives of each other, as are $(y_2 - y_1)$ and $(y_1 - y_2)$, and squaring a number or its negative gives us the same numerical result.

EXAMPLE 2 Given the points $A(-2, 3)$ and $B(4, 1)$, find **(a)** the coordinates of the midpoint of $\overline{AB}$ and **(b)** AB.

SOLUTION

(a) Using the midpoint formula with $(x_1, y_1) = (-2, 3)$ and $(x_2, y_2) = (4, 1)$, we have

$$\left(\frac{-2 + 4}{2}, \frac{3 + 1}{2} \right) = (1, 2).$$

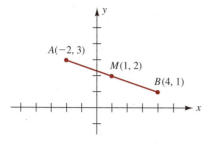

FIGURE 1.11
The coordinates of the midpoint M of $\overline{AB}$ are $(1, 2)$.

Thus, as shown in Figure 1.11, the coordinates of the midpoint M are $(1, 2)$.

(b) The notation AB designates the length of the line segment that joins the points A and B. Using the distance formula with $(x_1, y_1) = (-2, 3)$ and $(x_2, y_2) = (4, 1)$ we have

$$AB = \sqrt{[4 - (-2)]^2 + (1 - 3)^2}$$
$$= \sqrt{(6)^2 + (-2)^2}$$
$$= \sqrt{40} = 2\sqrt{10} \approx 6.32$$

PROBLEM 2 Given the points $A(-3, 1)$ and $B(1, -2)$, find **(a)** the coordinates of the midpoint of $\overline{AB}$ and **(b)** AB.

♦ **Analytic Geometry**

As illustrated in the next example, we can place geometric figures in the coordinate plane and solve related problems by using the distance and midpoint formulas. We refer to this branch of mathematics, which uses algebra to solve geometric problems, as **analytic geometry.**

EXAMPLE 3 Show that the triangle with vertices at the points $A(-2, 2)$, $B(3, 2)$, and $C(1, -2)$ is an isosceles triangle (with two sides of equal length). Then find the area of triangle ABC.

SOLUTION We begin by plotting the points and constructing the triangle, as shown in Figure 1.12. Now, using the distance formula, we have

$$AB = \sqrt{[3 - (-2)]^2 + (2 - 2)^2} = \sqrt{(5)^2 + (0)^2} = 5$$
$$BC = \sqrt{(1 - 3)^2 + (-2 - 2)^2} = \sqrt{(-2)^2 + (-4)^2} = \sqrt{20} = 2\sqrt{5}$$
$$AC = \sqrt{[1 - (-2)]^2 + (-2 - 2)^2} = \sqrt{(3)^2 + (-4)^2} = \sqrt{25} = 5$$

Since $AB = AC$, we conclude the triangle is isosceles with BC the nonequal side.

To find the area of triangle ABC, recall from geometry that the altitude to the nonequal side of an isosceles triangle bisects that side. Thus, if $\overline{AM}$ is the altitude to the base $\overline{BC}$, then M is the midpoint of $\overline{BC}$, as shown in Figure 1.13. By the midpoint formula, the coordinates of the midpoint M are

$$\left(\frac{1 + 3}{2}, \frac{-2 + 2}{2}\right) = (2, 0)$$

and, by the distance formula, the length of the altitude $\overline{AM}$ is

$$AM = \sqrt{[(2 - (-2)]^2 + (0 - 2)^2} = \sqrt{(4)^2 + (-2)^2} = \sqrt{20} = 2\sqrt{5}.$$

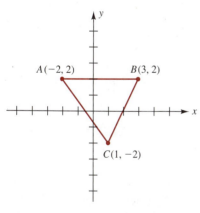

FIGURE 1.12
The triangle formed by the points $A(-2, 2)$, $B(3, 2)$, and $C(1, -2)$.

Hence, the area of triangle ABC is

$$\text{Area} = \tfrac{1}{2}(\text{base})(\text{height}) = \tfrac{1}{2}(BC)(AM)$$
$$= \tfrac{1}{2}(2\sqrt{5})(2\sqrt{5})$$
$$= 10 \text{ square units.}$$

As an alternative method for finding the area of triangle ABC, let $\overline{CN}$ be the altitude to the base $\overline{AB}$. Since $\overline{AB}$ is a horizontal line segment, $CN = 4$. Hence,

$$\text{Area} = \tfrac{1}{2}(AB)(CN) = \tfrac{1}{2}(5)(4) = 10 \text{ square units.} \quad \blacklozenge$$

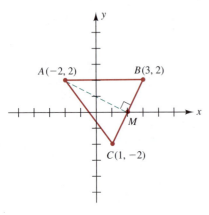

FIGURE 1.13
The altitude to the nonequal side of an isosceles triangle bisects that side. Thus, M is the midpoint of $\overline{BC}$.

The *converse* of the Pythagorean theorem is also true. If the square of the longest side of a triangle is equal to the sum of the squares of the two other sides, then the triangle is a right triangle, with the right angle opposite the longest side.

PROBLEM 3 Use the distance formula and the converse of the Pythagorean theorem to show that the triangle with vertices at the points $A(2, 5)$, $B(-2, 3)$, and $C(4, 1)$ is an isosceles right triangle. Then find the area of triangle ABC. $\quad \blacklozenge$

♦ **Graph of an Equation**

We may also use the coordinate plane to show the *graph* of an equation in two unknowns. The **graph** of an equation is the set of *all* ordered pairs (x, y) in the coordinate plane that are solutions of the given equation. Consider the equation

$$y = 2x + 1$$

with the two unknowns x and y. We say that the ordered pair $(0, 1)$ is a solution of this equation since replacing x with 0 and y with 1 yields a true statement.

$$y = 2x + 1$$
$$1 = 2(0) + 1$$
$$1 = 1 \quad \text{is true.}$$

To find other ordered pairs that are solutions of this equation, we arbitrarily choose values of x and then determine the corresponding values of y. Under these conditions we say that x is the **independent variable** and y the **dependent variable** in the equation. In order to determine other solutions of this equation, it is convenient to set up a *table of values* such as the one shown in Table 1.2.

Table 1.2
Table of values for $y = 2x + 1$

x	-2	-1	0	1	2
$y = 2x + 1$	-3	-1	1	3	5

Thus, along with (0, 1) we have (−2, −3), (−1, −1), (1, 3), and (2, 5) as four other solutions of the equation $y = 2x + 1$. By continuing this process, we can generate many other ordered pairs that satisfy this equation. The graph of the equation $y = 2x + 1$ is a straight line, as shown in Figure 1.14. The arrowhead at each end of the line indicates that the graph continues indefinitely in each direction. Note that every point on this line is a solution of the equation $y = 2x + 1$.

FIGURE 1.14
The graph of the equation $y = 2x + 1$ is a straight line.

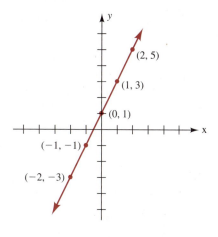

◆ Point-Plotting Method

The **point-plotting method** is a procedure used in elementary algebra courses to *sketch the graph* of an equation.

 Point-Plotting Method

To sketch the graph of an equation by the **point-plotting method,**

1. Set up a *table of values* and find a few ordered pairs that satisfy the equation.
2. *Plot* and label the corresponding points in the coordinate plane.
3. Look for a pattern, and *connect the plotted points* to form a smooth curve.

Note: Although this method works well for some simple equations in two unknowns, it is inadequate for sketching the graph of more complicated equations. Throughout this chapter, we will introduce various graphical aids that are useful for sketching an accurate graph while plotting as few points as possible.

EXAMPLE 4 Sketch the graph of each equation by using the point-plotting method.

(a) $y = |x|$ (b) $x = y^2$ (c) $2y - x^3 = 0$

SOLUTION

(a) We begin by selecting arbitrary values of *x,* and then finding their corresponding values of *y*. The following table of values organizes our work.

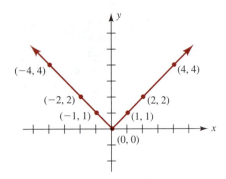

FIGURE 1.15
Graph of $y = |x|$.

x	−4	−2	−1	0	1	2	4
$y = |x|$	4	2	1	0	1	2	4

We now plot and label the points given by this table and connect them according to the suggested pattern. The graph of $y = |x|$ is the V-shaped curve shown in Figure 1.15. The arrowheads on the curve indicate that the graph continues upward to the left and upward to the right, according to the suggested pattern.

(b) For the equation $x = y^2$, it is easier to choose arbitrary values of y and then determine the corresponding values of x. In doing so, we are treating y as the independent variable and x as the dependent variable. The following table of values shows several values of y and their corresponding x-values.

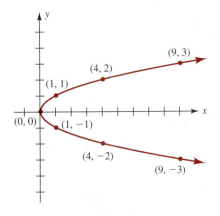

FIGURE 1.16
Graph of $x = y^2$.

y	−3	−2	−1	0	1	2	3
$x = y^2$	9	4	1	0	1	4	9

We plot and label the points given by this table, and connect them according to the suggested pattern. *Remember that the x-coordinate is always written first in an ordered pair.* The graph of $x = y^2$ is shown in Figure 1.16. It is a cup-shaped curve called a *parabola*. The arrowheads on the curve indicate that the curve continues indefinitely, according to the suggested pattern.

(c) It is best to solve the equation $2y - x^3 = 0$ for either x or y before setting up a table of values. For this equation, solving for y is easier than solving for x, and we obtain

$$y = \tfrac{1}{2}x^3.$$

The following table of values shows several values of x and their corresponding y-values.

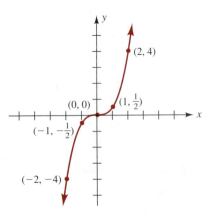

FIGURE 1.17
Graph of $2y - x^3 = 0$.

x	−2	−1	0	1	2
$y = \tfrac{1}{2}x^3$	−4	−$\tfrac{1}{2}$	0	$\tfrac{1}{2}$	4

We now plot and label the points given by this table and draw a smooth curve through them. The graph of $2y - x^3 = 0$ is shown in Figure 1.17. The arrowheads on the curve indicate that the graph continues indefinitely downward to the left and upward to the right. ◆

PROBLEM 4 Sketch the graph of the equation $4x + y = 8$.

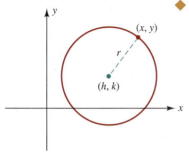

◆ Circles

A **circle** is the set of all points in a plane that lie a fixed distance from a given point. We call the fixed distance the **radius** and the given point the **center** of the circle. Shown in Figure 1.18 is a circle with radius r and center (h, k). The point (x, y) lies on this circle if and only if the distance from the center (h, k) to (x, y) is r. By using the distance formula, we have

$$r = \sqrt{(x - h)^2 + (y - k)^2}.$$

FIGURE 1.18
Circle with center (h, k) and radius r.

If we square both sides of this equation, we obtain the equation of a circle in standard form.

Equation of a Circle in Standard Form

> The **equation of a circle in standard form** with center (h, k) and radius r is
>
> $$(x - h)^2 + (y - k)^2 = r^2.$$

EXAMPLE 5 Determine the equation of a circle in standard form with center $(2, -1)$ and radius 3.

SOLUTION The equation of a circle in standard form with center $(2, -1)$ and radius 3 is

$$(x - h)^2 + (y - k)^2 = r^2$$
$$(x - 2)^2 + [y - (-1)]^2 = 3^2 \qquad \text{Substitute } h = 2, k = -1 \text{ and } r = 3$$
$$(x - 2)^2 + (y + 1)^2 = 9 \qquad \text{Simplify}$$

Conversely, we can state that the graph of any equation of the form

$$(x - h)^2 + (y - k)^2 = r^2 \quad \text{with } r > 0$$

is a circle with center (h, k) and radius r.

PROBLEM 5 Sketch the graph of the equation $(x + 2)^2 + (y - 1)^2 = 25$.

Squaring the expressions $(x - h)$ and $(y - k)$ in the equation

$$(x - h)^2 + (y - k)^2 = r^2,$$

gives us

$$x^2 - 2hx + h^2 + y^2 - 2ky + k^2 = r^2$$
$$x^2 + y^2 - 2hx - 2ky + (h^2 + k^2 - r^2) = 0.$$

Replacing the constants $-2h$ with D, $-2k$ with E, and $(h^2 + k^2 - r^2)$ with F, we have

$$\boxed{x^2 + y^2 + Dx + Ey + F = 0}$$

This equation is called the **equation of a circle in general form** and is characterized by the presence of x^2 and y^2 terms, each having a coefficient of 1. To sketch the graph of an equation of the form $x^2 + y^2 + Dx + Ey + F = 0$, it is best to write the equation in the standard form $(x - h)^2 + (y - k)^2 = r^2$. Once in standard form, we can easily draw its graph.

EXAMPLE 6 Sketch the graph of the equation $x^2 + y^2 = 25$.

SOLUTION The equation $x^2 + y^2 = 25$ is of the form

$$x^2 + y^2 + Dx + Ey + F = 0, \quad \text{where } D = E = 0, \text{ and } F = -25$$

By thinking of this equation as

$$\underbrace{(x - 0)^2 + (y - 0)^2 = 5^2}_{\text{Standard form of a circle}}$$

we know its graph is a circle with center $(0, 0)$ and radius 5. The graph of this equation is shown in Figure 1.19.

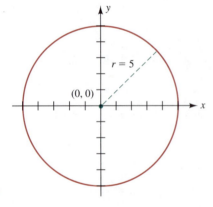

FIGURE 1.19
The graph of $x^2 + y^2 = 25$ is a circle with center $(0, 0)$ and radius 5.

PROBLEM 6 Sketch the graph of the equation $x^2 + y^2 = 4$.

As illustrated in the next example, we sometimes use the process of *completing the square* to write the equation $x^2 + y^2 + Dx + Ey + F = 0$ in the form $(x - h)^2 + (y - k)^2 = r^2$.

EXAMPLE 7 Sketch the graph of each equation.

(a) $x^2 + y^2 + 6x = 0$ (b) $4x^2 + 4y^2 - 8x + 20y + 13 = 0$.

SOLUTION

(a) The equation $x^2 + y^2 + 6x = 0$ appears to be the equation of a circle in general form with $D = 6$, $E = 0$, and $F = 0$. Using the process of completing the square, we have

$(x^2 + 6x +) + y^2 = 0$ Regroup

$(x^2 + 6x + 9) + y^2 = 9$ Complete the square by adding 9 to both sides

$(x + 3)^2 + y^2 = 9$ Factor the perfect square trinomial

$[x - (-3)]^2 + (y - 0)^2 = 3^2$ Write in standard form

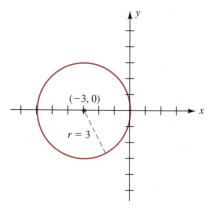

FIGURE 1.20
The graph of $x^2 + y^2 + 6x = 0$ is a circle with center $(-3, 0)$ and radius 3.

Thus, the center of the circle is $(-3, 0)$ and the radius is 3. The graph of this equation is shown in Figure 1.20.

(b) Dividing both sides by 4, we obtain

$$x^2 + y^2 - 2x + 5y + \tfrac{13}{4} = 0,$$

which appears to be the equation of a circle in general form with $D = -2$, $E = 5$, and $F = \tfrac{13}{4}$. Using the process of completing the square, we have

$(x^2 - 2x) + (y^2 + 5y \phantom{+\tfrac{25}{4}}) = -\dfrac{13}{4}$ Regroup

Add 1 to both sides.

$(x^2 - 2x + 1) + \left(y^2 + 5y + \dfrac{25}{4}\right) = -\dfrac{13}{4} + \left(1 + \dfrac{25}{4}\right)$ Complete the squares

Add $\tfrac{25}{4}$ to both sides.

$(x - 1)^2 + \left(y + \dfrac{5}{2}\right)^2 = 4$ Factor

$(x - 1)^2 + \left[y - \left(-\dfrac{5}{2}\right)\right]^2 = 2^2$ Write in standard form

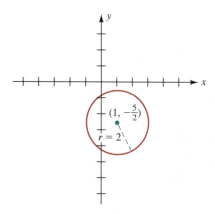

FIGURE 1.21
The graph of $4x^2 + 4y^2 - 8x + 20y + 13 = 0$ is a circle with center $(1, -\tfrac{5}{2})$ and radius 2.

Thus the center of this circle is $(1, -\tfrac{5}{2})$ and the radius is 2. The graph of this equation is shown in Figure 1.21. ◆

Caution Not every equation of the form $x^2 + y^2 + Dx + Ey + F = 0$ is a circle. If this equation is written in the standard form $(x - h)^2 + (y - k)^2 = r^2$ and $r^2 = 0$, then its graph is the single point (h, k). If $r^2 < 0$ when the equation is written in standard form, then it has no graph, since no point (x, y) with real coefficients can satisfy the equation.

PROBLEM 7 Is $x^2 + y^2 + 4x + 6y + 13 = 0$ the equation of a circle? Explain. ◆

Exercises 1.2

Basic Skills

In Exercises 1–6, fill in the blank to complete each statement.

1. If both coordinates are negative, the point is located in quadrant _____.

2. If the x-coordinate is _____ and the y-coordinate is _____, the point is located in quadrant II.

3. If the x-coordinate is _____ and the y-coordinate is _____, the point is located in quadrant IV.

4. If the x-coordinate is zero, the point is located on the _____.

5. If the y-coordinate is zero, the point is located on the _____.

6. If both coordinates are _____, the point is located at the origin.

In Exercises 7 and 8, name the ordered pair associated with each point.

7.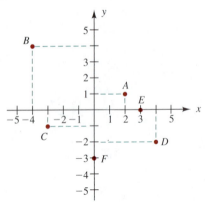

8.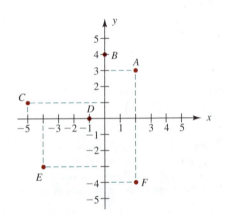

In Exercises 9–18, find

(a) the length of the line segment joining the points A and B, and

(b) the coordinates of the midpoint M of the line segment joining the points A and B.

9. $A(1, 2)$; $B(4, 6)$
10. $A(3, -6)$; $B(-2, 6)$
11. $A(-2, 3)$; $B(1, 8)$
12. $A(-1, -2)$; $B(-3, 2)$
13. $A(-5, 2)$; $B(3, -5)$
14. $A(-2, -1)$; $B(-6, -5)$
15. $A\left(-\frac{1}{2}, \frac{2}{3}\right)$; $B\left(-\frac{3}{4}, 1\right)$
16. $A\left(\frac{1}{2}, \frac{1}{2}\right)$; $B\left(\frac{2}{3}, \frac{5}{8}\right)$
17. $A(0, \sqrt{2})$; $B(-4, 0)$
18. $A(\sqrt{3}, -1)$; $B(0, 2)$

19. Find a formula for the distance d between the origin and the point (x, y) in the coordinate plane.

20. Find the coordinates of the midpoint M of a line segment joining the origin to the point (x, y) in the coordinate plane.

21. Find the perimeter of a triangle whose vertices are the points $A(-2, 1)$, $B(1, 3)$, and $C(4, -3)$.

22. Find the perimeter of a quadrilateral whose vertices are the points $A(-1, 0)$, $B(2, 4)$, $C(8, -4)$, and $D(4, -12)$.

23. Show that the points $A(1, 4)$, $B(2, -3)$, and $C(-1, -2)$ are the vertices of a right triangle. Then find the area of triangle ABC.

24. Show that the triangle whose vertices are the points $A(-5, 14)$, $B(1, 4)$, and $C(11, 10)$ is isosceles. Then find its area.

25. Show that the triangle whose vertices are $A(1, 1)$, $B(-1, -1)$, and $C(\sqrt{3}, -\sqrt{3})$ is an equilateral triangle. Then find its area.

26. Show that the triangle whose vertices are $A(0, 6)$, $B(2, 0)$, and $C(8, 2)$ is an isosceles right triangle. Then find its area.

27. A line segment joins the points $A(8, -12)$ and $B(-4, 6)$. Use the midpoint formula to find the coordinates of the three points that divide this line segment into four equal parts.

28. Use the distance formula to determine if the points $A(-2, -3)$, $B(1, 3)$ and $C(2, 5)$ are collinear.

29. A *median* of a triangle is a line segment that joins a vertex of the triangle to the midpoint of the opposite side.

Find the lengths of the medians of a triangle whose vertices are $A(-5, 4)$, $B(5, 2)$ and $C(-1, -4)$.

30. The midpoint of a line segment is $M(3, -2)$. One end point of the segment has coordinates $(6, 3)$. Find the coordinates of the other endpoint.

In Exercises 31–42, use the point-plotting method to sketch the graph of each equation.

31. $y = x$
32. $y = x^2$
33. $y = \sqrt{x}$
34. $y = x^{1/3}$
35. $xy = 1$
36. $y - 3x = 0$
37. $3x + 2y = 12$
38. $4x - 3y = 12$
39. $x - 2|y| = 0$
40. $x + 4y^3 = 0$
41. $|x| + |y| = 2$
42. $|x| - |y| = 2$

In Exercises 43–54, give the center and radius for each equation that defines a circle. Then sketch the graph.

43. $x^2 + (y - 2)^2 = 16$
44. $(x + 5)^2 + (y + 2)^2 = 4$
45. $x^2 + y^2 = 9$
46. $x^2 + y^2 - 36 = 0$
47. $x^2 + y^2 - 2y = 3$
48. $x^2 + y^2 + 5x = 0$
49. $x^2 + y^2 - 6x + 8y + 9 = 0$
50. $x^2 + y^2 - x - 4y = 2$
51. $4x^2 + 4y^2 - 8x + 12y = 3$
52. $4x^2 + 4y^2 + 20x - 16y + 41 = 0$
53. $3x^2 + 3y^2 - 4x - 8y + 24 = 0$
54. $2x^2 + 2y^2 - x + 20y + 40 = 0$

In Exercises 55–62, write the equation of a circle that has the given characteristics.

55. Center at $(0, 0)$; radius of 2
56. Center at $(-3, 0)$; radius of 3
57. Center at $(2, -3)$; radius of $\sqrt{2}$
58. Center at $(-1, 2)$; radius of $\frac{2}{3}$
59. Center at the origin; passes through the point $(-3, 4)$
60. Center at $(-1, 3)$; passes through the point $(1 -1)$
61. Line segment from $(2, 3)$ to $(-2, 5)$ is the diameter
62. Line segment from $(-1, -1)$ to $(3, 1)$ is the radius (*Hint:* Two answers are possible.)

Critical Thinking

For Exercises 63–66, refer to the right triangle with vertices A, 0, and B as shown:

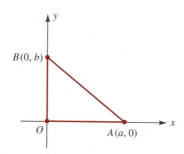

63. Find the coordinates of the midpoint M of the hypotenuse $\overline{AB}$.
64. Find AM and BM.
65. Find OM.
66. What conclusion can you make about the mipdoint of the hypotenuse in regard to the three vertices of the right triangle?

For Exercises 67–70, refer to the parallelogram with vertices A, B, C, and 0 as shown:

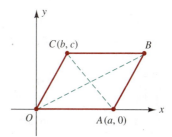

67. Find the coordinates of point B.
68. Find the midpoint of the diagonal $\overline{AC}$.
69. Find the midpoint of the diagonal $\overline{OB}$.
70. What conclusion can you make about the diagonals of a parallelogram?
71. Graph the equation $y = |x|$. Then, on the same set of axes, sketch the graphs of the equations $y = |x| + 1$,

EXERCISE 71 *(continued)*

$y = |x| + 2$, and $y = |x| + 3$. Compare the graphs of these equations. What effect does the constant c have on the graph of $y = |x| + c$ when $c > 0$?

72. Graph the equation $y = x^3$. Then, on the same set of axes, sketch the graphs of the equations $y = x^3 - 1$, $y = x^3 - 2$, and $y = x^3 - 3$. Compare the graphs of these equations. What effect does the constant c have on the graph of $y = x^3 + c$ when $c < 0$?

73. Graph the equation $y = \sqrt{x}$. Then, on the same set of axes, sketch the graphs of the equations $y = \sqrt{x} - 1$, $y = \sqrt{x} - 2$, and $y = \sqrt{x} - 3$. Compare the graphs of these equations. What effect does the constant c have on the graph of $y = \sqrt{x} + c$ when $c < 0$?

74. Graph the equation $y = x^2$. Then on the same set of axes, sketch the graphs of the equations $y = (x + 1)^2$, $y = (x + 2)^2$, and $y = (x + 3)^2$. Compare the graphs of these equations. What effect does the constant c have on the graph of $y = (x + c)^2$ when $c > 0$?

75. Two circles having the same center but different radii are called *concentric circles*.

    (a) If the ratio of the radii of two concentric circles is $2:1$ and the equation of the circle with the larger radii is $x^2 + y^2 - 10x + 6y = 2$, determine the equation of the other circle.

    (b) Find the area between the two circles described in part (a).

76. For the circle $x^2 + y^2 + Dx + Ey + F = 0$, use the method of completing the square to determine a formula in terms of D, E, and F for the radius of the circle.

  Calculator Activities

In Exercises 77–80, find **(a)** *the length of the line segment that joins the given point to the origin, and* **(b)** *the coordinates of the midpoint M of the line segment that joins the given point to the origin.*

77. $(2.56, 3.20)$ 78. $(-10.6, 14.8)$

79. $(-46.9, -76.8)$ 80. $(123.6, -457.9)$

81. Find the radius and area of a circle that passes through the point $A(2.24, 3.71)$ and has its center at the origin.

82. The line segment that joins the point $A(4.3, 2.5)$ to the point $B(-2.7, -4.9)$ is the diameter of a circle. Find the center, radius, and area of the circle.

In Exercises 83 and 84, determine the center and radius of the circle with the given equation. Round each answer to three significant digits.

83. $x^2 + y^2 - 7.22x - 4.84y + 12.34 = 0$

84. $4.1x^2 + 4.1y^2 - 24.1x - 16.3y + 25.2 = 0$

85. A castle entrance has the form of a Gothic arch, as shown in the figure.

    (a) Taking the axes as shown, find the equation of the circle that forms the left-hand side of the arch.

    (b) Use the equation in part (a) to find the height h of the arch, rounded to three significant digits.

86. The radius of a Ferris wheel is 10.1 meters, as shown in the figure.

    (a) Taking the axes as shown, find the equation of the circle that forms the Ferris wheel.

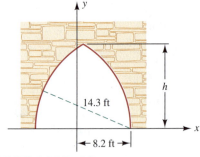

EXERCISE 85

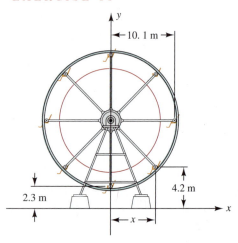

   (b) Use the equation in part (a) to find the distance x from a chair to the vertical axis of the wheel when the height of the chair is 4.2 meters above the ground. Round the answer to three significant digits.

1.3 Functions

◆ Introductory Comments

Given an equation in two unknowns x and y, suppose we *choose* values for x and *obtain* corresponding values for y. Hence, we are using x as the *independent variable* and y as the *dependent variable*. The set of all numbers from which we may choose is called the **domain** and the set of all numbers that we obtain is called the **range**. If for each value of x in the domain there corresponds one and only one value of y in the range, we say that y is a **function** of x.

◆ Definition of a Function

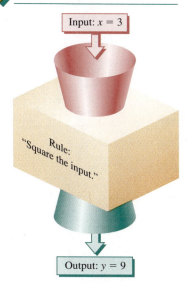

FIGURE 1.22
Function machine for $y = x^2$.

> A **function** from a set X to a set Y is a rule of correspondence that assigns to each element x in X *exactly one* element y in Y.

A function can be represented by a list or table, by a graph, or by a formula or equation. Many functions in this text are specified by equations. For example, consider the equation $y = x^2$. It is useful to think of the x-values as **inputs** and their corresponding y-values as **outputs**. We can think of this equation as a rule that says, "Square the input." A *function machine* for this equation is shown in Figure 1.22. If we place $x = 3$ in the input hopper, the machine follows the rule "square the input" and gives us an output of $y = 9$. The rule "square the input" assigns to each input value x one and only one output value y. Thus we say the equation $y = x^2$ *defines y as a function of x.*

Not every equation in two unknowns x and y defines y as a function of x. Consider the equation $y^2 = x$. If we choose the input $x = 9$, the equation becomes $y^2 = 9$ and, consequently, $y = \pm 3$. In this equation we obtain two outputs for the input $x = 9$. Since our definition of a function requires that there be exactly one output for each input, we know that the equation $y^2 = x$ does not define y as a function of x.

◆ Vertical Line Test

Suppose we are given the graph of an equation in two unknowns, x and y. If every vertical line that can be drawn in the coordinate plane intersects the graph *at most once*, then we can say the equation has exactly one output y for each input value x that we can assign. Hence, the equation defines y as a function of x. We refer to this graphical method of determining whether y is a function of x as the **vertical line test.**

◆ Vertical Line Test

> An equation defines y as a function of x if and only if every vertical line in the coordinate plane intersects the graph of the equation at most once.

The graph of $y = x^2$ (in Figure 1.23) passes the vertical line test, but the graph of $y^2 = x$ (in Figure 1.24) fails this test.

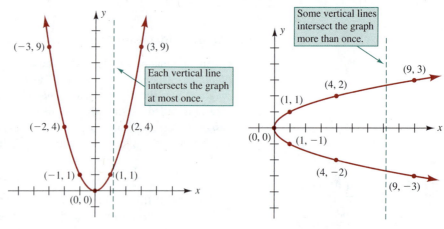

FIGURE 1.23
The graph of $y = x^2$ passes the vertical line test.

FIGURE 1.24
The graph of $y^2 = x$ fails the vertical line test.

EXAMPLE 1 Determine if the equation defines y as a function of x.

(a) $2y - x^3 = 0$ **(b)** $x^2 + y^2 = 4$

SOLUTION

(a) If we solve the equation $2y - x^3 = 0$ for y, we obtain

$$y = \tfrac{1}{2}x^3$$

In this form, it is easy to see that for each input value x we choose, one and only one output value y corresponds to it. Hence, we conclude that the equation $2y - x^3 = 0$ defines y as a function of x.

We can also sketch the graph of the equation $2y - x^3 = 0$ by using the point-plotting method [see Example 4(c) in Section 1.2]. We note that every vertical line intersects this graph *at most once* (see Figure 1.25).

FIGURE 1.25
The graph of $2y - x^3 = 0$ passes the vertical line test.

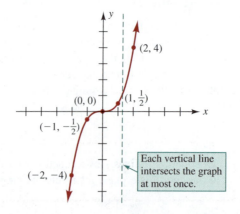

SECTION 1.3 Functions

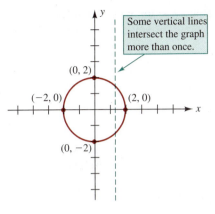

FIGURE 1.26
The graph of $x^2 + y^2 = 4$ fails the vertical line test.

Thus, we may also conclude by the vertical line test that the equation $2y - x^3 = 0$ defines y as a function of x.

(b) From Problem 6 in Section 1.2, we know that the graph of the equation $x^2 + y^2 = 4$ is a circle with center $(0, 0)$ and radius 2. The graph of this equation is shown in Figure 1.26. Since at least one vertical line intersects this graph twice, we know that the equation $x^2 + y^2 = 4$ does *not* define y as a function of x. Also, if we solve this equation for y, we obtain

$$y = \pm\sqrt{4 - x^2}.$$

This form of the equation makes it obvious that we have *two outputs* for each input value x in the interval $(-2, 2)$. ◆

Note: Taken separately, each of the equations $y = \sqrt{4 - x^2}$ and $y = -\sqrt{4 - x^2}$ defines y as a function of x, and the graph of each function is *half* a circle. Referring to Figure 1.26, we see that the graph of $y = \sqrt{4 - x^2}$ is the semicircle above the x-axis, and the graph of $y = -\sqrt{4 - x^2}$ is the semicircle below the x-axis. Note that each semicircle passes the vertical line test.

PROBLEM 1 Determine if the equation $2x + 3y = 6$ defines y as a function of x. ◆

◆ **Functional Notation**

The letters f, F, g, G, h and H are often used to represent functions. The *functional notation* $f(x)$ is read "f of x" and is defined as follows.

Functional Notation

> If f is a function and x is an input for the function, then the **functional notation**
>
> $$f(x), \quad \text{read "}f \text{ of } x\text{,"}$$
>
> denotes the corresponding output of the function.

The function f defined by the equation $y = x^2$ may be written as

$$f(x) = x^2,$$

where f specifies a rule for determining the value of an output $f(x)$ from a given input x. For instance, $f(-3)$ denotes the output for a given input of -3 using the rule "square the input." Hence,

$$f(-3) = (-3)^2 = 9.$$

Table 1.3 shows several other inputs x and outputs $f(x)$ for the function f defined by $f(x) = x^2$. We refer to the process of determining the value of $f(x)$ from a given input x as *computing the functional value*.

Table 1.3
Table of values for $f(x) = x^2$

x	-3	-2	-1	0	1	2	3
$f(x)$	9	4	1	0	1	4	9

EXAMPLE 2 Let G be a function defined by $G(x) = x^2 - 4x + 3$. Compute each functional value.

(a) $G(-2)$ (b) $G(2a)$ (c) $G(x-1)$

SOLUTION

(a) The notation $G(-2)$ denotes the output obtained from an input of -2. Replacing x with -2 in the rule for G, we obtain

$$G(-2) = (-2)^2 - 4(-2) + 3$$
$$= 4 + 8 + 3$$
$$= 15.$$

(b) The notation $G(2a)$ denotes the output obtained from an input of $2a$. Replacing x with $2a$ in the rule for G, we obtain

$$G(2a) = (2a)^2 - 4(2a) + 3$$
$$= 4a^2 - 8a + 3.$$

(c) The notation $G(x-1)$ denotes the output obtained from an input of $x-1$. Replacing x with $x-1$ in the rule for G, we obtain

$$G(x-1) = (x-1)^2 - 4(x-1) + 3$$
$$= x^2 - 2x + 1 - 4x + 4 + 3$$
$$= x^2 - 6x + 8.$$ ◆

PROBLEM 2 For the function G defined in Example 2, find $G(x - h)$. ◆

◆ Finding the Domain of a Function

We can think of the *domain* of a function as the set of all possible inputs. When a function is defined by an equation, the domain is assumed to be the set of all real numbers that give real number outputs. For the function f defined by $f(x) = x^2$, every real number assigned to x gives a corresponding real number for $f(x)$. Thus, the domain of the function f is the set of all real numbers. In this text, we express the domain of a function by using *interval notation* (see Section 1.1). Thus, we express the domain of f as $(-\infty, \infty)$.

The *range* of a function is the set of all outputs that we obtain from the elements in the domain. Consider the function f defined by $f(x) = x^2$. When real numbers are squared, we obtain *nonnegative* real numbers. Thus, using interval notation, we conclude that the range of f is $[0, \infty)$. For most functions, however, the range is quite difficult to find and is best obtained from the graph of the function, which we will discuss later in this section. As illustrated in the next example, we must be aware of two properties when determining the domain of a function.

(1) Division by zero   and   **(2)** Even roots of negative numbers.

EXAMPLE 3   Determine the domain of each function.

(a) $F(x) = \dfrac{2x + 5}{x^3 - 3x^2 + 2x - 6}$ **(b)** $g(x) = \dfrac{\sqrt{2 - x}}{x + 5}$

SOLUTION

(a) Remember that division by zero is undefined. Thus, the domain of this function is all real numbers x except those for which the denominator equals zero:

$$x^3 - 3x^2 + 2x - 6 = 0$$

$$x^2(x - 3) + 2(x - 3) = 0 \qquad \text{Factor by grouping terms}$$

$$(x - 3)(x^2 + 2) = 0$$

$$x = 3 \quad \text{or} \quad \underbrace{x^2 + 2 = 0}_{\text{No real solution}}$$

Hence, the domain of this function is all real numbers *except* $x = 3$. Using interval notation, we express the domain of this function as

$$(-\infty, 3) \cup (3, \infty).$$

(b) Remember that an even root of a negative number is *not* a real number. Thus, the radicand $2 - x$ must be *nonnegative*, that is,

$$2 - x \geq 0$$
$$-x \geq -2$$
$$x \leq 2.$$

Also, we must exclude $x = -5$ from the domain, since this value of x makes the denominator zero. Hence, the domain of this function is all real numbers less than or equal to 2 except $x = -5$. Using interval notation, we express the domain of this function as

$$(-\infty, -5) \cup (-5, 2].$$

PROBLEM 3 Determine the domain of the function H defined by $H(x) = \dfrac{x+5}{\sqrt{2-x}}$.

◆ **The Graph of a Function**

The graph of a function f is the same as the graph of the equation $y = f(x)$ and is defined as follows.

Graph of a Function

> The **graph of a function** f is the set of all points (x, y) in the coordinate plane such that x is in the domain of f and $y = f(x)$.

The point-plotting method we discussed in Section 1.2 may be used to sketch the graph of a simple function. We select various x-values from the domain of f and find the corresponding outputs $f(x)$. Using the inputs as the x-coordinates and the outputs as the y-coordinates, we plot the points $(x, f(x))$ in the coordinate plane. Connecting these points to form a smooth curve gives us the graph of f.

The graph of the function f defined by $f(x) = x^2$ is shown in Figure 1.27. The graph of this function is the same as the graph of the equation $y = x^2$ (see Figure 1.23). Referring to Figure 1.27, if we fold the coordinate plane along the y-axis, the right- and left-hand portions of the graph coincide. This means that for each point $(x, f(x))$ on the graph, there corresponds a point $(-x, f(x))$ that is also on the graph. We say that a graph with this characteristic is **symmetric with respect to the y-axis**. A function whose graph is symmetric with respect to the y-axis is called an **even function**. Hence, $f(x) = x^2$ is an example of an even function.

The graph of the function g defined by $g(x) = \tfrac{1}{2}x^3$ is shown in Figure 1.28. The graph of this function is the same as the graph of the equation $2y = x^3$ (see Figure 1.25). Referring to Figure 1.28, if we fold the coordinate plane along the x-axis and

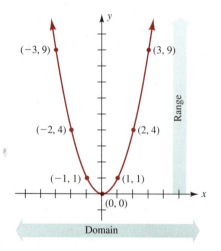

FIGURE 1.27
The graph of $f(x) = x^2$ is symmetric with respect to the y-axis. Hence, f is an even function.

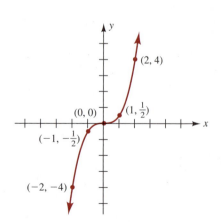

FIGURE 1.28
The graph of $g(x) = \tfrac{1}{2}x^3$ is symmetric with respect to the origin. Hence, g is an odd function.

then along the y-axis, the upper and lower portions of the graph coincide. This means that for each point $(x, f(x))$ on the graph, there corresponds a point $(-x, -f(x))$ that is also on the graph. We say a graph with this characteristic is **symmetric with respect to the origin.** A function whose graph is symmetric with respect to the origin is called an **odd function.** Hence, $g(x) = \frac{1}{2}x^3$ is an example of an odd function. The following tests can be used to determine if a function is even or odd.

Tests for Even and Odd Functions

1. A function f is an **even function** if $f(-x) = f(x)$ for every x in the domain of f.
2. A function f is an **odd function** if $f(-x) = -f(x)$ for every x in the domain of f.

EXAMPLE 4 Determine if the function is even, odd, or neither.

(a) $g(x) = \sqrt{x^2 - 4}$ (b) $h(x) = x - x^3$

SOLUTION

(a) Since
$$g(-x) = \sqrt{(-x)^2 - 4} = \sqrt{x^2 - 4} = g(x),$$
the function g is *even*. Thus, we know the graph of the function g is symmetric with respect to the y-axis.

(b) Since
$$h(-x) = (-x) - (-x)^3 = -x + x^3 = -(x - x^3) = -h(x),$$
the function h is *odd*. Thus, we know the graph of the function h is symmetric with respect to the origin. ◆

PROBLEM 4 Repeat Example 4 for the function h defined by $h(x) = x - x^2$. ◆

The y-coordinates of the points where a curve crosses the y-axis are called the **y-intercepts.** By the vertical line test, the graph of a function f can have *at most one* y-intercept. Since x is zero when the curve crosses the y-axis, the y-intercept can be found by evaluating $f(0)$.

The **x-intercepts** are the x-coordinates of the points where the curve crosses the x-axis. Since $f(x)$ is zero when the curve crosses the x-axis, the x-intercepts can be found by solving the equation $f(x) = 0$ for real values of x. Values of x for which $f(x) = 0$ are called the **zeros** of the function f.

EXAMPLE 5 Find the real zeros of each function, and then sketch the graph of the function:

(a) $g(x) = \sqrt{x^2 - 4}$ (b) $h(x) = x - x^3$

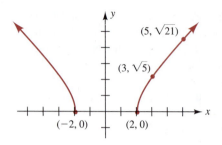

FIGURE 1.29

Graph of $g(x) = \sqrt{x^2 - 4}$, an *even* function with *zeros* ± 2.

SOLUTION

(a) The zeros of g are the roots of the equation $g(x) = \sqrt{x^2 - 4} = 0$. Squaring both sides, we have

$$x^2 - 4 = 0 \quad \text{or} \quad x = \pm 2.$$

Thus, the zeros of the function g are ± 2, and the x-intercepts of the graph of g are also ± 2. Using the point-plotting method, along with the fact that g is an even function with real zeros ± 2, we can sketch the graph of g. The graph of the function g is shown in Figure 1.29.

(b) The zeros of h are the roots of the equation $h(x) = x - x^3 = 0$. Factoring to form a product that equals zero, we have

$$x - x^3 = 0$$
$$x(1 - x^2) = 0$$
$$x = 0 \quad \text{or} \quad 1 - x^2 = 0$$
$$x = \pm 1$$

Thus, the zeros of the function h are 0 and ± 1, and the x-intercepts of the graph of h are also 0 and ± 1. Using the point-plotting method, along with the fact that h is an odd function with real zeros 0 and ± 1, we can sketch the graph of h. The graph of the function h is shown in Figure 1.30. ◆

The domain and range of a function are apparent from its graph. From the graph of the function g in Figure 1.29, we can state that the domain of g is $(-\infty, -2] \cup [2, \infty)$ and the range is $[0, \infty)$.

FIGURE 1.30

Graph of $h(x) = x - x^3$, an *odd* function with *zeros* 0 and ± 1.

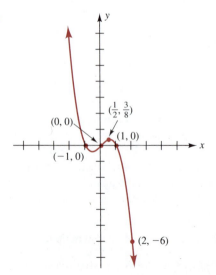

SECTION 1.3   Functions   31

PROBLEM 5   Use the graph of the function h in Figure 1.30 to find the domain and range of h. ◆

◆ **The Graphing Calculator**

Recent technology enables us to generate the graph of a function by using a personal computer with graphing software or a hand-held *graphing calculator*. Although you may not have access to a personal computer with graphing software, you can purchase a hand-held graphing calculator at a cost that is slightly more than a regular scientific calculator.

The key to using a graphing calculator is to select a viewing screen that shows the portion of the graph you wish to display. The viewing screen on a graphing calculator, which is often called the *viewing rectangle,* represents a portion of the coordinate plane. On most graphing calculators, the [RANGE] key is used to select a viewing rectangle. After pressing the [RANGE] key, you enter several values

x-min	x-axis minimum value	**y-min**	y-axis minimum value
x-max	x-axis maximum value	**y-max**	y-axis maximum value
x-scl	distance between scale marks on x-axis	**y-scl**	distance between scale marks on y-axis

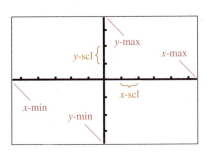

FIGURE 1.31
Viewing rectangle on a graphing calculator.

These values are shown in Figure 1.31.

For instance, to generate the graph of $g(x) = \sqrt{x^2 - 4}$ from Figure 1.29 on a graphing calculator, begin by pressing the [RANGE] key to choose a viewing rectangle. Although the choice of a viewing rectangle is arbitrary, Figure 1.29 suggests the viewing rectangle

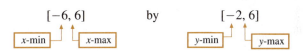

with x-scale and y-scale of 1 unit. After entering these values, we may now obtain the graph of $g(x) = \sqrt{x^2 - 4}$ in the viewing rectangle. For instance, on a *Texas Instrument TI-81,* we enter the keying sequence

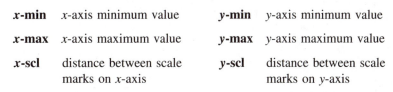

or, on a *CASIO fx-7000G,* we enter the keying sequence

The procedure is similar with other graphing calculators. If you have access to a graphing calculator, you may wish to consult the Graphing Calculator Supplement that accompanies this text for more detailed explanations.

The [TRACE] key on a graphing calculator can be used to display the x- and y-coordinates associated with each dot, or *pixel,* along the graph. With the [ZOOM] key we can magnify the graph n times. Look for the logo in the *Calculator Activities* section of each exercise set. This logo indicates that a problem is written for students who have access to a graphing calculator.

Do not use a graphing calculator to obtain a quick solution with no understanding of the fundamental mathematical ideas being presented. To become dependent upon a machine for solutions can produce disastrous effects. Instead, use it to explore, investigate, and check your mathematical problems. In this way, you will enhance your learning and understanding of college mathematics.

Exercises 1.3

 Basic Skills

In Exercises 1–14, determine if the given equation defines y as a function of x.

1. $x^2 + y - 1 = 0$
2. $3x + 2y = 6$
3. $x + y^2 = 1$
4. $x^2 + y - 2x = 0$
5. $y = \sqrt{x^2 - 4}$
6. $y = |x - 5|$
7. $x^2 + y^2 - 4x = 0$
8. $x^2 - y^2 - 4 = 0$
9. $|x|y = 1$
10. $|xy| = 1$
11. $y^3 + x - 1 = 0$
12. $x^3 + y^3 = 1$
13. $xy - x^2 = 1$
14. $x^2y + y = 1$

Given the functions f, g, and h, defined by
$$f(x) = 4x^2 - 2x + 1$$
$$g(x) = \sqrt{x} - 4, \quad \text{and} \quad h(x) = |2x + 3|,$$

respectively, compute the functional values in Exercises 15–34.

15. $f(2)$
16. $g(9)$
17. $h(-4)$
18. $g(\frac{1}{4}) + g(1)$
19. $f(\sqrt{2})$
20. $-h(\frac{3}{2})$
21. $h(ab)$
22. $f(3p)$
23. $g(t^2), \quad t \geq 0$
24. $f(-x)$
25. $f\left(\frac{n}{2}\right)$
26. $h(x^2 - 3)$
27. $g(x) - g(0)$
28. $f(x + 2)$
29. $g(1 + x^2)$
30. $f(x + h) - f(x)$
31. $f(\sqrt{x - 2})$
32. $h\left(-\frac{x}{2}\right)$
33. $f(g(x))$
34. $g(f(x))$

In Exercises 35–50, state the domain of each function using interval notation.

35. $f(x) = 3x + 2$
36. $g(x) = 1 - x^2$
37. $f(x) = x^3 + 2$
38. $f(x) = \sqrt[3]{2x - 1}$
39. $H(x) = \sqrt{4 - x}$
40. $h(x) = \sqrt{2x - 3}$
41. $g(x) = -\sqrt{16 - x^2}$
42. $F(x) = \sqrt{x^2 - 4}$
43. $f(x) = \dfrac{x}{x + 2}$
44. $f(x) = \dfrac{-3}{\sqrt{2 - x}}$
45. $G(x) = \dfrac{1}{x^2 - 4}$
46. $f(x) = \dfrac{2x}{x^2 + 3}$
47. $f(x) = \dfrac{-3}{x^2 + 3x - 10}$
48. $f(x) = \dfrac{2x - 3}{x^3 - 4x^2 + x - 4}$
49. $F(x) = \dfrac{\sqrt{2x - 1}}{4 - x}$
50. $H(x) = \sqrt{x^3 - 2x^2 - 8x}$

In Exercises 51–60, determine if the function is even, odd, or neither. If you have access to a graphing calculator, verify each answer by looking at the symmetry of the graph.

51. $F(x) = x$
52. $g(x) = x^3$
53. $f(x) = 3 - |x|$
54. $f(x) = 2x^2 - 4$
55. $f(x) = 4x - x^5$
56. $f(x) = x^4 + 2x^2 + 3$
57. $g(x) = \dfrac{3 - x^2}{2 + x^6}$
58. $H(x) = \dfrac{x + x^3}{2|x|}$
59. $h(x) = x^3 - 3x + 1$
60. $g(x) = (3x^2 - 2x)^2$

In Exercises 61–70, find the real zeros of each function. If you have access to a graphing calculator, verify each answer by tracing to the x-intercepts of the graph.

61. $g(x) = 3x - 5$
62. $f(x) = 4 + 5x$
63. $F(x) = x^2 - 3x - 40$
64. $g(x) = 2x^2 - 7x + 3$
65. $h(x) = x^2 - 8x + 4$
66. $f(x) = 2x^4 - 5x^2 - 12$
67. $g(x) = \dfrac{2x - 3}{x^2 + 4}$
68. $G(x) = \dfrac{9 - x^2}{2x + 3}$

69. $H(x) = 5 - \sqrt{x^4 + 9}$

70. $h(x) = x - \sqrt{4x^2 - x - 2}$

In Exercises 71–80, a function and its graph are given.

(a) Find the domain of the function.
(b) Find the range of the function.
(c) Find any real zeros of the function.
(d) Determine if the function is even, odd, or neither.

71. $f(x) = |x| - 2$

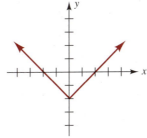

72. $g(x) = 3x - 2$

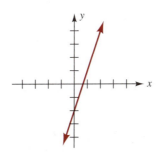

73. $g(x) = x^2 + 2x$

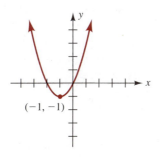

$(-1, -1)$

74. $H(x) = 2x^2 - x^4$

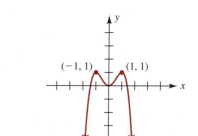

$(-1, 1)$ $(1, 1)$

75. $f(x) = \sqrt{5 - x^2}$

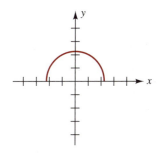

76. $g(x) = -\sqrt{x^2 - 3}$

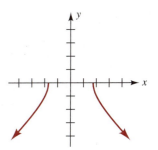

77. $f(x) = \dfrac{-2}{x^2 + 1}$

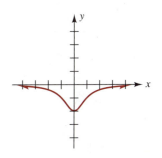

78. $g(x) = \dfrac{x^2 + 1}{x}$

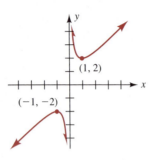

79. $g(x) = x^5 - 5x, \quad -2 \leq x \leq 2$

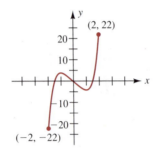

80. $f(x) = \begin{cases} x^2 & \text{if } -2 \leq x < 1 \\ 3 - 2x & \text{if } 1 \leq x \leq 3 \end{cases}$

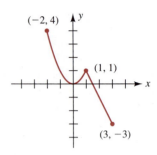

Critical Thinking

In calculus, it is necessary to evaluate the **difference quotient**

$$\dfrac{f(x + \Delta x) - f(x)}{\Delta x}$$

for a given function f. The symbol Δx is read "delta x" and is commonly used to denote a small change in x. In Exercises 81–88, find the difference quotient for the given function f and simplify the result.

81. $f(x) = 2x + 1$
82. $f(x) = 2 - 3x$
83. $f(x) = x^2$
84. $f(x) = x^3$
85. $f(x) = x^2 + 2x - 3$
86. $f(x) = 2x^2 - 3x$
87. $f(x) = \dfrac{1}{x}$
    (*Hint:* Subtract fractions in the numerator and eliminate Δx)
88. $f(x) = \sqrt{x}$
    (*Hint:* Rationalize the numerator and eliminate Δx.)

89. The radius r (in meters) of an oil spill is a function of the time t (in minutes) that the oil has been leaking. This function is given by

$$r(t) = \sqrt{2t - 1}, \quad t \geq 1.$$

(a) Evaluate and interpret $r(25)$.
(b) Find the time t when $r(t) = 5$ meters.

90. The bending moment M in pound-feet (lb-ft) along a simply supported 9-ft steel beam that carries a uniformly distributed load of 800 lb/ft is a function of the distance x (in feet) from one end of the beam and is given by

$$M(x) = \tfrac{1}{2}(9)(800)x - \tfrac{1}{2}(800)x^2$$
$$= 3600x - 400x^2, \quad 0 \leq x \leq 9.$$

(a) Evaluate and interpret $M(1)$.
(b) Find the distance when $M(x) = 8100$ lb-ft.

91. The population P of a certain organism is a function of the temperature t (in degrees Celsius) of the medium in which the organism exists and is given by

$$P(t) = 3t^2, \quad 0 \le t \le 40.$$

If t_1 is the present temperature of the medium, compare $P(t_1)$ with $P(\frac{1}{2}t_1)$, and state the effect that halving the temperature has on the population.

92. In a psychological experiment, a student is given electrical shocks of varying intensities and asked to rate the intensity of each shock in relation to an initial shock s_0, which is given a rating of 10. It is found that the response number R given by the student is a function of the magnitude (in milliamps) of the shock s and is defined by

$$R(s) = \frac{s^{3/2}}{40}.$$

Compare $R(s_0)$ with $R(4s_0)$, and state the effect that quadrupling the intensity has on the response number.

Calculator Activities

For Exercises 93–96, use the functions f and g defined by

$$f(x) = 14.6x^2 - 12.9x + 25.4$$

and

$$g(x) = \frac{26.5x - 92.4}{86.3 - 29.6x}$$

to compute the functional values to three significant digits.

93. $f(2.6)$

94. $f(-1.31)$

95. $g(-23.5)$

96. $g(0.035)$

In Exercises 97–100, find the real zeros of each function to three significant digits. If you have access to a graphing calculator, verify each answer by tracing to the x-intercept(s) of the graph.

97. $f(x) = 2.478x - 5.937$

98. $g(x) = 2.6x^2 - 3.7x - 8.9$

99. $h(x) = \dfrac{27.9x^2 - 93.2}{81.5x + 42.3}$

100. $F(x) = \sqrt{0.532x - 0.273} - 0.632$

101. The cost C (in dollars) for a taxicab fare is a function of the miles driven, x, and is given by

$$C(x) = 1.50 + 1.45x, \quad x \ge 0.$$

(a) Evaluate and interpret $C(12.4)$.

(b) Determine the miles driven when $C(x) = \$33.98$.

102. If \$5000 is borrowed at $10\frac{1}{2}\%$ simple interest per year, the amount A (in dollars) that must be paid back is a function of the time t (in years) over which it is borrowed and is given by

$$A(t) = 5000(1 + 0.105t), \quad t \ge 0.$$

(a) Evaluate and interpret $A(12)$.

(b) Find the time t if $A(t) = \$8150$.

1.4 Techniques of Graphing Functions

◆ Introductory Comments

Figure 1.32 shows the graphs of eight basic functions that occur frequently in mathematics: **constant function, identity function, absolute value function, squaring**

FIGURE 1.32
Graphs of eight basic functions with domains and ranges specified.

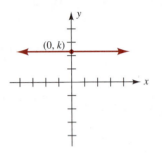

(a) Constant function
$f(x) = k$
Domain $(-\infty, \infty)$
Range $\{k\}$

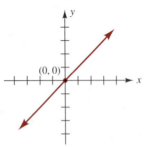

(b) Identity function
$f(x) = x$
Domain $(-\infty, \infty)$
Range $(-\infty, \infty)$

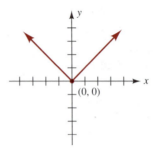

(c) Absolute value function
$f(x) = |x|$
Domain $(-\infty, \infty)$
Range $[0, \infty)$

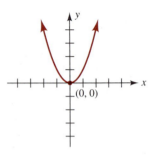

(d) Squaring function
$f(x) = x^2$
Domain $(-\infty, \infty)$
Range $[0, \infty)$

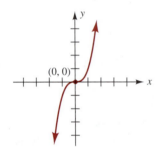

(e) Cubing function
$f(x) = x^3$
Domain $(-\infty, \infty)$
Range $(-\infty, \infty)$

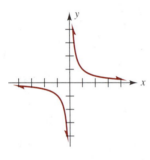

(f) Reciprocal function
$f(x) = 1/x$
Domain $(-\infty, 0) \cup (0, \infty)$
Range $(-\infty, 0) \cup (0, \infty)$

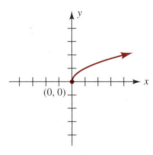

(g) Square root function
$f(x) = \sqrt{x}$
Domain $[0, \infty)$
Range $[0, \infty)$

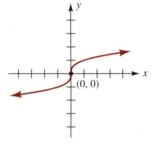

(h) Cube root function
$f(x) = \sqrt[3]{x}$
Domain $(-\infty, \infty)$
Range $(-\infty, \infty)$

function, cubing function, reciprocal function, square root function, and **cube root function**. By plotting points or using a graphing calculator, we can verify the graph of each of these function.

In this section, we use the graphs of these eight basic functions to sketch the graphs of many other related functions by

1. *shifting* a graph vertically or horizontally,
2. *reflecting* a graph about the *x*-axis or *y*-axis, and
3. *stretching* or *shrinking* a graph.

◆ Vertical Shift Rule

Suppose we wish to sketch the graph of the function F defined by

$$F(x) = |x| + 2.$$

We can set up a table of values and plot points, as shown in Figure 1.33. However, notice the graph of this function is the same as the graph of $f(x) = |x|$, the absolute value function shown in Figure 1.32(c), but *shifted vertically upward* 2 units. When we substitute the same input into the functions F and f, the output $F(x)$ is always 2 more than the output $f(x)$. Thus, *without plotting points,* we can sketch the graph of $F(x) = |x| + 2$ by shifting the graph of $f(x) = |x|$ vertically upward 2 units. In summary, we state the **vertical shift rule.**

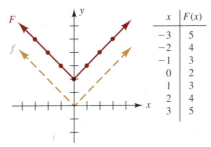

FIGURE 1.33
Comparison of the graphs of $F(x) = |x| + 2$ and $f(x) = |x|$.

x	F(x)
−3	5
−2	4
−1	3
0	2
1	3
2	4
3	5

Vertical Shift Rule

If f is a function and c is a constant, then the graph of the function F, defined by

$$F(x) = f(x) + c$$

is the same as the graph of f shifted *vertically upward* $|c|$ units, if $c > 0$ or shifted *vertically downward* $|c|$ units if $c < 0$.

To obtain an accurate sketch, it is a good practice to determine and label the *axis intercepts* of the graph. Remember, the *x-intercepts* of the graph of a function f are the real zeros of the function and may be found by solving the equation $f(x) = 0$. The *y-intercept* of the graph of a function f may be found by evaluating $f(0)$. The graph of a function can have at most one *y*-intercept.

EXAMPLE 1   Sketch the graph of each function and label the *x*-intercept and *y*-intercept.

(a) $F(x) = x^2 + 1$ (b) $G(x) = \sqrt{x} - 2$

SOLUTION

(a) We can think of this function as $F(x) = f(x) + 1$, where $f(x) = x^2$, the squaring function shown in Figure 1.32(d). Thus, by the vertical shift rule, the graph of $F(x) = x^2 + 1$ is the same as the graph of $f(x) = x^2$ shifted

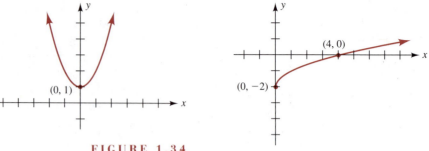

FIGURE 1.34
Graph of $F(x) = x^2 + 1$.

FIGURE 1.35
Graph of $G(x) = \sqrt{x} - 2$.

vertically *upward* 1 unit. The graph of the function F is shown in Figure 1.34.

x-intercept: $F(x) = 0$ *y*-intercept: $F(0) = (0)^2 + 1 = 1$

$$x^2 + 1 = 0$$

$$x^2 = -1$$

No real solution.
No *x*-intercept.

(b) We can think of this function as $G(x) = f(x) - 2$, where $f(x) = \sqrt{x}$, the squre root function shown in Figure 1.32(g). Thus, by the vertical shift rule, the graph of $G(x) = \sqrt{x} - 2$ is the same as the graph of $f(x) = \sqrt{x}$ shifted vertically *downward* 2 units. The graph of the function G is shown in Figure 1.35.

x-intercept: *y*-intercept:

$$G(x) = 0$$ $$G(0) = \sqrt{0} - 2 = -2$$

$$\sqrt{x} - 2 = 0$$

$$\sqrt{x} = 2$$

$$x = 4 \quad \text{Square both sides}$$

PROBLEM 1   Sketch the graph of $h(x) = x^3 - 1$ and label the *x*-intercept and *y*-intercept.

♦ Horizontal Shift Rule

Suppose we wish to sketch the graph of the function F defined by

$$F(x) = |x + 2|.$$

We can set up a table of values and plot points as shown in Figure 1.36. However, notice that the graph of this function is the same as the graph of $f(x) = |x|$ in Fig-

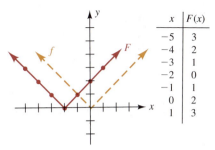

FIGURE 1.36
Comparison of the graphs of
$F(x) = |x + 2|$ and $f(x) = |x|$.

ure 1.32(c) *shifted horizontally to the left* 2 units. When we substitute an input into F that is 2 *less than* the input we use in f, the values of the outputs $F(x)$ and $f(x)$ are equal. Thus, *without plotting points,* we can sketch the graph of $F(x) = |x + 2|$ by shifting the graph of $f(x) = |x|$ horizontally to the left 2 units. In summary, we state the **horizontal shift rule.**

Horizontal Shift Rule

If f is a function and c is a constant, then the graph of the function F defined by

$$F(x) = f(x + c)$$

is the same as the graph f shifted *horizontally to the left* $|c|$ units if $c > 0$ or shifted *horizontally to the right* $|c|$ units if $c < 0$.

EXAMPLE 2 Sketch the graph of each function and label the x-intercept and y-intercept.

(a) $F(x) = (x + 1)^2$ (b) $G(x) = \sqrt[3]{x - 2}$

SOLUTION

(a) We can think of this function as $F(x) = f(x + 1)$, where $f(x) = x^2$, from Figure 1.32(d). Thus, by the horizontal shift rule, the graph of $F(x) = (x + 1)^2$ is the same as the graph of $f(x) = x^2$ shifted horizontally *to the left* 1 unit. The graph of the function F is shown in Figure 1.37

x-intercept: $F(x) = 0$ y-intercept: $F(0) = (0 + 1)^2 = 1$

$(x + 1)^2 = 0$

$x + 1 = 0$ **Take the square root of both sides**

$x = -1$

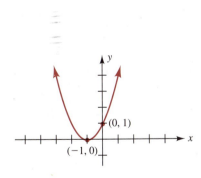

FIGURE 1.37
Graph of $F(x) = (x + 1)^2$.

(b) We can think of this function as $G(x) = f(x - 2)$, where $f(x) = \sqrt[3]{x}$, the cube root function shown in Figure 1.32(h). Thus, by the horizontal shift rule, the graph of $G(x) = \sqrt[3]{x - 2}$ is the same as the graph of $f(x) = \sqrt[3]{x}$ shifted horizontally *to the right* 2 units.

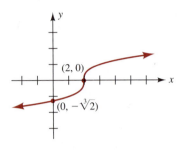

FIGURE 1.38
Graph of $G(x) = \sqrt[3]{x - 2}$.

x-intercept: $\quad G(x) = 0 \quad\quad$ *y*-intercept: $\quad G(0) = \sqrt[3]{(0) - 2} = \sqrt[3]{-2} \approx -1.26$

$$\sqrt[3]{x - 2} = 0$$

$$x - 2 = 0 \quad\quad \textbf{Cube both sides}$$

$$x = 2$$

The graph of the function G is shown in Figure 1.38. ◆

PROBLEM 2 Sketch the graph of $h(x) = (x - 1)^3$ and label the *x*-intercept and *y*-intercept. ◆

As illustrated in the next example, we can use both the vertical and horizontal shift rules to help sketch the graph of a function.

EXAMPLE 3 Sketch the graph of $F(x) = \dfrac{1}{x - 2} + 1$ and label the *x*-intercept and *y*-intercept.

SOLUTION We can think of this function as $F(x) = f(x - 2) + 1$, where $f(x) = 1/x$, the reciprocal function shown in Figure 1.32(f). Thus, by the horizontal and vertical shift rules, we obtain the graph of F by shifting the graph of $f(x) = 1/x$ *to the right* 2 units and *upward* 1 unit as shown in Figure 1.39.

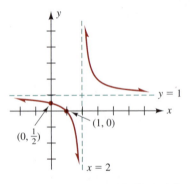

FIGURE 1.39
Graph of $F(x) = \dfrac{1}{x - 2} + 1$.

x-intercept: $\quad F(x) = 0 \quad\quad$ *y*-intercept: $\quad F(0) = \dfrac{1}{0 - 2} + 1 = \dfrac{1}{2}$

$$\dfrac{1}{x - 2} + 1 = 0$$

$$1 + (x - 2) = 0 \quad\quad \textbf{Multiply both sides by } (x - 2)$$

$$x = 1$$
◆

Referring to Figure 1.39, the line $x = 2$ is called a *vertical asymptote* for the graph of F, and the line $y = 1$ is called a *horizontal asymptote* for the graph of F. For a discussion of vertical and horizontal asymptotes, see Appendix B.

PROBLEM 3 Sketch the graph of $F(x) = (x + 1)^2 - 4$ and label the *x*-intercepts and *y*-intercept. ◆

◆ ***x*-Axis and *y*-Axis Reflection Rules**

Consider the functions f and F defined by

$$f(x) = \sqrt{x} \quad \text{and} \quad F(x) = -\sqrt{x},$$

FIGURE 1.40
Comparison of the graphs of
$f(x) = \sqrt{x}$ and $F(x) = -\sqrt{x}$.

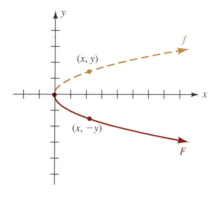

respectively. [As shown in Figure 1.32(g), f is the square root function.] If we substitute the same input into the functions f and F, the output $F(x)$ is always the *negative* of the output $f(x)$. Hence, as shown in Figure 1.40, the graph of F is the same as the graph of f *reflected about the x-axis*. In summary, we state the **x-axis reflection rule**.

 x-Axis Reflection Rule

> If f is a function, then the graph of the function F defined by
>
> $$F(x) = -f(x)$$
>
> is the same as the graph of f reflected about the x-axis.

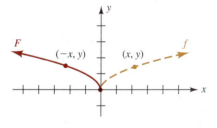

FIGURE 1.41
Comparison of the graphs of
$f(x) = \sqrt{x}$ and $F(x) = \sqrt{-x}$.

Consider the functions f and F defined by

$$f(x) = \sqrt{x} \quad \text{and} \quad F(x) = \sqrt{-x},$$

respectively. If we substitute an input into the function F that is the *opposite* of the input we use for f, the values of the outputs $F(x)$ and $f(x)$ are equal. Hence, as shown in Figure 1.41, the graph of F is the same as the graph of f *reflected about the y-axis*. In summary, we state the **y-axis reflection rule**.

 y-Axis Reflection Rule

> If f is a function, then the graph of the function F defined by
>
> $$F(x) = f(-x)$$
>
> is the same as the graph of f reflected about the y-axis.

EXAMPLE 4 Sketch the graph of each function and label the x-intercept and y-intercept.

(a) $F(x) = 2 - x^3$ (b) $h(x) = \sqrt{2-x}$

FIGURE 1.42
Graph of $F(x) = 2 - x^3$.

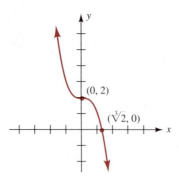

SOLUTION

(a) We can think of this function as $F(x) = -f(x) + 2$, where $f(x) = x^3$, the cubing function shown in Figure 1.32(e). Thus, by the x-axis reflection rule and the vertical shift rule, the graph of $F(x) = 2 - x^3$ is obtained by reflecting the graph of $y = x^3$ about the x-axis and then shifting this graph upward 2 units as shown in Figure 1.42.

x-intercept: $F(x) = 0$ y-intercept: $F(0) = 2 - (0)^3 = 2$
$2 - x^3 = 0$
$x^3 = 2$
$x = \sqrt[3]{2} \approx 1.26$

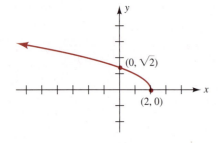

FIGURE 1.43
Graph of $h(x) = \sqrt{2-x}$.

(b) We can think of this function as $h(x) = f(x - 2)$, where $f(x) = \sqrt{-x}$. Thus, by the y-axis reflection rule and the horizontal shift rule, the graph of $h(x) = \sqrt{2 - x}$ is obtained by reflecting the graph of $y = \sqrt{x}$ about the y-axis and then shifting this graph to the right 2 units as shown in Figure 1.43.

x-intercept: $h(x) = 0$ y-intercept: $h(0) = \sqrt{2 + 0} = \sqrt{2} \approx 1.41$
$\sqrt{2 - x} = 0$
$2 - x = 0$ **Square both sides**
$x = 2$

◆

PROBLEM 4 Sketch the graph of $F(x) = -|x + 2|$ and label the x-intercept and y-intercept.

◆

◆ **Vertical Stretch and Shrink Rule**

Consider the functions f, F, and G defined by

$$f(x) = \sqrt{x}, \quad F(x) = 2\sqrt{x}, \quad \text{and} \quad G(x) = \tfrac{1}{2}\sqrt{x},$$

respectively. If we substitute the same input into the functions f, F, and G, the output $F(x)$ is *twice* the output $f(x)$, and the output $G(x)$ is *half* the output $f(x)$. As illustrated in Figure 1.44, the graphs of F and G are similar to the graph of f. The

FIGURE 1.44
Comparison of the graphs of $f(x) = \sqrt{x}$, $F(x) = 2\sqrt{x}$, and $G(x) = \frac{1}{2}\sqrt{x}$.

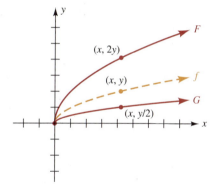

graph of F is *stretched vertically* by a factor of 2, and the graph of G is *shrunk vertically* by a factor of 2. In summary, we state the **vertical stretch and shrink rule.**

Vertical Stretch and Shrink Rule

> If f is a function and c is a real number with $c > 1$, then the graph of the function F defined by
>
> $$F(x) = cf(x)$$
>
> is similar to the graph of f *stretched vertically* by a factor of c, and the graph of the function G defined by
>
> $$G(x) = \frac{1}{c}f(x)$$
>
> is similar to the graph of f *shrunk vertically* by a factor of c.

EXAMPLE 5 Sketch the graph of each function and label the *x*-intercept and *y*-intercept.

(a) $G(x) = -\dfrac{|x|}{4}$ (b) $F(x) = 4(x - 1)$

SOLUTION

(a) We can think of this function as $G(x) = \frac{1}{4}f(x)$, where $f(x) = -|x|$. Thus, by the *x*-axis reflection rule and the vertical stretch and shrink rule, we obtain the graph of $G(x) = -|x|/4$ by reflecting the graph of $y = |x|$ about the *x*-axis, and then shrinking this graph by a factor of 4, as shown in Figure 1.45.

(b) We can think of this function as $F(x) = 4f(x)$, where $f(x) = x - 1$. By the vertical shift rule, the graph of f is the same as the graph of $y = x$, the identity function shown in Figure 1.32(b), shifted vertically downward 1 unit. The graph of $F(x) = 4(x - 1)$ is then obtained by stretching the graph of f by a factor of 4 as shown in Figure 1.46.

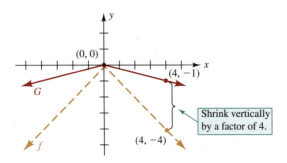

FIGURE 1.45
Graph of $G(x) = -|x|/4$ formed from the graph of $f(x) = -|x|$.

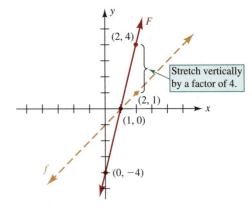

FIGURE 1.46
Graph of $F(x) = 4(x - 1)$ formed from the graph of $f(x) = x - 1$.

x-intercept: $\quad F(x) = 0 \qquad$ *y-intercept:* $\quad F(0) = 4(0 - 1) = -4$
$$4(x - 1) = 0$$
$$x - 1 = 0$$
$$x = 1$$

PROBLEM 5   Sketch the graph of $F(x) = 4x - 1$ and label the *x*-intercept and *y*-intercept.

◆ Increasing and Decreasing Functions

A function f is said to be an *increasing function* if as x increases, the value of $f(x)$ also *increases*. A function f is said to be a *decreasing function* if as x increases, $f(x)$ *decreases*. More precisely, we state the following definitions.

 Increasing and Decreasing Functions

1. A function f is an **increasing function** if for all a and b in the domain of f,

   $$f(a) < f(b) \quad \text{whenever} \quad a < b.$$

2. A function f is a **decreasing function** if for all a and b in the domain of f,

   $$f(a) > f(b) \quad \text{whenever} \quad a < b.$$

From the graph of a function, it is usually easy to determine if a function is increasing or decreasing.

EXAMPLE 6 Determine whether each function is increasing, decreasing, or neither.

(a) $F(x) = 4(x - 1)$ (b) $h(x) = \sqrt{2 - x}$ (c) $G(x) = -\dfrac{|x|}{4}$

SOLUTION

(a) The graph of $F(x) = 4(x - 1)$, shown in Figure 1.46, always *rises* as we move from left to right along the *x*-axis. Thus, as *x* increases, $F(x)$ also increases, and we conclude the function F is an increasing function.

(b) The graph of $h(x) = \sqrt{2 - x}$, shown in Figure 1.43, always *falls* as we move from left to right along the *x*-axis. Hence, as *x* increases, $h(x)$ decreases, and we conclude the function h is a decreasing function.

(c) The graph of $G(x) = -|x|/4$, shown in Figure 1.45, rises as we move from left to right until we reach the *y*-axis. The graph then falls as we move right along the *x*-axis, starting from the *y*-axis. Thus, this function is *neither* increasing nor decreasing over its entire domain. This situation is typical of many functions: A function may increase on some intervals in its domain and decrease on other intervals. For $G(x) = -|x|/4$, we say that G is *increasing on the interval* $(-\infty, 0)$ and *decreasing on the interval* $(0, \infty)$ ◆

PROBLEM 6 Referring to Figure 1.39, determine whether the function F defined by $F(x) = \dfrac{1}{x - 2} + 1$ is increasing, decreasing, or neither. ◆

Exercises 1.4

Basic Skills

In Exercises 1–42, sketch the graph of each function and label the x-intercept(s) and y-intercept. If you have access to a graphing calculator, verify each graph.

1. $f(x) = x + 3$
2. $g(x) = x^2 - 1$
3. $G(x) = x^3 - 2$
4. $H(x) = \dfrac{1}{x} + 2$
5. $F(x) = 3 + \sqrt{x}$
6. $f(x) = -2 + |x|$
7. $f(x) = |x - 1|$
8. $g(x) = \sqrt{x + 3}$
9. $H(x) = (x + 2)^2$
10. $F(x) = (x + 1)^3$
11. $F(x) = \dfrac{1}{x + 2}$
12. $f(x) = \sqrt[3]{x - 1}$
13. $h(x) = -\dfrac{1}{x}$
14. $g(x) = -x^3$
15. $f(x) = -x^2$
16. $H(x) = \sqrt[3]{-x}$
17. $g(x) = \sqrt{1 - x}$
18. $f(x) = -\sqrt{3 - x}$
19. $F(x) = (x - 1)^2 + 3$
20. $G(x) = |x - 4| - 1$
21. $f(x) = \sqrt{x + 2} - 1$
22. $h(x) = \dfrac{1}{x + 1} - 2$
23. $h(x) = -|x - 3|$
24. $f(x) = -(x + 1)$
25. $G(x) = 3 - x^2$
26. $H(x) = 2 - \sqrt{x}$
27. $F(x) = 2 - (x + 4)^3$
28. $h(x) = 1 - |x - 1|$
29. $f(x) = \sqrt[3]{x + 2} - 3$
30. $f(x) = 2 - \sqrt{4 - x}$
31. $H(x) = \dfrac{1}{2 - x}$
32. $g(x) = 3 + (2 - x)^2$
33. $g(x) = 3|x|$
34. $f(x) = -2x$
35. $f(x) = -\tfrac{1}{3}x^2$
36. $G(x) = \dfrac{x^3}{4}$
37. $G(x) = 2x - 1$
38. $F(x) = 3 - \tfrac{1}{2}x$
39. $F(x) = 2\sqrt{x - 4}$
40. $g(x) = 3 - 2|x|$
41. $F(x) = \dfrac{5}{x + 2}$
42. $h(x) = \dfrac{(x - 1)^2}{4}$

In Exercises 43–50, determine whether the function is increasing, decreasing, or neither.

43. The function G in Exercise 3.

44. The function g in Exercise 8.

45. The function g in Exercise 17.

46. The function g in Exercise 14.

47. The function G in Exercise 25.

48. The function G in Exercise 20.

49. The function f in Exercise 29.

50. The function f in Exercise 34.

  Critical Thinking

The graph of $y = f(x)$ is shown in the figure. Use this graph and the techniques of shifting and reflecting to sketch the graph of each of the equations in Exercises 51–54.

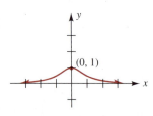

51. $y = -f(x)$

52. $y = f(x) - 3$

53. $y = f(x - 2)$

54. $y = 2 - f(x + 1)$

The graph of $y = \sqrt{4 - x^2}$ is shown in the figure. Use this graph to write an equation for each of the functions whose graphs are shown in Exercises 55–58.

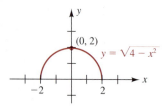

55.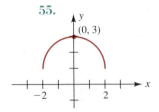

56.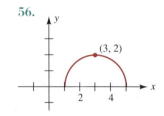

57.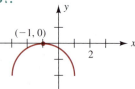

58.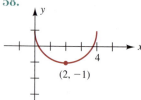

59. The cost C (in cents) to mail a letter with first-class postage is a function of a letter's weight w (in ounces) and is given by

$$C(w) = \begin{cases} 29 & \text{if } 0 < w \leq 1 \\ 52 & \text{if } 1 < w \leq 2 \\ 75 & \text{if } 2 < w \leq 3 \\ 98 & \text{if } 3 < w \leq 4 \\ 121 & \text{if } 4 < w \leq 5 \end{cases}$$

This function is called a *step function,* because its graph looks like a step of stairs.

(a) Sketch the graph of this function.

(b) Evaluate and interpret $C(2.4)$.

60. In mathematics, the notation $[\![x]\!]$ denotes the largest integer n such that $n \leq x$. For example,

$$[\![8.9]\!] = 8, \quad [\![5]\!] = 5, \quad [\![\pi]\!] = 3,$$
$$[\![-8.9]\!] = -9, \quad [\![-\sqrt{3}]\!] = -2,$$

and so on. The *greatest integer function* f defined by $f(x) = [\![x]\!]$ is another example of a step function (see Exercise 59). Sketch the graph of the function f.

Calculator Activities

In Exercises 61–65, sketch the graph of each function and determine to three significant digits the x-intercept(s) and y-intercept. If you have access to a graphing calculator, verify each answer by tracing to the x-intercept(s) and y-intercept of the graph.

61. $f(x) = 2.35x + 9.45$ **62.** $g(x) = 5.62 - 1.51x^3$

63. $h(x) = 0.85(x - 1.67)^2$

64. $F(x) = 1.34|x + 2.65| - 4.46$

65. $G(x) = 1.70 + \sqrt{2.32 - x}$

66. Use a graphing calculator to generate the graphs of the following equations in the same viewing rectangle:

$$f(x) = |x| \quad g(x) = |x| + 2 \quad h(x) = |x| + 1$$
$$G(x) = |x| - 1 \quad \text{and} \quad H(x) = |x| - 2$$

What rule in this section verifies the picture you observe?

67. Use a graphing calculator to generate the graphs of the following equations in the same viewing rectangle:

$$f(x) = x^3 \quad g(x) = (x + 2)^3 \quad h(x) = (x + 1)^3$$
$$G(x) = (x - 1)^3 \quad \text{and} \quad H(x) = (x - 2)^3$$

What rule in this section verifies the picture you observe?

68. Use a graphing calculator to generate the graphs of the following equations in the same viewing rectangle:

$$f(x) = x^2 \quad g(x) = 0.25x^2 \quad h(x) = 0.5x^2$$
$$G(x) = 2x^2 \quad \text{and} \quad H(x) = 4x^2$$

What rule in this section verifies the picture you observe?

1.5 Composite and Inverse Functions

If we select a number, cube it, and then take the cube root of the result, we obtain the number which we selected to cube. Reversing the procedure, that is, taking the cube root first and then cubing the result, also returns us to the original number. Since one operation "undoes" the other, we say that cubing and taking a cube root are *inverse operations*. In this section, we discuss functions that behave in a similar manner. We begin by discussing a way in which we may combine two functions.

◆ Composition of Functions

One method of combining two functions is called the *composition of functions*. If f and g are functions, then f composed with g, denoted $f \circ g$, is formed by using the output of g as the input of f. Similarly, g composed with f, denoted $g \circ f$, is formed by using the output of f as the input of g.

Composite Function

> If f and g are functions, then the **composite function** $f \circ g$ is defined by
>
> $$(f \circ g)(x) = f(g(x))$$
>
> and the **composite function** $g \circ f$ is defined by
>
> $$(g \circ f)(x) = g(f(x)).$$

Remember that the rule $f \circ g$ tells us to apply g first, then f, whereas the rule $g \circ f$ tells us to apply f first, then g.

EXAMPLE 1 Use the graphs of f and g in Figures 1.47 and 1.48 to compute each functional value.

(a) $(f \circ g)(4)$ (b) $(g \circ f)(0)$

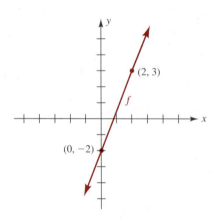

FIGURE 1.47   FIGURE 1.48

SOLUTION

(a) From the graph of g in Figure 1.48, we observe that $g(4) = 2$. Thus,

$$(f \circ g)(4) = f(g(4)) = f(2).$$

Now, from the graph of f in Figure 1.47, we observe that $f(2) = 3$. Hence,

$$(f \circ g)(4) = 3.$$

(b) From the graph of f in Figure 1.47, we observe that $f(0) = -2$. Thus,

$$(g \circ f)(0) = g(f(0)) = g(-2).$$

However, as shown in Figure 1.48, -2 is *not* in the domain of g. Hence, we conclude that

$$(g \circ f)(0) \quad \text{is undefined.} \qquad \blacklozenge$$

PROBLEM 1 Repeat Example 1 for $(f \circ g)(0)$.   ◆

Observe in Example 1(b) that for $g(f(x))$ to make sense, the values of $f(x)$ must be acceptable inputs for the function g. Remember, *the domain of $g \circ f$ is the set of all inputs x in the domain of f such that $f(x)$ is in the domain of g.* Likewise, *the domain of $f \circ g$ is the set of all inputs in the domain of g such that $g(x)$ is in the domain of f.*

SECTION 1.5 Composite and Inverse Functions

EXAMPLE 2 Given the functions f and g defined by $f(x) = \sqrt{x}$ and $g(x) = x^2 - 4$, find each of the following functions, and determine the domain of the new function.

(a) $f \circ g$ (b) $g \circ f$

SOLUTION

(a) The function $f \circ g$ is defined by

$$(f \circ g)(x) = f(g(x)) = f(x^2 - 4) = \sqrt{x^2 - 4}.$$

The domain of $f \circ g$ is all real numbers in the domain of g such that $g(x)$ is in the domain of f. The domain of g is $(-\infty, \infty)$, and the domain of f is $[0, \infty)$. Thus, from the interval $(-\infty, \infty)$ we can select only inputs x such that the outputs $g(x)$ are *nonnegative*:

$$g(x) \geq 0$$
$$x^2 - 4 \geq 0$$
$$x \leq -2 \quad \text{or} \quad x \geq 2.$$

Hence, the domain of $f \circ g$ is $(-\infty, -2] \cup [2, \infty)$.

(b) The function $g \circ f$ is defined by

$$(g \circ f)(x) = g(f(x)) = g(\sqrt{x}) = (\sqrt{x})^2 - 4 = x - 4.$$

The domain of $g \circ f$ is all real numbers in the domain of f such that $f(x)$ is in the domain of g. Since the domain of f is $[0, \infty)$, we can choose inputs only from this interval. Since each *nonnegative* input x gives us an output $f(x)$ that is in the domain of g, the domain of $g \circ f$ is also $[0, \infty)$. Hence we write

$$(g \circ f)(x) = x - 4, \quad x \geq 0. \qquad \blacklozenge$$

As illustrated in Example 2, the composite functions $f \circ g$ and $g \circ f$ are *not* necessarily the same.

PROBLEM 2 For the functions f and g given in Example 2, find $(g \circ f)(9)$ by

(a) first finding $f(9)$, and then evaluating $g(f(9))$.

(b) using the result of Example 2(b), $(g \circ f)(x) = x - 4$. $\qquad \blacklozenge$

◆ **Verifying Inverse Functions**

We now discuss a special class of functions in which

$$(f \circ g)(x) = (g \circ f)(x) = x.$$

Consider the *cubing function* f defined by $f(x) = x^3$ and the *cube root function* g defined by $g(x) = \sqrt[3]{x}$. The composite function $g \circ f$ represents cubing a number and then taking the cube root of the result, whereas the composite function $f \circ g$ represents taking the cube root of a number and then cubing the result. Note that

$$(g \circ f)(x) = g(f(x)) = g(x^3) = \sqrt[3]{x^3} = x$$

and

$$(f \circ g)(x) = f(g(x)) = f(\sqrt[3]{x}) = (\sqrt[3]{x})^3 = x.$$

In summary, composing f with g in either order is the *identity function,* the function that assigns each input to itself. Since one function "undoes" the other, we say that the cubing function and the cube root function are **inverse functions.**

Inverse Functions

> The functions f and g are **inverses** of each other if
>
> $$f(g(x)) = x \quad \text{for all } x \text{ in the domain of } g$$
>
> and
>
> $$g(f(x)) = x \quad \text{for all } x \text{ in the domain of } f.$$

If f is the inverse of g, then g is the inverse of f, that is, inverses come in *pairs.* Also, since inverse functions deal with the composition of functions, it is essential to check the domains and ranges to be certain that the domain of f equals the range of g and the domain of g equals the range of f.

EXAMPLE 3 Show that the functions f and g defined by

$$f(x) = 2x - 8 \quad \text{and} \quad g(x) = \tfrac{1}{2}x + 4$$

are inverses of each other.

SOLUTION Since the domain and range of both f and g are the set of all real numbers, we know the compositions of f and g exist. Thus, we proceed to show that $f(g(x)) = g(f(x)) = x$.

$$\begin{aligned}
f(g(x)) &= f(\tfrac{1}{2}x + 4) & g(f(x)) &= g(2x - 8) \\
&= 2(\tfrac{1}{2}x + 4) - 8 & &= \tfrac{1}{2}(2x - 8) + 4 \\
&= x + 8 - 8 & &= x - 4 + 4 \\
&= x & &= x
\end{aligned}$$

Since $f(g(x)) = g(f(x)) = x$, we conclude that the functions f and g are inverses of each other. ◆

SECTION 1.5 Composite and Inverse Functions

PROBLEM 3 Repeat Example 3 if $f(x) = (x - 2)^5$ and $g(x) = \sqrt[5]{x} + 2$.

◆ One-to-One Functions

There exists an inverse function for a given function f only if f is a *one-to-one function*. A function f is said to be **one-to-one** if each element in the range of f is associated with only one element in its domain. More precisely, we state the following definition.

◆ **One-to-One Function**

> A function f is a **one-to-one function** if, for a and b in the domain of f,
>
> $$f(a) = f(b) \text{ implies } a = b.$$

A simple method for determining whether a function is *one-to-one* is to look at its graph. If every horizontal line that can be drawn in the coordinate plane intersects the graph at most once, then each output of the function is associated with only one input. Hence, the function is *one-to-one*. We refer to this graphical method of determining whether a function is one-to-one as the **horizontal line test.**

◆ **Horizontal Line Test**

> A function f is one-to-one if no horizontal line in the coordinate plane intersects the graph of the function in more than one point.

EXAMPLE 4 Determine if the given function is one-to-one.

(a) $f(x) = (x - 1)^3$ **(b)** $h(x) = x^2 - 1$

SOLUTION

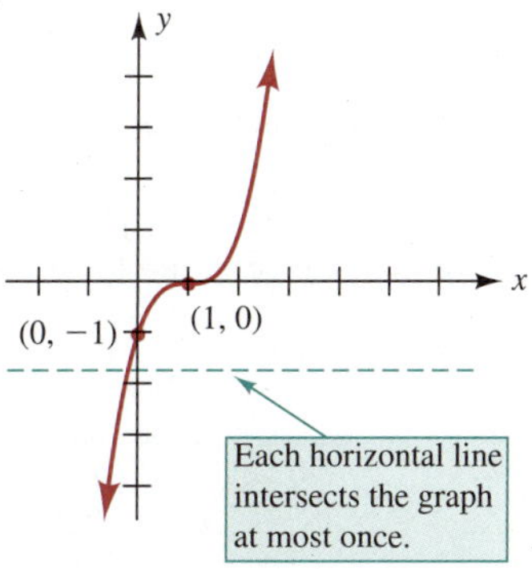

(a) By the horizontal shift rule (Section 1.4), the graph of $f(x) = (x - 1)^3$ is the same as the graph of $y = x^3$ shifted to the right 1 unit, as shown in Figure 1.49. No horizontal line drawn in the coordinate plane intersects the graph more than once. Thus, by the horizontal line test, the function f is a one-to-one function. Also, note that

$$f(a) = f(b)$$

implies $(a - 1)^3 = (b - 1)^3$

$a - 1 = b - 1$ **Take the cube root of both sides**

$a = b$ **Add 1 to both sides**

FIGURE 1.49
The graph of $f(x) = (x - 1)^3$ passes the horizontal line test. Hence, f is one-to-one.

(0, −1) (1, 0)

Each horizontal line intersects the graph at most once.

Since $f(a) = f(b)$ implies $a = b$, we conclude that f is one-to-one.

(b) By the vertical shift rule (Section 1.4), the graph of $h(x) = x^2 - 1$ is the same as the graph of $y = x^2$ shifted downward 1 unit, as shown in Figure 1.50. Notice the x-axis is one horizontal line that intersects the graph more

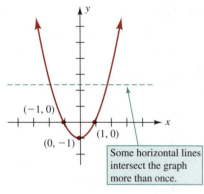

FIGURE 1.50
The graph of $h(x) = x^2 - 1$ fails the horizontal line test. Hence, h is *not* one-to-one.

than once. Thus, by the horizontal line test, the function h is *not* a one-to-one function. Also, note that

$$h(a) = h(b)$$

implies $\quad a^2 - 1 = b^2 - 1$

$\qquad\qquad a^2 = b^2 \qquad$ **Add 1 to both sides**

$\qquad\qquad a = \pm b \qquad$ **Take the square root of both sides**

Since $h(a) = h(b)$ does *not* imply $a = b$, we may also conclude that h is *not* one-to-one. ◆

Note: The function f defined in Example 4(a) is an increasing function (see Section 1.4). *Every function that is either an increasing function or a decreasing function is also one-to-one.* If a function is *not* one-to-one, we can often find a suitable restriction on its domain in order to form a new function that is one-to-one. For example, if we restrict the domain of the function h defined in Example 4(b) to *nonnegative numbers,* the new function that is formed is one-to-one.

PROBLEM 4 Show that the function H defined by $H(x) = x^2 - 1$ with $x \geq 0$ is a one-to-one function. ◆

◆ **Finding the Inverse of a Function**

As we previously stated, there exists an inverse function for a given function f only if f is a one-to-one function. This is because each output of a one-to-one function comes from just one input. Thus, when we reverse the roles, we can assign to each output the input from which it came. We usually denote the inverse function of the function f by using the notation f^{-1} (read "f inverse") and write

$$f(f^{-1}(x)) = f^{-1}(f(x)) = x$$

In the notation f^{-1}, the superscript -1 denotes an inverse and should not be confused with an exponent. That is, f^{-1} does not mean $1/f$.

The following procedure may be used to find the inverse of a one-to-one function.

Finding the Inverse of a One-to-One Function

If f is a one-to-one function, then the inverse function f^{-1} may be found by

1. Replacing each x with $f^{-1}(x)$ in the equation that defines f, and then solving for $f^{-1}(x)$.
2. Adjusting the domain of f^{-1} (if necessary) so that it equals the range of f.

SECTION 1.5 Composite and Inverse Functions

This procedure is illustrated in the next example.

EXAMPLE 5 Find the inverse of the one-to-one function defined by each rule.

(a) $f(x) = x^3 - 1$ (b) $g(x) = \sqrt{x - 1}$

SOLUTION

(a) We can find the inverse of this one-to-one function as follows:

$$f(x) = x^3 - 1$$

$$f(f^{-1}(x)) = [f^{-1}(x)]^3 - 1 \quad \text{Replace each } x \text{ with } f^{-1}(x)$$

$$x = [f^{-1}(x)]^3 - 1 \quad \text{By definition, } f(f^{-1}(x)) = x$$

$$x + 1 = [f^{-1}(x)]^3 \quad \text{Solve for } f^{-1}(x)$$

$$f^{-1}(x) = \sqrt[3]{x + 1}$$

Since the domains and ranges of both f and f^{-1} are the set of all real numbers, we conclude that the inverse function f^{-1} is defined by

$$f^{-1}(x) = \sqrt[3]{x + 1}.$$

FIGURE 1.51
Graph of $g(x) = \sqrt{x - 1}$.

(b) As shown in Figure 1.51, the function g is a one-to-one function with domain $[1, \infty)$ and range $[0, \infty)$. To find g^{-1}, we proceed as follows:

$$g(x) = \sqrt{x - 1}$$

$$g(g^{-1}(x)) = \sqrt{g^{-1}(x) - 1} \quad \text{Replace each } x \text{ with } g^{-1}(x)$$

$$x = \sqrt{g^{-1}(x) - 1} \quad \text{By definition, } g(g^{-1}(x)) = x$$

$$x^2 = g^{-1}(x) - 1 \quad \text{Square both sides and solve for } g^{-1}(x)$$

$$g^{-1}(x) = x^2 + 1$$

Since the domain of g^{-1} must be equal to the range of g, we must restrict the domain of g^{-1} to the interval $[0, \infty)$. Hence, we conclude that g^{-1} is defined by

$$g^{-1}(x) = x^2 + 1 \quad x \geq 0. \qquad \blacklozenge$$

Referring to Example 5(b), the result of composing g and g^{-1} in either order must be the identity function. Thus, to check Example 5(b), we show that $g(g^{-1}(x)) = g^{-1}(g(x)) = x$:

$$g(g^{-1}(x)) = g(x^2 + 1) \qquad\qquad g^{-1}(g(x)) = g^{-1}(\sqrt{x - 1})$$

$$= \sqrt{(x^2 + 1) - 1} \qquad\qquad\qquad = (\sqrt{x - 1})^2 + 1$$

$$= \sqrt{x^2} \qquad\qquad\qquad\qquad\quad = (x - 1) + 1$$

$$= x \quad \text{since } x \geq 0 \qquad\qquad\qquad = x$$

PROBLEM 5 Check Example 5(a) by showing that $f(f^{-1}(x)) = f^{-1}(f(x)) = x$.

Alternatively, the inverse of a one-to-one function f may be found by interchanging x and y in the equation $y = f(x)$ and solving for y. The procedure is illustrated in the next example.

EXAMPLE 6 Find the inverse of the one-to-one function f defined by $f(x) = 2x - 3$.

SOLUTION To find the inverse, we can replace $f(x)$ with y and proceed as follows:

$$y = 2x - 3$$

$$x = 2y - 3 \qquad \text{Interchange } x \text{ and } y$$

$$x + 3 = 2y \qquad \text{Solve for } y$$

$$y = \frac{x + 3}{2}$$

Hence, the inverse of $y = f(x) = 2x - 3$ is $y = f^{-1}(x) = \dfrac{x + 3}{2}$.

PROBLEM 6 Find the inverse of the one-to-one function f defined by $f(x) = \dfrac{1}{x + 1}$.

◆ Graphs of Inverse Functions

The graphs of g and g^{-1}, the functions defined in Example 5(b), are shown in Figure 1.52. Notice that for every point (a, b) on the graph of g there corresponds a point (b, a) on the graph of g^{-1}. Thus, if we were to fold the coordinate plane along the dotted line $y = x$, the graphs of g and g^{-1} would coincide. In other words, the graphs of g and g^{-1} are *reflections of one another in the line $y = x$*. This special relationship between the graphs of g and g^{-1} is true for any function and its inverse.

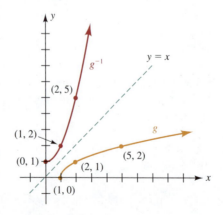

FIGURE 1.52
The graphs of $g(x) = \sqrt{x - 1}$ and $g^{-1}(x) = x^2 + 1$, $x \geq 0$, are reflections of one another in the line $y = x$.

SECTION 1.5 Composite and Inverse Functions

EXAMPLE 7 Suppose the function f is defined by $f(x) = 4 - x^2$ for $x \geq 0$. Sketch the graph of f^{-1}.

SOLUTION We first sketch the graph of $f(x) = 4 - x^2$ for $x \geq 0$ by using the graphing techniques discussed in Section 1.4. The graph of f^{-1} is the same as this graph, but reflected in the line $y = x$, as shown in Figure 1.53. From the graphs of f and f^{-1}, we can observe that the domain of f^{-1} and the range of f are both $(-\infty, 4]$. Also the range of f^{-1} and the domain of f are both $[0, \infty)$. ◆

FIGURE 1.53
Graph of f^{-1} formed from the graph of $f(x) = 4 - x^2$, $x \geq 0$, by reflecting about the line $y = x$.

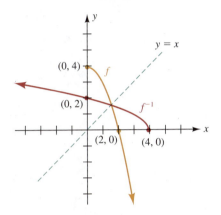

PROBLEM 7 Referring to Example 7, find the equation that defines f^{-1}. ◆

Exercises 1.5

  Basic Skills

In Exercises 1–8, use the given graphs of f, g, and h to compute (if possible) the indicated functional value.

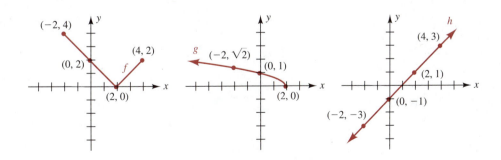

1. $(g \circ f)(0)$
2. $(h \circ f)(-2)$
3. $(f \circ h)(-2)$
4. $(h \circ g)(3)$
5. $(f \circ g)(2)$
6. $(f \circ f)(-2)$
7. $[h \circ (f \circ g)](2)$
8. $[h \circ (g \circ f)](0)$

In Exercises 9–20, given that f, g, h, F, G, and H are defined as follows:

$$f(x) = 2x + 1 \qquad F(x) = \sqrt{x^2 - 4}$$

$$g(x) = x^2 + 1 \qquad G(x) = \frac{1}{x}$$

$$h(x) = \sqrt{x} \qquad H(x) = \frac{2}{x - 1}$$

(a) Find the indicated function.
(b) Find the domain of the indicated function.

9. $f \circ g$
10. $g \circ f$
11. $G \circ f$
12. $G \circ F$
13. $F \circ h$
14. $g \circ h$
15. $G \circ H$
16. $H \circ G$
17. $H \circ f$
18. $G \circ G$
19. $f \circ f$
20. $H \circ H$

21. Use the results of Exercise 15 to evaluate $(G \circ H)(2)$.

22. Use the results of Exercise 16 to evaluate $(H \circ G)(2)$.

In Exercises 23–32, show that the functions f and g are inverses of each other by showing that $f(g(x)) = g(f(x)) = x$.

23. $f(x) = 3x - 1$; $g(x) = \dfrac{x + 1}{3}$

24. $f(x) = 5 - 2x$; $g(x) = \dfrac{5 - x}{2}$

25. $f(x) = \dfrac{1}{x}$; $g(x) = \dfrac{1}{x}$

26. $f(x) = \sqrt[5]{x - 6}$; $g(x) = x^5 + 6$

27. $f(x) = x^3 - 8$; $g(x) = \sqrt[3]{x + 8}$

28. $f(x) = (x - 2)^3$; $g(x) = \sqrt[3]{x} + 2$

29. $f(x) = \dfrac{1}{x - 1}$; $g(x) = \dfrac{1}{x} + 1$

30. $f(x) = \dfrac{x}{x + 3}$; $g(x) = \dfrac{-3x}{x - 1}$

31. $f(x) = (x - 1)^2$, $x \geq 1$; $g(x) = \sqrt{x} + 1$

32. $f(x) = x^2 + 2$, $x \geq 0$; $g(x) = \sqrt{x - 2}$

In Exercises 33–42,

(a) Determine if the function is one-to-one.

(b) If the function is not one-to-one, find a suitable restriction on its domain in order to form a new function that is one-to-one. There is no unique choice for this new function.

33. $f(x) = 2x$
34. $f(x) = \dfrac{x^3}{3}$
35. $f(x) = x^2$
36. $f(x) = |x| - 3$
37. $f(x) = \sqrt{9 - x}$
38. $f(x) = \dfrac{1}{x - 3}$
39. $f(x) = |x + 2|$
40. $f(x) = 1 - \sqrt[3]{x}$
41. $f(x) = (x - 2)^2 - 1$
42. $f(x) = (x - 1)^{2/3}$

Each function f in Exercises 43–60 is a one-to-one function.

(a) Find the inverse function f^{-1}.

(b) Sketch the graph of f and f^{-1} on the same coordinate plane and verify that the graphs of f and f^{-1} are symmetric with respect to the line $y = x$. If you have access to a graphing calculator, check each graph.

43. $f(x) = 2x$
44. $f(x) = \dfrac{-x}{4}$
45. $f(x) = 3 - x$
46. $f(x) = x + 1$
47. $f(x) = 2 - x^3$
48. $f(x) = (x + 1)^3$
49. $f(x) = \sqrt[3]{x - 4}$
50. $f(x) = 3 + \sqrt[3]{x}$
51. $f(x) = \dfrac{1}{x + 1}$
52. $f(x) = \dfrac{1}{x} - 2$
53. $f(x) = \sqrt{x - 4} + 1$
54. $f(x) = \sqrt{x} + 3$
55. $f(x) = (x - 1)^2$, $x \geq 1$
56. $f(x) = (x - 3)^2 + 2$, $x \geq 3$
57. $f(x) = \sqrt{9 - x^2}$, $0 \leq x \leq 3$
58. $f(x) = \sqrt{x^2 - 9}$, $x \geq 3$
59. $f(x) = x^2 + 3$, $x \leq 0$
60. $f(x) = 2 - x^2$, $x \leq 0$

Critical Thinking

For Exercises 61–66, refer to the following table of values.

x	f(x)	g(x)
−1	4	3
0	7	2
2	3	−1
3	−1	−5

61. Given that $h(x) = (f \circ g)(x)$, find $h(2)$.
62. Given that $h(x) = (g \circ f)(x)$, find $h(2)$.
63. Given that $g(x) = (f \circ h)(x)$, find $h(2)$.
64. Given that $f(x) = (g \circ h)(x)$, find $h(2)$
65. Given that $f(x) = (g \circ h)(x)$, find $h(3)$
66. Given that $g(x) = (f \circ h)(x)$, find $h(-1)$.

In Exercises 67–70, the graph of a function f is shown. Determine if the function f has an inverse. If so, sketch the graph of f^{-1} and complete this table of values.

x	−2	0	3
$f^{-1}(x)$			

67.

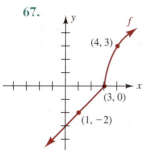

68.

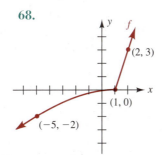

69.

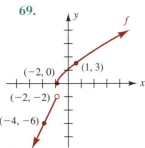

70.

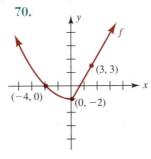

For Exercises 71–76, assume the functions f and g are defined as follows:

$$f(x) = \frac{x - 2}{3} \quad \text{and} \quad g(x) = 3x + 1$$

71. Find $f^{-1}(4)$.
72. Find $g^{-1}(7)$.
73. Find $(f^{-1} \circ g)(-2)$.
74. Find $(f \circ g^{-1})(25)$.
75. Compare $(f \circ g)^{-1}(x)$ with $(g^{-1} \circ f^{-1})(x)$.
76. Compare $(g \circ f)^{-1}(x)$ with $(f^{-1} \circ g^{-1})(x)$.

In Exercises 77–80, determine the range of the given function by examining the domain of its inverse function.

77. $f(x) = \dfrac{1}{2x - 1}$

78. $f(x) = \dfrac{x}{6 - 3x}$

79. $f(x) = \dfrac{2x + 1}{3x - 2}$

80. $f(x) = \dfrac{9x - 3}{3x + 1}$

81. Find $f(x)$ if $(g \circ f)(x) = 2x - 8$ and $g(x) = 6x + 4$.

82. A spherical balloon is being inflated with air. The volume V of the balloon is a function of its radius r and is given by $V(r) = \frac{4}{3}\pi r^3$. In turn, the radius r (in inches) is a function of the time t (in seconds) after the inflation process begins and is given by $r(t) = 3(t + 1)$.

   (a) Determine $(V \circ r)(t)$.
   (b) What does the composition function $V \circ r$ describe?
   (c) Find the volume of the balloon after inflating it for 3 seconds.
   (d) Find the time at which the volume is $36{,}000\pi$ cubic inches.

For Exercises 83–86, use the following facts:

(i) *The volume V of a sphere is a function of its radius r and is given by $V = f(r) = \frac{4}{3}\pi r^3$.*

(ii) *The surface area S of a sphere is also a function of its radius r and is given by $S = g(r) = 4\pi r^2$.*

83. Find the inverse function g^{-1}.
84. Find the inverse function f^{-1}.
85. Find an interpret $(f \circ g^{-1})(S)$.
86. Find and interpret $(g \circ f^{-1})(V)$.

Calculator Activities

Given the functions f and g defined by

$$f(x) = 2x - 1 \quad \text{and} \quad g(x) = x^2 + 2x,$$

compute to three significant digits the functional values given in Exercises 87 and 88.

87. $(f \circ g)(0.076)$ **88.** $(g \circ f)(4.57)$

Given the functions f and g defined by

$$f(x) = 3.24x - 3.46 \quad \text{and} \quad g(x) = \frac{2.72}{x - 8.65},$$

compute to three significant digits the functional values indicated in Exercises 89–92.

89. $f^{-1}(2.75)$ **90.** $g^{-1}(9.87)$

91. $(f \circ f^{-1})(0.173)$ **92.** $(g \circ g^{-1})(4.33)$

 93. Use a graphing calculator to generate the graph of $y = x$ and the graphs of the given pair of functions in the same viewing rectangle. From the picture, determine if the given functions appear to be inverses of each other. Verify each answer.

(a) $f(x) = \dfrac{3x + 1}{x}$ and $g(x) = \dfrac{1}{x - 3}$

(b) $f(x) = \dfrac{x}{x - 2}$ and $g(x) = \dfrac{2x}{x - 1}$

94. In a ski factory, the daily cost C (in dollars) of producing n pairs of skis is given by

$$C(n) = 225n - 0.8n^2, \quad 0 \le n \le 40$$

and the number of pairs produced in t hours is given by

$$n(t) = 4.5t, \quad t \le 10.$$

(a) Determine $(C \circ n)(t)$.

(b) What does $(C \circ n)(t)$ describe?

(c) Find the daily production cost to the nearest dollar if the factory runs for 8.2 hours per day.

(d) How many hours does the factory operate if the daily production cost is $4228?

1.6 Applied Functions and Variation

To solve many types of word problems—especially those that appear in calculus—we must begin by setting up an equation that defines a function. We may then analyze the function, draw its graph, and answer questions concerning the functional relationship between the quantities. In this section we practice setting up functions from words.

◆ Applied Functions

For many applied problems, we begin with an established formula and then use *substitution* to obtain a functional relationship between the desired variables. The procedure is illustrated in the next two examples.

EXAMPLE 1   Suppose 200 feet of fencing is needed to enclose a rectangular garden.

(a) Express the area A of the rectangular garden as a function of its length l.

(b) Give the domain of this function.

FIGURE 1.54
Rectangular garden of length l and width w.

SOLUTION

(a) Drawing a rectangular garden, we let

l = the length and w = the width

as shown in Figure 1.54.

Recall that the area A of a rectangle is the product of its length and width:

$$A = lw.$$

For the area A to be a function of its length l, we need to write A in terms of just l. Since 200 feet of fencing is needed to enclose the garden, we have

$$200 = 2l + 2w.$$

Solving this equation for w gives us

$$w = 100 - l.$$

Now, we substitute $100 - l$ for w in the area formula, $A = lw$, and obtain

$$A = lw = l(100 - l) = 100l - l^2.$$

The equation $A = 100l - l^2$ defines the area A of this rectangle as a function of its length l. If we wish to emphasize this functional relationship, we may use functional notation to write

$$A(l) = 100l - l^2.$$

(b) Algebraically, the domain of this function is the set of all real numbers but, geometrically, this domain does not make sense. This is because l represents the length of the rectangle, and lengths can only be *positive*. Thus, we must have $l > 0$. Also, since the perimeter is 200 feet, we must have $l < 100$, otherwise, the width w would be zero or negative. Because of the geometric restrictions placed on this function, the domain is (0, 100). ◆

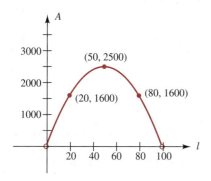

FIGURE 1.55
Graph of the function defined by the equation $A = 100l - l^2$ with $0 < l < 100$.

To sketch the graph of $A = 100l - l^2$, $0 < l < 100$, we can choose a convenient scale for each axis and plot some points. The graph is shown in Figure 1.55. The vertical axis is labeled as A instead of y and the horizontal axis is labeled as l instead of x. From the graph, we can determine that the range of the function is (0, 2500].

PROBLEM 1 Use the graph of $A = 100l - l^2$ in Figure 1.55 to determine the dimensions of the rectangle whose area is the *largest* possible. ◆

EXAMPLE 2 Water is flowing into a conical funnel. The diameter of the base of the funnel is 8 inches and its height is 12 inches.

(a) Find the volume V of water in the funnel as a function of the height h of water in the funnel.

(b) Give the domain of this function, then sketch its graph.

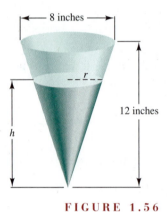

FIGURE 1.56
Conical funnel 8 inches wide and 12 inches deep.

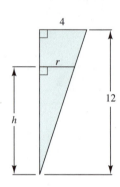

FIGURE 1.57
Similar triangles formed from the cross section of the funnel.

SOLUTION

(a) We draw a conical funnel as in Figure 1.56, showing that

$$h = \text{the height of water in the funnel}$$

and $\qquad r = \text{the radius of the surface of the water.}$

From elementary geometry, we know that the volume of water in the funnel is given by

$$V = \tfrac{1}{3}\pi r^2 h.$$

For the volume V to be a function of h, we must write the volume in terms of just h. To do this, we use similar triangles from the cross section of the funnel, as shown in Figure 1.57. Since corresponding sides of similar triangles are proportional, we have

$$\frac{4}{12} = \frac{r}{h} \qquad \text{or} \qquad r = \frac{h}{3}.$$

Now we substitute $h/3$ for r in the volume formula $V = \tfrac{1}{3}\pi r^2 h$ and obtain

$$V = \tfrac{1}{3}\pi \left(\frac{h}{3}\right)^2 h \qquad \text{or} \qquad V = \tfrac{1}{27}\pi h^3.$$

If we wish to emphasize this functional relationship, we use functional notation to write

$$V(h) = \tfrac{1}{27}\pi h^3.$$

(b) Since the height of the funnel is 12 inches, the water level must be between 0 and 12 inches. Thus, the domain of the function defined by the equation $V = \tfrac{1}{27}\pi h^3$ is [0, 12]. Choosing convenient scales for the h-axis and V-axis, we sketch the graph by plotting some points. The graph of the function defined by the equation

$$V = \tfrac{1}{27}\pi h^3, \quad 0 \leq h \leq 12,$$

is shown in Figure 1.58. Note that the range of the function is [0, 64π]. ◆

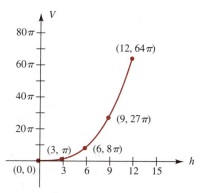

FIGURE 1.58
Graph of the function defined by the equation $V = \tfrac{1}{27}\pi h^3$ with $0 \leq h \leq 12$.

PROBLEM 2 Suppose water is flowing into the funnel described in Example 2 at the rate of 8 cubic inches per minute. Find the volume V of water (in cubic inches) in the funnel as a function of time t (in minutes), and state the domain of this function. ◆

◆ **Variation**

In business, engineering, and science, the functional relationship between two quantities is often given in terms of *variation*. When we state that **y varies directly as x** or that **y is directly proportional to x,** we mean that the ratio of y to x is always the same. In other words,

$$\frac{y}{x} = k \quad \text{or} \quad y = kx$$

where k is called the *variation constant*.

EXAMPLE 3 The real estate tax T on a property varies directly as its assessed value V.

(a) Express T as a function of V if T = $2800 when V = $112,000.

(b) State the domain of the function defined in part (a), then sketch its graph.

SOLUTION

(a) Since T varies directly as V, we write

$$T = kV$$

where k is the variation constant to be determined. To find k, we replace T with 2800 and V with 112,000 as follows:

$$T = kV$$
$$2800 = k(112{,}000)$$
$$k = \frac{2800}{112{,}000} = \frac{1}{40}$$

Thus, $T = \frac{1}{40}V$ or $T = \frac{V}{40}$.

(b) Since the assessed value V must be greater than or equal to zero, the domain is $[0, \infty)$. By the vertical stretch and shrink rule (Section 1.4), we know that the graph of $T = V/40$ is part of a straight line. Choosing convenient scales for the V-axis and T-axis, we show the graph in Figure 1.59. From the graph, we can see that the range is also $[0, \infty)$. ◆

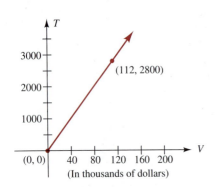

FIGURE 1.59

Graph of the function defined by the equation $T = \frac{1}{40}V$ with $V \geq 0$.

PROBLEM 3 Using the results from Example 3(a), find the tax T on a piece of property with an assessed value of $168,000. Check your answer by using the graph in Figure 1.59. ◆

When we state that y **varies inversely as** x or that y **is inversely proportional to** x, we mean that the product of y and x is always the same. In other words,

$$yx = k \quad \text{or} \quad y = \frac{k}{x}$$

where k is the *variation constant*.

EXAMPLE 4 The time t it takes a person to travel a fixed distance varies inversely as the rate of speed r at which the person travels.

(a) Express t as a function of r if $t = 4$ hours when $r = 60$ miles per hour.

(b) State the domain of the function defined in part (a) then sketch its graph.

SOLUTION

(a) Since t varies inversely as r, we write

$$t = \frac{k}{r},$$

where k is the variation constant to be determined. To find k, we replace t with 4 and r with 60 as follows:

$$t = \frac{k}{r}$$

$$4 = \frac{k}{60}$$

$$k = 240$$

Thus, $t = \dfrac{240}{r}$.

(b) Since rates of speed are assumed to be positive, the domain of the function defined by $t = 240/r$ is $(0, \infty)$. Choosing convenient scales for the r-axis and t-axis, we show the graph in Figure 1.60. ◆

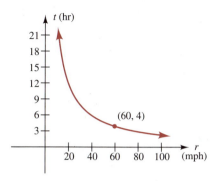

FIGURE 1.60
Graph of the function defined by the equation $t = 240/r$ with $r > 0$.

PROBLEM 4 Using the results of Example 4(a), find the time t it takes to travel a fixed distance if the rate of speed is 50 miles per hour. Check your answer by using the graph in Figure 1.60. ◆

The following table gives several other types of variation that occur frequently. In each case, k denotes the variation constant.

Statement	Formula
y varies directly as the nth power of x.	$y = kx^n$
y varies inversely as the nth power of x.	$y = \dfrac{k}{x^n}$
y varies directly as x and inversely as z.	$y = \dfrac{kx}{z}$
y varies jointly as x and z.	$y = kxz$

EXAMPLE 5 The volume V of a right circular cylinder (see Figure 1.61) varies jointly as its height h and the square of its radius r. Describe what happens to the volume of a cylinder if its height and radius are doubled.

SOLUTION Since V varies jointly as h and the square of r, we write

$$V = kr^2h,$$

where k is the variation constant. If we double both the height and the radius, we obtain

$$V = k(2r)^2(2h) = 8kr^2h.$$

FIGURE 1.61
A right circular cylinder.

From this equation we can see that the volume of the cylinder is *8 times the original volume*. Thus, when the height and radius of a cylinder are doubled, its volume becomes 8 times as large. ◆

PROBLEM 5 The electrical resistance R of a wire varies directly as its length l and inversely as the square of its radius r. Describe what happens to the resistance of a wire if its length and radius are doubled. ◆

Exercises 1.6

Basic Skills

1. Express the diameter d of a circle as a function of its circumference C.

2. The length of a rectangle is 5 cm. Express its width w as a function of its perimeter P.

3. Suppose a leasing company charges $40 per day plus $0.20 per mile to rent a car. Express in dollars the daily cost C of renting a car as a function of the number n of miles driven.

4. Suppose a salesperson earns $200 per week plus 25% commission on all sales. Express in dollars the weekly earnings E of a salesperson as a function of the amount A of merchandise that she sells.

5. A square with side s and diagonal d is shown in the figure.

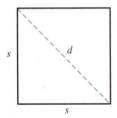

   (a) Express s as a function of d.
   (b) Express the area A of the square as a function of d.
   (c) Express the perimeter P of the square as a function of d.

6. An equilateral triangle with side s and height h is shown in the figure.

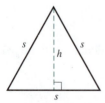

   (a) Express h as a function of s.
   (b) Express the area A of the triangle as a function of s.
   (c) Express the perimeter P of the triangle as a function of h.

7. A small computer company has a present net worth of $125,000. It is estimated that the future weekly income and weekly expenses for the company will be $30,000 and $26,000, respectively. Express the net worth W of the company at the end of t weeks as a function of t.

8. A $15,000 automobile depreciates 20% of its original value each year. Express the value V of the automobile at the end of t years as a function of t.

9. The radius r of a pile of sand in the shape of a cone is twice its height h.
   (a) Express the volume V of sand as a function of h.
   (b) Express the volume V of sand as a function of r.

10. The height h of a tin can in the shape of a cylinder is equal to its diameter d.
    (a) Express the volume V of the tin can as a function of h.
    (b) Express the surface area S of the tin can as a function of d.

11. The point $P(x, y)$ lies on the graph of the circle $x^2 + y^2 = 1$ as shown in the figure.

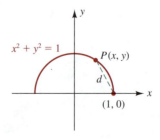

(a) Express the distance d from point P to the point $(1, 0)$ as a function of the x-coordinate of P.

(b) State the domain of the function in part (a).

12. The point $P(x, y)$ lies on the graph of the parabola $2y = x^2$ shown in the figure.

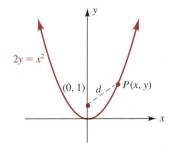

(a) Express the distance d from the point P to the point $(0, 1)$ as a function of the y-coordinate of P.

(b) State the domain of the function in part (a).

13. The point $Q(x, y)$ lies on the graph of $y = \sqrt{x}$ shown in the figure.

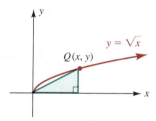

(a) Express the area A of the shaded right triangle as a function of the x-coordinate of Q.

(b) Express the perimeter P of the shaded right triangle as a function of the x-coordinate of Q.

14. The point $Q(x, y)$ lies on the graph of the semicircle $y = \sqrt{25 - x^2}$ shown in the figure.

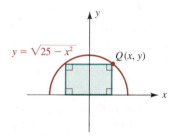

(a) Express the area A of the shaded rectangle as a function of the x-coordinate of Q.

(b) Express the perimeter P of the shaded rectangle as a function of the x-coordinate of Q.

15. The volume of a rectangular box with a square base is 64 cubic inches. Express its surface area S as a function of the width x of its base.

16. The volume of a rectangular box with a square base and an open top is 100 cubic meters. The material used to construct the base costs $4 per square meter, and the material for the sides cost $2.50 per square meter. Express the cost C to construct the rectangular box as a function of the width x of its base.

17. The ends of a water trough are isosceles triangles, as shown in the figure.

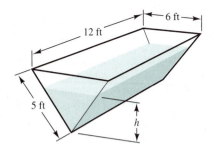

(a) Express the volume V of water in the trough as a function of the height h of the water in the trough.

(b) Give the domain of the function in part (a), then sketch its graph.

18. Suppose the trough described in Exercise 17 is initially empty, and begins to fill with water at the rate of 1 cubic foot per minute. Express the height h of the water in the trough as a function of the time t (in minutes) that the water flows into the trough.

19. A baseball diamond is a square with 90-foot base paths, as shown in the figure. Suppose a runner on first base runs toward second base at the rate of 30 feet per second as soon as the pitcher throws the ball to home plate.

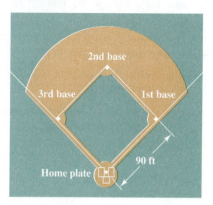

EXERCISE 19 *(continued)*

(a) Express the runner's distance d from home plate as a function of the time t (in seconds) after the pitcher throws the ball.

(b) Describe the domain and range of the function in part (a).

20. The fixed cost to run a wood-stove company is $25,000 per month, and the variable cost to produce each stove is $200 per unit. If the stoves sell for x dollars each, then the number of stoves sold per month is estimated to be $1000 - 2x$. Express the monthly profit of the company as a function of x. (*Hint:* Profit = Total revenue − total cost.)

In Exercises 21–32, the functional relationship between the quantities is given in terms of variation.

21. **Hooke's law** states that the distance d a spring stretches varies directly as the force F applied to the spring as long as the elastic limit of the spring is not exceeded. Assume that the elastic limit of a spring occurs at 2400 Newtons.

    (a) Express d as a function of F if $d = 8$ cm when $F = 400$ Newtons.

    (b) State the domain of the function in part (a), then sketch its graph.

22. The electrical resistance R of a wire is directly proportional to the length l of the wire.

    (a) Express R as a function of l if $R = 2$ ohms when $l = 30$ meters.

    (b) State the domain of the function in part (a), then sketch its graph.

23. The weight W of a steel beam varies directly as its length l.

    (a) Express W as a function of l if $W = 1500$ pounds when $l = 12$ feet.

    (b) State the domain of the function in part (a), then sketch its graph.

24. The weekly payroll P of a company varies directly as the number n of workers assigned to the job.

    (a) Express P as a function of n if $P = \$5200$ when $n = 13$ workers.

    (b) State the domain of the function in part (a), then sketch its graph.

25. The number N of long-distance phone calls per day between two towns, each with a population of approximately 25,000 people varies inversely as the distance d between the towns.

    (a) Express N as a function of d if $N = 20$ calls when $d = 500$ miles.

    (b) State the domain of the function in part (a), then sketch its graph.

26. **Boyle's law** states that when the temperature of a confined gas remains constant, the pressure P it exerts varies inversely as the volume V it occupies.

    (a) Express P as a function of V if $P = 4$ pounds per square inch when $V = 30$ cubic inches.

    (b) State the domain of the function in part (a), then sketch its graph.

27. The wavelength w of a radio wave varies inversely as its frequency f.

    (a) Express w as a function of f if $w = 40$ meters when $f = 7.5$ megahertz.

    (b) State the domain of the function in part (a), then sketch its graph.

28. If the voltage in an electrical circuit is constant, the current I through a resistor varies inversely as its resistance R.

    (a) Express I as a function of R if $I = 2$ milliamps when $R = 100$ ohms.

    (b) State the domain of the function in part (a), then sketch its graph.

29. The gravitational force F between two objects varies inversely as the square of the distance d between them.

    (a) Express F as a function of d if $F = 100$ Newtons when $d = 40$ meters.

    (b) State the domain of the function in part (a), then sketch its graph.

30. The lift L of an airplane wing varies directly as the square of the speed v of air flowing over it.

    (a) Express L as a function of v if $L = 225$ pounds per square foot when $v = 150$ miles per hour.

    (b) State the domain of the function in part (a), then sketch its graph.

31. The centrifugal force acting on an object traveling in a circular path varies directly as the square of its velocity and inversely as the radius of the circle. Describe what happens to the centrifugal force acting on the object if the velocity and radius are both tripled.

32. The power produced by an electric generator varies jointly as the load resistance and the square of the current. Describe what happens to the power produced by the generator if the load resistance is doubled and the current is halved.

Critical Thinking

33. A church window has the shape of a rectangle surmounted by a semicircle, as shown in the figure. Express the area A of the window as a function of the radius r of the semicircle if the perimeter of the window is 30 feet.

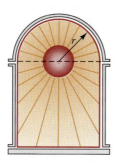

34. A football stadium has the shape of a rectangle with semicircular ends, as shown in the figure. Express the area A of the stadium as a function of the radius r of the semicircle if the perimeter of the stadium is 1 mile.

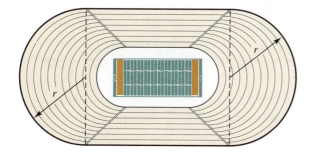

35. A baseball player hits the ball to the outfield and tries for a double running directly on the base paths (see the figure in Exercise 19). Suppose he can run at the rate of 30 feet per second. Express the straight-line distance d from home plate to the runner as a function of the time t (in seconds) after he hits the ball.

36. An author is paid a royalty of $3.00 per book for the first 1000 books sold. The royalty increases to $3.05 per book for each book sold in excess of 1000. Express the author's royalty R as a function of the number n of books sold.

37. The time required for an elevator to lift its passengers varies jointly as the weight of the passengers and the distance they are lifted, and inversely as the horsepower of the motor that is used. Suppose it takes 10 seconds for an elevator to lift 800 pounds to a height of 40 feet with a 20 horsepower motor. Determine the time it takes this elevator to lift 1000 pounds to a height of 80 feet with the same motor.

38. The safe-load capacity of a wooden rectangular beam supported at both ends varies jointly as its width and the square of its depth, and inversely as the distance between the supports. Suppose the safe-load capacity of a beam 4 inches wide and 12 inches deep is 3600 pounds when the distance between the supports is 16 feet. Determine the safe-load capacity of a similar beam that is 8 inches wide and 8 inches deep if the distance between the supports remains 16 feet.

Calculator Activities

39. The length of the hypotenuse of a right triangle is 16.2 inches.
  (a) Express the area A of the triangle as a function of the length x of one of the legs.
  (b) Find the area when $x = 12.4$ inches.

40. Two joggers start from the same place at the same time. One runs due east at 8.2 miles per hour (mph) and the other runs due north at 6.3 mph.
  (a) Determine the distance d between the joggers as a function of the time t (in hours) that they have been running.
  (b) Find the distance between the joggers after they have been running 1 hour 24 minutes.

41. One thousand feet of fencing is to be used to enclose a rectangular pasture along a river, as shown in the figure. No fencing is needed along the river.

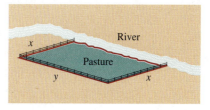

  (a) Express the area A of the pasture as a function of the length x of the pasture.
  (b) State the domain of the function defined in part (a).

EXERCISE 41 *(continued)*

(c) Use a graphing calculator to generate the graph of the function defined in part (a). Trace to the peak of this curve, then state the dimensions of the rectangular pasture whose area is the *largest* possible.

42. Three adjacent rectangular corrals are to be built with 120 feet of fencing, as shown in the figure.

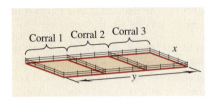

(a) Express the total enclosed area A as a function of the common length x.

(b) State the domain of the function defined in part (a).

(c) Use a graphing calculator to generate the graph of the function defined in part (a). Trace to the peak of this curve, then state the overall dimensions of the corrals so that the total enclosed area is the *largest* possible.

43. The period T of a pendulum is directly proportional to the square root of the length l of the pendulum.

(a) Express T as a function of l if $T = 2.1$ seconds when $l = 1.095$ meters.

(b) Find the period when the length of the pendulum is 2.405 meters.

44. The horsepower H required to propel a motorboat through the water varies directly as the cube of the speed s of the boat.

(a) Express H as a function of s if $H = 48$ horsepower when $s = 11.2$ knots.

(b) Find the horsepower when the speed is 15.6 knots.

Chapter 1 Review

Questions for Group Discussion

1. Explain the difference between the *interval* (a, b) and the *ordered pair* (a, b).
2. Explain in words how to find the *midpoint* of a line segment.
3. What is the difference between the *independent variable* and the *dependent variable* in an equation?
4. What is the advantage of writing the equation of a circle in *standard form*?
5. Is the graph of $x^2 + y^2 + Dx + Ey + F = 0$ always a *circle*? Explain.
6. What is a *function*? What is meant by its *domain* and *range*?
7. Describe the *point-plotting method* of sketching the graph of a simple function.
8. How can the *vertical line test* be used to determine if an equation defines y as a function of x?
9. How can the *zeros* of a function be determined? Graphically, what do the real zeros represent?
10. What is the difference between an *even function* and an *odd function*? Give an example of each.
11. Which of the eight basic functions in Figure 1.32 are neither *increasing* nor *decreasing* functions?
12. From the graph of a function, how is it possible to determine if the function is one-to-one?
13. Does every function have an *inverse function* associated with it? Explain.
14. How does the graph of a function relate to the graph of its inverse?

15. Explain the algebraic procedure for finding the inverse of a one-to-one function.
16. Can the graph of a function be symmetric with respect to the x-axis? Explain.
17. How are the graphs of the following functions related to the graph of $y = x^2$?
    (a) $y = x^2 + 3$ (b) $y = (x - 2)^2$ (c) $y = -x^2$ (d) $y = (x + 2)^2 - 1$
18. What is the difference between $g \circ f$ and $f \circ g$? Illustrate with examples.
19. How is *composition* used to show that the functions f and g are inverse functions?
20. If f is an even function, what can you conclude about f^{-1}? Explain.
21. Can a function be its own inverse? If so, give an example.
22. Explain how the *distance formula* might be used to determine if three points are collinear.
23. Suppose *y varies inversely as x* and *x varies inversely as t*. What can you conclude about y and t?
24. Give an example of how a *composite function* might be used in a practical work situation.

Review Exercises

In Exercises 1–8, use interval notation to describe the set of all real numbers designated by each graph.

1.   (endpoints 3 and 7)
2. (open at -2, closed at 8)
3. (open at -1, extending right)
4. (open at 5)
5. (closed at -6, open at -1)
6. (open at 4, open at 12)
7. (closed at 0, open at 4)
8. (open at -9, closed at 1)

In Exercises 9–14, find

(a) The length of the line segment joining the points A and B.

(b) the coordinates of the midpoint M of the line segment joining the points A and B.

(c) the standard form of the equation of a circle whose diameter is the line segment joining the points A and B.

9. $A(0, 3)$; $B(4, 6)$ 10. $A(2, 0)$; $B(7, 12)$

11. $A(-2, 1)$; $B(2, 3)$ 12. $A(4, -1)$; $B(1, -4)$
13. $A(-3, -5)$; $B(6, -1)$ 14. $A(7, -3)$; $B(-3, -5)$

15. Show that the quadrilateral that joins the points $A(2, 1)$, $B(4, 3)$, $C(2, 5)$, and $D(0, 3)$ is a square. Then find its perimeter and area.

16. Show that the triangle that joins the points $A(-1, 8)$, $B(5, -2)$, and $C(15, 4)$ is an isosceles triangle. Then find its perimeter and area.

For Exercises 17–20, refer to the triangle with vertices A, O, and B as shown:

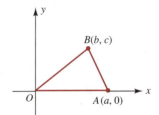

17. Find the length of the line segment joining the midpoints of the sides $\overline{OB}$ and $\overline{AB}$. Compare this length to OA, the length of the third side of the triangle.

18. Find the length of the line segment joining the midpoints of the sides $\overline{OA}$ and $\overline{AB}$. Compare this length to OB, the length of the third side of the triangle.

19. Find the length of the line segment joining the midpoints of the sides $\overline{OA}$ and $\overline{OB}$. Compare this length to AB, the length of the third side of the triangle.

20. What conclusion can be made about the length of a line segment joining the midpoints of any two sides of a triangle with respect to the length of the third side of the triangle.

In Exercises 21–30, sketch the graph of each equation.

21. $3x - |y| = 0$ **22.** $4x = y^2$

23. $x^2 y = 1$ **24.** $|x + y| = 1$

25. $x^2 + y^2 - 16 = 0$ **26.** $x^2 + y^2 = 1$

27. $x^2 + y^2 - 6x = 7$

28. $x^2 + y^2 - 6y - 16 = 0$

29. $x^2 + y^2 - 10x - 8y + 16 = 0$

30. $4x^2 + 4y^2 - 8x + 12y = 3$

Given the functions f, g, and h, defined by

$$f(x) = 3x + 4, \quad g(x) = 2x^2 - x, \quad \text{and} \quad h(x) = \frac{x + 1}{x - 3},$$

compute the functional values in Exercises 31–50.

31. $f(-1)$ **32.** $g(3)$

33. $h(4)$ **34.** $f(2x)$

35. $g(x - 3)$ **36.** $h\left(\dfrac{a}{2}\right)$

37. $f\left(\dfrac{1}{3}\right) - f(2)$ **38.** $g(x) \cdot f(x)$

39. $\dfrac{f(7)}{h(4)}$ **40.** $(f \circ h)(2)$

41. $(g \circ f)(x^2)$ **42.** $(f \circ f)(-5)$

43. $h^{-1}(x)$ **44.** $f(x) + f^{-1}(x)$

45. $f^{-1}(x + 2)$ **46.** $h^{-1}(-x)$

47. $(h \circ h^{-1})(2)$ **48.** $(g \circ f^{-1})(x)$

49. $\dfrac{g(x + \Delta x) - g(x)}{\Delta x}$ **50.** $\dfrac{f^{-1}(a + h) - f^{-1}(a)}{h}$

In Exercises 51–70,

(a) find the domain of the function.

(b) find any real zeros of the function.

(c) determine if the function is even, odd, or neither.

(d) sketch the graph of the function (if it is not given).

(e) determine the intervals where the function is increasing, decreasing, or constant.

(f) determine if the function is one-to-one.

(g) use the graph to find the range of the function.

51. $f(x) = x + 3$ **52.** $g(x) = -2x$

53. $h(x) = \sqrt[3]{2x}$ **54.** $F(x) = \sqrt[3]{x - 1}$

55. $G(x) = x^2 - 9$ **56.** $H(x) = |x| + 5$

57. $g(x) = (x + 3)^2$ **58.** $F(x) = \sqrt{3 - x}$

59. $G(x) = x^3 + 8$ **60.** $h(x) = \dfrac{1}{x - 4}$

61. $F(x) = |x - 3| - 1$ **62.** $H(x) = 8 - (x + 1)^3$

63. $H(x) = 6x - x^2$

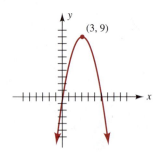

64. $f(x) = |x + 1| + |x - 1|$

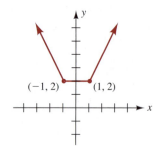

65. $f(x) = \sqrt{100 - x^2}$

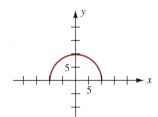

66. $g(x) = \sqrt{x^2 - 49}$

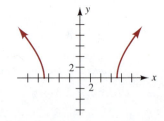

67. $g(x) = 2x^2 - x^4$

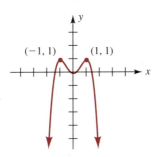

68. $H(x) = 2x^3 - 6x$

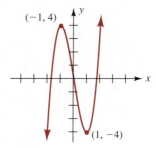

69. $F(x) = \begin{cases} x^2 + 1 & \text{if } x \leq 0 \\ 1 & \text{if } 0 < x \leq 3 \end{cases}$

70. $h(x) = \begin{cases} \sqrt{x-1} & \text{if } 1 \leq x \leq 5 \\ x - 3 & \text{if } x > 5 \end{cases}$

In Exercises 71–76, use the graph of $y = x^{2/3}$ in the figure and the techniques of shifting, reflecting, shrinking, and stretching to sketch the graph of the function defined by each equation.

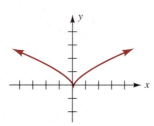

71. $y = x^{2/3} + 2$ **72.** $y = (x - 3)^{2/3}$

73. $y = (x + 1)^{2/3} - 4$ **74.** $y = 3 - x^{2/3}$

75. $y = -3x^{2/3}$ **76.** $y = \dfrac{x^{2/3}}{3}$

In Exercises 77–84, each function f is one-to-one.
(a) Find the inverse function f^{-1}.

(b) Sketch the graphs of f and f^{-1} on the same coordinate plane.

77. $f(x) = 2x$ **78.** $f(x) = \sqrt[3]{x - 8}$

79. $f(x) = 3 - x^3$ **80.** $f(x) = \dfrac{1}{x} - 3$

81. $f(x) = \sqrt{x - 2}$ **82.** $f(x) = x + 3$

83. $f(x) = 4 - x^2$, $x \leq 0$

84. $f(x) = (x + 1)^2 - 4$, $x \geq -1$

In Exercises 85–92, refer to the graph of the function f that is shown in the figure.

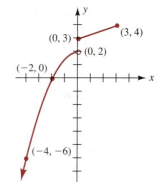

85. Specify the domain of f.

86. Specify the range of f.

87. Find $f(0)$.

88. Find $f(-2)$.

89. Find $f(f(0))$.

90. Find $f(f(-2))$.

91. Does f possess an inverse? If so, determine each value:
(a) $f^{-1}(0)$ (b) $f^{-1}(f^{-1}(4))$ (c) $f(f^{-1}(-6))$

92. Sketch the graph of f^{-1}, if it exists.

93. If the function f is one-to-one and defined by

$$f(x) = \dfrac{x + a}{x + b},$$

where a and b are constants and $a \neq b$, find the inverse function f^{-1}.

94. A function f is its own inverse if $f(f(x)) = x$. Show that the function f defined by

$$f(x) = \dfrac{x + k}{x - 1},$$

EXERCISE 94 *(continued)*

where k is constant, is its own inverse for any real number k if $k \neq -1$.

95. The cost C (in dollars) for a daily truck rental is a function of the miles x driven and is given by

$$C(x) = 32 + 0.25x.$$

(a) Evaluate $C(125.7)$, rounding to the nearest cent.

(b) Determine the miles driven when $C(x) = \$138.40$.

96. The volume of a rectangular box with a square base and open top is 9 cubic feet. Express the total surface area S as a function of the width x of its base.

97. The monthly charge for water in a small town is \$0.015 per gallon for the first 1000 gallons that are used and \$0.02 per gallon for each gallon in excess of 1000 gallons used. Express the charge C for the water as a function of the number n of gallons used.

98. The fixed cost to run a company that manufactures picnic tables is \$10,000 per month, and the variable cost to produce each table is \$100 per unit. If each table sells for x dollars, then the number of tables sold per month is estimated to be $800 - 2x$. Express the monthly profit P of the company as a function of x. (*Hint:* Profit = Total revenue − total cost.)

99. The point $P(x, y)$ lies on the right portion of the parabola $y = 3 - x^2$ with $y \geq 0$, as shown in the figure.

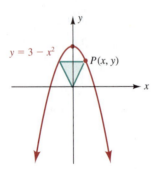

(a) Express the area A of the shaded isosceles triangle as a function of the x-coordinate of P.

(b) State the domain of the function defined in part (a).

100. A street light 30 feet above level ground casts the shadow of a man 6 feet tall.

(a) Express the shadow length s as a function of the man's distance d from the base of the light pole.

(b) If the man stands 8 feet from the base of the light pole and then walks away from the pole at the rate of 4 feet per second, express the shadow length s (in feet) as a function of the time t (in seconds).

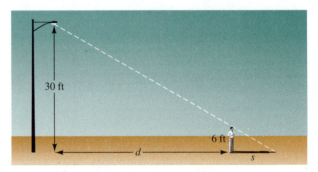

101. The excise tax T on an automobile varies directly as its actual value V.

(a) Express T as a function of V if $T = \$200$ when $V = \$5000$.

(b) State the domain of the function defined in part (a), then sketch its graph.

102. The intensity I of a light source varies inversely as the square of the distance d from the light source.

(a) Express I as a function of d if $I = 4$ candlepower when $d = 5$ feet.

(b) State the domain of the function defined in part (a), then sketch its graph.

103. A large field is capable of supporting life for a maximum of 2000 field mice. Suppose the rate of growth G of the field mouse population varies jointly as the number n of field mice present and the difference between the maximum number supportable by the field and the number present.

(a) Express G as a function of n if $G = 50$ field mice per week when $n = 1200$ mice.

(b) Find the rate of growth when 1500 field mice are living in the field.

104. The time t required for the excavation of a sewer line varies jointly as its length l, width w, and square of its depth d, and inversely as the number n of backhoes used. Describe what happens to the time for excavation if the depth is halved and the number of backhoes is doubled.

CHAPTER 2

Trigonometric Functions

During a baseball game, a runner on first base attempts to steal second base. Given that a baseball diamond is a square with 90-foot base paths (sides) and that θ is the angle between the first-base path and a line from home plate to the runner,
(a) express angle θ as a function of the runner's distance d from home plate, and
(b) use the function described in part (a) to find the measure of angle θ when $d = 100$ ft.

(For the solution, see Example 7 in Section 2.6.)

- 2.1 Angles and Their Measures
- 2.2 Trigonometric Functions of Angles
- 2.3 Evaluating Trigonometric Functions of Angles and Real Numbers
- 2.4 Properties and Graphs of the Sine and Cosine Functions
- 2.5 Properties and Graphs of the Other Trigonometric Functions
- 2.6 The Inverse Trigonometric Functions

2.1 Angles and Their Measures

◆ Introductory Comments

Any function that can be expressed as sums, differences, products, quotients, powers, or roots of polynomials is classified as an **algebraic function**. All the functions that we discussed in Chapter 1 are algebraic functions. Any function that goes beyond the limits of, or transcends an algebraic function is a **transcendental function**. In this chapter we study a special group of transcendental functions, the **trigonometric functions.**

Early Greek mathematicians used the trigonometric functions to solve many types of problems involving sides and angles of triangles. The domains of the trigonometric functions, when used in this manner, are angles. After the advent of calculus, mathematicians discovered that the trigonometric functions could also be used to describe cyclic patterns that vary with time. The domains of the trigonometric functions, when used in this manner, are real numbers. In this chapter, we will discuss both approaches to trigonometry and will use trigonometry to find the missing parts of a triangle as well as to describe patterns such as weather patterns, predator-prey relationships, and electric voltage.

We begin with a discussion of angles and various ways to measure them. Then, in Sections 2.2 and 2.3, we define the trigonometric functions of angles and the trigonometric functions of real numbers.

◆ Angles

A *ray* with endpoint A that continues indefinitely through another point B is designated $\overrightarrow{AB}$. One such ray is shown in Figure 2.1. When we rotate this ray about its endpoint, we form an **angle** with *vertex A, initial side $\overrightarrow{AB}$* and *terminal side $\overrightarrow{AB'}$*. One method of naming this angle is to place a Greek letter, such as theta (θ), next to the curved arrow that designates the rotation (see Figure 2.2). Angles formed by rotation are usually placed in a Cartesian coordinate system using *standard position*. **Standard position** means that the vertex of the angle is at the origin and its initial side is along the positive x-axis, as shown in Figure 2.3.

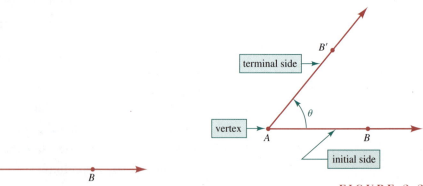

FIGURE 2.1
A ray with endpoint A, denoted $\overrightarrow{AB}$

FIGURE 2.2
Angle θ, formed by rotating $\overrightarrow{AB}$ about its endpoint

SECTION 2.1 Angles and Their Measures

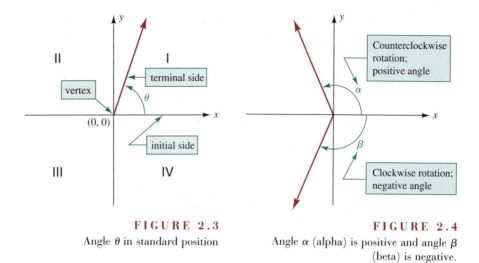

FIGURE 2.3
Angle θ in standard position

FIGURE 2.4
Angle α (alpha) is positive and angle β (beta) is negative.

By convention, an angle formed by *counterclockwise* rotation is said to be a **positive angle** and an angle formed by *clockwise* rotation is said to be a **negative angle**. Figure 2.4 illustrates a positive angle and a negative angle in standard position.

◆ Degree Measure

The *measure* of an angle is determined by the direction as well as the amount of the rotation from the initial side to the terminal side. One way to measure an angle is by using a unit of measure called a **degree** (°). One degree (1°) is formed when a ray is rotated $\frac{1}{360}$ of a full counterclockwise revolution about the origin of a coordinate system. Thus, one full counterclockwise revolution measures 360°. By convention, we write $\theta = 360°$ to indicate that the measure of angle θ is 360°. In Figure 2.5, several angles are shown in standard position. Each angle is measured in degrees.

Notice in Figure 2.5 that although angles with measures 0°, 360°, and −720° differ in the amount and direction of their rotation, they have the same terminal side. Angles in standard position that have the same terminal side are called **coterminal**

FIGURE 2.5
Angles measured in degrees in standard position

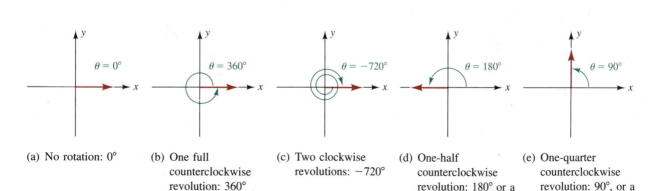

(a) No rotation: 0°

(b) One full counterclockwise revolution: 360°

(c) Two clockwise revolutions: −720°

(d) One-half counterclockwise revolution: 180° or a *straight angle*

(e) One-quarter counterclockwise revolution: 90°, or a *right angle*

angles. We can find the measures of angles that are coterminal with an angle measured in degrees by adding and subtracting multiples of 360°.

EXAMPLE 1 Determine the degree measures of a positive angle and a negative angle that are coterminal with an angle having the given measure.

(a) 390°   (b) −420°

SOLUTION

(a) We find angles coterminal with 390° by adding and subtracting multiples of 360°. Thus, all angles of the form

$$(390 + 360n)°, \quad \text{where } n \text{ is an integer,}$$

are coterminal with 390°.

Letting $n = -1$, we obtain   Letting $n = -2$, we obtain

$$(390 - 360)° = 30°. \qquad (390 - 720)° = -330°.$$

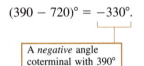

A *positive* angle coterminal with 390°   A *negative* angle coterminal with 390°

FIGURE 2.6
30°, 390°, and −330° are coterminal angles.

Figure 2.6 shows angles of 30°, 390° and −330° in standard position. Note that these angles have the same terminal side.

(b) We may find angles coterminal with −420° by adding and subtracting multiples of 360°. Thus, all angles of the form

$$(-420 + 360n)°, \quad \text{where } n \text{ is an integer,}$$

are coterminal with −420°.

Letting $n = 1$, we obtain   Letting $n = 2$, we obtain

$$(-420 + 360)° = -60°. \qquad (-420 + 720)° = 300°.$$

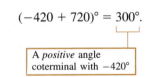

A *negative* angle coterminal with −420°   A *positive* angle coterminal with −420°

FIGURE 2.7
−420°, 300°, and −60° are coterminal angles.

Figure 2.7 shows angles of −420°, 300°, and −60° in standard position. Note that these angles have the same terminal side. ◆

PROBLEM 1 Repeat Example 1 for 600°. ◆

When greater precision is needed to measure an angle, we divide each degree into 60 equal parts, each called a **minute** (′). If still greater precision is needed to measure an angle, then each minute is divided into 60 equal parts, each called a **second** (″). For example, an angle written as 16°24′27″ is read "16 degrees, 24

SECTION 2.1   Angles and Their Measures

minutes, 27 seconds." Many types of calculators have keys that convert angle measures written in degrees, minutes, seconds into decimal degrees, and vice versa. If such a calculator is not available, we can use the following facts to make the conversion.

$$1° = 60', \quad 1' = 60'', \quad 1° = 3600''$$

EXAMPLE 2   Convert the measure of the angle in part (a) to decimal degrees and the measure of the angle in part (b) to degrees, minutes, seconds.

(a) $16°24'27''$ (b) $123.423°$

SOLUTION

(a)   To convert to decimal degrees, we apply two conversion factors:

$$16°24'27'' = 16° + 24'\left(\frac{1°}{60'}\right) + 27''\left(\frac{1°}{3600''}\right)$$

conversion factors

$$= 16° + 0.4° + 0.0075°$$
$$= 16.4075°$$

(b)   First, we convert the decimal part of the degrees (0.423°) to minutes:

$$0.423° = 0.423°\left(\frac{60'}{1°}\right) = 25.38'$$

conversion factor

Thus, $123.423° = 123°25.38'$

Now, we must convert the decimal part of the minutes (0.38′) to seconds:

$$0.38' = 0.38'\left(\frac{60''}{1'}\right) = 22.8''$$

conversion factor

Thus, $123.423° \approx 123°25'23''$ (to the nearest second).   ◆

PROBLEM 2   Check the results of Example 2 by converting each measure as indicated.

(a) $16.4075°$ to degrees, minutes, seconds

(b) $123°25'23''$ to decimal degrees   ◆

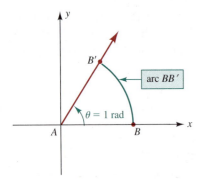

FIGURE 2.8
When the length of arc BB' equals AB, an angle θ with radian measure 1 is formed.

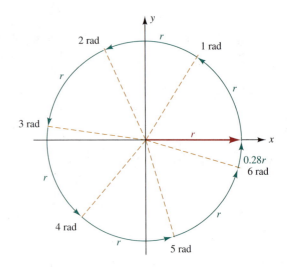

FIGURE 2.9
The radius can be marked off 2π (≈ 6.28) times along its circumference.

◆ Radian Measure

Another way to measure an angle is by using a unit of measure called a *radian*. An angle whose measure is one radian (1 rad) is formed when we rotate $\overrightarrow{AB}$ counterclockwise such that the *length of the arc BB'* equals the *length of the line segment $\overline{AB}$*, as shown in Figure 2.8.

Recall from elementary geometry that the circumference C of a circle is given by the formula $C = 2\pi r$, where r is the radius of the circle. Thus, for any circle, its radius r can be marked off 2π (≈ 6.28) times along its circumference, as shown in Figure 2.9. Thus, one full counterclockwise revolution corresponds to 2π rad (≈ 6.28 rad). By convention, if no unit of measure is specified with an angle, then the angle is in radians. Thus, if we write $\theta = 2\pi$, we mean that θ measures 2π rad. In Figure 2.10, several angles are shown in standard position. Each angle is measured in radians.

FIGURE 2.10
Angles measured in radians in standard position

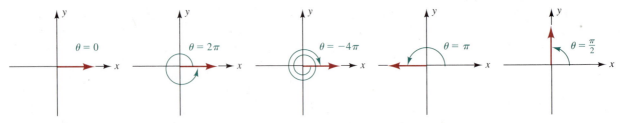

(a) No rotation: 0 radian
(b) One full counterclockwise revolution: 2π radians
(c) Two clockwise revolutions: -4π radians
(d) One-half counterclockwise revolution: π radians
(e) One-quarter counterclockwise revolution: $\pi/2$ radians

One complete revolution measures 2π radians. One complete revolution also measures 360°. Thus,

$$2\pi \text{ rad} = 360° \quad \text{or} \quad \pi \text{ rad} = 180°.$$

From this relationship, we obtain

$$1 \text{ rad} = \frac{180°}{\pi} \approx 57.3° \quad \text{and} \quad 1° = \frac{\pi}{180} \text{ rad} \approx 0.01745 \text{ rad}$$

These equations give us the following conversion factors.

Conversion Factors: Radians to Degrees and Degrees to Radians

To convert radians to degrees, multiply the radian measure by $\dfrac{180°}{\pi}$.

To convert degrees to radians, multiply the degree measure by $\dfrac{\pi}{180°}$.

EXAMPLE 3 Convert the degree measures to radians, and convert the radian measures to degrees.

(a) $\theta = 30°$ (b) $\theta = -240°$ (c) $\theta = \dfrac{9\pi}{4}$ (d) $\theta = 30$

SOLUTION The angular measures in parts (a) and (b) are in degrees. Thus, we convert to radian measure as follows:

(a) $30° = 30°\left(\dfrac{\pi}{180°}\right) = \dfrac{\pi}{6}$ (b) $-240° = -240°\left(\dfrac{\pi}{180°}\right) = -\dfrac{4\pi}{3}$

The angular measures in parts (c) and (d) are in radians, since no unit of measure is specified. We convert radians to degrees as follows:

(c) $\dfrac{9\pi}{4} = \dfrac{9\pi}{4}\left(\dfrac{180°}{\pi}\right) = 405°$

(d) $30 = 30\left(\dfrac{180°}{\pi}\right) = \dfrac{5400°}{\pi} \approx 1719°$ ◆

We can find the measures of angles that are coterminal with an angle measured in radians by adding and subtracting multiples of 2π. For example, all angles of the form

$$\frac{\pi}{6} + 2\pi n, \quad \text{where } n \text{ is an integer}$$

are coterminal with $\pi/6$. Thus,

$$\frac{\pi}{6} + 2\pi = \frac{13\pi}{6}, \qquad \frac{\pi}{6} + 4\pi = \frac{25\pi}{6}, \qquad \frac{\pi}{6} - 2\pi = -\frac{11\pi}{6}$$

are all coterminal with $\pi/6$.

PROBLEM 3 Determine the radian measure of the smallest positive angle that is coterminal with $9\pi/4$.

Application: Arc Length

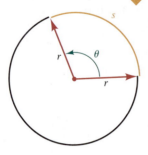

FIGURE 2.11
A circle of radius r with central angle θ that intercepts an arc of length s.

Shown in Figure 2.11 is a circle of radius r with *central angle* θ that intercepts an arc of length s. For a central angle $\theta = 2\pi$, the arc length s is the circumference of the entire circle. Hence $s = 2\pi r$ (recall that circumference $= 2\pi r$). For a central angle of $\theta = \pi$, the arc length s is one-half the circumference of the circle. Hence $s = \pi r$. For a central angle of $\theta = \pi/2$, the arc length s is one-quarter the circumference of the circle. Hence $s = \pi r/2$. Note that in each case the ratio of the central angle θ to the arc length s is always $1/r$, that is,

$$\frac{\theta}{s} = \frac{1}{r}$$

$s = \theta r$ **Cross product property**

Arc Length Formula

If a central angle θ (in radians) intercepts an arc of length s on a circle with radius r, then

$$s = \theta r.$$

EXAMPLE 4 Referring to the intersecting highways shown in Figure 2.12, determine the length of the circular arc from the point of tangency, PT, to the point of curvature, PC.

SOLUTION Before applying the arc length formula, we convert 115° to radians, as follows:

$$115° = 115°\left(\frac{\pi}{180°}\right) = \frac{23\pi}{36}.$$

Thus,

$$s = \theta r = \frac{23\pi}{36} \cdot 48.0 \text{ ft} \approx 96.3 \text{ ft}.$$

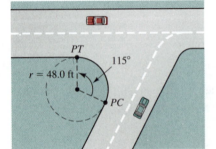

FIGURE 2.12 Hence, the arc length from PT to PC is approximately 96.3 ft.

PROBLEM 4 Repeat Example 4 if the central angle θ measures 125°.

Application: Angular Speed

Linear speed v is defined as the distance s traveled per unit of time t, that is

$$v = \frac{s}{t}$$

For example, if a boy walks a distance of 6 miles in 2 hours, then his linear speed is

$$v = \frac{s}{t} = \frac{6 \text{ mi}}{2 \text{ h}} = 3 \text{ mph}.$$

Angular speed ω (denoted by the Greek letter *omega*) is defined as the amount of rotation θ per unit of time t, that is,

$$\omega = \frac{\theta}{t}$$

For example, if a wheel rotates 6 rad in 2 seconds, then its angular speed is

$$\omega = \frac{\theta}{t} = \frac{6 \text{ rad}}{2 \text{ s}} = 3 \text{ rad}/s.$$

To develop a relationship between linear speed and angular speed, we proceed as follows:

$$s = \theta r$$

$$\frac{s}{t} = \frac{\theta}{t} r \qquad \textcolor{red}{\textbf{Divide both sides by } t}$$

$$v = \omega r \qquad \textcolor{red}{\textbf{Substitute}}$$

Relationship between Linear Speed and Angular Speed

If v is the linear speed of an object, ω its angular speed in radians per unit of time, and r the radius of rotation, then

$$v = \omega r.$$

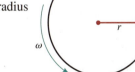

EXAMPLE 5 Each tire on an automobile has a radius of 1.25 ft. How many revolutions per minute (rpm) does a tire make when the automobile is traveling at a speed of 88.0 ft/s?

SOLUTION We first find the angular speed ω of a tire:

$$v = \omega r$$
$$88.0 \text{ ft/s} = \omega(1.25 \text{ ft})$$
$$\omega = 70.4 \text{ rad/s} \quad \textbf{Divide both sides by 1.25 ft}$$

Therefore, the angle through which a tire rotates in *one minute* is

$$\theta = \omega t = (70.4 \text{ rad/s})(60 \text{ s}) = 4224 \text{ rad}.$$

Since every 2π rad equals one revolution (rev), we have

$$\theta = 4224 \text{ rad}\left(\frac{1 \text{ rev}}{2\pi \text{ rad}}\right) \approx 672 \text{ rev}.$$

Thus, the tire is rotating at 672 rpm when the automobile travels at 88 ft/s. ◆

PROBLEM 5 Repeat Example 5 if the automobile is traveling at a speed of 45 mph. ◆

Exercises 2.1

Basic Skills

In Exercises 1–8, draw each angle in standard position, and give its degree measure and radian measure.

1. One-eighth of a revolution counterclockwise
2. Two-thirds of a revolution counterclockwise
3. Five-sixths of a revolution clockwise
4. One-fourth of a revolution clockwise
5. One and one-half revolutions counterclockwise
6. Two and one-sixth revolutions counterclockwise
7. Three revolutions clockwise
8. One and three-fourths revolutions clockwise

In Exercises 9–16, find the degree measure of the smallest positive angle that is coterminal with the given angular measure.

9. 580°
10. 695°
11. −1824°
12. 1200°
13. 1000°
14. −906°
15. 414°38′
16. 628°22′15″

In Exercises 17–24, find the radian measure of the smallest positive angle that is coterminal with the given angular measure.

17. 7π
18. $\dfrac{5\pi}{2}$
19. $-\dfrac{15\pi}{4}$
20. $\dfrac{24\pi}{5}$
21. $\dfrac{13\pi}{3}$
22. $\dfrac{11\pi}{4}$
23. 8
24. −12

In Exercises 25–32, convert the degree measure to radian measure. Write the answer as a reduced fraction in terms of π.

25. 60°
26. 45°
27. 225°
28. −150°
29. 432°
30. 520°
31. −375°
32. 774°

In Exercises 33–40, convert each radian measure to degree measure.

33. $\dfrac{5\pi}{4}$
34. $\dfrac{7\pi}{6}$
35. $-\dfrac{\pi}{18}$
36. $\dfrac{5\pi}{3}$
37. $\dfrac{13\pi}{8}$
38. $-\dfrac{22\pi}{5}$
39. 12π
40. $\dfrac{25\pi}{12}$

41. The radius of a circle is 12 cm. Find the length of an arc intercepted on its circumference by the given central angle.
    (a) $\pi/2$ (b) $7\pi/6$ (c) $60°$ (d) $135°$

42. Find the diameter of a circle if the length of an arc intercepted on its circumference is 21π inches when the central angle is θ.
    (a) $\theta = \pi/2$ (b) $\theta = 7\pi/6$
    (c) $\theta = 60°$ (d) $\theta = 135°$

43. On a carousel the horses are 15 feet from the center. How many feet does a passenger travel when riding a horse if the carousel makes 20 revolutions before stopping?

44. Winnipeg, Canada, is approximately due north of Austin, Texas. Austin is at latitude 30° North and Winnipeg is at latitude 50° North, as shown in the sketch. If the radius of the earth is approximately 4000 miles, what is the distance between the two cities?

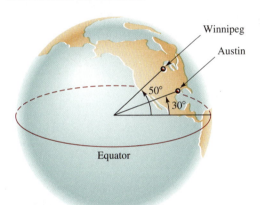

45. A freight train rounds a curve of radius 1320 feet at the speed of 30 mph. Through how many radians does a point on the train turn in 30 seconds?

46. A lawn mower blade turns at 1800 revolutions per minute (rpm). What is the linear speed (in ft/s) of the tip of the blade if the diameter of the blade is 18 inches?

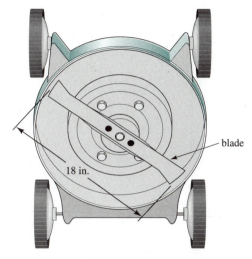

47. What is the angular speed (in rad/s) of each clock part?
    (a) the minute hand   (b) the hour hand

48. A space shuttle orbits the moon once every 130 minutes and maintains a constant altitude of 70 miles above the surface of the moon. If the radius of the moon is approximately 1100 miles, what is the linear speed (in mph) of the satellite?

49. A stereo record has a diameter of 12 inches and rotates on a turntable at $33\frac{1}{3}$ revolutions per minute (rpm).
    (a) Find the angular speed of the record (in rad/min).
    (b) Find the linear speed of the record (in ft/s).

50. A gear in a piece of machinery makes 12 revolutions per minute (rpm). How many seconds does it take for the gear to turn through 144°?

Critical Thinking

51. Which is larger, an angle that measures 1 rad or an angle that measures 1°? Explain.

52. The wheels on a bicycle are turning at the rate of 30 revolutions per minute (rpm). Assuming the radius of each wheel is 1 foot, find the time necessary for the cyclist to travel a distance of $600\pi \approx 1885$ ft?

53. A particle moves along a circular path given by the equation $x^2 + y^2 = 144$, where x and y are in centimeters. How far has the particle moved after sweeping through a central angle of 150°?

54. A belt connects two pulleys with radii 8 cm and 12 cm, as shown in the sketch. If the larger pulley turns through an angle of 10 rad, find the angle through which the smaller pulley turns.

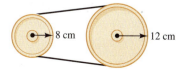

55. A *sector* of a circle is the region bounded by the arc PQ and the two radii $\overline{OP}$ and $\overline{OQ}$ shown in the figure.

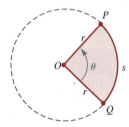

(a) Using the fact that the area A of a sector is directly proportional to the measure of its central angle θ (in radians), show that

$$A = \tfrac{1}{2}\theta r^2.$$

(b) Show that if s is the length of the arc from P to Q, then the area A of the sector is given by

$$A = \tfrac{1}{2}sr.$$

(c) Using the formulas in parts (a) and (b), find the area of each sector.

(i)   (ii)

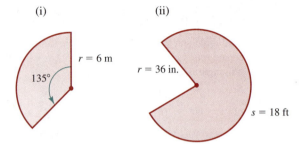

56. Referring to Exercise 55, suppose the perimeter of a sector is 12 cm.

(a) Express the area A of the sector as a function of r.

(b) Sketch the graph of the function defined in part (a).

(c) Find the value of the radius r that gives the maximum area of the sector. What is the maximum area?

(d) What is the measure of the central angle θ when the area of the sector is the largest possible?

Calculator Activities

In Exercises 57–60, convert each angular measure to (a) decimal degrees and (b) radians (to the nearest hundredth).

57. 22°36′   58. 178°54′

59. 215°28′48″   60. −6°2′49.2″

In Exercises 61–64, convert each angular measure to (a) degrees, minutes, seconds and (b) radians (to the nearest hundredth).

61. 32.31°   62. −112.325°

63. 306.1225°   64. 258.0008°

In Exercises 65–68, convert each angular measure to (a) decimal degrees (to the nearest ten-thousandth) and (b) degrees, minutes, (nearest) seconds.

65. 1.5   66. −2.36   67. 10   68. 1.571

69. A pendulum swinging through a central angle of 6°45′ sweeps an arc of length 10.0 cm. What is the length of the pendulum?

70. Referring to the intersecting highways shown in the sketch, find the central angle θ in degrees, minutes, seconds if the arc length from the point of tangency PT to the point of curvature PC is 162.35 ft.

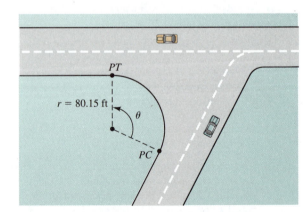

71. As the drum in the figure rotates counterclockwise, the rope is wound around the drum and the weight moves upward. After 1.5 seconds, how high is the weight off the ground if the angular speed of the drum is 2.25 rad/s and the radius of the drum is 25.6 cm?

72. Referring to Exercise 71, find the radius of the drum if the weight is raised 30.2 cm when the drum rotates 85°.

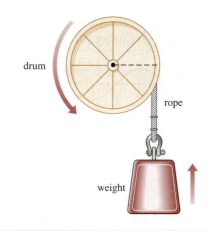

EXERCISE 71

2.2 Trigonometric Functions of Angles

◆ **Introductory Comments**

Consider an angle θ in standard position with $P(x, y)$ a point other than $(0, 0)$ on the terminal side of θ as shown in Figure 2.13. By the *distance formula* (Section 1.2), the distance r from the origin to the point $P(x, y)$ is always positive and is given by

$$r = \sqrt{x^2 + y^2}.$$

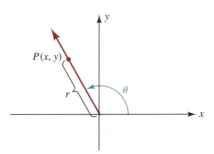

FIGURE 2.13
The values of x, y, and r determine the six trigonometric ratios for angle θ.

From the values of x, y, and r, we define the six *trigonometric ratios* for angle θ. These six ratios are called the *sine* (sin), *cosine* (cos), *tangent* (tan), *cosecant* (csc), *secant* (sec), and *cotangent* (cot).

Trigonometric Ratios

If θ is an angle in standard position with $P(x, y)$ a point other than $(0, 0)$ on the terminal side of θ, then the six **trigonometric ratios** of angle θ are defined as follows:

1. $\sin \theta = \dfrac{y}{r}$
2. $\cos \theta = \dfrac{x}{r}$
3. $\tan \theta = \dfrac{y}{x}, \quad x \neq 0$

4. $\csc \theta = \dfrac{r}{y}, \quad y \neq 0$
5. $\sec \theta = \dfrac{r}{x}, \quad x \neq 0$
6. $\cot \theta = \dfrac{x}{y}, \quad y \neq 0$

where $r = \sqrt{x^2 + y^2}$.

The value of each trigonometric ratio is determined by the angle θ, not by the particular point $P(x, y)$ that we choose on the terminal side of θ. To show this, consider an angle θ in standard position with points $P(x, y)$ and $P_1(x_1, y_1)$ on its terminal side, as shown in Figure 2.14. Since the triangles OPQ and OP_1Q_1 are similar, it follows from elementary geometry that their corresponding sides are proportional. Hence,

$$\sin \theta = \frac{y}{r} = \frac{y_1}{r_1}, \qquad \cos \theta = \frac{x}{r} = \frac{x_1}{r_1}, \qquad \tan \theta = \frac{y}{x} = \frac{y_1}{x_1}, \qquad \text{and so on.}$$

Since the value of each ratio depends only on angle θ, we can say that the ratios are *functions* of angle θ. In turn, we call these ratios the **trigonometric functions** of angle θ. In this section, we work with the trigonometric functions of angles and discuss some of the fundamental relationships between these functions.

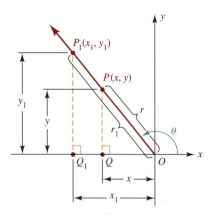

FIGURE 2.14
When the points $P(x, y)$ and $P_1(x_1, y_1)$ are on the terminal side of angle θ, the triangles OPQ and OP_1Q_1 are similar.

◆ Angle Domains of the Trigonometric Functions

The *domain* of each trigonometric function is all angles θ such that the denominator of its corresponding ratio is not zero. If the terminal side of an angle θ is along the *x*-axis and $P(x, y)$ is a point on its terminal side, then $y = 0$ and, therefore,

$$\csc \theta = \frac{r}{y} \quad \text{and} \quad \cot \theta = \frac{x}{y} \quad \text{are undefined.}$$

If the terminal side of θ is along the *y*-axis and $P(x, y)$ is a point on its terminal side, then $x = 0$ and, therefore,

$$\tan \theta = \frac{y}{x} \quad \text{and} \quad \sec \theta = \frac{r}{x} \quad \text{are undefined.}$$

However, since r is always positive,

$$\sin \theta = \frac{y}{r} \quad \text{and} \quad \cos \theta = \frac{x}{r} \quad \text{are defined for all angles } \theta.$$

Table 2.1 summarizes our results.

Table 2.1
Angle domains of the trigonometric functions

Trigonometric Function	Domain
sine or cosine	All angles
cosecant or cotangent	All angles except those whose terminal side is on the *x*-axis
tangent or secant	All angles except those whose terminal side is on the *y*-axis

Our definitions of the trigonometric functions as ratios allow us to find the following products:

$$\sin \theta \csc \theta = \left(\frac{y}{r}\right)\left(\frac{r}{y}\right) = 1, \quad y \neq 0,$$

$$\cos \theta \sec \theta = \left(\frac{x}{r}\right)\left(\frac{r}{x}\right) = 1, \quad x \neq 0,$$

$$\tan \theta \cot \theta = \left(\frac{y}{x}\right)\left(\frac{x}{y}\right) = 1, \quad x \neq 0, y \neq 0.$$

When the product of two numbers equals 1, the two numbers are said to be *reciprocals* of each other. Thus, to find the six trigonometric functions of an angle θ whose terminal side passes through a given point, we begin by evaluating $\sin \theta$, cos θ, and tan θ. For csc θ, sec θ, and cot θ, we simply use the fact that these are reciprocals of sin θ, cos θ, and tan θ, respectively:

$$\csc\theta = \frac{1}{\sin\theta} \qquad \sec\theta = \frac{1}{\cos\theta} \qquad \cot\theta = \frac{1}{\tan\theta}$$

EXAMPLE 1 Given that $P(-1, 2)$ is a point on the terminal side of an angle θ in standard position, find the values of the six trigonometric functions of θ.

SOLUTION Referring to Figure 2.15, we have $x = -1$, $y = 2$, and

$$r = \sqrt{x^2 + y^2} = \sqrt{(-1)^2 + (2)^2} = \sqrt{5}.$$

Thus,

$$\sin\theta = \frac{y}{r} = \frac{2}{\sqrt{5}} \qquad\qquad \csc\theta = \frac{1}{\sin\theta} = \frac{\sqrt{5}}{2}$$

$$\cos\theta = \frac{x}{r} = \frac{-1}{\sqrt{5}} = -\frac{1}{\sqrt{5}} \qquad \sec\theta = \frac{1}{\cos\theta} = -\sqrt{5}$$

$$\tan\theta = \frac{y}{x} = \frac{2}{-1} = -2 \qquad\qquad \cot\theta = \frac{1}{\tan\theta} = -\frac{1}{2}$$

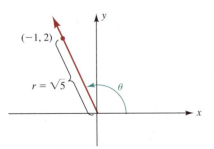

FIGURE 2.15
An angle θ in standard position with terminal side passing through the point $(-1, 2)$

PROBLEM 1 Repeat Example 1 given that $(-2, 4)$ is a point on the terminal side of an angle θ.

◆ **Algebraic Signs of the Trigonometric Functions**

When the terminal side of an angle θ lies in a particular quadrant, we say that angle θ lies in that quadrant as well. Selecting $P(x, y)$ as a point on the terminal side of angle θ (see Figure 2.16), and remembering that r is always positive, we can show that the algebraic sign of a trigonometric function of θ depends entirely on the quadrant in which θ lies.

FIGURE 2.16
Values of x and y, when θ is in each quadrant

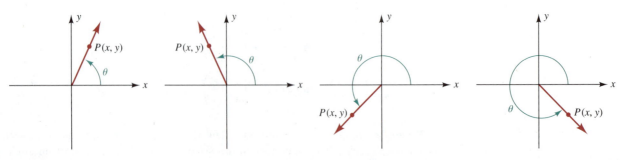

(a) θ in quadrant I; $x > 0$, $y > 0$

(b) θ in quadrant II; $x < 0$, $y > 0$

(c) θ in quadrant III; $x < 0$, $y < 0$

(d) θ in quadrant IV; $x > 0$, $y < 0$

1. If θ is in quadrant I [as shown in part (a) of Figure 2.16], then $x > 0$ and $y > 0$. Thus

$$\sin\theta = \frac{y}{r} = \frac{+}{+} = + \qquad \cos\theta = \frac{x}{r} = \frac{+}{+} = + \qquad \tan\theta = \frac{y}{x} = \frac{+}{+} = +$$

Hence, *all the trigonometric functions are positive when θ is in quadrant I.*

2. If θ is in quadrant II [part (b) of the figure], then $x < 0$ and $y > 0$. Thus

$$\sin\theta = \frac{y}{r} = \frac{+}{+} = + \qquad \cos\theta = \frac{x}{r} = \frac{-}{+} = - \qquad \tan\theta = \frac{y}{x} = \frac{+}{-} = -$$

Hence, *only $\sin\theta$ and its reciprocal $\csc\theta$ are positive when θ is in quadrant II.*

3. If θ is in quadrant III [part (c) of Figure 2.16], then $x < 0$ and $y < 0$. Thus

$$\sin\theta = \frac{y}{r} = \frac{-}{+} = - \qquad \cos\theta = \frac{x}{r} = \frac{-}{+} = - \qquad \tan\theta = \frac{y}{x} = \frac{-}{-} = +$$

Hence, *only $\tan\theta$ and its reciprocal $\cot\theta$ are positive when θ is in quadrant III.*

4. If θ is in quadrant IV [part (d) of the figure], then $x > 0$ and $y < 0$. Thus

$$\sin\theta = \frac{y}{r} = \frac{-}{+} = - \qquad \cos\theta = \frac{x}{r} = \frac{+}{+} = + \qquad \tan\theta = \frac{y}{x} = \frac{-}{+} = -$$

Hence, *only $\cos\theta$ and its reciprocal $\sec\theta$ are positive when θ is in quadrant IV.*

To help remember the algebraic signs of the trigonometric functions of θ, we can use the phrase

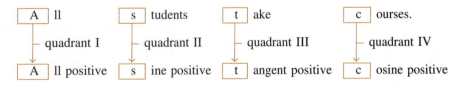

As illustrated in the next example, if the value of a trigonometric function is given and the quadrant containing the angle is known, then we can determine the values of the other five trigonometric functions.

EXAMPLE 2 The value of one trigonometric function is given, along with the sign of another one. Use the trigonometric ratios to find the values of the other five trigonometric functions of θ.

(a) $\tan\theta = -\frac{3}{4}$ and $\sin\theta > 0$ (b) $\cos\theta = 0.3$ and $\sin\theta < 0$

SOLUTION

(a) If θ is in either the first or third quadrant then $\tan\theta > 0$. Thus, for $\tan\theta = -\frac{3}{4} < 0$, we know that θ is in either quadrant II or quadrant IV.

In quadrant II, $\sin\theta > 0$, and in quadrant IV, $\sin\theta < 0$. Since we want $\sin\theta > 0$, we conclude that θ is in quadrant II. Now in quadrant II, we know that $x < 0$ and $y > 0$. Since

$$\tan\theta = -\frac{3}{4} = \frac{y}{x},$$

we choose $P(-4, 3)$ as a point on the terminal side of θ. For $x = -4$ and $y = 3$, we have

$$r = \sqrt{x^2 + y^2} = \sqrt{(-4)^2 + (3)^2} = \sqrt{25} = 5,$$

as illustrated in Figure 2.17.
Thus,

$$\sin\theta = \frac{y}{r} = \frac{3}{5} \quad \text{and} \quad \cos\theta = \frac{x}{r} = \frac{-4}{5} = -\frac{4}{5}.$$

Therefore,

$$\csc\theta = \frac{1}{\sin\theta} = \frac{5}{3}, \quad \sec\theta = \frac{1}{\cos\theta} = -\frac{5}{4}, \quad \cot\theta = \frac{1}{\tan\theta} = -\frac{4}{3}.$$

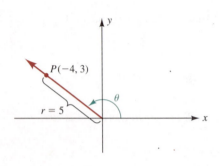

FIGURE 2.17
If we have $\tan\theta = -\frac{3}{4}$ and $\sin\theta > 0$, then we can choose the point $P(-4, 3)$ on the terminal side of θ.

(b) For $\cos\theta = 0.3 > 0$, we know that θ is in either the first or fourth quadrant. In quadrant I, we have $\sin\theta > 0$ and in quadrant IV, $\sin\theta < 0$. Since we want $\sin\theta < 0$, we conclude that θ is in quadrant IV. Now, in quadrant IV, we know that $x > 0$ and $y < 0$. Thus, for

$$\cos\theta = 0.3 = \frac{3}{10} = \frac{x}{r},$$

we choose a point on the terminal side of θ such that $x = 3$ and $r = 10$. To find $y < 0$, we proceed as follows:

$$r = \sqrt{x^2 + y^2}$$
$$10 = \sqrt{(3)^2 + y^2} \quad \text{Substitute for } x \text{ and } r$$
$$100 = 9 + y^2 \quad \text{Square both sides}$$
$$y^2 = 91 \quad \text{Subtract 9 from both sides}$$
$$y = \pm\sqrt{91} \;\boxed{\text{Choose}} \to y = -\sqrt{91}$$

$\boxed{\text{Angle } \theta \text{ in quadrant IV implies } y < 0.}$

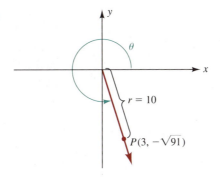

FIGURE 2.18
If we have $\cos \theta = 0.3$ and $\sin \theta < 0$, then we can choose the point $P(3, -\sqrt{91})$ on the terminal side of θ.

Now, referring to Figure 2.18, we have

$$\sin \theta = \frac{y}{r} = \frac{-\sqrt{91}}{10} = -\frac{\sqrt{91}}{10}, \qquad \tan \theta = \frac{y}{x} = \frac{-\sqrt{91}}{3} = -\frac{\sqrt{91}}{3},$$

and

$$\csc \theta = \frac{1}{\sin \theta} = -\frac{10}{\sqrt{91}}, \qquad \sec \theta = \frac{1}{\cos \theta} = \frac{10}{3},$$

$$\cot \theta = \frac{1}{\tan \theta} = -\frac{3}{\sqrt{91}}.$$

◆

PROBLEM 2 Repeat Example 2 given that $\cot \theta = \frac{3}{4}$ and $\cos \theta < 0$. ◆

◆ **The Fundamental Trigonometric Identities**

Several relationships between the trigonometric functions are important for our future work with these functions. We have already discussed the following reciprocal relationships:

$$\sin \theta \csc \theta = 1 \qquad \cos \theta \sec \theta = 1 \qquad \tan \theta \cot \theta = 1$$

Now, consider the quotients $\dfrac{\sin \theta}{\cos \theta}$ and $\dfrac{\cos \theta}{\sin \theta}$. Using the trigonometric ratios, we have

$$\frac{\sin \theta}{\cos \theta} = \frac{y/r}{x/r} = \frac{y}{r} \cdot \frac{r}{x} = \frac{y}{x} = \tan \theta$$

and

$$\frac{\cos \theta}{\sin \theta} = \frac{x/r}{y/r} = \frac{x}{r} \cdot \frac{r}{y} = \frac{x}{y} = \cot \theta.$$

In summary, we can state two identities involving quotients:

$$\tan \theta = \frac{\sin \theta}{\cos \theta} \qquad \cot \theta = \frac{\cos \theta}{\sin \theta}$$

Next, we consider the expression $(\sin \theta)^2 + (\cos \theta)^2$. Using the trigonometric ratios, we have

$$(\sin \theta)^2 + (\cos \theta)^2 = \left(\frac{y}{r}\right)^2 + \left(\frac{x}{r}\right)^2 = \frac{x^2 + y^2}{r^2}.$$

However, since $r = \sqrt{x^2 + y^2}$, we have $r^2 = x^2 + y^2$. Hence,

$$(\sin \theta)^2 + (\cos \theta)^2 = \frac{x^2 + y^2}{r^2} = \frac{r^2}{r^2} = 1.$$

To avoid using parentheses, we usually write $(\sin \theta)^2$ as $\sin^2 \theta$ and $(\cos \theta)^2$ as $\cos^2 \theta$. This convention is applied to other positive powers of the trigonometric functions as well. In summary, we state an identity involving powers:

$$\sin^2 \theta + \cos^2 \theta = 1$$

Do not confuse $\sin^2 \theta$ with $\sin \theta^2$. Remember,

$$\sin^2 \theta = (\sin \theta)(\sin \theta) = (\sin \theta)^2,$$

whereas

$$\sin \theta^2 = \sin (\theta \cdot \theta).$$

In general, $\sin^2 \theta \neq \sin \theta^2$.

Starting with the relationship $(\sin \theta)^2 + (\cos \theta)^2 = 1$, if we divide both sides by $(\cos \theta)^2$, we obtain

$$\frac{(\sin \theta)^2}{(\cos \theta)^2} + \frac{(\cos \theta)^2}{(\cos \theta)^2} = \frac{1}{(\cos \theta)^2}$$

$$\left(\frac{\sin \theta}{\cos \theta}\right)^2 + 1 = \left(\frac{1}{\cos \theta}\right)^2$$

$$(\tan \theta)^2 + 1 = (\sec \theta)^2$$

In a similar manner, starting with the same relationship $(\sin \theta)^2 + (\cos \theta)^2 = 1$, and dividing both sides by $(\sin \theta)^2$, we obtain $1 + (\cot \theta)^2 = (\csc \theta)^2$. In summary, we have

$$1 + \tan^2\theta = \sec^2\theta \quad \text{and} \quad 1 + \cot^2\theta = \csc^2\theta$$

These basic relationships between the trigonometric functions, which lay the foundation for more advanced work in trigonometry, are often referred to as the **fundamental trigonometric identities.**

Fundamental Trigonometric Identities

If θ is in the domain of the trigonometric function, then

1. $(\sin\theta)(\csc\theta) = 1$
2. $(\cos\theta)(\sec\theta) = 1$
3. $(\tan\theta)(\cot\theta) = 1$
4. $\tan\theta = \dfrac{\sin\theta}{\cos\theta}$
5. $\cot\theta = \dfrac{\cos\theta}{\sin\theta}$
6. $\sin^2\theta + \cos^2\theta = 1$
7. $1 + \tan^2\theta = \sec^2\theta$
8. $1 + \cot^2\theta = \csc^2\theta$

EXAMPLE 3 Use the fundamental trigonometric identities to find the value of the indicated trigonometric function from the given information.

(a) Find $\tan\theta$ if $\sin\theta = \dfrac{12}{13}$ and θ is in quadrant I.

(b) Find $\sin\theta$ if $\cot\theta = \sqrt{3}$ and $\sin\theta < 0$.

SOLUTION

(a) From the trigonometric identity $\sin^2\theta + \cos^2\theta = 1$, we have

$$\cos^2\theta = 1 - \sin^2\theta$$
$$\cos\theta = \pm\sqrt{1 - \sin^2\theta}$$

However, since we are given that θ is in quadrant I, we know that all trigonometric functions must be positive. Thus,

$$\cos\theta = \sqrt{1 - \sin^2\theta} = \sqrt{1 - \left(\frac{12}{13}\right)^2} = \sqrt{\frac{25}{169}} = \frac{5}{13}.$$

Now, using the trigonometric identity $\tan\theta = \sin\theta/\cos\theta$, we have

$$\tan\theta = \frac{\frac{12}{13}}{\frac{5}{13}} = \frac{12}{5}.$$

(b) For $\cot\theta = \sqrt{3} > 0$, we know that θ is in either the first or third quadrant. We know that in quadrant I, $\sin\theta > 0$ and in quadrant III, $\sin\theta < 0$. Since we want $\sin\theta < 0$, we conclude that θ is in quadrant III. Now, from the trigonometric identity $1 + \cot^2\theta = \csc^2\theta$, we obtain

$$\csc\theta = \pm\sqrt{1 + \cot^2\theta}$$

However, since θ is in quadrant III, we know that $\csc\theta < 0$. Thus,

$$\csc\theta = -\sqrt{1 + \cot^2\theta} = -\sqrt{1 + \left(\sqrt{3}\right)^2} = -\sqrt{4} = -2.$$

From the identity $\sin\theta \csc\theta = 1$, we find

$$\sin\theta = \frac{1}{\csc\theta} = -\frac{1}{2}.$$

◆

PROBLEM 3 Given that $\tan\theta = -1$ and $\cos\theta < 0$, use the fundamental trigonometric identities to find $\cos\theta$. ◆

◆ Trigonometric Ratios for Right Triangles

Recall from elementary geometry that a **right triangle** is a triangle in which one of the interior angles is a *right angle.* Since the sum of the interior angles in a right triangle is 180°, and the right angle measures 90°, it follows that the other two angles must be positive angles that each measure less than 90°. We refer to a positive angle with measure less than 90° as an *acute angle.* Figure 2.19 shows a right triangle with acute angles A and B and right angle C. By convention, we call the side opposite the right angle the *hypotenuse* and assign it the lowercase letter c. The sides opposite the angles A and B are called *legs,* and are denoted by lowercase letters a and b, respectively. By placing the vertex of angle A at the origin and side b along the positive x-axis, as shown in Figure 2.20, we can express the trigonometric functions of the acute angle A as ratios of the sides of the right triangle:

$$\sin A = \frac{a}{c} = \frac{\text{side opposite } A}{\text{hypotenuse}} \qquad \csc A = \frac{c}{a} = \frac{\text{hypotenuse}}{\text{side opposite } A}$$

$$\cos A = \frac{b}{c} = \frac{\text{side adjacent to } A}{\text{hypotenuse}} \qquad \sec A = \frac{c}{b} = \frac{\text{hypotenuse}}{\text{side adjacent to } A}$$

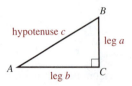

FIGURE 2.19
A right triangle with hypotenuse c and legs a and b

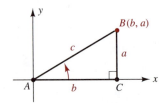

FIGURE 2.20
A right triangle with acute angle A in standard position

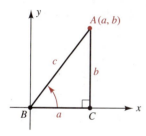

$$\tan A = \frac{a}{b} = \frac{\text{side opposite } A}{\text{side adjacent to } A} \qquad \cot A = \frac{b}{a} = \frac{\text{side adjacent to } A}{\text{side opposite } A}$$

Similarly, by placing the vertex of angle B at the origin and side a along the positive x-axis, as shown in Figure 2.21, we can express the trigonometric functions of the acute angle B as ratios of the sides of the right triangle:

$$\sin B = \frac{b}{c} = \frac{\text{side opposite } B}{\text{hypotenuse}} \qquad \csc B = \frac{c}{b} = \frac{\text{hypotenuse}}{\text{side opposite } B}$$

$$\cos B = \frac{a}{c} = \frac{\text{side adjacent to } B}{\text{hypotenuse}} \qquad \sec B = \frac{c}{a} = \frac{\text{hypotenuse}}{\text{side adjacent to } B}$$

$$\tan B = \frac{b}{a} = \frac{\text{side opposite } B}{\text{side adjacent to } B} \qquad \cot B = \frac{a}{b} = \frac{\text{side adjacent to } B}{\text{side opposite } B}$$

FIGURE 2.21
A right triangle with acute angle B in standard position

In summary, we state the **trigonometric ratios for right triangles:**

◆ Trigonometric Ratios for Right Triangles

If θ is an acute angle in a right triangle, then

$$\sin \theta = \frac{\text{opp}}{\text{hyp}}, \qquad \cos \theta = \frac{\text{adj}}{\text{hyp}}, \qquad \tan \theta = \frac{\text{opp}}{\text{adj}},$$

$$\csc \theta = \frac{\text{hyp}}{\text{opp}}, \qquad \sec \theta = \frac{\text{hyp}}{\text{adj}}, \qquad \cot \theta = \frac{\text{adj}}{\text{opp}},$$

where *opp* is the length of the side opposite θ, *adj* is the length of the side adjacent to θ, and *hyp* is the length of the hypotenuse of the right triangle.

EXAMPLE 4 Given the right triangle in Figure 2.22, find the value of each trigonometric function:

(a) $\sin A$ **(b)** $\cos A$ **(c)** $\tan A$

SOLUTION First, we apply the Pythagorean theorem to find the length of the leg opposite angle A. Letting a be the length of this leg, we find

$$12^2 = a^2 + 8^2$$
$$12^2 - 8^2 = a^2$$
$$a^2 = 80$$
$$a = \sqrt{80} = 4\sqrt{5}$$

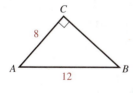

FIGURE 2.22

Now, using the trigonometric ratios for right triangles, we obtain the values of the indicated trigonometric functions.

(a) $\sin A = \dfrac{\text{opp}}{\text{hyp}} = \dfrac{4\sqrt{5}}{12} = \dfrac{\sqrt{5}}{3}$ (b) $\cos A = \dfrac{\text{adj}}{\text{hyp}} = \dfrac{8}{12} = \dfrac{2}{3}$

(c) $\tan A = \dfrac{\text{opp}}{\text{adj}} = \dfrac{4\sqrt{5}}{8} = \dfrac{\sqrt{5}}{2}$ ◆

PROBLEM 4 Referring to the right triangle in Example 4, find the value of each trigonometric function.

(a) $\sin B$ (b) $\cos B$ (c) $\tan B$ ◆

Since the sum of the interior angles in a right triangle is 180°, and the right angle measures 90°, it follows that the sum of the two acute angles is 180° − 90° = 90°. Any two acute angles whose sum is 90° are called **complementary angles.** Now, referring to Figures 2.20 and 2.21, observe that

$\cos A = \dfrac{b}{c} = \sin B,$ $\cos B = \dfrac{a}{c} = \sin A$ — The cosine of an acute angle is the sine of its complementary angle.

$\cot A = \dfrac{b}{a} = \tan B,$ $\cot B = \dfrac{a}{b} = \tan A$ — The cotangent of an acute angle is the tangent of its complementary angle.

$\csc A = \dfrac{c}{a} = \sec B,$ $\csc B = \dfrac{c}{b} = \sec A$ — The cosecant of an acute angle is the secant of its complementary angle.

Because of these relationships, we say that the *sine* and *cosine* are *cofunctions,* the *tangent* and *cotangent* are cofunctions, and the *secant* and *cosecant* are cofunctions. A trigonometric function of any acute angle θ is the same as the cofunction of its complementary angle 90° − θ, that is,

$$\sin \theta = \cos(90° - \theta) \qquad \cos \theta = \sin(90° - \theta)$$
$$\tan \theta = \cot(90° - \theta) \qquad \cot \theta = \tan(90° - \theta)$$
$$\sec \theta = \csc(90° - \theta) \qquad \csc \theta = \sec(90° - \theta)$$

EXAMPLE 5 Express each trigonometric function as the cofunction of its complementary angle.

(a) $\cos 42°$ (b) $\tan \dfrac{\pi}{8}$

SOLUTION

(a) $\cos 42° = \sin(90° - 42°) = \sin 48°$

(b) $\tan \dfrac{\pi}{8} = \cot\left(\dfrac{\pi}{2} - \dfrac{\pi}{8}\right) = \cot \dfrac{3\pi}{8}$

PROBLEM 5 Repeat Example 5 for $\sec 12°32'$.

Exercises 2.2

Basic Skills

In Exercises 1–6, a point P on the terminal side of θ is shown in the figure. Evaluate (if possible) the six trigonometric functions of θ.

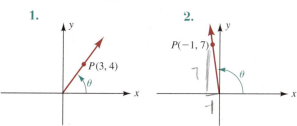

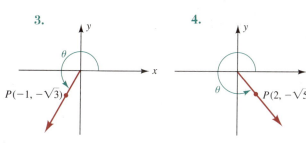

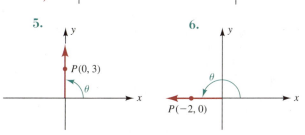

In Exercises 7–14, find the quadrant in which θ lies if the given conditions are satisfied.

7. $\tan\theta > 0$ and $\sin\theta < 0$
8. $\sin\theta > 0$ and $\tan\theta < 0$
9. $\sec\theta < 0$ and $\tan\theta < 0$
10. $\cos\theta < 0$ and $\cot\theta > 0$
11. $\cot\theta > 0$ and $\csc\theta > 0$
12. $\sec\theta > 0$ and $\sin\theta < 0$
13. $\sin\theta < 0$ and $\cos\theta > 0$
14. $\cos\theta < 0$ and $\csc\theta < 0$

In Exercises 15–24, the value of one of the trigonometric functions is given, along with some additional information. Use the trigonometric ratios to find the values of the other five trigonometric functions of θ.

15. $\sin\theta = \tfrac{3}{5}$, θ is in quadrant I
16. $\cos\theta = \tfrac{5}{13}$, θ is in quadrant I
17. $\tan\theta = -\tfrac{2}{3}$, θ is in quadrant II
18. $\sec\theta = -\tfrac{3}{2}$, θ is in quadrant III
19. $\cos\theta = \dfrac{\sqrt{3}}{2}$, $\sin\theta < 0$
20. $\sin\theta = \tfrac{1}{2}$, $\tan\theta < 0$
21. $\cot\theta = -1$, $\cos\theta > 0$
22. $\csc\theta = 2$, $\tan\theta > 0$
23. $\sec\theta = -1.2$, $\tan\theta > 0$
24. $\tan\theta = -0.8$, $\cos\theta > 0$

In Exercises 25–38, use the fundamental trigonometric identities to find the value of the indicated trigonometric function from the given information.

25. $\csc\theta = 3$; find $\sin\theta$
26. $\sec\theta = -\tfrac{5}{2}$; find $\cos\theta$
27. $\tan\theta = -0.2$; find $\cot\theta$
28. $\sin\theta = -0.25$; find $\csc\theta$
29. $\sin\theta = \dfrac{3}{\sqrt{10}}$, $\cos\theta = \dfrac{1}{\sqrt{10}}$; find $\tan\theta$
30. $\sin\theta = \tfrac{4}{5}$, $\cot\theta = \tfrac{3}{4}$; find $\cos\theta$
31. $\sec\theta = \tfrac{13}{12}$, $\tan\theta = \tfrac{5}{12}$; find $\sin\theta$
32. $\sin\theta = \dfrac{3}{\sqrt{13}}$, $\tan\theta = \dfrac{3}{2}$; find $\sec\theta$
33. $\cos\theta = \tfrac{3}{5}$, θ is in quadrant I; find $\sin\theta$
34. $\sin\theta = -\tfrac{1}{2}$, θ is in quadrant III; find $\cos\theta$

35. $\tan\theta = -\sqrt{2}$, $\sin\theta > 0$; find $\sec\theta$
36. $\sec\theta = 3$, $\tan\theta < 0$; find $\sin\theta$
37. $\csc\theta = -\sqrt{3}$, $\cos\theta > 0$; find $\tan\theta$
38. $\tan\theta = -1$, $\sec\theta < 0$; find $\sin\theta$

In Exercises 39 and 40, use the right triangle shown in the sketch to find the values of the indicated trigonometric functions.

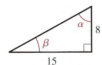

39. (a) $\sin\alpha$ (b) $\cos\alpha$ (c) $\cot\beta$
40. (a) $\tan\alpha$ (b) $\csc\beta$ (c) $\sec\beta$

In Exercises 41 and 42, use the right triangle sketched in the accompanying figure to find the values of the indicated trigonometric functions.

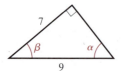

41. (a) $\tan\beta$ (b) $\csc\alpha$ (c) $\sec\alpha$
42. (a) $\sin\beta$ (b) $\cos\beta$ (c) $\cot\alpha$

In Exercises 43 and 44, use the right triangle shown in the sketch to find the values of the indicated trigonometric functions.

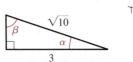

43. (a) $\sin\beta$ (b) $\cos\alpha$ (c) $\tan\beta$
44. (a) $\cot\alpha$ (b) $\sec\beta$ (c) $\csc\alpha$

In Exercises 45–50, express the trigonometric function as the cofunction of its complementary angle.

45. $\sin 30°$ 46. $\cos 63°$ 47. $\tan\dfrac{\pi}{4}$

48. $\sec\dfrac{5\pi}{12}$ 49. $\cot 34°43'$ 50. $\csc 56°17'45''$

In Exercises 51–54, sketch and label the sides of a right triangle corresponding to the given trigonometric function of the acute angle θ.

51. $\sin\theta = \tfrac{1}{2}$ 52. $\cos\theta = \tfrac{3}{4}$
53. $\tan\theta = 2$ 54. $\csc\theta = 3$

Critical Thinking

55. For a given angle θ, is it possible to have $\sin\theta > 0$ and $\csc\theta < 0$? Explain.

56. For a given angle θ, is it possible to have $\tan\theta < 0$ and $\cot\theta > 0$? Explain.

57. Use the fundamental trigonometric identities to express $\cos\theta$, $\tan\theta$, $\csc\theta$, $\sec\theta$, and $\cot\theta$ in terms of $\sin\theta$ only.

58. In each of the eight fundamental trigonometric identities, replace each function with its cofunction. Are the resulting equations identities? Explain.

59. Suppose angle θ is in standard position with the point (a,b) on its terminal side. Compare $\sin\theta$ with the value of the sine of an angle in standard position that passes through the point $(2a,2b)$. Explain.

60. The terminal side of angle θ passes through the intersection point of the given curves. Find the trigonometric functions of θ, if they exist.

(a) $2x - y = -10$ and $3x + y = -5$
(b) $y = x^2 - 4x$ and $y = 4x - 16$

Calculator Activities

In Exercises 61–66, the value of one of the trigonometric functions is given, along with some additional information. Use the trigonometric ratios and a calculator to find the values of the other five trigonometric functions of θ. Round each answer to three significant digits.

61. $\cos \theta = 0.853$, θ in quadrant IV

62. $\sin \theta = -0.934$, θ in quadrant III

63. $\csc \theta = 2.252$, $\cot \theta < 0$

64. $\cot \theta = 1.552$, $\sin \theta > 0$

65. $\tan \theta = 0.773$, $\cos \theta < 0$

66. $\sec \theta = -3.014$, θ in quadrant II

In Exercises 67–72, use the fundamental trigonometric identities and a calculator to find the value of the indicated trigonometric function from the given information. Round each answer to three significant digits.

67. $\cos \theta = 0.8875$, $\tan \theta > 0$; find $\sin \theta$.

68. $\sin \theta = -0.3290$, $\cot \theta > 0$; find $\cos \theta$.

69. $\cot \theta = 2.334$, $\sec \theta < 0$; find $\sin \theta$.

70. $\sec \theta = -1.973$, $\sin \theta > 0$; find $\cot \theta$.

71. $\sin \theta = -0.7254$, θ in quadrant IV; find $\cot \theta$.

72. $\cot \theta = 0.4591$, θ in quadrant I; find $\cos \theta$.

In Exercises 73 and 74, use the right triangle shown in the sketch and a calculator to find the values of the indicated trigonometric functions to three significant digits.

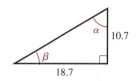

73. (a) $\sin \alpha$ (b) $\cos \alpha$ (c) $\cot \beta$

74. (a) $\tan \alpha$ (b) $\csc \beta$ (c) $\sec \beta$

2.3 Evaluating Trigonometric Functions of Angles and Real Numbers

In this section, we evaluate

1. the trigonometric functions of any angle whose measure is known and
2. the trigonometric functions of real numbers.

For some angles and real numbers, we can find the *exact value* of each trigonometric function. For other angles and real numbers, however, we use the SIN, COS, or TAN keys on a calculator to help find the *approximate value* of each trigonometric function. If you do not have access to a calculator with these trigonometric keys, refer to the trigonometric table in Appendix C when working through this section.

◆ Quadrantal Angles

A **quadrantal angle** is any angle whose terminal side is on either the *x*-axis or the *y*-axis. To evaluate the trigonometric functions of a quadrantal angle, we simply select an arbitrary point on the terminal side of the angle and then apply the trigonometric ratios from Section 2.2.

EXAMPLE 1 Evaluate each trigonometric function, if it is defined.

(a) $\tan 0°$ (b) $\sin 90°$

(c) $\cos \pi$ (d) $\tan \dfrac{3\pi}{2}$

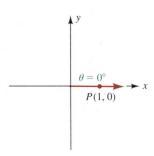

FIGURE 2.23
The point $P(1, 0)$ is on the terminal side of $\theta = 0°$.

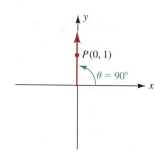

FIGURE 2.24
The point $P(0, 1)$ is on the terminal side of $\theta = 90°$.

SOLUTION

(a) We may select any point on the terminal side of $\theta = 0°$ to evaluate $\tan 0°$. Selecting the point $P(1, 0)$, as shown in Figure 2.23, we find

$$\tan 0° = \frac{y}{x} = \frac{0}{1} = 0.$$

(b) Any point on the terminal side of $\theta = 90°$ may be selected to evaluate $\sin 90°$. Selecting the point $P(0, 1)$, as shown in Figure 2.24, we find

$$r = \sqrt{x^2 + y^2} = \sqrt{(0)^2 + (1)^2} = 1.$$

Thus,

$$\sin 90° = \frac{y}{r} = \frac{1}{1} = 1.$$

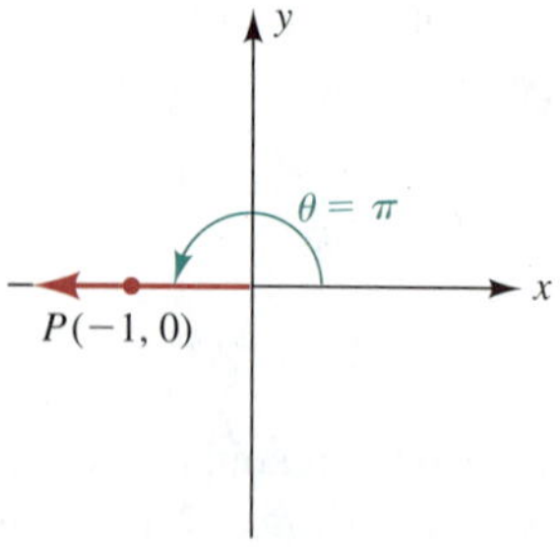

(c) We may select any point on the terminal side of $\theta = \pi = 180°$ to evaluate $\cos \pi$. Selecting the point $P(-1, 0)$, as shown in Figure 2.25, we find

$$r = \sqrt{x^2 + y^2} = \sqrt{(-1)^2 + (0)^2} = 1.$$

Thus,

$$\cos \pi = \frac{x}{r} = \frac{-1}{1} = -1.$$

FIGURE 2.25
The point $P(-1, 0)$ is on the terminal side of $\theta = \pi$.

(d) Recall from Section 2.2 that the domain of the tangent function is all angles except those having a terminal side on the y-axis. Hence, $\theta = 3\pi/2 = 270°$ is not in the domain of the tangent function, and therefore

$$\tan \frac{3\pi}{2} \text{ is undefined.}$$

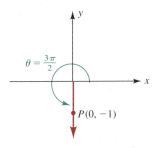

FIGURE 2.26
The point $P(0, -1)$ is on the terminal side of $\theta = 3\pi/2$.

Note that if we select the point $P(0, -1)$ on the terminal side of $\theta = 3\pi/2$, as shown in Figure 2.26, we obtain

$$\tan \frac{3\pi}{2} = \frac{y}{x} = \frac{-1}{0} \quad \text{(undefined)}.$$

Table 2.2 gives the values of the trigonometric functions for 0°, 90°, 180°, and 270°. By using the procedure shown in Example 1, we can verify each entry.

Table 2.2
Values of the trigonometric functions for 0°, 90°, 180°, and 270°

θ			$\sin \theta$	$\cos \theta$	$\tan \theta$	$\csc \theta$	$\sec \theta$	$\cot \theta$
0°	or	0	0	1	0	undefined	1	undefined
90°	or	$\pi/2$	1	0	undefined	1	undefined	0
180°	or	π	0	-1	0	undefined	-1	undefined
270°	or	$3\pi/2$	-1	0	undefined	-1	undefined	0

We may obtain values of the trigonometric functions for other quadrantal angles from Table 2.2 by using the idea of coterminal angles (Section 2.1). For example, since 360° is coterminal with 0°, we have

$$\sin 360° = \sin 0° = 0, \quad \cos 360° = \cos 0° = 1, \quad \tan 360° = \tan 0° = 0,$$

and so on.

PROBLEM 1 Evaluate each trigonometric function, if it is defined.

(a) $\sin 450°$ (b) $\cos 7\pi$ (c) $\tan\left(-\frac{\pi}{2}\right)$ (d) $\sec(-1080°)$

◆ **Special Angles: 30°, 45°, and 60°**

By using the relationships between the sides of an isosceles right triangle and a 30°-60°-90° triangle, we can find the exact values of the trigonometric functions of 30°, 45°, and 60°. Recall from elementary geometry that in an *isosceles right triangle,* the legs are equal in length and the interior angles opposite the legs measure 45°. Referring to Figure 2.27, if the length of one leg is 1, then the length of the

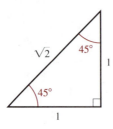

FIGURE 2.27
Relationship between the sides of an isosceles right triangle

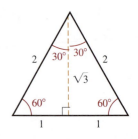

FIGURE 2.28
Relationship between the sides of a 30°-60°-90° triangle

other leg is also 1 and, by the Pythagorean theorem, the length of the hypotenuse is $\sqrt{2}$.

Also recall from elementary geometry that in an *equilateral triangle* all sides have the same length and all interior angles measure 60°. If an altitude is drawn to one of the sides of an equilateral triangle, the vertex angle and base are both cut in half, and two identical 30°-60°-90° triangles are formed. Referring to Figure 2.28, if the length of the leg opposite the 30° angle is 1, then the hypotenuse is 2 and, by the Pythagorean theorem, the length of the leg opposite the 60° angle is $\sqrt{3}$.

EXAMPLE 2 Find the exact value of each trigonometric function.

(a) $\cos 45°$ (b) $\sin 60°$ (c) $\tan \dfrac{\pi}{6}$

FIGURE 2.29
The point $P(1, 1)$ is on the terminal side of $\theta = 45°$.

SOLUTION

(a) By the relationships between the sides of an isosceles right triangle, we must have $x = 1$, $y = 1$, and $r = \sqrt{2}$, as shown in Figure 2.29. Thus,

$$\cos 45° = \frac{x}{r} = \frac{1}{\sqrt{2}} \quad \text{or} \quad \frac{\sqrt{2}}{2}.$$

(b) By the relationships between the sides of a 30°-60°-90° triangle, we must have $x = 1$, $y = \sqrt{3}$, and $r = 2$, as shown in Figure 2.30. Thus,

$$\sin 60° = \frac{y}{r} = \frac{\sqrt{3}}{2}.$$

(c) Remember that $\pi/6 = 30°$. By the relationships between the sides of a 30°-60°-90° triangle, we must have $x = \sqrt{3}$, $y = 1$, and $r = 2$, as shown in Figure 2.31. Thus,

$$\tan \frac{\pi}{6} = \frac{y}{x} = \frac{1}{\sqrt{3}} \quad \text{or} \quad \frac{\sqrt{3}}{3}.$$

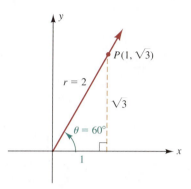
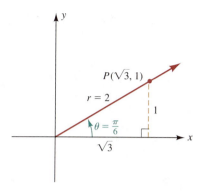

FIGURE 2.30
The point $P(1, \sqrt{3})$ is on the terminal side of $\theta = 60°$.

FIGURE 2.31
The point $P(\sqrt{3}, 1)$ is on the terminal side of $\theta = 30°$.

Table 2.3 gives the values of the trigonometric functions for 30°, 45°, and 60°. By using the procedure shown in Example 2, we can verify these entries.

Table 2.3
Values of the trigonometric functions for the angle of 30°, 45°, and 60°

θ	$\sin \theta$	$\cos \theta$	$\tan \theta$	$\csc \theta$	$\sec \theta$	$\cot \theta$
30° or $\pi/6$	1/2	$\sqrt{3}/2$	$1/\sqrt{3}$	2	$2/\sqrt{3}$	$\sqrt{3}$
45° or $\pi/4$	$1/\sqrt{2}$	$1/\sqrt{2}$	1	$\sqrt{2}$	$\sqrt{2}$	1
60° or $\pi/3$	$\sqrt{3}/2$	1/2	$\sqrt{3}$	$2/\sqrt{3}$	2	$1/\sqrt{3}$

We can use Table 2.3 to obtain values of the trigonometric functions for angles coterminal with 30°, 45°, and 60°. For example, since 420° is coterminal with 60°, we have

$$\sin 420° = \sin 60° = \frac{\sqrt{3}}{2}, \qquad \cos 420° = \cos 60° = \tfrac{1}{2},$$

$$\tan 420° = \tan 60° = \sqrt{3},$$

and so on.

PROBLEM 2 Find the exact value of each trigonometric function.

(a) $\sin 750°$ (b) $\tan \dfrac{25\pi}{4}$ (c) $\cos (-660°)$ (d) $\csc \left(-\dfrac{5\pi}{3}\right)$ ◆

◆ Reference Angles

To evaluate the trigonometric functions of other special angles, we introduce the idea of a *reference angle*.

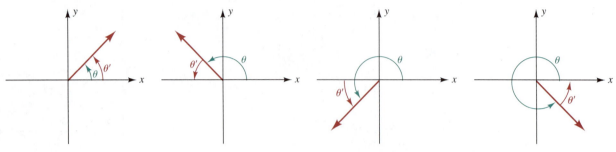

(a) θ in quadrant I;
θ = θ′

(b) θ in quadrant II;
θ′ = 180° − θ,
θ′ = π − θ

(c) θ in quadrant III;
θ′ = θ − 180°,
θ′ = θ − π

(d) θ in quadrant IV;
θ′ = 360° − θ,
θ′ = 2π − θ

FIGURE 2.32 Relationship between an angle θ and its reference angle θ′ for θ in each of the four quadrants

 **Reference Angle**

> If θ is an angle in standard position, then its **reference angle** is the acute angle θ′ that is formed with the terminal side of θ and the x-axis.

Figure 2.32 gives the relationship between an angle θ and its reference angle θ′ for θ in each quadrant.

EXAMPLE 3 Find the reference angle θ′ for each angle θ.

(a) $\theta = 225°$ (b) $\theta = 117°35'$ (c) $\theta = 5$ (d) $\theta = \dfrac{7\pi}{3}$

SOLUTION

(a) The angle $\theta = 225°$ is between 180° and 270°. Hence, θ lies in quadrant III. Thus, referring to Figure 2.32(c), the reference angle is

$$\theta' = 225° - 180° = 45°.$$

(b) The angle $\theta = 117°35'$ is between 90° and 180°. Hence, θ lies in quadrant II. Thus, referring to Figure 2.32(b), the reference angle is

$$\theta' = 180° - 117°35' = 179°60' - 117°35' = 62°25'.$$

(c) The angle $\theta = 5$ is between $\dfrac{3\pi}{2} \approx 4.71$ and $2\pi \approx 6.28$. Hence, θ lies in quadrant IV. Thus, referring to Figure 2.32(d), the reference angle is

$$\theta' = 2\pi - 5 \approx 1.28.$$

(d) The angle $\theta = \dfrac{7\pi}{3} = 2\pi + \dfrac{\pi}{3}$ is coterminal with $\dfrac{\pi}{3}$. Since $\dfrac{\pi}{3}$ lies in quadrant I, and the reference angle for any first-quadrant angle is itself, we conclude that the reference angle for $\theta = \dfrac{7\pi}{3}$ is $\theta' = \dfrac{\pi}{3}$. ◆

PROBLEM 3 Repeat Example 3 for (a) $\theta = \dfrac{5\pi}{6}$ and (b) $\theta = -495°$. ◆

Suppose angle θ lies in either the second, third, or fourth quadrant and its corresponding reference angle θ' is drawn in standard position. Furthermore, suppose P is a point on the terminal side of θ' and Q is a point on the terminal side of θ such that both points are r units from the origin, as shown in Figure 2.33.

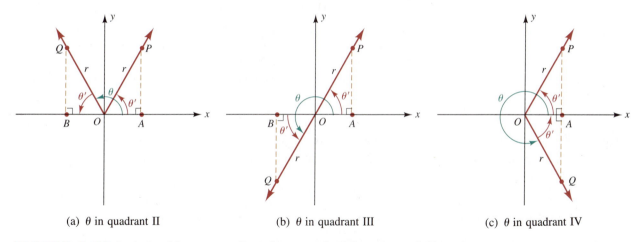

(a) θ in quadrant II   (b) θ in quadrant III   (c) θ in quadrant IV

FIGURE 2.33 Angle θ and its corresponding reference angle θ' shown in standard position.

Using Figure 2.33, we can show a special relationship that exists between the trigonometric functions of θ and θ'. This relationship enables us to state a procedure for evaluating a trigonometric function of any nonquadrantal angle.

1. Referring to Figure 2.33(a), if θ is in quadrant II, then triangles OAP and OBQ are congruent (identical in shape and size). Hence, if the coordinates of point P are (x, y), then the coordinates of point Q must be $(-x, y)$. Therefore,

$$\underbrace{\sin \theta = \dfrac{y}{r} = \sin \theta'}_{\text{Same value}}, \quad \underbrace{\cos \theta = \dfrac{-x}{r} = -\dfrac{x}{r} = -\cos \theta'}_{\text{Same value, but different sign}}, \quad \underbrace{\tan \theta = \dfrac{y}{-x} = -\dfrac{y}{x} = -\tan \theta'}_{\text{Same value, but different sign}}.$$

2. Referring to Figure 2.33(b), if θ is in quadrant III, then triangles *OAP* and *OBQ* are congruent. Hence, if the coordinates of point *P* are (x, y), then the coordinates of point *Q* must be (−x, −y). Therefore,

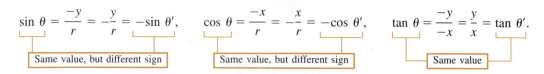

3. Referring to Figure 2.33(c), if θ is in quadrant IV, then triangles *OAP* and *OAQ* are congruent. Hence, if the coordinates of point *P* are (x, y), then the coordinates of point *Q* must be (x, −y). Therefore,

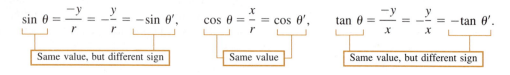

From these examples we conclude that *the values of the trigonometric functions of angle θ are the same as the values of the trigonometric functions of its reference angle θ′, except possibly for a difference in algebraic sign.* The following procedure may be used to evaluate the trigonometric functions of any nonquadrantal angle θ.

Procedure for Evaluating a Trigonometric Function of Any Nonquadrantal Angle

To find the value of a trigonometric function of any nonquadrantal angle θ,

Step 1: Find the reference angle θ′ associated with θ.

Step 2: Evaluate the given trigonometric function for θ′.

Step 3: Affix the appropriate sign to the function value found in step 2 by considering the quadrant in which θ lies.

EXAMPLE 4 Find the exact value of each trigonometric function.

(a) $\cos 150°$ (b) $\sin \dfrac{7\pi}{4}$ (c) $\cot(-480°)$

SOLUTION

(a) *Step 1:* The reference angle for 150° is 30°, as illustrated in Figure 2.34.

Step 2: From Table 2.3, we have $\cos 30° = \dfrac{\sqrt{3}}{2}$.

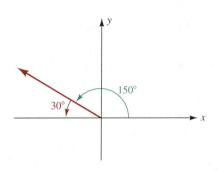

FIGURE 2.34
The reference angle for 150° is 30°.

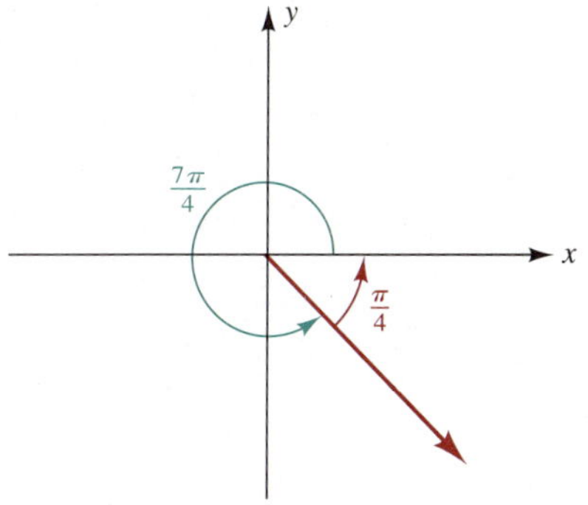

FIGURE 2.35
The reference angle for $7\pi/4$ is $\pi/4$.

Step 3: Since 150° is a second-quadrant angle and the value of the cosine is negative in quadrant II, we have

$$\cos 150° = -\cos 30° = -\frac{\sqrt{3}}{2}.$$

(b) *Step 1:* The reference angle for $\dfrac{7\pi}{4}$ is $\dfrac{\pi}{4}$, as illustrated in Figure 2.35.

Step 2: From Table 2.3, we find $\sin \dfrac{\pi}{4} = \dfrac{1}{\sqrt{2}}$.

Step 3: Since $\dfrac{7\pi}{4}$ is a fourth-quadrant angle and the value of the sine is negative in quadrant IV, we have

$$\sin \frac{7\pi}{4} = -\sin \frac{\pi}{4} = -\frac{1}{\sqrt{2}}.$$

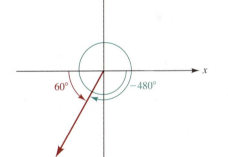

FIGURE 2.36
The reference angle for −480° is 60°.

(c) *Step 1:* The reference angle for −480° is 60°, as illustrated in Figure 2.36.

Step 2: From Table 2.3, we have $\cot 60° = \dfrac{1}{\sqrt{3}}$.

Step 3: Since −480° is a third-quadrant angle and the value of the cotangent is positive in quadrant III, we have

$$\cot(-480°) = \cot 60° = \frac{1}{\sqrt{3}}. \qquad \blacklozenge$$

PROBLEM 4 Repeat Example 4 for each trigonometric function.

(a) $\sin 150°$ (b) $\tan \dfrac{3\pi}{4}$ (c) $\sec(-420°)$ $\blacklozenge$

Defining the Trigonometric Functions of Real Numbers

Several applications of the trigonometric functions do not involve angles but, instead, describe periodic phenomena that vary with respect to a real number t representing time. To define the trigonometric functions of a real number t, we may imagine a real number line wrapped continuously around a **unit circle**—a circle with center at the origin and radius 1 unit—as shown in Figure 2.37. By placing the origin of the real number line at the point $(1, 0)$ on the circle, we are able to wrap the positive part of the line around the circle in a counterclockwise direction, wrap the negative part of the line around the circle in a clockwise direction, and have an angle of 0 rad correspond to the number 0 on the line.

Now for each real number t on the line, there corresponds a central angle θ, as shown in Figure 2.38. Recall from Section 2.1 that if a central angle θ (in radians) intercepts an arc of length t on a circle with radius r, then

$$t = \theta r.$$

Since the unit circle has radius 1 unit, we have

$$t = \theta.$$

Hence, we may find the values of the trigonometric functions for a real number t by considering the real number as the radian measure of the central angle θ.

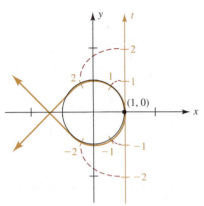

FIGURE 2.37
A real number line being wrapped around the unit circle.

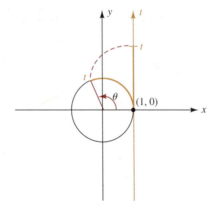

FIGURE 2.38
For each real number t, there corresponds a central angle θ.

◆ Trigonometric Functions of Real Numbers

If t is a real number, then

1. $\sin t = \sin \theta$
2. $\cos t = \cos \theta$
3. $\tan t = \tan \theta$
4. $\csc t = \csc \theta$
5. $\sec t = \sec \theta$
6. $\cot t = \cot \theta$

where θ is an angle whose measure is t radians.

SECTION 2.3 Evaluating Trigonometric Functions

Note: As we stated in Section 2.1, if no unit of measure is specified with an angle, then the angle is in radians. Thus, a trigonometric expression such as cos π means either

1. the cosine of the real number π or

2. the cosine of an angle with radian measure π.

The context in which cos π appears determines which meaning is intended, but both meanings have the same numerical value, that is, cos $\pi = -1$.

Table 2.4 shows the angle domains and the corresponding real number domains of the trigonometric functions.

Table 2.4
Angle domains and real number domains of the trigonometric functions

Trigonometric Function	Angle Domain	Real Number Domain
sine or cosine	All angles	All real numbers
cosecant or cotangent	All angles except those whose terminal side is on the x-axis	All real numbers except πn, where n is an integer
tangent or secant	All angles except those whose terminal side is on the y-axis	All real numbers except $\frac{\pi}{2} + \pi n$, where n is an integer.

EXAMPLE 5 Given that $f(t) = 1 + \cos 2t$, and $g(t) = 2 \tan \left[\frac{\pi}{4}(t-1) \right]$ find each functional value.

(a) $f\left(\frac{\pi}{6}\right)$ (b) $g(4)$

SOLUTION

(a) Replacing t with the real number $\pi/6$, we obtain

$$f\left(\frac{\pi}{6}\right) = 1 + \cos\left(2 \cdot \frac{\pi}{6}\right) = 1 + \cos\frac{\pi}{3}.$$

Now, the value of cos ($\pi/3$) is the same as the value of cos ($\pi/3$) rad, which in turn has the same value as cos 60°. From Table 2.3, we conclude that

$$f\left(\frac{\pi}{6}\right) = 1 + \cos\frac{\pi}{3} = 1 + \frac{1}{2} = \frac{3}{2}.$$

(b) Replacing t with the real number 4, we obtain

$$g(4) = 2 \tan\left[\frac{\pi}{4}(4-1)\right] = 2 \tan\frac{3\pi}{4}.$$

Now, the value of tan $(3\pi/4)$ is the same as the value of tan $(3\pi/4)$ rad, which in turn has the same value as tan $135°$. Since the angle $3\pi/4$ rad lies in quadrant II and has reference angle $\pi/4$, and since the tangent function is negative in quadrant II, we conclude that

$$g(4) = 2\tan\frac{3\pi}{4} = -2\tan\frac{\pi}{4} = -2(1) = -2.$$

PROBLEM 5 Given the function g in Example 5, find $g(5)$.

◆ **Approximate Values of the Trigonometric Functions**

We can use a calculator to find the decimal value of a trigonometric function of an angle or of a real number. The following is a guideline for using a calculator to evaluate trigonometric functions.

Guideline for Calculator Usage

1. Set the calculator in *degree mode* to find the trigonometric functions of angles in degrees or in *radian mode* to find the trigonometric functions of real numbers or angles in radians.

2. Press the appropriate key, [SIN], [COS], or [TAN], to evaluate the sine, cosine, or tangent of a given angle or real number.

3. For cosecant, secant, or cotangent, use the [SIN], [COS], or [TAN] key, respectively. Then press the reciprocal key, [1/x], to obtain the value of the cosecant, secant, or cotangent of a given angle or real number.

EXAMPLE 6 Use a calculator to find the approximate value of each trigonometric function. Round each answer to four significant digits.

(a) sin $215°$ (b) sec 8   (c) cot $32°15'45''$

SOLUTION

(a) To find sin $215°$ using a calculator, we set the calculator in degree mode and use the [SIN] key. (With some calculators, we first enter the angle and then press the [SIN] key. With others, first press the [SIN] key and then enter the angle. Both procedures yield the same result.) Rounding the calculator display to four significant digits, we obtain

$$\sin 215° \approx -0.5736.$$

The calculator automatically displays the correct sign.

(b) To find sec 8, we set the calculator in radian mode and use the $\boxed{\text{COS}}$ key to find the value of cos 8. Next, we press the reciprocal key, $\boxed{1/x}$, to find the value of sec 8. Rounding the calculator display to four significant digits, we obtain

$$\sec 8 \approx -6.873.$$

(c) For angles written in degrees, minutes, and seconds (see Section 2.1), we must first convert the measure to decimal degrees. If a calculator does not have a conversion key, then we can make the conversion using our procedure from Section 2.1:

$$32°15'45'' = 32° + \left(\frac{15}{60}\right)° + \left(\frac{45}{3600}\right)° = 32.2625°.$$

Now, to find cot 32.2625°, we set the calculator in degree mode and use the $\boxed{\text{TAN}}$ key to find the value of tan 32.2625°. Then we press the reciprocal key, $\boxed{1/x}$, to find the value of cot 32.2625°. Rounding the calculator display to four significant digits, we obtain

$$\cot 32°15'45'' \approx 1.584. \qquad \blacklozenge$$

We may also obtain the approximate values of the trigonometric functions of angles and real numbers from Table 4 in Appendix C. For example, to find the approximate value of sin 215°, we proceed with the three-step procedure used in Example 4.

Step 1: The reference angle for 215° is 35°, as illustrated in Figure 2.39.

Step 2: From the trigonometric table in Appendix C we read

$$\sin 35° \approx 0.5736 \qquad \text{\textcolor{red}{\textbf{four significant digits}}}$$

Step 3: Since 215° is a third-quadrant angle and the sine is negative in quadrant III, we have

$$\sin 215° = -\sin 35° \approx -0.5736,$$

which agrees with the answer we obtained in Example 6(a).

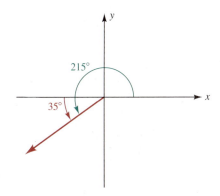

FIGURE 2.39
The reference angle for 215° is 35°.

PROBLEM 6 Use a calculator or Table 4 in Appendix C to find the approximate value of each trigonometric function. Round each answer to four significant digits.

(a) csc 237°   **(b)** tan 121°30'   **(c)** $\sec \dfrac{43\pi}{24}$ $\blacklozenge$

Exercises 2.3

Basic Skills

In Exercises 1–16, draw each angle θ in standard position, label a point on the terminal side of θ, and find the exact values of the six trigonometric functions of θ that are defined. Do not use a calculator.

1. $\theta = 180°$
2. $\theta = 360°$
3. $\theta = 450°$
4. $\theta = 630°$
5. $\theta = 4\pi$
6. $\theta = \dfrac{9\pi}{2}$
7. $\theta = -\dfrac{\pi}{2}$
8. $\theta = -900°$
9. $\theta = 30°$
10. $\theta = \dfrac{\pi}{4}$
11. $\theta = \dfrac{13\pi}{3}$
12. $\theta = 780°$
13. $\theta = -315°$
14. $\theta = -1020°$
15. $\theta = \dfrac{9\pi}{4}$
16. $\theta = -\dfrac{11\pi}{6}$

In Exercises 17–28, find the reference angle θ' for each angle θ.

17. $\theta = 240°$
18. $\theta = 324°$
19. $\theta = -234°$
20. $\theta = 753°$
21. $\theta = \dfrac{11\pi}{6}$
22. $\theta = \dfrac{2\pi}{3}$
23. $\theta = \dfrac{16\pi}{5}$
24. $\theta = -\dfrac{12\pi}{7}$
25. $\theta = 2$
26. $\theta = 4$
27. $\theta = 306°22'$
28. $\theta = 198°21'36''$

In Exercises 29–52, use the three-step procedure illustrated in Example 4 to find the exact value of each trigonometric function. Do not use a calculator.

29. $\cos 120°$
30. $\tan 225°$
31. $\sin 210°$
32. $\sec 330°$
33. $\cot 570°$
34. $\sin 660°$
35. $\cot 1035°$
36. $\csc 960°$
37. $\csc \dfrac{7\pi}{4}$
38. $\cot \dfrac{5\pi}{6}$
39. $\sin \dfrac{2\pi}{3}$
40. $\cos \dfrac{5\pi}{4}$
41. $\sin \dfrac{11\pi}{4}$
42. $\cos \dfrac{11\pi}{3}$
43. $\sec \dfrac{19\pi}{6}$
44. $\cot \dfrac{31\pi}{6}$
45. $\tan(-210°)$
46. $\csc(-330°)$
47. $\cos(-870°)$
48. $\tan(-945°)$
49. $\sec\left(-\dfrac{5\pi}{3}\right)$
50. $\cot\left(-\dfrac{3\pi}{4}\right)$
51. $\csc\left(-\dfrac{41\pi}{6}\right)$
52. $\sin\left(-\dfrac{14\pi}{3}\right)$

In Exercises 53–68, assume the functions f, g, and h are defined by

$$f(t) = 3 + \tan \dfrac{\pi t}{4}, \qquad g(t) = -2 \sin 3\pi t,$$

and

$$h(t) = 3 \cos\left(\pi t - \dfrac{\pi}{3}\right).$$

Compute the functional value, if it is defined. Do not use a calculator.

53. $f(0)$
54. $f(1)$
55. $f(-3)$
56. $f(-8)$
57. $g(\tfrac{2}{3})$
58. $g(\tfrac{3}{2})$
59. $g(-\tfrac{1}{6})$
60. $g(\tfrac{7}{18})$
61. $h(0)$
62. $h(3)$
63. $h(\tfrac{1}{2})$
64. $h(-\tfrac{2}{3})$
65. $g(f(1))$
66. $f(h(\tfrac{1}{3}))$
67. $h(g(-\tfrac{1}{18}))$
68. $h(f(2))$

69. The displacement d (in centimeters) of an oscillating spring from its equilibrium position is a function of time t (in seconds) and is given by

$$d(t) = 12 \cos \pi t.$$

Find the displacement at each value of t.

(a) 0 s (b) 5 s

(c) $\frac{1}{2}$ s (d) $\frac{3}{4}$ s

70. The population P of mice in a certain field is a function of time t in years and is given by

$$P(t) = 350 + 200 \sin \frac{2\pi}{3} t.$$

(a) Find the initial mouse population.

(b) Find the mouse population after 3 years 9 months.

Critical Thinking

71. Choose various values of θ and evaluate $\sin \theta$ and $\sin(-\theta)$. How does the value of $\sin \theta$ compare to the value of $\sin(-\theta)$?

72. Choose various values of θ and evaluate $\cos \theta$ and $\cos(-\theta)$. How does the value of $\cos \theta$ compare to the value of $\cos(-\theta)$?

73. Only four of the eight fundamental trigonometric identities (Section 2.2) hold when θ is replaced by π. Which four do *not* hold? Explain.

74. Only four of the eight fundamental trigonometric identities (Section 2.2) hold when θ is replaced by $\pi/2$. Which four do *not* hold? Explain.

75. What are the real numbers x, $0 \leq x < 2\pi$, for which $\sin x = \csc x$? $\cos x = \sec x$? $\tan x = \cot x$?

76. Compare the slope m of a line that passes through the origin and the point $P(x, y)$ to the value of $\tan \theta$, where θ is the positive angle in standard position whose terminal side passes through the point $P(x, y)$. What can you conclude about m and $\tan \theta$? Use your observation to find the slope of a line that is formed by the terminal side of each angle θ.

(a) $\theta = 60°$ (b) $\theta = 150°$

(c) $\theta = 225°$ (d) $\theta = 300°$

77. For the central angle θ there corresponds a point P on the unit circle, as shown in the figure.

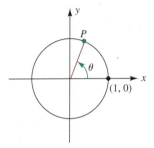

(a) Express the x-coordinate of P as a trigonometric function of θ.

(b) Express the y-coordinate of P as a trigonometric function of θ.

(c) Use the distance formula (Section 1.2) to verify that the distance from point P to the origin is 1 unit. What trigonometric identity did you use?

78. Usually, for a real number t, $\sin t \neq \sin t°$. For instance, $\sin 1 \approx 0.8415$, but $\sin 1° \approx 0.01752$. However, if t radians and t degrees have the same terminal side, as shown in the figure, then $\sin t = \sin t°$. Find all real numbers t for which $\sin t = \sin t°$.

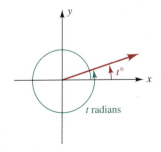

Calculator Activities

In Exercises 79–98, use a calculator to find the approximate value of each trigonometric function. Round each answer to four significant digits.

79. sin 25°
80. cos 76°
81. sin 195°
82. tan 306°
83. cos 617°
84. csc 1224°
85. sec (−216°)
86. cot (−500°)
87. sec $\dfrac{\pi}{5}$
88. cot $\dfrac{3\pi}{10}$
89. tan $\left(-\dfrac{11\pi}{12}\right)$
90. cos $\dfrac{17\pi}{10}$
91. tan 1
92. cos 0.4
93. csc 4.28
94. sec 7.6
95. sin 46°40′
96. cot (−1674°30′)
97. sin 29°20′45″
98. cos 600°00′30″

99. In calculus, when finding the derivatives of the sine and cosine functions, it is necessary to evaluate the quotient (sin t)/t as t approaches zero. Use a calculator to complete the accompanying table and then state the value that (sin t)/t seems to approach as $t \to 0^+$.

t	1	0.1	0.01	0.001	0.0001
$\dfrac{\sin t}{t}$					

100. In calculus, when finding the derivatives of the sine and cosine functions, it is necessary to evaluate the quotient (cos t − 1)/t as t approaches zero. Use a calculator to complete the accompanying table and then state the value that (cos t − 1)/t seems to approach as $t \to 0^+$.

t	1	0.1	0.01	0.001	0.0001
$\dfrac{\cos t - 1}{t}$					

101. The voltage v (in volts) in an electric circuit is a function of time t (in seconds) and is given by

$$v(t) = 20 \sin 4t.$$

Find the voltage at each value of t.

(a) 1 s (b) 3 s
(c) 4.5 s (d) 6.3 s

102. The expected daily high temperature T (in degrees Fahrenheit) for a certain city can be approximated by

$$T = 52 - 32 \cos\left[\dfrac{2\pi}{365}(t - 25)\right],$$

where t is the time in days with $t = 1$ corresponding to January 1. Find the expected daily high temperature on each day.

(a) January 1 (b) January 25
(c) July 4 ($t = 185$) (d) September 1 ($t = 244$)

2.4 Properties and Graphs of the Sine and Cosine Functions

In Section 2.3 we stated that the domains of both the sine function and the cosine function are the set of all real numbers. In this section, we discuss other important properties of these two trigonometric functions and then use this information to help sketch the graphs of equations of the form

$$y = a \sin (bx + c) \quad \text{and} \quad y = a \cos (bx + c),$$

where a, b, and c are constants.

Properties of the Sine and Cosine Functions

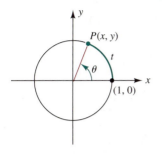

FIGURE 2.40
A unit circle where the arc length from $(1,0)$ to $P(x, y)$ is t units

Consider a unit circle where the arc length from $(1, 0)$ to the point $P(x, y)$ is t units, as shown in Figure 2.40. On the unit circle, $r = 1$. Thus, it follows that angle θ in Figure 2.40 has radian measure t, and the sine and cosine of the real number t can be defined as follows:

$$\sin t = \sin \theta = \frac{y}{r} = \frac{y}{1} = y \quad \text{and} \quad \cos t = \cos \theta = \frac{x}{r} = \frac{x}{1} = x$$

Hence, the coordinates of point P on the unit circle are

$$(x, y) = (\cos t, \sin t),$$

as illustrated in Figure 2.41.

Referring to Figure 2.41, if t increases by a multiple of 2π, the coordinates of point P remain unchanged, that is,

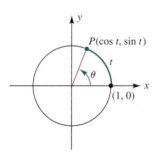

FIGURE 2.41
Renaming the coordinates of point P in Figure 2.40.

$$\boxed{\sin(t + 2\pi n) = \sin t}$$

and

$$\boxed{\cos(t + 2\pi n) = \cos t}$$

for any integer n. When a function repeats its values in a repetitive manner, as do $\sin t$ and $\cos t$, we say that the function is *periodic* and, in this case, has a *period* of 2π. In general, we have the following definition for a *periodic function*.

◆ Periodic Function

A function f is said to be **periodic** if there exists a positive real number c such that

$$f(t + c) = f(t)$$

for all t in the domain of f. The least such positive number is called the **period** of f.

Now, referring to Figure 2.41, observe the changes in the x-coordinate $\cos t$ and the y-coordinate $\sin t$ as we let t increase from 0 to 2π. Table 2.5 summarizes these changes.

Table 2.5

Changes in cos t and sin t as t increases from 0 to 2π

t	x-coordinate cos t	y-coordinate sin t
0 to $\pi/2$	1 to 0	0 to 1
$\pi/2$ to π	0 to -1	1 to 0
π to $3\pi/2$	-1 to 0	0 to -1
$3\pi/2$ to 2π	0 to 1	-1 to 0

Using the information in Table 2.5 and plotting a few points, we are able to graph $f(t) = \sin t$ and $g(t) = \cos t$ on the interval $[0, 2\pi]$. When we graph a periodic function over one period, we say that we have graphed *one cycle* of the function. The graphs of one cycle of $f(t) = \sin t$ and $g(t) = \cos t$ are shown in Figures 2.42 and 2.43, respectively.

The graph of one cycle of $f(t) = \sin t$ has five key points:

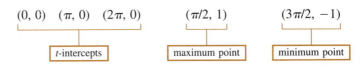

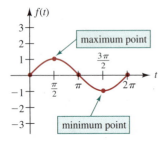

FIGURE 2.42
Graph of one cycle of $f(t) = \sin t$

The graph of one cycle of $g(t) = \cos t$ also has five key points:

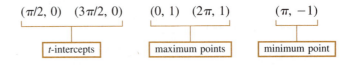

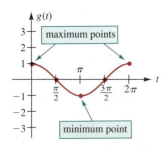

FIGURE 2.43
Graph of one cycle of $g(t) = \cos t$

Since the sine and cosine functions are periodic with period 2π, we can obtain the entire graphs of $f(t) = \sin t$ and $g(t) = \cos t$ simply by repeating the portion of the graph from 0 to 2π, as illustrated in Figures 2.44 and 2.45, respectively.

The following properties are evident from the graphs of $f(t) = \sin t$ and $g(t) = \cos t$.

 Properties of the Sine and Cosine Functions

Sine function: $f(t) = \sin t$

1. The *domain* is the set of all real numbers.
2. The *range* is all real numbers in the closed interval $[-1, 1]$.
3. The *period* is 2π.
4. The *zeros* are $0, \pm\pi, \pm2\pi, \pm3\pi, \ldots$.
5. The function is *odd*, that is, $\sin(-t) = -\sin t$ for all t.

Cosine function: $g(t) = \cos t$

1. The *domain* is the set of all real numbers.
2. The *range* is all real numbers in the closed interval $[-1, 1]$.
3. The *period* is 2π.
4. The *zeros* are $\pm\dfrac{\pi}{2}, \pm\dfrac{3\pi}{2}, \pm\dfrac{5\pi}{2}, \ldots$.
5. The function is *even*, that is, $\cos(-t) = \cos t$ for all t.

SECTION 2.4   Properties and Graphs of the Sine and Cosine

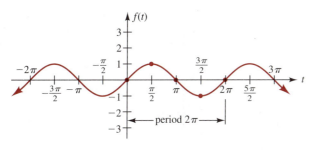

FIGURE 2.44
Graph of the sine function $f(t) = \sin t$

FIGURE 2.45
Graph of the cosine function $g(t) = \cos t$

Note: We may use any letter to denote the independent or dependent variable for the sine and cosine functions. It is convenient to choose x and y for these variables so that the graphs are shown in the usual xy-coordinate system. However, when writing $y = \sin x$ or $y = \cos x$, do not confuse the variables x and y with the coordinates of the point $P(x, y)$ on the unit circle.

◆ **Graphs of $y = a \sin x$ and $y = a \cos x$**

For $y = \sin x$ and $y = \cos x$, the values of y range from a minimum value of -1 to a maximum value of 1. Thus, by the vertical stretch and shrink rules given in Section 1.4, we know that for $y = a \sin x$ and $y = a \cos x$ with $a > 0$, the values of y range from a minimum of $-a$ to a maximum of a. The *maximum value* that y attains is called the *amplitude*.

◆ **Amplitude**

> The **amplitude** of $y = a \sin x$ and $y = a \cos x$ is $|a|$ for $a \neq 0$.

We can sketch the graphs of $y = a \sin x$ and $y = a \cos x$ by knowing the basic shapes of the sine and cosine curves (Figures 2.44 and 2.45) and noting their amplitudes.

EXAMPLE 1 Sketch the graph of each function.

(a) $y = 3 \sin x$ (b) $y = \frac{1}{2} \cos x$

SOLUTION

(a) The amplitude of $y = 3 \sin x$ is $|3| = 3$. Thus, the values of y range from a maximum of 3 to a minimum of -3, as shown in Figure 2.46. On the interval $[0, 2\pi]$, the five key points are

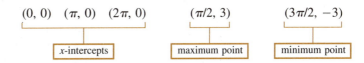

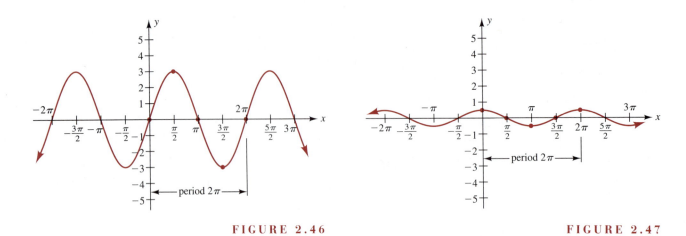

FIGURE 2.46
Graph of $y = 3 \sin x$

FIGURE 2.47
Graph of $y = \frac{1}{2} \cos x$

(b) The amplitude of $y = \frac{1}{2} \cos x$ is $|\frac{1}{2}| = \frac{1}{2}$. Thus, the values of y range from a maximum of $\frac{1}{2}$ to a minimum of $-\frac{1}{2}$, as shown in Figure 2.47. On the interval $[0, 2\pi]$, the five key points are

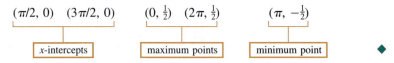

By the x-axis reflection rule (in Section 1.4), we know that the graph of $y = -a \sin x$ is the same as the graph of $y = a \sin x$ reflected about the x-axis. Thus, to graph $y = -3 \sin x$, we sketch $y = 3 \sin x$ as in Example 1(a), and then reflect this graph about the x-axis. The graph of $y = -3 \sin x$ is shown in Figure 2.48. Note that the amplitude of $y = -3 \sin x$ is $|-3| = 3$.

PROBLEM 1 Sketch the graph of $y = -\frac{1}{2} \cos x$. ◆

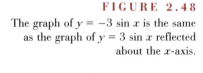

FIGURE 2.48
The graph of $y = -3 \sin x$ is the same as the graph of $y = 3 \sin x$ reflected about the x-axis.

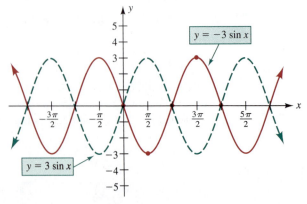

◆ Graphs of $y = a \sin bx$ and $y = a \cos bx$

The graph of $y = a \sin x$ and $y = a \cos x$ each has period 2π. Each graph completes one cycle from $x = 0$ to $x = 2\pi$. Therefore, the graphs of $y = a \sin bx$ and $y = a \cos bx$ must each complete one cycle from

$$bx = 0 \quad \text{to} \quad bx = 2\pi$$

$$x = 0 \quad \text{to} \quad x = \frac{2\pi}{b} \qquad \text{Divide both sides by } b$$

Hence, the period of $y = a \sin bx$ and $y = a \cos bx$ is the length of the interval from $x = 0$ to $x = 2\pi/b$.

▶ **Period Formula: Sine and Cosine**

The **period** P of $y = a \sin bx$ and $y = a \cos bx$ is given by

$$P = \frac{2\pi}{|b|} \quad \text{for } b \neq 0.$$

We can sketch the graph of $y = a \sin bx$ or $y = a \cos bx$ by noting its amplitude and period. As illustrated in the next example, the value of b has the effect of stretching or shrinking the sine or cosine curve horizontally.

EXAMPLE 2 Sketch the graph of each function.

(a) $y = 3 \sin 2x$ (b) $y = \frac{1}{2} \cos \frac{\pi}{4} x$

SOLUTION

(a) For $y = 3 \sin 2x$, we have

$$\text{Amplitude: } |a| = |3| = 3 \qquad \text{Period: } \frac{2\pi}{|b|} = \frac{2\pi}{|2|} = \pi$$

Thus, this sine curve completes one cycle from $x = 0$ to $x = \pi$ and has a maximum y-value of 3. Dividing the interval $[0, \pi]$ into four equal parts gives us the five key points on this cycle:

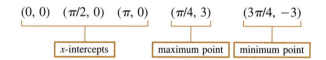

(0, 0) ($\pi/2$, 0) (π, 0)   ($\pi/4$, 3)   ($3\pi/4$, −3)

x-intercepts   maximum point   minimum point

The graph of $y = 3 \sin 2x$ is shown in Figure 2.49.

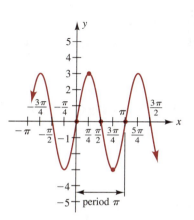

FIGURE 2.49
Graph of $y = 3 \sin 2x$

(b) For $y = \frac{1}{2} \cos \frac{\pi}{4} x$, we have

$$\text{Amplitude: } |a| = \left|\frac{1}{2}\right| = \frac{1}{2} \qquad \text{Period: } \frac{2\pi}{|b|} = \frac{2\pi}{|\pi/4|} = 8$$

Thus, this cosine curve completes one cycle from $x = 0$ to $x = 8$ and has a maximum y-value of $\frac{1}{2}$. Dividing the interval $[0, 8]$ into four equal parts gives us the five key points on this cycle:

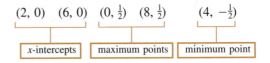

The graph of $y = \frac{1}{2} \cos \frac{\pi}{4} x$ is shown in Figure 2.50. ◆

FIGURE 2.50

Graph of $y = \frac{1}{2} \cos \frac{\pi}{4} x$

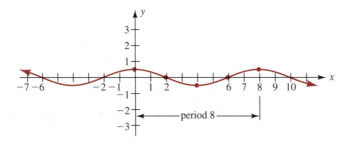

If b is negative in $y = a \sin bx$ or $y = a \cos bx$, then we use the fact that

$$\sin(-x) = -\sin x \quad \text{and} \quad \cos(-x) = \cos x$$

to rewrite the function before sketching the graph. For example, to graph $y = 3 \sin(-2x)$, we write

$$y = 3 \sin(-2x) = -3 \sin 2x.$$

Now the graph of $y = 3 \sin(-2x) = -3 \sin 2x$ is the same as the graph of $y = 3 \sin 2x$ reflected about the x-axis.

PROBLEM 2 How does the graph of $y = \frac{1}{2} \cos\left(-\frac{\pi}{4} x\right)$ compare to the graph of $y = \frac{1}{2} \cos \frac{\pi}{4} x$?

◆

◆ **Graphs of $y = a \sin(bx + c)$ and $y = a \cos(bx + c)$**

Next we consider the graphs of the equations $y = a \sin(bx + c)$ and $y = a \cos(bx + c)$. By factoring, we can rewrite these equations as follows:

$$y = a \sin b\left(x + \frac{c}{b}\right) \quad \text{and} \quad y = a \cos b\left(x + \frac{c}{b}\right).$$

Now, by the horizontal shift rule (in Section 1.4), we know that the graphs of these equations are the same as the graphs of $y = a \sin bx$ and $y = a \cos bx$, respectively, shifted horizontally $|c/b|$ units. If $c/b < 0$, the shift is to the right, and if $c/b > 0$ the shift is to the left. We refer to the absolute value of c/b as the *phase shift*.

Phase Shift

> The **phase shift** of $y = a \sin(bx + c)$ and $y = a \cos(bx + c)$ is $\left|\frac{c}{b}\right|$ units to the right if $\frac{c}{b} < 0$ and $\left|\frac{c}{b}\right|$ units to the left if $\frac{c}{b} > 0$.

We can sketch the graph of $y = a \sin(bx + c)$ or $y = a \cos(bx + c)$ by noting its amplitude, period, and phase shift.

EXAMPLE 3 Sketch the graph of each function.

(a) $y = 3 \sin(2x + \pi)$ **(b)** $y = \frac{1}{2} \cos\left(\frac{\pi}{4}x - \frac{\pi}{4}\right)$

SOLUTION

(a) For $y = 3 \sin(2x + \pi)$, we have $a = 3$, $b = 2$, and $c = \pi$. Therefore,

$$\text{Amplitude: } |a| = |3| = 3 \qquad \text{Period: } \frac{2\pi}{|b|} = \frac{2\pi}{|2|} = \pi$$

$$\text{Phase shift: } \left|\frac{c}{b}\right| = \left|\frac{\pi}{2}\right| = \frac{\pi}{2} \text{ units to the left, since } \frac{c}{b} > 0.$$

Thus, the graph of $y = 3 \sin(2x + \pi) = 3 \sin 2\left(x + \frac{\pi}{2}\right)$ is the same as the graph of $y = 3 \sin 2x$ shifted horizontally to the left $\pi/2$ units. Hence, it completes one cycle from

$$x = -\frac{\pi}{2} \quad \text{to} \quad x = -\frac{\pi}{2} + \pi = \frac{\pi}{2},$$

[Add the period]

as shown in Figure 2.51. Dividing the interval $[-\pi/2, \pi/2]$ into four equal parts gives us the five key points on this cycle:

$(-\pi/2, 0) \quad (0, 0) \quad (\pi/2, 0) \qquad (-\pi/4, 3) \qquad (\pi/4, -3)$

x-intercepts — maximum point — minimum point

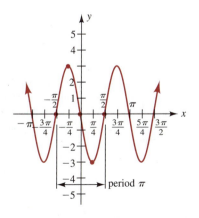

FIGURE 2.51
Graph of $y = 3 \sin(2x + \pi)$

(b) For $y = \frac{1}{2} \cos\left(\frac{\pi}{4}x - \frac{\pi}{4}\right)$, we have $a = \frac{1}{2}$, $b = \pi/4$, and $c = -\pi/4$. Therefore,

$$\text{Amplitude: } |a| = \left|\frac{1}{2}\right| = \frac{1}{2} \qquad \text{Period: } \frac{2\pi}{|b|} = \frac{2\pi}{|\pi/4|} = 8$$

$$\text{Phase shift: } \left|\frac{c}{b}\right| = \left|\frac{-\pi/4}{\pi/4}\right| = |-1| = 1 \text{ unit to the right, since } \frac{c}{b} < 0.$$

Thus, the graph of $y = \frac{1}{2} \cos\left(\frac{\pi}{4}x - \frac{\pi}{4}\right) = \frac{1}{2} \cos\frac{\pi}{4}(x - 1)$ is the same as the graph of $y = \frac{1}{2} \cos\frac{\pi}{4}x$ shifted horizontally to the right 1 unit. Hence, it completes one cycle from

$$x = 1 \quad \text{to} \quad x = 1 + 8 = 9,$$

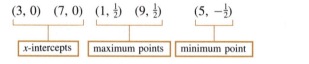

as shown in Figure 2.52. Dividing the interval [1, 9] into four equal parts gives us the five key points on this cycle:

$$(3, 0) \quad (7, 0) \quad (1, \tfrac{1}{2}) \quad (9, \tfrac{1}{2}) \quad (5, -\tfrac{1}{2})$$

FIGURE 2.52

Graph of $y = \frac{1}{2} \cos\left(\frac{\pi}{4}x - \frac{\pi}{4}\right)$

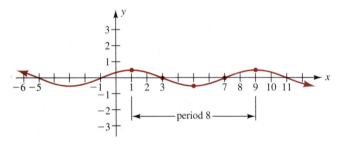

Do not confuse $y = 3 \sin(2x + \pi)$ with $y = 3 \sin 2x + \pi$. As illustrated in Example 3(a), the graph of $y = 3 \sin(2x + \pi)$ is the same as the graph of $y = 3 \sin 2x$ shifted horizontally to the left $\pi/2$ units. However, for $y = 3 \sin 2x + \pi = \pi + 3 \sin 2x$, we follow the vertical shift rule given in Section 1.4. That is, the graph of $y = 3 \sin 2x + \pi$ is the same as the graph of $y = 3 \sin 2x$ shifted vertically upward π units.

PROBLEM 3 Sketch the graph of $y = 3 \sin 2x + \pi$.

In the next example, we reverse the preceding procedure to determine the equation of a sine or cosine curve from its graph.

SECTION 2.4 Properties and Graphs of the Sine and Cosine

EXAMPLE 4   Write the equation of the graph shown in Figure 2.53 in the form

$$y = a \sin(bx + c),$$

using the least nonnegative real number c with $a > 0$ and $b > 0$.

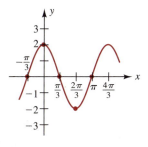

FIGURE 2.53

SOLUTION   From the graph, we note the following facts:

1. The amplitude is 2. Hence, we have $a = 2$.

2. Since the sine curve completes one cycle from $x = -\pi/3$ to $x = \pi$, the period is $P = \pi - (-\pi/3) = 4\pi/3$. Now, using the formula $P = 2\pi/b$, we solve for b:

$$\frac{4\pi}{3} = \frac{2\pi}{b}$$

$$4\pi b = 6\pi \qquad \text{\color{red}Cross product property}$$

$$b = \frac{3}{2} \qquad \text{\color{red}Divide both sides by } 4\pi$$

3. The phase shift c/b is $\pi/3$ units to the left. Thus,

$$\frac{c}{b} = \frac{\pi}{3}$$

$$\frac{c}{\frac{3}{2}} = \frac{\pi}{3} \qquad \text{\color{red}Replace } b \text{ with } \tfrac{3}{2}$$

$$c = \frac{\pi}{2} \qquad \text{\color{red}Multiply both sides by } \tfrac{3}{2}$$

Hence, the equation of the curve is $y = 2 \sin\left(\frac{3}{2}x + \frac{\pi}{2}\right)$. ◆

PROBLEM 4   We may think of the curve shown in Figure 2.53 as a cosine curve that completes one cycle from $x = 0$ to $x = 4\pi/3$. Write the equation of this graph in the form $y = a \cos(bx + c)$ using the least nonnegative real number c with $a > 0$ and $b > 0$. ◆

◆ **Application: Predator-Prey Relationships**

Consider the relationship between mice (prey) and hawks (predators) in a certain field. If the number of mice is relatively large, the number of hawks in the area starts to increase, since the mice provide a plentiful food supply for the hawks. Now, as the number of hawks in the area increases, their food supply (the mice) begins to be depleted. This in turn causes the hawks to search elsewhere for food, which in turn allows the mouse population to begin to increase again. This cyclic pattern repeats itself over and over again with both populations oscillating in a

FIGURE 2.54
Mouse and hawk populations in a predator-prey relationship oscillating in a periodic manner about their average values.

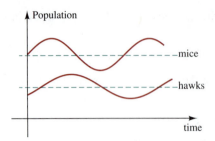

periodic manner about their average values, as shown in Figure 2.54. Hence, the population in a **predator-prey relationship** may be described by using the periodic function

$$f(x) = k + a \sin(bx + c).$$

EXAMPLE 5 Rangers have determined that the population P of red-tailed hawks in a certain area of a state forest can be described by the periodic function

$$P = 70 + 50 \sin\left(\frac{\pi t}{6} - \frac{\pi}{3}\right)$$

where t is the time in months with $t = 1$ corresponding to January 1.

(a) Sketch the graph of this function over the period of one year.

(b) Determine the greatest number of red-tailed hawks in this area during one year and the month in which this occurs.

SOLUTION

(a) For $P = 70 + 50 \sin\left(\frac{\pi t}{6} - \frac{\pi}{3}\right)$, we have $a = 50$, $b = \frac{\pi}{6}$, and $c = -\frac{\pi}{3}$. Therefore,

Amplitude: $|a| = |50| = 50$ Period: $\frac{2\pi}{|b|} = \frac{2\pi}{|\pi/6|} = 12$

Phase shift: $\left|\frac{c}{b}\right| = \left|\frac{-\pi/3}{\pi/6}\right| = |-2| = 2$ units to the right, since $\frac{c}{b} < 0$.

Now, the graph of $P = 70 + 50 \sin\left(\frac{\pi t}{6} - \frac{\pi}{3}\right)$ is the same as the graph of $P = 50 \sin\left(\frac{\pi t}{6} - \frac{\pi}{3}\right)$ shifted vertically upward 70 units. It completes one cycle from

$t = 2$ to $t = 2 + 12 = 14$,
                        └─────┬─────┘
                        Add the period

as shown in Figure 2.55.

FIGURE 2.55

The graph of
$$P = 70 + 50 \sin\left(\frac{\pi t}{6} - \frac{\pi}{3}\right)$$
completes one cycle from $t = 2$ to $t = 14$.

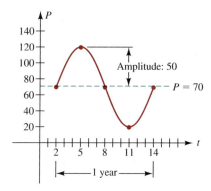

(b) Dividing the interval [2, 14] into four equal parts gives us the five key points on this cycle:

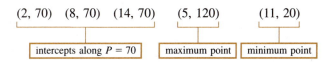

(2, 70) (8, 70) (14, 70)   (5, 120)   (11, 20)

intercepts along $P = 70$; maximum point ; minimum point

Therefore, the greatest number of red-tailed hawks is 120, and this occurs in the fifth month, May. ◆

PROBLEM 5 Referring to Example 5, find the red-tailed hawk population on September 1 ($t = 9$). ◆

Exercises 2.4

Basic Skills

1. By completing the accompanying table of values and then plotting points, graph one cycle of $f(x) = \sin x$ and $g(x) = \cos x$.

x	0	$\frac{\pi}{4}$	$\frac{\pi}{2}$	$\frac{3\pi}{4}$	π	$\frac{5\pi}{4}$	$\frac{3\pi}{2}$	$\frac{7\pi}{4}$	2π
$f(x)$									
$g(x)$									

2. By completing the accompanying table of values and then plotting points, graph one cycle of $f(x) = \sin x$ and $g(x) = \cos x$.

x	0	$\frac{\pi}{6}$	$\frac{\pi}{3}$	$\frac{\pi}{2}$	$\frac{2\pi}{3}$	$\frac{5\pi}{6}$	π	$\frac{7\pi}{6}$	$\frac{4\pi}{3}$	$\frac{3\pi}{2}$	$\frac{5\pi}{3}$	$\frac{11\pi}{6}$	2π
$f(x)$													
$g(x)$													

In Exercises 3–10, use the results of Exercises 1 and 2 and the fact that the sine and cosine functions are periodic with period 2π to find all x in the interval $[-4\pi, 4\pi]$ that satisfy the given equation.

3. $\sin x = 0$
4. $\cos x = 0$
5. $\cos x = -1$
6. $\sin x = 1$
7. $\cos x = \dfrac{1}{\sqrt{2}}$
8. $\sin x = -\dfrac{1}{\sqrt{2}}$
9. $\sin x = -\frac{1}{2}$
10. $\cos x = \dfrac{\sqrt{3}}{2}$

In Exercises 11–22, state the amplitude and period for the graph of each function. Then sketch the graph.

11. $y = 5 \sin x$
12. $y = \frac{2}{3} \cos x$
13. $y = 4 \cos 3x$
14. $y = 3 \sin 0.1x$
15. $y = -\frac{1}{2} \sin \frac{1}{6}x$
16. $y = -\cos 6x$
17. $y = \dfrac{4}{3} \cos(-\pi x)$
18. $y = \sqrt{3} \sin\left(-\dfrac{\pi}{3}x\right)$
19. $y = -\sin(-1.5x)$
20. $y = -\frac{5}{4} \cos \dfrac{\pi x}{5}$
21. $y = 1 + \cos \dfrac{3\pi x}{8}$
22. $y = \sin 2x - 3$

In Exercises 23–36, state the amplitude, period, and phase shift for the graph of each function. Then sketch the graph.

23. $y = \sin\left(x + \dfrac{\pi}{6}\right)$
24. $y = 2\cos\left(x - \dfrac{\pi}{4}\right)$
25. $y = 4 \cos\left(\dfrac{x}{2} - \dfrac{3\pi}{8}\right)$
26. $y = \frac{1}{3} \sin\left(2x + \dfrac{\pi}{3}\right)$
27. $y = -\frac{1}{2} \sin\left(\pi x + \dfrac{3\pi}{4}\right)$
28. $y = -4 \cos \dfrac{\pi}{3}(x - 1)$
29. $y = \cos 2\left(\dfrac{\pi}{6} - x\right)$
30. $y = 3 \sin\left(\dfrac{2\pi}{3} - \pi x\right)$
31. $y = 1.5 \sin(0.75x + \pi)$
32. $y = 5 \cos 0.4(x - \pi)$
33. $y = 3 + 2 \cos 3\left(x - \dfrac{\pi}{12}\right)$
34. $y = 2 - 3 \sin\left(\dfrac{x}{3} + \dfrac{\pi}{9}\right)$
35. $y = \sin(6x + \pi) - 3$
36. $y = -\left[1 + \cos\left(3x - \dfrac{\pi}{2}\right)\right]$

In Exercises 37–42, write the equation of the graph in the form

(a) $y = a \sin(bx + c)$ 　(b) $y = a \cos(bx + c)$

for the least nonnegative real number c, with $a > 0$ and $b > 0$.

37.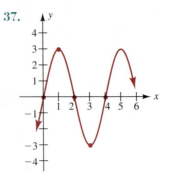

38.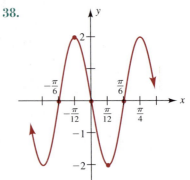

39.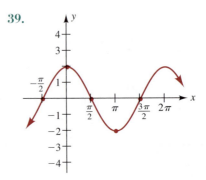

40.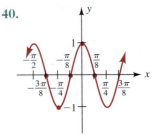

41.

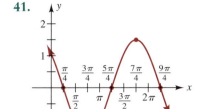

42.

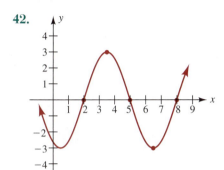

43. Ordinary household voltage v is given by

$$v = 170 \sin 377t,$$

where t is time (in seconds).

(a) Determine the amplitude and period, and then graph this equation for $t \geq 0$.

(b) The number of cycles that the voltage completes in one second is called its *frequency*. The frequency of household voltage, in cycles per second, or Hertz (Hz), is the reciprocal of the period. Find the frequency for household voltage.

44. Rangers have determined that the population P of black bears in a state forest in Maine can be approximated by

$$P = 75 + 40 \sin \frac{\pi t}{4},$$

where t is the time (in years) with $t = 0$ corresponding to January 1, 1980.

(a) Graph this equation for $t \geq 0$.

(b) Find the first year after the year 2000 when the population is at its maximum.

45. Over a period of years, the owner of a ski shop has found that the amount of monthly sales S (in dollars) can be approximated by

$$S = 4500 + 3200 \cos \frac{\pi}{6}(t - 2),$$

where t is the time (in months) with $t = 1$ corresponding to January.

(a) Determine the month when sales are greatest and the amount of sales for that month.

(b) Determine the month when the sales are least and the amount of sales for that month.

46. Meteorologists project that the daily high temperature T (in degrees Fahrenheit) in Buffalo, New York, for 1996 may be found by using the equation

$$T = 42 - 55 \cos\left(\frac{\pi t}{183} + \frac{2\pi}{61}\right),$$

where t is the time (in days) with $t = 1$ corresponding to January 1.

(a) Sketch the graph of this equation over the period of one year.

(b) Determine the greatest temperature and the day that it occurs.

47. The cross section of an aquifer is in the shape of a sine wave, as illustrated in the sketch.

(a) Determine the depth d of the aquifer below level ground as a function of the distance x for $0 \leq x \leq 480$.

(b) Find the depth below level ground when $x = 280$ ft.

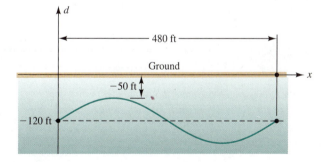

48. A portion of a roller coaster is built in the shape of a cosine wave, as sketched in the figure.

(a) Determine the height h of the roller-coaster car above level ground as a function of the distance x for $0 \leq x \leq 120$.

(b) Find the height above level ground when $x = 60$ ft.

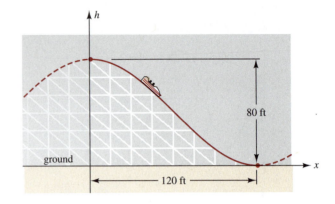

Critical Thinking

If the graph of $y = \sin x$ is shifted to the left $\pi/2$ units, we obtain the graph of $y = \cos x$. Based upon this observation, we conclude that

$$\sin\left(x + \frac{\pi}{2}\right) = \cos x$$

for all real numbers x. In Exercises 49–54, sketch the graph of the function f and compare it with the graphs of $y = \sin x$, $y = -\sin x$, $y = \cos x$, and $y = -\cos x$. Based upon your comparisons, state a trigonometric identity that relates $f(x)$ to either $\sin x$, $-\sin x$, $\cos x$, or $-\cos x$.

49. $f(x) = \sin\left(x - \dfrac{\pi}{2}\right)$
50. $f(x) = \cos\left(x + \dfrac{\pi}{2}\right)$
51. $f(x) = \cos\left(x + \dfrac{3\pi}{2}\right)$
52. $f(x) = \sin\left(x - \dfrac{3\pi}{2}\right)$
53. $f(x) = \sin(x - \pi)$
54. $f(x) = \cos(x + \pi)$

55. Find a function f of the form $f(x) = k + a \sin bx$ if its graph has a maximum point at $(2, 3)$ and a minimum point at $(6, -5)$.

56. By carefully sketching the graphs of $y = \sin x$ and $y = \cos x$ on the same set of coordinate axes, determine all x in the interval $[-4\pi, 4\pi]$ for which $\sin x = \cos x$.

57. Is it possible for a periodic function to be a one-to-one function? Explain.

58. Sketch the graphs of the functions $f(x) = \sin |x|$ and $g(x) = |\sin x|$, and then determine whether these functions are periodic. If either is periodic, state its period.

Calculator Activities

In Exercises 59–64, state the amplitude, period, and phase shift for the graph of each function, rounding each answer to three significant digits.

59. $y = 125 \sin 342x$

60. $y = 9.87 + 5.46 \cos 3.65x$

61. $y = 26.6 \cos (0.223x - 1.24)$

62. $y = 2.24 \sin (1.24 - 19.1x)$

63. $y = 5.46 + \sin (12.5x + 10.3)$

64. $y = 113 - 105 \cos [0.765(x + 3.95)]$

65. The silver fox population P in a national forest in Canada is given by

$$P = 540 + 220 \sin 0.449t,$$

where t is the time (in years) with $t = 0$ corresponding to January 1, 1990. What is the first year after 1990 when the fox population is expected to reach its maximum value? minimum value?

66. It is projected that the amount of snowfall S (in inches) during the year in the Boston area is

$$S = 18.2 + 16.5 \cos [0.785(t + 1)],$$

where t is the time (in years) with $t = 0$ corresponding to January 1, 1990. What is the first year after 1998 when the snowfall is expected to reach its maximum value? minimum value?

67. Use a graphing calculator to generate the graphs of $y = x$ and $y = \cos x$ in the same viewing rectangle. Then use the calculator's trace and zoom features to estimate the solution of the equation $x = \cos x$. Round the answer to three significant digits.

68. Use a graphing calculator to generate the graphs of $y = \sin 2x$ and $y = \cos x$ in the same viewing rectangle. Then use the calculator's trace and zoom features to estimate the solution of the equation $\sin 2x = \cos x$ in the interval $[0, \pi/2]$. Round the answer to three significant digits.

In calculus, it is shown that certain polynomial functions represent good approximations of the trigonometric functions. In Exercises 69 and 70, use a graphing calculator or a computer with graphing capabilities to generate the graph of the polynomial function P over the interval $[-\pi, \pi]$.

(a) *What trigonometric function does P seem to approximate?*

(b) *Compare P(1) to the value of the trigonometric function found in part (a) when $x = 1$. Round each answer to four significant digits.*

69. $P(x) = x - \dfrac{x^3}{6} + \dfrac{x^5}{120} - \dfrac{x^7}{5040}$

70. $P(x) = 1 - \dfrac{x^2}{2} + \dfrac{x^4}{24} - \dfrac{x^6}{720}$

2.5 Properties and Graphs of the Other Trigonometric Functions

As we stated in Section 2.3, the domain of the tangent and secant functions is all real numbers except odd multiples of $\pi/2$, and the domain of the cotangent and cosecant functions is all real numbers except multiples of π. Thus, the graphs of $f(t) = \tan t$ and $g(t) = \sec t$ must each possess vertical asymptotes (see Appendix B) at

$$t = \pm \frac{\pi}{2}, \pm \frac{3\pi}{2}, \pm \frac{5\pi}{2}, \pm \frac{7\pi}{2}, \ldots$$

and the graphs of $F(t) = \cot t$ and $G(t) = \csc t$ must each possess vertical asymptotes at

$$t = 0, \pm \pi, \pm 2\pi, \pm 3\pi, \pm 4\pi, \ldots$$

In this section, we gather some additional information about these four trigonometric functions and then use this information to help sketch the graphs of these functions. We begin with the tangent function.

◆ Properties of the Tangent Function

Recall from Section 2.4 that if the arc length from (1, 0) to a point P on the unit circle is t units, then the coordinates of point P are ($\cos t$, $\sin t$), as shown in Figure 2.56.

Now, by one of our fundamental trigonometric identities (Section 2.2), the ratio of the y-coordinate ($\sin t$) to the x-coordinate ($\cos t$) is

$$\frac{\sin t}{\cos t} = \tan t.$$

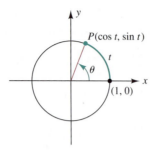

FIGURE 2.56
The coordinates of a point on the unit circle corresponding to an arc length of t are $(\cos t, \sin t)$

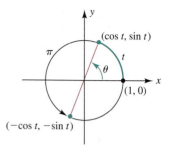

FIGURE 2.57
If t increases by π units, the coordinates of the point are $(-\cos t, -\sin t)$.

Note that $\tan t$ is undefined when $\cos t = 0$, that is, when

$$t = \pm \frac{\pi}{2}, \pm \frac{3\pi}{2}, \pm \frac{5\pi}{2}, \pm \frac{7\pi}{2}, \text{ and so on.}$$

If t increases by π units, as shown in Figure 2.57, then the coordinates of the point are

$$(\cos(t + \pi), \sin(t + \pi)) = (-\cos t, -\sin t).$$

Hence,

$$\tan(t + \pi) = \frac{-\sin t}{-\cos t} = \frac{\sin t}{\cos t} = \tan t.$$

In general, for any integer n, we have

$$\boxed{\tan(t + \pi n) = \tan t}$$

Thus, we conclude the tangent function is periodic with period π. Hence, it must complete one period between $-\pi/2$ and $\pi/2$.

Referring to Figure 2.56, as t approaches $\pi/2$ through values less than $\pi/2$, then $\sin t$ approaches 1 and $\cos t$ approaches 0. Thus, $(\sin t)/(\cos t)$ becomes very large. Hence, we say that $\tan t$ *increases without bound*. Symbolically, we write

$$\tan t \to \infty \quad \text{as} \quad t \to \frac{\pi}{2}^-.$$

Again, referring to Figure 2.56, as t approaches $-\pi/2$ through values greater than $-\pi/2$, then $\sin t$ approaches -1 and $\cos t$ approaches 0. Thus, $(\sin t)/(\cos t)$ is

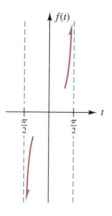
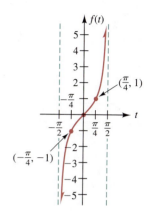

FIGURE 2.58
Appearance of the graph of $f(t) = \tan t$ as t approaches $-\pi/2$ from the right and $\pi/2$ from the left.

FIGURE 2.59
Graph of one period of $f(t) = \tan t$ on the interval $(-\pi/2, \pi/2)$

negative and very large in absolute value. Hence, we say that tan *t decreases without bound.* Symbolically, we write

$$\tan t \to -\infty \quad \text{as} \quad t \to -\frac{\pi}{2}^+.$$

In summary, the lines $t = \pi/2$ and $t = -\pi/2$ are *vertical asymptotes* for the graph of the tangent function $f(t) = \tan t$. The appearance of the graph of $f(t) = \tan t$ as t approaches $-\pi/2$ from the right and $\pi/2$ from the left is shown in Figure 2.58. Now, plotting a few points on the interval $(-\pi/2, \pi/2)$ gives us the graph of one period of $f(t) = \tan t$, as shown in Figure 2.59. We can obtain the entire graph of $f(t) = \tan t$ by repeating the pattern in Figure 2.59 over successive intervals of length π, as illustrated in Figure 2.60.

The following properties are evident from the graph of $f(t) = \tan t$.

FIGURE 2.60
Graph of the tangent function $f(t) = \tan t$

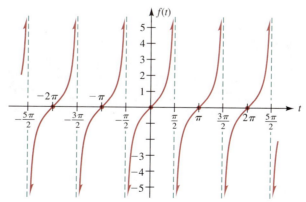

Properties of the Tangent Function: $f(t) = \tan t$

1. The *domain* is the set of all real numbers except $t = \pm \frac{\pi}{2}, \pm \frac{3\pi}{2}, \pm \frac{5\pi}{2}, \ldots$.
2. The *range* is the set of all real numbers.
3. The *period* is π.
4. The *zeros* are $0, \pm \pi, \pm 2\pi, \pm 3\pi, \ldots$.
5. The function is *odd*, that is, $\tan(-t) = -\tan t$ for all t in the domain of f.

Note: We may use any letter as the independent or dependent variable for the tangent function. Throughout the remainder of this section, we choose x and y for these variables so that graphs are shown in the usual xy-coordinate system.

◆ Graph of $y = a \tan(bx + c)$

By the vertical stretch and shrink rules (in Section 1.4), we know that the graph of $y = a \tan x$ with $a > 1$ is similar to the graph of $y = \tan x$ stretched vertically by a factor of a, as shown in Figure 2.61. Unlike the sine and cosine functions, the tangent function has no "largest" value. Thus, we do not refer to the value of a as the amplitude of the tangent function.

The graph of $y = a \tan x$ has five key values on the interval $[-\pi/2, \pi/2]$:

$$\underbrace{x = -\pi/2, \quad x = \pi/2}_{\text{vertical asymptotes}} \quad \underbrace{(0, 0)}_{x\text{-intercept}} \quad \underbrace{(-\pi/4, -a) \quad (\pi/4, a)}_{\text{stretch points}}$$

The graph of $y = a \tan x$ has period π and completes one cycle from $x = -\pi/2$ to $x = \pi/2$. Therefore, the graph of $y = a \tan bx$ completes one cycle from

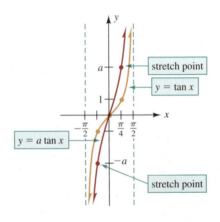

FIGURE 2.61
Comparison of the graphs of $y = \tan x$ and $y = a \tan x$ (with $a > 1$) on the interval $(-\pi/2, \pi/2)$

SECTION 2.5 Properties and Graphs of the Other Trig Functions

$$bx = -\frac{\pi}{2} \quad \text{to} \quad bx = \frac{\pi}{2}$$

$$x = -\frac{\pi}{2b} \quad \text{to} \quad x = \frac{\pi}{2b}$$

⎵ ⎵
A pair of consecutive vertical asymptotes for the graph of $y = a \tan bx$

The period of $y = a \tan bx$ is the length of the interval from $x = -\pi/(2b)$ to $x = \pi/(2b)$.

◆ **Period Formula: Tangent**

The **period** P of $y = a \tan bx$ is given by

$$P = \frac{\pi}{|b|} \quad \text{for } b \neq 0.$$

EXAMPLE 1   Find the period and sketch the graph of each function.

(a) $y = 3 \tan \frac{1}{2}x$ (b) $y = \frac{2}{3} \tan\left(\pi x + \frac{\pi}{4}\right)$

SOLUTION

(a) The graph of $y = 3 \tan \frac{1}{2}x$ completes one cycle from

$$bx = -\frac{\pi}{2} \quad \text{to} \quad bx = \frac{\pi}{2}$$

$$\tfrac{1}{2}x = -\frac{\pi}{2} \quad \text{to} \quad \tfrac{1}{2}x = \frac{\pi}{2} \qquad \text{Replace } bx \text{ with } \tfrac{1}{2}x$$

$$x = -\pi \quad \text{to} \quad x = \pi \qquad \text{Solve for } x$$

Thus, the period is 2π. Dividing the interval $[-\pi, \pi]$ into four equal parts gives us the five key values on this cycle:

$$x = -\pi, \quad x = \pi \qquad (0, 0) \qquad (-\pi/2, 3) \quad (\pi/2, 3)$$

⎵ ⎵ ⎵
vertical asymptotes   x-intercept   stretch points

The graph of $y = 3 \tan \frac{1}{2}x$ is shown in Figure 2.62.

(b) By the horizontal shift rule (in Section 1.4), we know that the graph of $y = \frac{2}{3} \tan\left(\pi x + \frac{\pi}{4}\right) = \frac{2}{3} \tan \pi(x + \frac{1}{4})$ is the same as the graph of $y = \frac{2}{3} \tan \pi x$ shifted horizontally to the left $\frac{1}{4}$ unit. The graph of

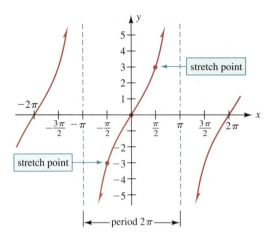

FIGURE 2.62
Graph of $y = 3 \tan \frac{1}{2}x$

$y = \frac{2}{3} \tan \left(\pi x + \frac{\pi}{4} \right)$ completes one cycle from

$$bx = -\frac{\pi}{2} \quad \text{to} \quad bx = \frac{\pi}{2}$$

$$\pi x + \frac{\pi}{4} = -\frac{\pi}{2} \quad \text{to} \quad \pi x + \frac{\pi}{4} = \frac{\pi}{2} \qquad \text{Replace } bx \text{ with } \pi x + \frac{\pi}{4}$$

$$x = -\frac{3}{4} \quad \text{to} \quad x = \frac{1}{4} \qquad \text{Solve for } x$$

Thus, the period is 1. Dividing the interval $[-\frac{3}{4}, \frac{1}{4}]$ into four equal parts gives us the five key values on this cycle:

$$x = -\tfrac{3}{4}, \; x = \tfrac{1}{4} \qquad (-\tfrac{1}{4}, 0) \qquad (-\tfrac{1}{2}, -\tfrac{2}{3}) \quad (0, \tfrac{2}{3})$$

vertical asymptotes   x-intercept   shrink points

The graph of $y = \frac{2}{3} \tan \left(\pi x + \frac{\pi}{4} \right)$ is shown in Figure 2.63. ◆

If b is negative in $y = a \tan bx$, then we use the fact that

$$\tan(-x) = -\tan x$$

to rewrite the function before sketching the graph. For example, to graph $y = \tan(-2x)$, we write

$$y = \tan(-2x) = -\tan 2x.$$

Now the graph of $y = \tan(-2x) = -\tan 2x$ is the same as the graph of $y = \tan 2x$ reflected about the x-axis.

FIGURE 2.63

Graph of $y = \frac{2}{3}\tan\left(\pi x + \frac{\pi}{4}\right)$

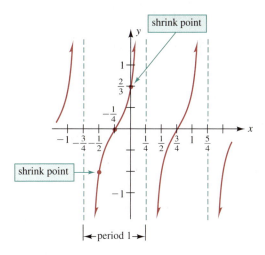

PROBLEM 1 Find the period and sketch the graph of $y = \tan(-2x)$.

◆ Graphing the Cotangent Function

Since we know that

$$\cot t = \frac{1}{\tan t},$$

we can obtain the graph of the cotangent function $f(t) = \cot t$ by finding, whenever possible, the reciprocals of the y-coordinates on the graph of $y = \tan t$. Values of t for which $\tan t = 0$ lead to vertical asymptotes on the graph of $f(t) = \cot t$, and values of t for which $\tan t$ is undefined lead to t-intercepts on the graph of $f(t) = \cot t$. The graph of $f(t) = \cot t$ is shown in Figure 2.64.

The following properties are evident from the graph of $f(t) = \cot t$.

FIGURE 2.64

Graph of the cotangent function $f(t) = \cot t$

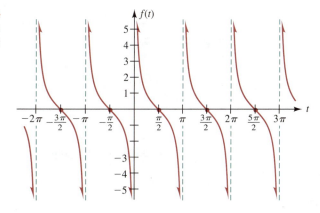

Properties of the Cotangent Function: $f(t) = \cot t$

1. The *domain* is the set of all real numbers except $0, \pm\pi, \pm 2\pi, \pm 3\pi, \ldots$.
2. The *range* is the set of all real numbers.
3. The *period* is π.
4. The *zeros* are $t = \pm\dfrac{\pi}{2}, \pm\dfrac{3\pi}{2}, \pm\dfrac{5\pi}{2}, \ldots$.
5. The function is *odd*, that is, $\cot(-t) = -\cot t$ for all t in the domain of f.

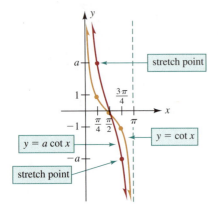

FIGURE 2.65
Comparison of the graphs of $y = \cot x$ and $y = a \cot x$ (with $a > 1$) on the interval $(0, \pi)$

By the vertical stretch and shrink rules (in Section 1.4), we know that the graph of $y = a \cot x$ with $a > 1$ is similar to the graph of $y = \cot x$ stretched vertically by a factor of a, as shown in Figure 2.65. Unlike the sine and cosine functions, the cotangent function has no "largest" value. Thus, we do not refer to the value of a as the amplitude of the cotangent function.

The graph of $y = a \cot x$ has five key values on the interval $[0, \pi]$:

$$x = 0, \quad x = \pi \qquad (\pi/2, 0) \qquad (\pi/4, a) \quad (3\pi/4, -a)$$

    vertical asymptotes   *x*-intercept   stretch points

The graph of $y = a \cot x$ has period π and completes one cycle from $x = 0$ to $x = \pi$. Therefore, the graph of $y = a \cot bx$ completes one cycle from

$$bx = 0 \quad \text{to} \quad bx = \pi$$

$$x = 0 \quad \text{to} \quad x = \dfrac{\pi}{b}$$

A pair of consecutive vertical asymptotes for the graph of $y = a \cot bx$

The period of $y = a \cot bx$ is the length of the interval from $x = 0$ to $x = \pi/b$.

Period Formula: Cotangent

The **period** P of $y = a \cot bx$ is given by

$$P = \dfrac{\pi}{|b|} \quad \text{for } b \neq 0.$$

EXAMPLE 2 Find the period and sketch the graph of each function.

    (a) $y = 2 \cot 3x$ **(b)** $y = \cot\left(2x - \dfrac{\pi}{2}\right)$

SOLUTION

(a) The graph of $y = 2 \cot 3x$ completes one cycle from

$$bx = 0 \quad \text{to} \quad bx = \pi$$
$$3x = 0 \quad \text{to} \quad 3x = \pi \quad \text{\textbf{Replace } } bx \text{ \textbf{with} } 3x$$
$$x = 0 \quad \text{to} \quad x = \frac{\pi}{3} \quad \text{\textbf{Solve for } } x$$

Thus, the period is $\pi/3$. Dividing the interval $[0, \pi/3]$ into four equal parts gives us the five key values on this cycle:

$$\underbrace{x = 0 \quad x = \pi/3}_{\text{vertical asymptotes}} \quad \underbrace{(\pi/6, 0)}_{x\text{-intercept}} \quad \underbrace{(\pi/12, 2) \quad (\pi/4, -2)}_{\text{stretch points}}$$

The graph of $y = 2 \cot 3x$ is shown in Figure 2.66.

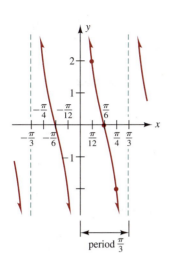

FIGURE 2.66
Graph of $y = 2 \cot 3x$

(b) By the horizontal shift rule (in Section 1.4), we know that the graph of

$$y = \cot\left(2x - \frac{\pi}{2}\right) = \cot 2\left(x - \frac{\pi}{4}\right)$$

is the same as the graph of $y = \cot 2x$ shifted horizontally to the right $\pi/4$ unit. The graph of $y = \cot\left(2x - \frac{\pi}{2}\right)$ completes one cycle from

$$bx = 0 \quad \text{to} \quad bx = \pi$$
$$2x - \frac{\pi}{2} = 0 \quad \text{to} \quad 2x - \frac{\pi}{2} = \pi \quad \text{\textbf{Replace } } bx \text{ \textbf{with} } 2x - \frac{\pi}{2}$$
$$x = \frac{\pi}{4} \quad \text{to} \quad x = \frac{3\pi}{4} \quad \text{\textbf{Solve for } } x$$

Thus, the period is $\pi/2$. Dividing the interval $[\pi/4, 3\pi/4]$ into four equal parts gives us the five key values on this cycle:

$$\underbrace{x = \pi/4 \quad x = 3\pi/4}_{\text{vertical asymptotes}} \quad \underbrace{(\pi/2, 0)}_{x\text{-intercept}} \quad \underbrace{(3\pi/8, 1) \quad (5\pi/8, -1)}_{\text{stretch points}}$$

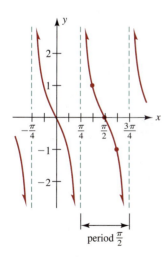

FIGURE 2.67
Graph of $y = \cot\left(2x - \frac{\pi}{2}\right)$

The graph of $y = \cot\left(2x - \frac{\pi}{2}\right)$ is shown in Figure 2.67. ◆

If b is negative in $y = a \cot bx$, then we use the fact that

$$\cot(-x) = -\cot x$$

to rewrite the function before sketching the graph. For example, to graph $y = 3 \cot(-\pi x)$, we write

$$y = 3 \cot(-\pi x) = -3 \cot \pi x.$$

Now the graph of $y = 3 \cot(-\pi x) = -3 \cot \pi x$ is the same as the graph of $y = 3 \cot \pi x$ reflected about the x-axis.

PROBLEM 2   Find the period and sketch the graph of $y = 3 \cot(-\pi x)$. ◆

◆ **Graphing the Cosecant and Secant Functions**

Since

$$\csc t = \frac{1}{\sin t} \quad \text{and} \quad \sec t = \frac{1}{\cos t},$$

we can obtain the graphs of the cosecant function $f(t) = \csc t$ and the secant function $g(t) = \sec t$ by finding, whenever possible, the reciprocals of the y-coordinates on the graphs of $y = \sin t$ and $y = \cos t$, respectively. Values of t for which $\sin t = 0$ lead to vertical asymptotes on the graph of $f(t) = \csc t$, and values of t for which $\cos t = 0$ lead to vertical asymptotes on the graph of $g(t) = \sec t$. The graphs of $f(t) = \csc t$ and $g(t) = \sec t$ are shown in Figures 2.68 and 2.69, respectively.

The following properties are evident from the graphs of $f(t) = \csc t$ and $g(t) = \sec t$.

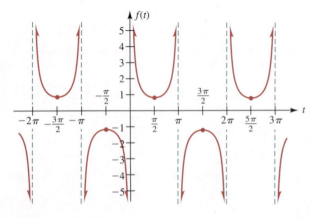

FIGURE 2.68
Graph of the cosecant function
$f(t) = \csc t$

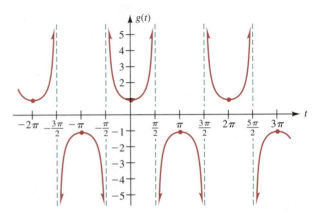

FIGURE 2.69
Graph of the secant function
$g(t) = \sec t$

SECTION 2.5   Properties and Graphs of the Other Trig Functions   139

Properties of the Cosecant and Secant Functions

Cosecant function: $f(t) = \csc t$

1. The *domain* is the set of all real numbers except

   $t = 0, \pm \pi, \pm 2\pi, \pm 3\pi, \ldots$

2. The *range* is $(-\infty, -1] \cup [1, \infty)$.

3. The *period* is 2π.

4. The function has no *zeros*.

5. The function is *odd*, that is, $\csc(-t) = -\csc t$ for all t in the domain of f.

Secant function: $g(t) = \sec t$

1. The *domain* is the set of all real numbers except

   $t = \pm \dfrac{\pi}{2}, \pm \dfrac{3\pi}{2}, \pm \dfrac{5\pi}{2}, \ldots$

2. The *range* is $(-\infty, -1] \cup [1, \infty)$.

3. The *period* is 2π.

4. The function has no *zeros*.

5. The function is *even*, that is, $\sec(-t) = \sec t$ for all t in the domain of g.

For $y = \csc x$ and $y = \sec x$, the range is $(-\infty, -1] \cup [1, \infty)$. Thus, for $y = a \csc x$ and $y = a \sec x$ with $a > 0$ the range is $(-\infty, -a] \cup [a, \infty)$, as shown in Figures 2.70 and 2.71, respectively.

The graph of $y = a \csc x$ has five key points on the interval $[0, 2\pi]$:

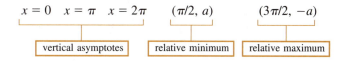

The graph of $y = a \sec x$ has five key points on the interval $[-\pi/2, 3\pi/2]$:

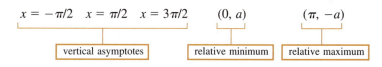

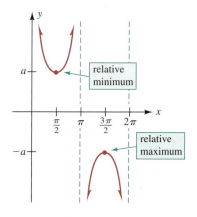

FIGURE 2.70

Graph of $y = a \csc x$ with $a > 0$ on the interval $[0, 2\pi]$

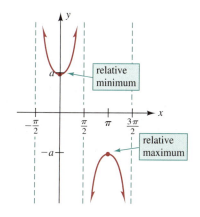

FIGURE 2.71

Graph of $y = a \sec x$ with $a > 0$ on the interval $[-\pi/2, 3\pi/2]$

The graph of $y = a \csc x$ has period 2π and completes one cycle from $x = 0$ to $x = 2\pi$. Therefore, the graph of $y = a \csc bx$ completes one cycle from

$$bx = 0 \quad \text{to} \quad bx = 2\pi$$
$$x = 0 \quad \text{to} \quad x = \frac{2\pi}{b}$$

The graph of $y = a \sec x$ has period 2π and completes one cycle from $x = -\pi/2$ to $x = 3\pi/2$. Therefore, the graph of $y = a \sec bx$ completes one cycle from

$$bx = -\frac{\pi}{2} \quad \text{to} \quad bx = \frac{3\pi}{2}$$
$$x = -\frac{\pi}{2b} \quad \text{to} \quad x = \frac{3\pi}{2b}$$

The period of $y = a \csc bx$ is the length of the interval from $x = 0$ to $x = 2\pi/b$, and the period of $y = a \sec bx$ is the length of the interval from $x = -\pi/(2b)$ to $x = 3\pi/(2b)$.

Period Formula: Cosecant and Secant

The **period** P of $y = a \csc bx$ and $y = a \sec bx$ is given by

$$P = \frac{2\pi}{|b|} \quad \text{for } b \neq 0.$$

EXAMPLE 3 Find the period and sketch the graph of each function.

(a) $y = 3 \csc 4x$ (b) $y = \sec\left(\pi x + \frac{\pi}{2}\right)$

SOLUTION

(a) The graph of $y = 3 \csc 4x$ completes one cycle from

$$bx = 0 \quad \text{to} \quad bx = 2\pi$$
$$4x = 0 \quad \text{to} \quad 4x = 2\pi \qquad \text{\color{red}Replace bx with $4x$}$$
$$x = 0 \quad \text{to} \quad x = \frac{\pi}{2} \qquad \text{\color{red}Solve for x}$$

Thus, the period is $\pi/2$. Dividing the interval $[0, \pi/2]$ into four equal parts gives us the five key values on this cycle:

SECTION 2.5 Properties and Graphs of the Other Trig Functions 141

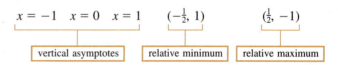

vertical asymptotes: $x = 0 \quad x = \pi/4 \quad x = \pi/2$
relative minimum: $(\pi/8, 3)$
relative maximum: $(3\pi/8, -3)$

The graph of $y = 3 \csc 4x$ is shown in Figure 2.72.

(b) By the horizontal shift rule (in Section 1.4), we know that the graph of

$$y = \sec\left(\pi x + \frac{\pi}{2}\right) = \sec \pi(x + \tfrac{1}{2})$$

is the same as the graph of $y = \sec \pi x$ shifted horizontally to the left $\tfrac{1}{2}$ unit. The graph of $y = \sec\left(\pi x + \frac{\pi}{2}\right)$ completes one cycle from

$$bx = -\frac{\pi}{2} \quad \text{to} \quad bx = \frac{3\pi}{2}$$

$$\pi x + \frac{\pi}{2} = -\frac{\pi}{2} \quad \text{to} \quad \pi x + \frac{\pi}{2} = \frac{3\pi}{2} \qquad \textbf{Replace } bx \textbf{ with } \pi x + \frac{\pi}{2}$$

$$x = -1 \quad \text{to} \quad x = 1 \qquad \textbf{Solve for } x$$

Thus, the period is 2. Dividing the interval $[-1, 1]$ into four equal parts gives us the five key points on this cycle:

vertical asymptotes: $x = -1 \quad x = 0 \quad x = 1$
relative minimum: $(-\tfrac{1}{2}, 1)$
relative maximum: $(\tfrac{1}{2}, -1)$

The graph of $y = \sec\left(\pi x + \frac{\pi}{2}\right)$ is shown in Figure 2.73. ◆

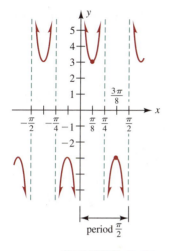

FIGURE 2.72
Graph of $y = 3 \csc 4x$

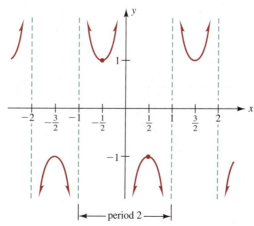

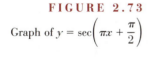

FIGURE 2.73
Graph of $y = \sec\left(\pi x + \frac{\pi}{2}\right)$

If b is negative in $y = a \csc bx$ or $y = a \sec bx$, then we use the facts that

$$\csc(-x) = -\csc x \quad \text{and} \quad \sec(-x) = \sec x$$

to rewrite the function before sketching the graph. For example, to graph $y = \sec(-x/3)$, we begin by writing

$$y = \sec\left(-\frac{x}{3}\right) = \sec\left(\frac{x}{3}\right).$$

Thus, the graph of $y = \sec(-x/3)$ is the same as the graph of $y = \sec(x/3)$.

PROBLEM 3 Find the period and sketch the graph of $y = \sec\left(-\frac{x}{3}\right)$. ◆

Exercises 2.5

Basic Skills

In Exercises 1–4, complete the table of values and then plot points to graph one cycle of the given function.

1. $f(x) = \tan x$

x	$-\frac{\pi}{2}$	$-\frac{\pi}{3}$	$-\frac{\pi}{4}$	$-\frac{\pi}{6}$	0	$\frac{\pi}{6}$	$\frac{\pi}{4}$	$\frac{\pi}{3}$	$\frac{\pi}{2}$
$f(x)$									

2. $f(x) = \cot x$

x	0	$\frac{\pi}{6}$	$\frac{\pi}{4}$	$\frac{\pi}{3}$	$\frac{\pi}{2}$	$\frac{2\pi}{3}$	$\frac{3\pi}{4}$	$\frac{5\pi}{6}$	π
$f(x)$									

3. $f(x) = \csc x$

x	0	$\frac{\pi}{4}$	$\frac{\pi}{2}$	$\frac{3\pi}{4}$	π	$\frac{5\pi}{4}$	$\frac{3\pi}{2}$	$\frac{7\pi}{4}$	2π
$f(x)$									

SECTION 2.5 Properties and Graphs of the Other Trig Functions 143

4. $f(x) = \sec x$

x	$-\frac{\pi}{2}$	$-\frac{\pi}{3}$	$-\frac{\pi}{6}$	0	$\frac{\pi}{6}$	$\frac{\pi}{3}$	$\frac{\pi}{2}$	$\frac{2\pi}{3}$	$\frac{5\pi}{6}$	π	$\frac{7\pi}{6}$	$\frac{4\pi}{3}$	$\frac{3\pi}{2}$
$f(x)$													

In Exercises 5–12, use the results of Exercises 1 and 2 and the fact that the tangent and cotangent functions are periodic with period π to find all x in the interval $[-4\pi, 4\pi]$ that satisfy the given equation.

5. $\tan x = 0$

6. $\cot x = 0$

7. $\cot x = -1$

8. $\tan x = 1$

9. $\tan x = \dfrac{1}{\sqrt{3}}$

10. $\cot x = -\dfrac{1}{\sqrt{3}}$

11. $\tan x = -\sqrt{3}$

12. $\cot x = \sqrt{3}$

In Exercises 13–20, use the results of Exercises 3 and 4 and the fact that the cosecant and secant functions are periodic with period 2π to find all x in the interval $[-4\pi, 4\pi]$ that satisfy the given equation.

13. $\csc x = 1$

14. $\sec x = -1$

15. $\sec x = -2$

16. $\csc x = \sqrt{2}$

17. $\sec x = \dfrac{2}{\sqrt{3}}$

18. $\csc x = -\sqrt{2}$

19. $\csc x = 0$

20. $\sec x = \dfrac{1}{2}$

In Exercises 21–48, find the period and sketch the graph of each function.

21. $y = \tan 3x$

22. $y = 2 \tan \dfrac{x}{3}$

23. $y = 2 \tan (-\tfrac{3}{4}x)$

24. $y = -\tfrac{1}{3} \tan 4x$

25. $y = \tfrac{1}{2} \cot 4x$

26. $y = 2 \cot 2x$

27. $y = -\cot \tfrac{2}{3}x$

28. $y = 3 \cot \left(-\dfrac{\pi x}{2}\right)$

29. $y = \csc \pi x$

30. $y = \csc \tfrac{2}{5}x$

31. $y = 2 \csc (-3x)$

32. $y = -0.5 \csc 1.2x$

33. $y = \sec 0.8\, x$

34. $y = \tfrac{2}{3} \sec 4x$

35. $y = -4 \sec \tfrac{3}{8}x$

36. $y = \tfrac{1}{2} \sec (-2x)$

37. $y = 3 + \tan x$

38. $y = 4 - \cot \dfrac{\pi}{12} x$

39. $y = 1 - \csc 2x$

40. $y = 2[1 + \sec (-2x)]$

41. $y = \tan (3x + \pi)$

42. $y = 3 \cot \left(2\pi x - \dfrac{\pi}{4}\right)$

43. $y = \tfrac{1}{3} \cot \left(\pi x - \dfrac{\pi}{6}\right)$

44. $y = \tan \left(2x + \dfrac{\pi}{8}\right)$

45. $y = \csc \left(\dfrac{\pi}{3} - x\right)$

46. $y = 2 \sec (4x + \pi)$

47. $y = \tfrac{3}{4} \sec \pi(x - \tfrac{1}{4})$

48. $y = 2 + 3 \csc \left(x - \dfrac{\pi}{8}\right)$

In Exercises 49 and 50, write the equation of the given graph in the form $y = a \tan bx$ or $y = a \cot bx$ with $a > 0$ and $b > 0$.

49.

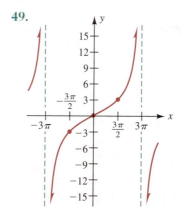

50.

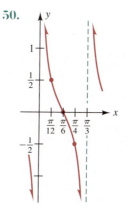

In Exercises 51 and 52, write the equation of the graph in the form $y = a \csc bx$ or $y = a \sec bx$ with $a > 0$ and $b > 0$.

51.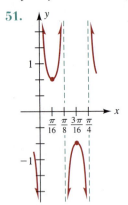

52.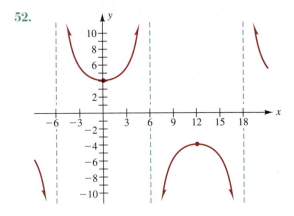

Critical Thinking

If the graph of $y = \tan x$ is shifted to the right $\pi/2$ units, we obtain the graph of $y = \cot x$ reflected about the x-axis. Based upon this observation, we conclude that

$$\tan\left(x - \frac{\pi}{2}\right) = -\cot x$$

for all real numbers x. In Exercises 53–56, sketch the graph of the function f and compare it with the graphs of $y = \tan x$, $y = -\tan x$, $y = \cot x$, and $y = -\cot x$. Based upon your comparisons, state a trigonometric identity that relates f(x) to either $\tan x$, $-\tan x$, $\cot x$, or $-\cot x$.

53. $f(x) = \tan\left(x + \dfrac{\pi}{2}\right)$

54. $f(x) = \tan(x + \pi)$

55. $f(x) = \cot(\pi - x)$

56. $f(x) = \cot\left(x + \dfrac{3\pi}{2}\right)$

If the graph of $y = \sec x$ is shifted to the right $\pi/2$ units, we obtain the graph of $y = \csc x$. Based upon this observation, we conclude that

$$\sec\left(x - \frac{\pi}{2}\right) = \csc x$$

for all real numbers x. In Exercises 57–60, sketch the graph of the function f and compare it with the graphs of $y = \csc x$, $y = -\csc x$, $y = \sec x$, and $y = -\sec x$. Based upon your observations, state a trigonometric identity that relates f(x) to either $\csc x$, $-\csc x$, $\sec x$, or $-\sec x$.

57. $f(x) = \csc\left(x + \dfrac{\pi}{2}\right)$

58. $f(x) = \csc(\pi - x)$

59. $f(x) = \sec\left(x - \dfrac{3\pi}{2}\right)$

60. $f(x) = \sec(x - \pi)$

61. Find a function f of the form $f(x) = k + a \csc bx$ if its graph has a relative minimum point at (2, 5) and a relative maximum point at (6, −1).

62. Sketch the graphs of the functions $f(x) = \tan |x|$ and $g(x) = |\tan x|$ and determine whether each of these functions is periodic. If the function is periodic, state its period.

Calculator Activities

In Exercises 63–68, state the period of each function, rounding each answer to three significant digits.

63. $y = 1.50 \tan 4.33x$

64. $y = 0.46 \cot 3.65x$

65. $y = 2.67 \csc 0.87x$

66. $y = 2.12 \sec(-1.91x)$

67. $y = \cot(11.2 - 1.32x)$

68. $y = 1.2 \csc(1.09x - 0.52)$

69. Use a graphing calculator to generate the graphs of $y = x$ and $y = \csc x$ in the same viewing rectangle. Then use the calculator's trace and zoom features to estimate the solution of the equation $x = \csc x$ in the interval $[-\pi, \pi]$. Round the answer to three significant digits.

70. Use a graphing calculator or a computer with graphing capabilities to generate the graph of the polynomial function

$$P(x) = x + \frac{x^3}{3} + \frac{2x^5}{15} + \frac{17x^7}{315} + \frac{62x^9}{2835}$$

over the interval $(-\pi/2, \pi/2)$.

(a) What trigonometric function does P seem to approximate?

(b) Compare $P(1)$ to the value of that trigonometric function when $x = 1$. Round each answer to four significant digits.

2.6 The Inverse Trigonometric Functions

Recall that in order to have an inverse, a function must be one-to-one. By the horizontal line test (Section 1.5), the sine function $y = \sin x$ is not a one-to-one function since, as shown in Figure 2.74, at least one horizontal line intersects its graph in more than one point. Similarly, the cosine function $y = \cos x$ and the tangent function $y = \tan x$ are not one-to-one functions since each graph is intercepted by horizontal lines in more than one point. Thus, we conclude that neither the sine, cosine, nor tangent function has an inverse over its entire domain.

In this section, we restrict the domain of each of these trigonometric functions so that the new function which is formed for each restricted domain is a one-to-one function and, therefore, does have an inverse. We begin by restricting the domain of the sine function.

◆ Restricted Sine Function and Its Inverse

Consider the *restricted sine function*

$$y = \sin x, \quad -\frac{\pi}{2} \leq x \leq \frac{\pi}{2}$$

whose graph is shown in Figure 2.75. Note that on the interval $[-\pi/2, \pi/2]$, the restricted sine function attains every value of the sine function once and only once.

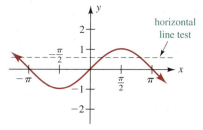

FIGURE 2.74
The graph of the sine function, $y = \sin x$, fails the horizontal line test. Thus, it is not one-to-one and does not have an inverse.

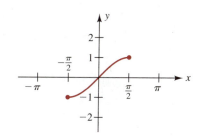

FIGURE 2.75
The restricted sine function $y = \sin x$, $-\pi/2 \leq x \leq \pi/2$, is one-to-one and has an inverse.

Thus, the restricted sine function is one-to-one and has an inverse. We refer to the inverse of the restricted sine function as the *inverse sine function*.

Using the method suggested in Section 1.5, we can attempt to find this inverse sine function by interchanging x and y in $y = \sin x$, $-\pi/2 \leq x \leq \pi/2$, and then solving for y. If we proceed in this manner, we obtain

$$x = \sin y, \quad -\frac{\pi}{2} \leq y \leq \frac{\pi}{2}.$$

However, we have no algebraic procedure enabling us to solve for y in this equation. By convention, we solve this equation for y by writing

$$y = \sin^{-1} x, \quad -\frac{\pi}{2} \leq y \leq \frac{\pi}{2}.$$

Hence, $\sin^{-1} x$ denotes the real number (or angle in radians) between $-\pi/2$ and $\pi/2$ whose sine is x.

Inverse Sine Function

> The **inverse sine function** is defined as
>
> $$y = \sin^{-1} x \quad \text{if and only if} \quad x = \sin y, \quad -\frac{\pi}{2} \leq y \leq \frac{\pi}{2}.$$

In the notation $\sin^{-1} x$, the superscript -1 denotes an inverse, not an exponent, that is,

$$\sin^{-1} x \quad \text{does } not \text{ mean} \quad \frac{1}{\sin x}.$$

Note: The inverse sine function is also referred to as the **arcsine function** and the notation $\arcsin x$ is used frequently in place of $\sin^{-1} x$. The arcsine notation comes from the fact that we can interpret the real number whose sine is x as the length of that arc on a unit circle whose sine is x.

Recall from Section 1.5 that the graphs of a function and its inverse are symmetric with respect to the line $y = x$. Thus, the graph of the inverse sine function (or arcsine function) is the same as the graph of the restricted sine function (Figure 2.75) reflected in the line $y = x$. Figure 2.76 shows the graph of $y = \sin^{-1} x$. Note that the domain of $y = \sin^{-1} x$ is $[-1, 1]$ and the range is $[-\pi/2, \pi/2]$. Also note that the graph of $y = \sin^{-1} x$ is symmetric with respect to the origin, which indicates that it is an odd function. Thus, we have

$$\sin^{-1}(-x) = -\sin^{-1} x$$

for every real number x in the interval $[-1, 1]$.

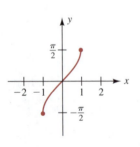

FIGURE 2.76
Graph of the inverse sine function, $y = \sin^{-1} x$. Its domain is $[-1, 1]$ and its range is $[-\pi/2, \pi/2]$.

EXAMPLE 1 Find the exact value of each expression, if it is defined.

(a) $\sin^{-1} \dfrac{1}{\sqrt{2}}$ 　　(b) $\arcsin(-1)$ 　　(c) $\sin^{-1} 2$

SOLUTION

(a) Letting $y = \sin^{-1} \dfrac{1}{\sqrt{2}}$, we have

$$\sin y = \frac{1}{\sqrt{2}}, \quad -\frac{\pi}{2} \le y \le \frac{\pi}{2}.$$

From Table 2.3 in Section 2.3, we know that $\sin(\pi/4) = 1/\sqrt{2}$. Since $\pi/4$ is in the interval $[-\pi/2, \pi/2]$, we conclude that

$$\sin^{-1} \frac{1}{\sqrt{2}} = \frac{\pi}{4}.$$

(b) Letting $y = \arcsin(-1)$, we have

$$\sin y = -1, \quad -\frac{\pi}{2} \le y \le \frac{\pi}{2}.$$

From Table 2.2 in Section 2.3, we have $\sin(\pi/2) = 1$. Therefore, using the fact that $\sin^{-1}(-x) = -\sin^{-1}x$, we write

$$\arcsin(-1) = -\arcsin 1 = -\frac{\pi}{2}.$$

Comment: Although $\sin(3\pi/2) = -1$, note that $\arcsin(-1) \ne 3\pi/2$ because the range of the inverse sine function is restricted to $[-\pi/2, \pi/2]$.

(c) The domain of the inverse sine function $y = \sin^{-1}x$ is $[-1, 1]$. Thus, for $x > 1$ or $x < -1$, the inverse sine function is undefined. Hence

$$\sin^{-1} 2 \text{ is undefined.} \quad \blacklozenge$$

In Example 1, we were able to find the *exact values* of $\sin^{-1}(1/\sqrt{2})$ and $\arcsin(-1)$ by relying on the information given in Table 2.2 and Table 2.3. To find the *approximate value* of $\arcsin x$, or $\sin^{-1}x$, for $-1 \le x \le 1$, we set a calculator in radian mode and use the inverse sine key, $\boxed{\text{SIN}^{-1}}$. (To access the inverse trig functions on some types of calculators, we must use the $\boxed{\text{INV}}$ or $\boxed{\text{2ND } f}$ key in conjunction with the trig key. If a calculator is not available, use Table 4 in Appendix C, finding the value in the function column, then reading the corresponding radian value.)

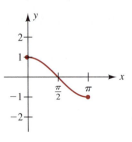
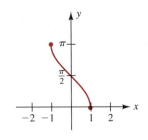

FIGURE 2.77
The restricted cosine function
$y = \cos x$, $0 \leq x \leq \pi$, is one-to-one and has an inverse.

FIGURE 2.78
Graph of the inverse cosine function, $y = \cos^{-1} x$. Its domain is $[-1, 1]$ and the range is $[0, \pi]$.

PROBLEM 1 Use a calculator to find the approximate value of each expression. Round each answer to four significant digits.

(a) arcsin (0.53)   (b) $\sin^{-1}(-0.766)$

◆ Restricted Cosine Function and Its Inverse

Like the sine function, the cosine function $y = \cos x$ is not one-to-one and does not have an inverse. However, consider the *restricted cosine function*

$$y = \cos x, \quad 0 \leq x \leq \pi$$

whose graph is shown in Figure 2.77. Note that on the interval $[0, \pi]$, the restricted cosine function attains every value of the cosine function once and only once. Thus, the restricted cosine function is one-to-one and has an inverse. We refer to the inverse of the restricted cosine function as the *inverse cosine function* and define it as follows.

The **inverse cosine function** is defined as

$$y = \cos^{-1} x \quad \text{if and only if} \quad x = \cos y, \quad 0 \leq y \leq \pi.$$

Note: The inverse cosine function is also referred to as the **arccosine function** and the notation arccos x is used frequently in place of $\cos^{-1} x$. Both $\cos^{-1} x$ and arccos x denote the real number (or angle in radians) between 0 and π whose cosine is x.

The graph of the inverse cosine function (or arccosine function) is the same as the graph of the restricted cosine function (Figure 2.77) reflected in the line $y = x$. Figure 2.78 shows the graph of $y = \cos^{-1} x$. Note that the domain of $y = \cos^{-1} x$ is $[-1, 1]$ and the range is $[0, \pi]$. Since the graph of $y = \cos^{-1} x$ is not symmetric with respect to the origin nor with respect to the y-axis, the inverse cosine function is neither even nor odd. However, we do have the following relationship between positive and negative values of x in the interval $[-1, 1]$:

$$\cos^{-1}(-x) = -\cos^{-1}x + \pi$$

In Exercise 61, we are asked to prove this relationship.

EXAMPLE 2 Find the exact value of each expression, if it is defined.

(a) $\cos^{-1} 1$ (b) $\cos^{-1}(-\frac{1}{2})$ (c) $\arccos(-2)$

SOLUTION

(a) Letting $y = \cos^{-1} 1$, we have

$$\cos y = 1, \quad 0 \le y \le \pi.$$

From Table 2.2 in Section 2.3, we have $\cos 0 = 1$. Since 0 is in the interval $[0, \pi]$, we conclude that

$$\cos^{-1} 1 = 0.$$

(b) Letting $y = \cos^{-1}(-\frac{1}{2})$, we have

$$\cos y = -\frac{1}{2}, \quad 0 \le y \le \pi.$$

From Table 2.3 in Section 2.3, we know that $\cos(\pi/3) = \frac{1}{2}$. Therefore, using the fact that $\cos^{-1}(-x) = -\cos^{-1}x + \pi$, we write

$$\cos^{-1}(-\tfrac{1}{2}) = -\cos^{-1}\tfrac{1}{2} + \pi = -\frac{\pi}{3} + \pi = \frac{2\pi}{3}.$$

(c) Remember that the domain of the arccosine function $y = \arccos x$ is $[-1, 1]$. Thus, for $x > 1$ or $x < -1$, the arccosine function is undefined. Hence,

$$\arccos(-2) \text{ is undefined.} \quad \blacklozenge$$

To find the approximate value of $\arccos x$ or $\cos^{-1}x$ for $-1 \le x \le 1$, we set a calculator in radian mode and use the inverse cosine key, $\boxed{\cos^{-1}}$.

PROBLEM 2 Use a calculator to find the approximate value of each expression. Round each answer to four significant digits.

(a) $\arccos(0.809)$ (b) $\cos^{-1}(-0.42)$ $\blacklozenge$

Restricted Tangent Function and Its Inverse

Like the sine and cosine functions, the tangent function $y = \tan x$ is not one-to-one and does not have an inverse. However, consider the *restricted tangent function*

$$y = \tan x, \quad -\frac{\pi}{2} < x < \frac{\pi}{2}$$

whose graph is shown in Figure 2.79. Note that on the interval $(-\pi/2, \pi/2)$, the restricted tangent function attains every value of the tangent function once and only once. Thus, the restricted tangent function is one-to-one and has an inverse. We refer to the inverse of the restricted tangent function as the *inverse tangent function* and define it as follows.

Inverse Tangent Function

The **inverse tangent function** is defined as

$$y = \tan^{-1} x \quad \text{if and only if} \quad x = \tan y, \quad -\frac{\pi}{2} < y < \frac{\pi}{2}.$$

Note: The inverse tangent function is also referred to as the **arctangent function** and the notation arctan x is used frequently in place of $\tan^{-1} x$. Both $\tan^{-1} x$ and arctan x denote the real number (or angle in radians) between $-\pi/2$ and $\pi/2$ whose tangent is x.

The graph of the inverse tangent function (or arctangent function) is the same as the graph of the restricted tangent function (Figure 2.79) reflected in the line $y = x$. Figure 2.80 shows the graph of $y = \tan^{-1} x$. Note that the domain of $y = \tan^{-1} x$ is $(-\infty, \infty)$ and the range is $(-\pi/2, \pi/2)$. Also note that the graph of $y = \tan^{-1} x$ is

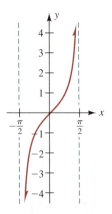

FIGURE 2.79
The restricted tangent function $y = \tan x, -\pi/2 < x < \pi/2$, is one-to-one and has an inverse.

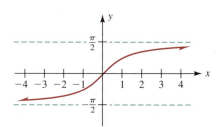

FIGURE 2.80
Graph of the inverse tangent function, $y = \tan^{-1} x$. The domain is $(-\infty, \infty)$ and the range is $(-\pi/2, \pi/2)$.

symmetric with respect to the origin, which indicates that it is an odd function. Thus, we have

$$\tan^{-1}(-x) = -\tan^{-1} x$$

for every real number x.

EXAMPLE 3 Find the exact value of each expression, if it is defined.

(a) $\arctan 0$ (b) $\tan^{-1} 1$ (c) $\tan^{-1}(-\sqrt{3})$

SOLUTION

(a) Letting $y = \arctan 0$, we have

$$\tan y = 0, \quad -\frac{\pi}{2} < y < \frac{\pi}{2}.$$

In Table 2.2 in Section 2.3, we read that $\tan 0 = 0$. Since 0 is in the interval $(-\pi/2, \pi/2)$, we conclude that

$$\arctan 0 = 0.$$

(b) Letting $y = \tan^{-1} 1$, we have

$$\tan y = 1, \quad -\frac{\pi}{2} < y < \frac{\pi}{2}.$$

From Table 2.3 in Section 2.3, we know $\tan(\pi/4) = 1$. Since $\pi/4$ is in the interval $(-\pi/2, \pi/2)$, we conclude that

$$\tan^{-1} 1 = \frac{\pi}{4}.$$

(c) Letting $y = \tan^{-1}(-\sqrt{3})$, we have

$$\tan y = -\sqrt{3}, \quad -\frac{\pi}{2} < y < \frac{\pi}{2}.$$

From Table 2.3 in Section 2.3, we know that $\tan(\pi/3) = \sqrt{3}$. Therefore, using the fact that $\tan^{-1}(-x) = -\tan^{-1} x$, we write

$$\tan^{-1}(-\sqrt{3}) = -\tan^{-1}(\sqrt{3}) = -\frac{\pi}{3}. \qquad \blacklozenge$$

To find the approximate value of $\arctan x$ or $\tan^{-1} x$, we set a calculator in radian mode and use the inverse tangent key, $\boxed{\text{TAN}^{-1}}$.

PROBLEM 3 Use a calculator to find the approximate value of each expression. Round each answer to four significant digits.

(a) arctan (0.977)   (b) $\tan^{-1}(-1.6)$ ◆

Note: The inverse cosecant, inverse secant, and inverse cotangent functions are rarely used. For a brief discussion of these three inverse trigonometric functions, see Exercises 57–60.

◆ Composing Trigonometric and Inverse Trigonometric Functions

Recall from Section 1.5 that if f and f^{-1} are a pair of inverse functions, then the result of composing f with f^{-1} in either order is the identity function, that is,

$$f(f^{-1}(x)) = x \quad \text{for all } x \text{ in the domain of } f^{-1}$$

and

$$f^{-1}(f(x)) = x, \quad \text{for all } x \text{ in the domain of } f.$$

If we apply this fact to the restricted trigonometric functions and their inverses, we obtain the following **trigonometric composition rules.**

Trigonometric Composition Rules

1. $\sin(\sin^{-1} x) = x$ for all x in the domain of the inverse sine function
2. $\sin^{-1}(\sin x) = x$ for all x in the domain of the restricted sine function
3. $\cos(\cos^{-1} x) = x$ for all x in the domain of the inverse cosine function
4. $\cos^{-1}(\cos x) = x$ for all x in the domain of the restricted cosine function
5. $\tan(\tan^{-1} x) = x$ for all x in the domain of the inverse tangent function
6. $\tan^{-1}(\tan x) = x$ for all x in the domain of the restricted tangent function

---

When using these composition rules, it is important to adhere to the domain restrictions. As illustrated in the next example, it may be necessary to rewrite the composition in an equivalent form before applying these rules.

EXAMPLE 4 Find the exact value of each expression, if it is defined.

(a) $\tan[\tan^{-1}(-6)]$ (b) $\sin(\arcsin \frac{3}{2})$

(c) $\cos^{-1}(\cos 3\pi)$ (d) $\arcsin\left(\sin \frac{5\pi}{4}\right)$

SOLUTION

(a) The domain of the inverse tangent function is $(-\infty, \infty)$. Since -6 is in the domain, we apply trigonometric composition rule 5, and write

$$\tan [\tan^{-1}(-6)] = -6.$$

(b) The domain of the arcsine function is $[-1, 1]$. Since $\frac{3}{2}$ is not in the domain, $\arcsin \frac{3}{2}$ is undefined. Hence,

$$\sin (\arcsin \tfrac{3}{2}) \quad \text{is undefined.}$$

(c) The domain of the restricted cosine function is $[0, \pi]$. Since 3π is not in the domain, we cannot apply trigonometric composition rule 4, that is,

$$\cos^{-1}(\cos 3\pi) \neq 3\pi.$$

However, we do know that $\cos 3\pi = \cos \pi = -1$, and since π is in the domain of the restricted cosine function, we may rewrite the expression and then apply trigonometric composition rule 4 as follows:

$$\cos^{-1}(\cos 3\pi) = \cos^{-1}(\cos \pi) = \pi.$$

(d) The domain of the restricted sine function is $[-\pi/2, \pi/2]$. Since $5\pi/4$ is not in the domain, we cannot apply trigonometric composition rule 2, that is

$$\arcsin \left(\sin \frac{5\pi}{4} \right) \neq \frac{5\pi}{4}.$$

However, we do know that $\sin (5\pi/4) = \sin (-\pi/4) = -1/\sqrt{2}$, and since $-\pi/4$ is in the domain of the restricted sine function, we may rewrite the expression and then apply trigonometric composition rule 2 as follows:

$$\arcsin \left(\sin \frac{5\pi}{4} \right) = \arcsin \left[\sin \left(-\frac{\pi}{4} \right) \right] = -\frac{\pi}{4}. \qquad \blacklozenge$$

PROBLEM 4 Repeat Example 4 for each expression.

(a) $\tan (\arctan 2)$ (b) $\sin^{-1}\left(\sin \dfrac{3\pi}{2} \right)$ $\blacklozenge$

We may use the trigonometric ratios for right triangles (in Section 2.2) to evaluate composite functions involving the trigonometric and inverse trigonometric functions. The procedure is illustrated in the next example.

EXAMPLE 5 Find the exact value of each expression.

(a) $\sin (\arctan \tfrac{1}{2})$ (b) $\sin [\arctan (-\tfrac{1}{2})]$

SOLUTION

(a) Letting $\theta = \arctan \frac{1}{2}$, we have

$$\tan \theta = \tfrac{1}{2}, \quad -\frac{\pi}{2} < \theta < \frac{\pi}{2}.$$

Since $\tan \theta > 0$, we may regard θ as an acute angle in a right triangle whose tangent is $\frac{1}{2}$. Now, by the trigonometric ratios for right triangles (Section 2.2), we have

$$\tan \theta = \frac{\text{opp}}{\text{adj}} = \frac{1}{2},$$

as shown in Figure 2.81. By the Pythagorean theorem, the length of the hypotenuse of this right triangle is $\sqrt{5}$ units. Thus,

$$\sin(\arctan \tfrac{1}{2}) = \sin \theta = \frac{\text{opp}}{\text{hyp}} = \frac{1}{\sqrt{5}}.$$

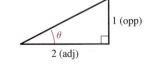

FIGURE 2.81
A right triangle with $\tan \theta = \frac{1}{2}$.

(b) To evaluate $\sin[\arctan(-\tfrac{1}{2})]$, we use the facts that

$$\arctan(-x) = -\arctan x \quad \text{and} \quad \sin(-x) = -\sin x$$

to obtain

$$\sin[\arctan(-\tfrac{1}{2})] = \sin(-\arctan \tfrac{1}{2}) = -\sin(\arctan \tfrac{1}{2}).$$

Using the result from part (a), we have

$$\sin[\arctan(-\tfrac{1}{2})] = -\sin(\arctan \tfrac{1}{2}) = -\frac{1}{\sqrt{5}}. \quad \blacklozenge$$

PROBLEM 5   Find the exact value of $\cos[\arctan(-\tfrac{1}{2})]$. $\quad \blacklozenge$

In calculus, it is necessary to express composite functions involving trigonometric and inverse trigonometric functions in x as algebraic expressions in x. We may use either the right-triangle approach illustrated in Example 5 or the fundamental trigonometric identities (Section 2.2) to accomplish this task, as illustrated in the following example.

EXAMPLE 6   Express $\cos(\sin^{-1} 2x)$ as an algebraic expression in x.

SOLUTION   Letting $y = \sin^{-1} 2x$, we have

$$\sin y = 2x, \quad -\frac{\pi}{2} \leq y \leq \frac{\pi}{2}.$$

SECTION 2.6 The Inverse Trigonometric Functions

If $-\pi/2 \leq y \leq \pi/2$, then y is in quadrant I or IV. Hence, it follows that $\cos y \geq 0$. Now using the trigonometric identity $\sin^2 y + \cos^2 y = 1$, we have

$$\cos y = \sqrt{1 - \sin^2 y}$$
$$= \sqrt{1 - (2x)^2}$$
$$= \sqrt{1 - 4x^2}$$

Hence,

$$\cos(\sin^{-1} 2x) = \cos y = \sqrt{1 - 4x^2}, \quad -\tfrac{1}{2} \leq x \leq \tfrac{1}{2} \quad \blacklozenge$$

PROBLEM 6 Use the right-triangle approach illustrated in Example 5 to express $\cos(\sin^{-1} 2x)$ as an algebraic expression in x. You should obtain the same result found in Example 6. ◆

◆ **Application: Stealing Second Base**

We conclude this section with an applied problem that involves an inverse trigonometric function. We will discuss additional applications of the trigonometric functions and their inverses in Chapter 3.

EXAMPLE 7 During a baseball game, a runner on first base attempts to steal second base. Given that a baseball diamond is a square with 90-foot base paths (sides) and that θ is the angle between the first-base path and a line from home plate to the runner,

(a) express angle θ as a function of the runner's distance d from home plate, and

(b) use the function described in part (a) to find the measure of angle θ when $d = 100$ ft.

SOLUTION

(a) By making a sketch of the information, we see that a right triangle is formed, as shown in Figure 2.82. Note that in relation to angle θ, the first-base path is the adjacent side of the right triangle (with length 90 ft) and d is the hypotenuse. Thus, using the trigonometric ratios for right triangles, we can state that

$$\cos \theta = \frac{\text{adj}}{\text{hyp}} = \frac{90}{d}.$$

Since θ is an acute angle $\left(0 < \theta < \dfrac{\pi}{2}\right)$, θ is in the domain of the restricted cosine function. Hence, we may use its inverse to write

$$\theta = \arccos \frac{90}{d}.$$

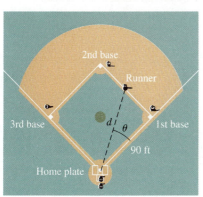

FIGURE 2.82
The quantities d, θ, and 90 are parts of a right triangle and $\theta = \arccos \dfrac{90}{d}$.

Exercises 2.6

Basic Skills

In Exercises 1–16, find the exact value of each expression, if it is defined.

1. $\sin^{-1} \frac{1}{2}$
2. $\cos^{-1} \frac{1}{2}$
3. $\arctan \sqrt{3}$
4. $\arcsin 1$
5. $\arccos \left(\frac{1}{\sqrt{2}}\right)$
6. $\cos^{-1} \frac{\sqrt{3}}{2}$
7. $\sin^{-1}\left(-\frac{\sqrt{3}}{2}\right)$
8. $\tan^{-1}\left(-\frac{1}{\sqrt{3}}\right)$
9. $\tan^{-1}(-1)$
10. $\arcsin\left(-\frac{1}{2}\right)$
11. $\cos^{-1}\left(-\frac{1}{\sqrt{2}}\right)$
12. $\cos^{-1}\left(-\frac{\sqrt{3}}{2}\right)$
13. $\cos^{-1} 0$
14. $\arcsin \sqrt{3}$
15. $\arccos(-5)$
16. $\cos^{-1}(-1)$

In Exercises 17–40, find the exact value of each expression, if it is defined.

17. $\sin(\sin^{-1} \frac{2}{3})$
18. $\tan(\tan^{-1} 1.6)$
19. $\cos[\arccos(-12)]$
20. $\cos[\arccos(-\frac{3}{5})]$
21. $\sin^{-1}\left(\sin \frac{\pi}{8}\right)$
22. $\arccos(\cos 2)$
23. $\tan^{-1}[\tan(-1)]$
24. $\arctan\left(\tan \frac{\pi}{2}\right)$
25. $\arccos(\cos 5\pi)$
26. $\arctan\left(\tan \frac{5\pi}{3}\right)$
27. $\sin^{-1}\left(\sin \frac{3\pi}{4}\right)$
28. $\cos^{-1}\left[\cos\left(-\frac{\pi}{2}\right)\right]$
29. $\sin^{-1}\left(\sin \frac{9\pi}{10}\right)$
30. $\cos^{-1}\left(\cos \frac{9\pi}{5}\right)$
31. $\tan^{-1}\left(\tan \frac{5\pi}{8}\right)$
32. $\cos^{-1}\left(\cos \frac{11\pi}{9}\right)$
33. $\sin(\arctan 3)$
34. $\sin(\arccos \frac{12}{13})$
35. $\cos(\sin^{-1} \frac{2}{3})$
36. $\cos(\tan^{-1} 0.3)$
37. $\csc[\tan^{-1}(-\frac{2}{3})]$
38. $\sec\left[\sin^{-1}\left(-\frac{\sqrt{7}}{4}\right)\right]$
39. $\cot[\arccos(-\frac{1}{2})]$
40. $\tan[\sin^{-1}(-\sqrt{2})]$

In Exercises 41–48, write each trigonometric expression as an algebraic expression in x.

41. $\sin(\cos^{-1} x)$
42. $\cos(\sin^{-1} 3x)$
43. $\tan\left(\arcsin \frac{1}{x}\right)$
44. $\sec\left(\arctan \frac{x}{2}\right)$
45. $\sec[\cos^{-1}(x-1)]$
46. $\cot[\tan^{-1}(x+2)]$
47. $\cos\left[\arctan\left(\frac{1}{x-1}\right)\right]$
48. $\csc\left[\arccos\left(\frac{x}{\sqrt{x^2+4}}\right)\right]$

In Exercises 49–54, use the shift rules, axis reflection rules, and stretch-and-shrink rules (in Section 1.4) to sketch the graph of each function.

49. $f(x) = -\arcsin x$
50. $f(x) = 2 \arctan(-x)$
51. $f(x) = \tan^{-1}(x-2)$
52. $f(x) = \cos^{-1}(x-1)$
53. $f(x) = \frac{1}{2} \arccos x - \frac{\pi}{2}$
54. $f(x) = \pi - \sin^{-1}(x+1)$

55. An observer is 550 feet from the launchpad when a rocket is fired vertically upward, as shown in the sketch.

   (a) Express angle θ as a function of the height h of the rocket.

   (b) Determine the domain and range of the function described in part (a).

   (c) Use the answer from part (a) to find the exact value of θ (in radians) when $h = 550$ ft.

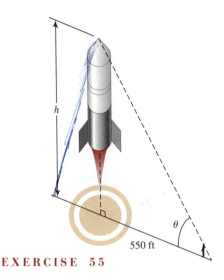

56. A 32-foot ladder is leaned against the side of a building, with the foot of the ladder and the base of the building on level ground.

(a) Express the acute angle θ between the ladder and the ground as a function of the distance x from the foot of the ladder to the base of the building.

(b) Determine the domain and range of the function described in part (a).

(c) Use the answer from part (a) to find the exact value of θ (in radians) when $x = 16$ ft.

EXERCISE 55

 Critical Thinking

57. The inverse of the *restricted cosecant function*

$$y = \csc x, \quad -\frac{\pi}{2} \leq x \leq \frac{\pi}{2}, \quad x \neq 0,$$

is called the *inverse cosecant function* and is defined as

$$y = \csc^{-1} x$$

if and only if $\quad \csc x = y, \quad -\frac{\pi}{2} \leq y \leq \frac{\pi}{2}, \quad y \neq 0.$

(a) Sketch the graph of the restricted cosecant function, and show that it is one-to-one.

(b) Sketch the graph of the inverse cosecant function, and state its domain and range.

58. The inverse of the *restricted secant function*

$$y = \sec x, \quad 0 \leq x \leq \pi, \quad x \neq \frac{\pi}{2},$$

is called the *inverse secant function* and is defined as

$$y = \sec^{-1} x$$

if and only if $\quad \sec x = y, \quad 0 \leq y \leq \pi, \quad y \neq \frac{\pi}{2}.$

(a) Sketch the graph of the restricted secant function, and show that it is one-to-one.

(b) Sketch the graph of the inverse secant function, and state its domain and range.

59. The inverse of the *restricted cotangent function*

$$y = \cot x, \quad 0 < x < \pi,$$

is called the *inverse cotangent function* and is defined as

$$y = \cot^{-1} x$$

if and only if $\quad \cot x = y, \quad 0 < y < \pi.$

(a) Sketch the graph of the restricted cotangent function, and show that it is one-to-one.

(b) Sketch the graph of the inverse cotangent function, and state its domain and range.

60. Use the definitions in Exercises 57–59 to find the exact value of each expression.

(a) $\csc^{-1} 2$ (b) $\cot^{-1}(-1)$

(c) $\sec(\sec^{-1} 2.63)$ (d) $\csc^{-1}\left(\csc \dfrac{5\pi}{3}\right)$

61. Sketch the graph of $f(t) = \cos(-t + \pi)$, and compare it to the graphs of $y = \sin t$, $y = -\sin t$, $y = \cos t$, and $y = -\cos t$.

(a) Based upon your comparisons, state a trigonometric identity that relates $f(t)$ to either

$$y = \sin t, \ y = -\sin t, \ y = \cos t, \text{ or } y = -\cos t.$$

EXERCISE 61 *(continued)*

(b) Let $t = \cos^{-1} x$ in the identity found in part (a), and show that

$$\cos^{-1}(-x) = -\cos^{-1} x + \pi,$$

provided x is in the interval $[-1, 1]$.

62. Sketch the graph of $f(t) = \cos\left(-t + \dfrac{\pi}{2}\right)$, and compare it to the graphs of $y = \sin t$, $y = -\sin t$, $y = \cos t$, and $y = -\cos t$.

   (a) Based upon your comparisons, state a trigonometric identity that relates $f(t)$ to either

   $$y = \sin t, \; y = -\sin t, \; y = \cos t, \; \text{or } y = -\cos t.$$

   (b) Let $t = \sin^{-1} x$ in the identity from part (a) and show that

   $$\sin^{-1} x + \cos^{-1} x = \dfrac{\pi}{2},$$

   provided x is in the interval $[-1, 1]$.

63. Find a value of x for which $\tan^{-1}(\tan x) \neq x$. Does this contradict the inverse function concept that states that $f^{-1}(f(x)) = x$? Explain.

64. From the fundamental trigonometric identities (Section 2.2) we know that

$$\sin^2 x + \cos^2 x = 1 \quad \text{and} \quad \tan x = \dfrac{\sin x}{\cos x}.$$

Do these identities hold when we replace $\sin x$, $\cos x$, and $\tan x$ with their corresponding inverse trigonometric functions? Use numerical examples to support your conclusion.

65. Explain the difference between the notations $\sin^{-1} x$ and $(\sin x)^{-1}$.

66. Given that $\csc x = k$ with $k > 1$, express x in terms of the inverse sine function.

In Exercises 67–70, solve each equation for x.

67. $\sin^{-1}(3x - 2) = \dfrac{\pi}{6}$

68. $\cos^{-1}(x^2 - 5x + 3) = \pi$

69. $4 \arctan (2x^2 - 3x) = -\pi$

70. $2 \arcsin (x - 3) = 3\pi$

Calculator Activities

In Exercises 71–86, use a calculator to find the approximate value of each expression, if it exists. Round each answer to four significant digits.

71. arcsin 0.28
72. arccos 0.92
73. $\tan^{-1} 8$
74. $\sin^{-1}(-0.6)$
75. $\cos^{-1}(-0.135)$
76. arctan 2.876
77. arcsin 0.9501
78. $\cos^{-1}(-0.045)$
79. arctan $\left(-\dfrac{5}{12}\right)$
80. $\cos^{-1} \dfrac{\sqrt{5}}{3}$
81. arcsin $\left(-\dfrac{4}{3}\right)$
82. $\cos^{-1} \pi$
83. $\sin [\sin^{-1}(0.887)]$
84. arccos (cos 3.156)
85. arctan [tan (−7)]
86. $\cos [\sin^{-1}(-.6781)]$

87. For the flight of stairs shown in the sketch,

   (a) express the acute angle θ as a function of the tread length t,

   (b) determine the domain and range of the function, defined in part (a), and

   (c) use the answer from part (a) and a calculator to find the approximate value of θ (in degrees) if $t = 12$ inches.

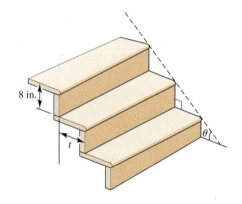

88. A pendulum hangs on a 1-meter string that is attached to the ceiling of a room. When the pendulum is pulled to one side, an angle θ is formed between the new position and its original, vertical position, as shown in the sketch.

   (a) Express angle θ as a function of the height h above its lowest position.

   (b) Determine the domain and range of the function defined in part (a).

   (c) Use the answer from part (a) and a calculator to find the approximate value of θ (in radians) when $h = 0.46$ meter.

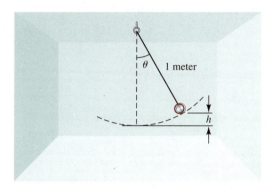

89. Use a graphing calculator to generate the graphs of $y = 1/x$ and $y = \arctan x$ in the same viewing rectangle. Then use the calculator's trace and zoom features to find all solutions of the equation $\arctan x = 1/x$. Round each answer to three significant digits.

90. Use a graphing calculator or a computer with graphing capabilities to generate the graph of the polynomial function

$$P(x) = x - \frac{x^3}{3} + \frac{x^5}{5} - \frac{x^7}{7} + \frac{x^9}{9}$$

   over the interval $[-1, 1]$.

   (a) What inverse trigonometric function does P seem to approximate?

   (b) Compare $P(0.5)$ to the value of that inverse trigonometric function when $x = 0.5$. Round each answer to four significant digits.

Chapter 2 Review

Questions for Group Discussion

1. When is an angle in *standard position*?
2. How are *negative angles* developed?
3. State the procedure for converting *radians* to *degrees*. Illustrate with an example.
4. State the procedure for finding the measures of angles that are *coterminal* with an angle measured in radians.
5. Discuss the procedure for converting angles measured in *degrees, minutes, seconds* to decimal degrees.
6. What is the measure of a *central angle* that intercepts an arc of length r on a circle with radius r?
7. Discuss the *algebraic signs of the trigonometric functions* of angle θ when θ lies in quadrant I, quadrant II, quadrant III, and quadrant IV.
8. List the eight *fundamental trigonometric identities*. Using $\theta = 30°$, verify each identity.
9. Explain the difference between $\sin^2 \theta$ and $\sin \theta^2$.
10. For a given angle θ, is it possible to have $\cos \theta > 0$ and $\sec \theta < 0$? Explain.
11. Explain the procedure for finding the reference angle θ' for angle θ given that $0 \leq \theta \leq 2\pi$.

12. Given that θ is an acute angle, express the trigonometric functions of θ in terms of their *cofunctions*.
13. What is a *quadrantal angle*? How can the trigonometric functions of a quadrantal angle be found without using a calculator?
14. Discuss the procedure for finding the *trigonometric functions of a real number* by using a calculator.
15. What are the *real number domains* of the trigonometric functions?
16. State the *range* of each trigonometric function.
17. Which trigonometric functions are *even* and which are *odd*?
18. How is the graph of $y = \sin x$ related to the graph of $y = \cos x$?
19. How is the graph of $y = \tan x$ related to the graph of $y = \cot x$?
20. Is each trigonometric function *periodic*? Give the *period* of each trigonometric function that is periodic.
21. As $t \to \frac{\pi}{2}^-$, discuss what happens to each function:
    (a) $\sin t$ (b) $\cos t$ (c) $\tan t$ (d) $\sec t$
22. Discuss *amplitude*, *period*, and *phase shift* in relation to the graph of $y = a \sin(bx + c)$.
23. Why must the domain of the sine, cosine, and tangent functions be restricted in order to define inverses for these functions?
24. Discuss the domain and range of the *inverse sine function*, the *inverse cosine function*, and the *inverse tangent function*. List some properties of these inverse functions.
25. State the conditions under which each of these relationships is true:
    $\sin(\arcsin x) = x$, $\arccos(\cos x) = x$
26. Which pair of trigonometric functions are *reciprocals* as well as *cofunctions* of each other?
27. Try using a calculator to evaluate $\tan 90°$. Explain the result.
28. If $-2\pi \leq x \leq 2\pi$, how many different values of x are there for which $\sin x = \frac{1}{2}$? for which $\sin x = 2$? Explain.

Review Exercises

In Exercises 1–6, an angle is described in terms of revolutions.

(a) Draw the given angle in standard position.
(b) Give the degree measure and radian measure of the given angle.
(c) Find the degree measure and radian measure of the smallest positive angle that is coterminal with the given angle.
(d) Find the degree measure and radian measure of the reference angle for the given angle.

1. one and two-thirds revolutions counterclockwise
2. one and one-ninth revolutions counterclockwise
3. two and two-fifths revolutions counterclockwise
4. one and five-sixths revolutions counterclockwise

CHAPTER 2 Review — 4 graphing problems

5. one and one-twelfth revolutions clockwise

6. three and three-eighths revolutions clockwise

In Exercises 7–10, convert the degree measure to radian measure. Write the answer as a reduced fraction in terms of π.

7. 200° **8.** 75° **9.** 780° **10.** −405°

In Exercises 11–14, convert the radian measure to degree measure.

11. $\dfrac{11\pi}{6}$ **12.** $\dfrac{5\pi}{18}$ **13.** -8π **14.** $\dfrac{31\pi}{10}$

15. Convert 123°42′18″ to (a) decimal degrees and (b) radians (to the nearest thousandth).

16. Convert 57.975° to (a) degrees, minutes, seconds and (b) radians (to the nearest thousandth).

17. Convert 2.65 radians to (a) decimal degrees (to the nearest thousandth) and (b) degrees, minutes, seconds.

18. Find the radian measure of the smallest positive angle that is coterminal with 456°36′.

In Exercises 19–22, a point on the terminal side of θ is shown. Find the values of the six trigonometric functions of θ.

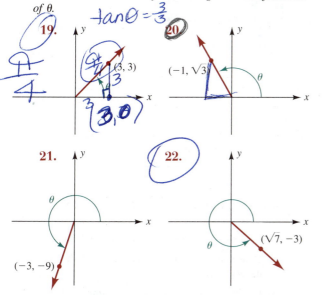

In Exercises 23–30, find the exact value of the indicated trigonometric function from the given information.

23. $\tan\theta = \tfrac{1}{2}$; find $\cot\theta$.

24. $\sec\theta = -4$; find $\cos\theta$.

25. $\sin\theta = \tfrac{4}{5}$, θ in quadrant II; find $\cos\theta$.

26. $\cos\theta = -\tfrac{2}{3}$, $\tan\theta > 0$; find $\sin\theta$.

27. $\tan\theta = \sqrt{2}$, $\sin\theta < 0$; find $\sec\theta$.

28. $\csc\theta = 2$, $\tan\theta < 0$; find $\cot\theta$.

29. $\sin\theta = -\tfrac{12}{13}$, $\cos\theta > 0$; find $\tan\theta$.

30. $\cot\theta = -\sqrt{3}$, θ in quadrant IV; find $\cos\theta$.

In Exercises 31–34, use the right triangle in the sketch to find the exact value of the indicated trigonometric function.

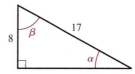

31. $\sin\alpha$ **32.** $\cos\alpha$ **33.** $\tan\beta$ **34.** $\sec\beta$

In Exercises 35–50, find the exact value of each expression, if it is defined.

35. sin 210° **36.** cos 225°

37. tan 420° **38.** cot (−540°)

39. $\sec\dfrac{7\pi}{4}$ **40.** $\csc\dfrac{2\pi}{3}$

41. $\cos(-3\pi)$ **42.** $\sin\dfrac{13\pi}{6}$

43. $\cos^{-1}\tfrac{1}{2}$ **44.** $\arctan(-1)$

45. $\arcsin\tfrac{3}{2}$ **46.** $\sin^{-1}\left(-\dfrac{1}{\sqrt{2}}\right)$

47. $\arctan\left(\tan\dfrac{\pi}{3}\right)$ **48.** $\sin^{-1}\left(\sin\dfrac{5\pi}{6}\right)$

49. $\sin(\cos^{-1}\tfrac{1}{3})$ **50.** $\cos(\arctan 4)$ = .24

In Exercises 51–62, find the approximate value of each expression. Round to four significant digits.

51. cos 72° **52.** sin 305°

53. cot 132°27′ **54.** tan (−24°22′36″)

55. csc 1.24 **56.** sec 12

57. $\sin\left(-\dfrac{7\pi}{8}\right)$ **58.** $\cos\dfrac{17\pi}{12}$

59. $\tan^{-1} 2$ **60.** $\sin^{-1}(-\tfrac{2}{3})$

61. arccos (−0.4561) **62.** arctan 1.254

In Exercises 63–66, compute the functional value given that

$$f(t) = \sin t, \quad g(t) = 2 \cos 3t, \quad h(t) = \tan (t - 1).$$

Round each answer to four significant digits.

63. $f(2)$
64. $g(-\frac{5}{3})$
65. $h(4)$
66. $g(h(2))$

Given the functions f and g in Exercises 67–70, define the composite function $f \circ g$ by using an algebraic expression.

67. $f(x) = \sin x, \quad g(x) = \cos^{-1} 4x$
68. $f(x) = \cos x, \quad g(x) = \tan^{-1} \frac{x}{2}$
69. $f(x) = \tan x, \quad g(x) = \arcsin \frac{1}{\sqrt{x}}$
70. $f(x) = \csc x, \quad g(x) = \arccos (x - 2)$

In Exercises 71–90, sketch a graph of each function.

71. $y = 3 \cos 4x$
72. $y = 2 \sin \frac{2x}{5}$
73. $y = 2 \tan \frac{\pi x}{2}$
74. $y = \sec 2x$
75. $y = -\csc \frac{3x}{4}$
76. $y = -3 \cot \pi x$
77. $y = \frac{5}{4} \sin (-3x)$
78. $y = 2 \cos (-0.3x)$
79. $y = \cos \left(x - \frac{\pi}{3} \right)$
80. $y = \frac{1}{3} \sin \left(\pi x + \frac{\pi}{12} \right)$
81. $y = 0.5 \cot (2x + 0.125\pi)$
82. $y = -2 \tan \left(3x - \frac{2\pi}{3} \right)$
83. $y = 3 + \sec \frac{\pi x}{6}$
84. $y = 1 - \csc 3x$
85. $y = 2 - 3 \sin \left(2x - \frac{\pi}{4} \right)$
86. $y = 1 + \cos \left(\frac{x}{2} + \pi \right)$
87. $y = -\arccos x$
88. $y = 2 \arcsin \frac{x}{3}$
89. $y = \tan^{-1}(x + 1)$
90. $y = \frac{\pi}{4} - \cos^{-1} x$

91. Montreal, Canada, is approximately due north of Santiago, Chile. Montreal is at latitude 45° North and Santiago is at latitude 33° South, as shown in the figure. If the radius of the earth is approximately 4000 miles, what is the distance between the two cities?

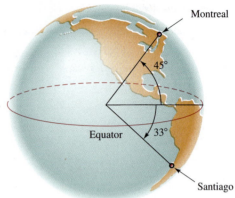

EXERCISE 91

92. What is the linear speed (in miles per hour) of the earth in its orbit around the sun? Assume that the earth's orbit is circular with a radius of 9.3×10^7 miles.

93. Express the length L of the *chord* in the sketch as a function of the central angle θ. Then complete the following table:

θ	1	$\frac{\pi}{12}$		
L			5	8

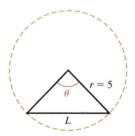

94. Three drill holes are located in an aluminum plate, as shown in the sketch.

(a) Express θ as a function of the distance x.

(b) Determine the domain and range of the function described in part (a).

(c) Use the answer from part (a) to find θ when $x = 18\sqrt{3}$ cm.

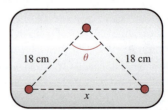

95. Four straight roads intersect, as shown in the sketch.

(a) Express α as a function of the distance x.

(b) Express β as a function of the distance x.

(c) Express θ as a function of the distance x.

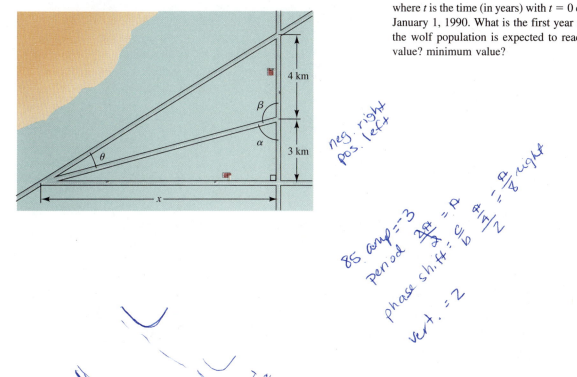

96. The wolf population P in a national forest in Canada is given by

$$P = 340 + 180 \cos 0.785t,$$

where t is the time (in years) with $t = 0$ corresponding to January 1, 1990. What is the first year after 1990 when the wolf population is expected to reach its maximum value? minimum value?

CHAPTER 3

Applications of the Trigonometric Functions

On a golf course, the distance from a tee to a hole is 355 yards. Suppose a golfer drives the ball 190 yd down the right side of the fairway at an angle of 20° from the center of the fairway. How far is the ball from the hole?

(For the solution, see Example 5 in Section 3.3.)

3.1 **Applications Involving Right Triangles and Harmonic Motion**
3.2 **Law of Sines**
3.3 **Law of Cosines**
3.4 **Vectors**

3.1 Applications Involving Right Triangles and Harmonic Motion

The trigonometric functions have several applications. In this section we discuss applied problems involving right triangles and *harmonic motion,* or periodic motion about an equilibrium position. For the right-triangle problems we use the trigonometric functions of angles, and for the harmonic motion problems we use the trigonometric functions of real numbers. In the remainder of this chapter, we discuss additional applications of the trigonometric functions, such as those involving oblique triangles and vectors.

◆ Solving Right Triangles

We begin with problems requiring that we find the unknown parts of a right triangle when given either one side and one acute angle or two sides of the right triangle. The procedure is called *solving a right triangle.* The Pythagorean theorem and the trigonometric ratios for right triangles (Section 2.2) are used to solve a right triangle. For our convenience, these ratios are listed once again.

Trigonometric Ratios for Right Triangles

If θ is an acute angle in a right triangle, then

$$\sin \theta = \frac{\text{opp}}{\text{hyp}}, \quad \cos \theta = \frac{\text{adj}}{\text{hyp}}, \quad \tan \theta = \frac{\text{opp}}{\text{adj}},$$

$$\csc \theta = \frac{\text{hyp}}{\text{opp}}, \quad \sec \theta = \frac{\text{hyp}}{\text{adj}}, \quad \cot \theta = \frac{\text{adj}}{\text{opp}},$$

where *opp* is the length of the side opposite θ, *adj* is the length of the side adjacent to θ, and *hyp* is the length of the hypotenuse of the right triangle.

In the first example, we solve a right triangle in which one of the sides and one of the acute angles are known. It is best to use the sine, cosine, or tangent function to solve a right triangle, since each of these functions can be evaluated directly by a key on a calculator.

EXAMPLE 1 Solve the right triangle shown in Figure 3.1. Round x and y to three significant digits.

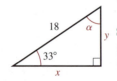

FIGURE 3.1

SOLUTION To solve the right triangle means to find the missing parts—in this case, angle α and sides x and y. Since the acute angles in a right triangle are complementary, we know that

$$\alpha = 90° - 33° = 57°.$$

In relation to the given angle (33°), x is the adjacent side, y is the opposite side, and 18 is the hypotenuse. Thus, using the trigonometric ratios for right triangles,

$$\cos \theta = \frac{\text{adj}}{\text{hyp}} \quad \text{and} \quad \sin \theta = \frac{\text{opp}}{\text{hyp}},$$

we have

$$\cos 33° = \frac{x}{18} \qquad \sin 33° = \frac{y}{18}$$

$$x = 18 \cos 33° \qquad y = 18 \sin 33° \qquad \textbf{Multiply both sides by 18}$$

Using the sine key, $\boxed{\text{SIN}}$, and cosine key, $\boxed{\text{COS}}$, on a calculator set in degree mode, we obtain

$$x = 18 \cos 33° \approx 15.1 \quad \text{and} \quad y = 18 \sin 33° \approx 9.80.$$

It is good practice to check the values of x and y by applying the Pythagorean theorem. Use a calculator to show that $18^2 \approx 15.1^2 + 9.80^2$. ◆

PROBLEM 1 Solve the right triangle shown in Figure 3.2. Round x and z to three significant digits.

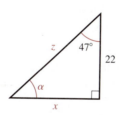

FIGURE 3.2

In the next example, we solve a right triangle in which two of the sides are known. We may apply the inverse trigonometric functions to solve problems of this type. ◆

EXAMPLE 2 Solve the right triangle shown in Figure 3.3. Round α and β to the nearest tenth of a degree.

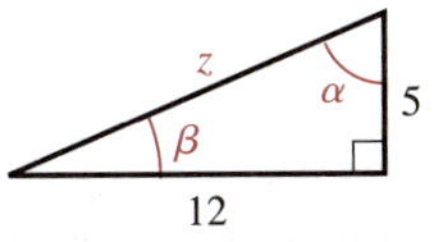

SOLUTION To solve the right triangle means to find the missing parts—in this triangle, angles α and β and side z. By the Pythagorean theorem, we have

$$z^2 = 5^2 + 12^2$$
$$z^2 = 169$$
$$z = \sqrt{169} = 13$$

FIGURE 3.3

In relation to angle α, 12 is the opposite side and 5 is the adjacent side. Thus, by the trigonometric ratios for right triangles,

$$\tan \alpha = \frac{\text{opp}}{\text{adj}} = \frac{12}{5}.$$

Since α is an acute angle, α is in the domain of the restricted tangent function. Hence, we may write

$$\alpha = \arctan \tfrac{12}{5}$$

Using the inverse tangent key, $\boxed{\text{TAN}^{-1}}$, on a calculator set in degree mode, we obtain

$$\alpha = \arctan \tfrac{12}{5} \approx 67.4°$$

Since the acute angles in a right triangle are complementary, we find

$$\beta = 90° - \alpha \approx 22.6°. \qquad \blacklozenge$$

PROBLEM 2 Solve the right triangle shown in the Figure 3.4. Round α and β to the nearest tenth of a degree and round x to three significant digits. $\qquad \blacklozenge$

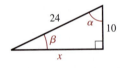

FIGURE 3.4

Applied Problems Involving Right Triangles

The next three examples illustrate applied problems that are solved by finding missing parts of right triangles.

EXAMPLE 3 To sharpen the cutting teeth of the saw blade shown in Figure 3.5, we must find the measure of angle β. Find β to the nearest tenth of a degree.

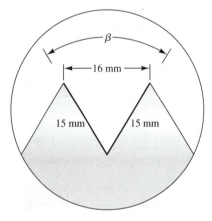

FIGURE 3.5

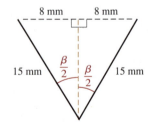

FIGURE 3.6

The altitude to the base of an isosceles triangle cuts both the vertex angle and base in half.

SOLUTION Note that the open part of the tooth forms an isosceles triangle. Recall from elementary geometry that if we construct an altitude from the vertex angle to the base of an isosceles triangle, both the vertex angle β and the base (16 mm) are cut in half, as shown in Figure 3.6. Working with the right triangle in Figure 3.6 we find that, in relation to angle $\beta/2$, 8 is the opposite side and 15 is the hypotenuse. Thus, by the trigonometric ratios for right triangles,

$$\sin \frac{\beta}{2} = \frac{\text{opp}}{\text{hyp}} = \frac{8}{15}.$$

Since $\beta/2$ is an acute angle, $\beta/2$ is in the domain of the restricted sine function. Hence, we may write

$$\frac{\beta}{2} = \arcsin \frac{8}{15} \quad \text{or} \quad \beta = 2 \arcsin \frac{8}{15}.$$

Using the inverse sine key, $\boxed{\text{SIN}^{-1}}$, on a calculator set in degree mode, we obtain

$$\beta = 2 \arcsin \tfrac{8}{15} \approx 64.4°.$$

PROBLEM 3 For the cutting teeth of the saw blade shown in Figure 3.7, find the distance x. Round to three significant digits.

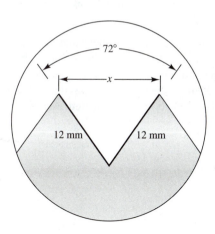

FIGURE 3.7

EXAMPLE 4 The chair lift at a ski area rises at an angle of 19.2° and attains a vertical rise of 1250 ft in 15 minutes. What is the speed (in mph) of the chair lift?

SOLUTION Letting

s = the distance (in feet) the lift travels,

170   CHAPTER 3   Applications of the Trigonometric Functions

we set up a sketch of the given information as shown in Figure 3.8. In relation to the given angle (19.2°), 1250 ft is the opposite side and s is the hypotenuse. Thus, using the fact that sin θ = opp/hyp, we have

$$\sin 19.2° = \frac{1250}{s}$$

$$s \sin 19.2° = 1250 \qquad \text{Multiply both sides by } s$$

$$s = \frac{1250}{\sin 19.2°} \text{ ft} \qquad \text{Divide both sides by sin 19.2°}$$

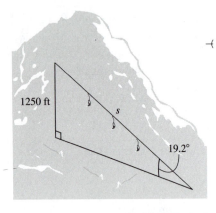

FIGURE 3.8
A right triangle in which the side opposite the given acute angle is known and the hypotenuse is to be determined.

Recall from Section 2.1 that linear speed v is defined as the distance s travels per unit of time t. Thus, the speed of the chair lift is

$$v = \frac{s}{t} = \frac{(1250/\sin 19.2°) \text{ ft}}{15 \text{ min}} \cdot \frac{60 \text{ min}}{1 \text{ hr}} \cdot \frac{1 \text{ mi}}{5280 \text{ ft}} \approx 2.9 \text{ mph}.$$

conversion factors ◆

PROBLEM 4   If the chair lift in Example 4 rises at an angle of 15.5° and has a speed of 3 mph, find the time (in minutes) required to attain a vertical rise of 1800 ft. ◆

A *transit* is an instrument that can be used to measure an angle from the horizontal to a point that is either above or below the horizontal. If the point sighted is above the horizontal, as shown in Figure 3.9(a), the angle is called an **angle of elevation.** If the point sighted is below the horizontal, as shown in Figure 3.9(b), the angle is called an **angle of depression.**

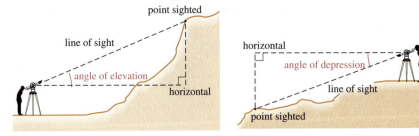

FIGURE 3.9
An angle of elevation and an angle of depression

EXAMPLE 5   The horizontal distance from a transit to a flagpole is 32.6 meters. The *angle of elevation* from the transit to the top of the flagpole is 38°27′ and the angle of depression from the transit to the bottom of the flagpole is 6°30′. Find the height of the flagpole.

SOLUTION   Letting

$h =$ the height (in meters) of the flagpole,

we set up a sketch of the given information as shown in Figure 3.10. In relation to the angle of elevation (38°27′), a is the opposite side and 32.6 is the adjacent

FIGURE 3.10
The height h of the flagpole is the sum of the legs a and b.

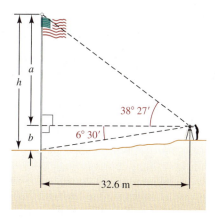

side. In relation to the angle of depression (6°30′), b is the opposite side and 32.6 is the adjacent side. Thus, using the fact that tan θ = opp/adj, we have

$$\tan 38°27' = \frac{a}{32.6} \quad \text{and} \quad \tan 6°30' = \frac{b}{32.6}$$

$$a = 32.6 \tan 38°27' \qquad b = 32.6 \tan 6°30'$$

Thus, the height h of the flagpole is

$$\begin{aligned} h &= a + b \\ &= 32.6 \tan 38°27' + 32.6 \tan 6°30' \\ &\approx 29.6 \text{ m}. \end{aligned}$$ ◆

PROBLEM 5 If the horizontal distance from the transit to the flagpole in Example 5 is 42.8 meters and the angles of elevation to the bottom and the top of the flagpole are 5°36′ and 22°45′, respectively, find the height of the pole. ◆

◆ **Simple Harmonic Motion**

Thus far we have looked at applied problems involving the trigonometric functions of angles. We now turn our attention to applied problems involving the trigonometric functions of real numbers. Specifically, we look at periodic motion that is symmetric about an equilibrium position. We refer to motion of this type as **simple harmonic motion.**

Consider a spring with an attached weight, as shown in Figure 3.11. If the weight is pulled down and then released, the weight will oscillate back and forth through the equilibrium position. If we ignore frictional forces, this oscillating motion will repeat itself over and over again.

From Figure 3.11 it appears that we may be able to describe the displacement of the weight from its equilibrium position by a sine or cosine function. Suppose we consider the maximum displacement of the weight from its equilibrium position as the amplitude. Also, suppose we consider the time required for the weight to travel from its maximum displacement below the equilibrium position to its maximum displacement above the equilibrium position and back again to its maximum displacement below the equilibrium position as the period. Now, we can express the

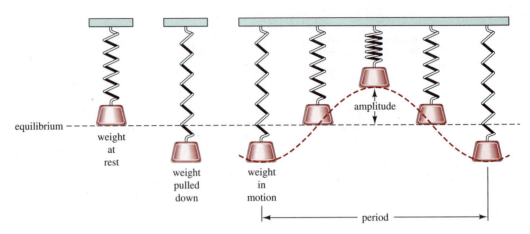

FIGURE 3.11 An illustration of simple harmonic motion

displacement d of the weight from its equilibrium position as a function of time t by writing

$$d = a \sin bt \quad \text{or} \quad d = a \cos bt,$$

where a is the amplitude and b is related to the period P by the formula $b = 2\pi/P$. Since the period P is the time required for one complete oscillation, the reciprocal of P must represent the number of oscillations per unit of time. We refer to $1/P$ as the **frequency** f. Hence,

$$b = \frac{2\pi}{P} = 2\pi \cdot \frac{1}{P} = 2\pi f.$$

In summary, the mathematical models for simple harmonic motion are

$$d = a \sin 2\pi ft \quad \text{or} \quad d = a \cos 2\pi ft$$

where

- d is the displacement from the equilibrium position after t units of time,
- $|a|$ is the maximum displacement, and
- f is the frequency, the number of oscillations per unit of time, $f = 1/P$.

Which model we use depends on the displacement d of the weight at $t = 0$. If $d = 0$ at $t = 0$ we use $d = a \sin 2\pi ft$, and if $d = a$ at $t = 0$ we use $d = a \cos 2\pi ft$.

EXAMPLE 6 A weight attached to a spring is pulled down 15 cm from its equilibrium position and then released at $t = 0$ (seconds). The weight completes one period in $\frac{1}{4}$ second.

(a) Write the equation for the simple harmonic motion.

(b) Sketch the graph of the motion.

(c) Find the time when the weight moves downward for the second time passing through the equilibrium position.

SOLUTION

(a) We shall assume that displacements below the equilibrium position are negative and displacements above the equilibrium position are positive. Since $d = -15$ cm at $t = 0$, we begin with the model

$$d = -15 \cos 2\pi ft.$$

Moreover, since the weight completes one period in $\frac{1}{4}$ second, we have

$$f = \frac{1}{P} = \frac{1}{\frac{1}{4}} = 4.$$

Hence, the equation that describes the motion is

$$d = -15 \cos 2\pi(4)t = -15 \cos 8\pi t, \quad t \geq 0.$$

(b) For $d = -15 \cos 8\pi t$, we have

Amplitude: $|a| = |-15| = 15$ Period: $\dfrac{2\pi}{|b|} = \dfrac{2\pi}{|8\pi|} = \dfrac{1}{4}$

Thus, this cosine curve completes one cycle from $t = 0$ to $t = \frac{1}{4}$ second and has a maximum displacement of 15 cm. Dividing the interval $[0, \frac{1}{4}]$ into four equal parts gives us the five key points on this cycle:

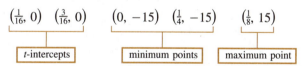

The graph of $d = -15 \cos 8\pi t$ for $t \geq 0$ is shown in Figure 3.12.

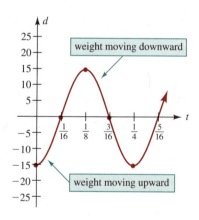

FIGURE 3.12

The graph of $d = -15 \cos 8\pi t$ for $t \geq 0$ illustrates the motion of the spring.

(c) Referring to Figure 3.12, the weight moves upward on the interval $(0, \frac{1}{8})$, passing through the equilibrium position at $t = \frac{1}{16}$ s. It then moves downward on the interval $(\frac{1}{8}, \frac{1}{4})$, passing through the equilibrium position again at $t = \frac{3}{16}$ s. Because of the periodic nature of this function, the weight moves downward for the second time on the interval

$$\left(\tfrac{1}{8} + \tfrac{1}{4}, \tfrac{1}{4} + \tfrac{1}{4}\right) = \left(\tfrac{3}{8}, \tfrac{1}{2}\right)$$

passing through the equilibrium position at

$$t = \tfrac{3}{16} + \tfrac{1}{4} = \tfrac{7}{16} \text{ s.} \qquad \blacklozenge$$

PROBLEM 6 Referring to Example 6, find the displacement at $t = 2$ seconds. ◆

Exercises 3.1

Basic Skills

In Exercises 1–6, solve each right triangle. Round x and y to three significant digits.

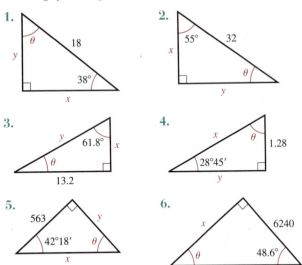

In Exercises 7–12, solve each right triangle. Round α and β to the nearest tenth of a degree and round x to three significant digits.

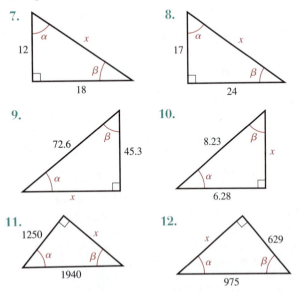

In Exercises 13–32, solve each problem. Round angles to the nearest tenth of a degree and all other measurements to three significant digits.

13. Find the vertical height of a kite if 350 ft of string are out and the angle from the ground to the kite string is 75°. Assume the string is taut.

14. An inclined ramp leading into a parking garage is 170 m long and rises 18 m. What is the angle of incline of the ramp?

15. When a helicopter is 1.2 km above one end of an island, the angle of depression to the other end of the island is 22.7°. What is the length of the island?

16. As illustrated in the sketch, the length of the shadow of a telephone pole is 62 ft when the angle of elevation of the sun is 32°. What is the height of the pole?

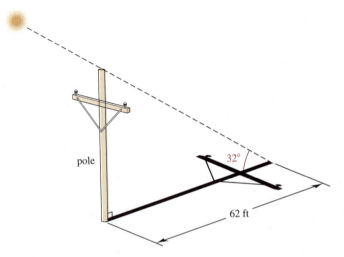

17. A guy wire from the top of an antenna is anchored 43.9 m from the base of the antenna and makes an angle of 72.6° with the ground, as shown in the sketch. What is the length of the guy wire?

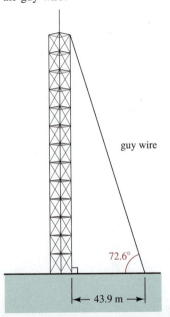

18. A security camera is used by a bank to monitor a teller's counter. The camera is mounted on a wall at a height of 12 ft. If the line-of-sight distance from the camera to the counter is 20 ft and the counter is 4 ft high, what is the angle of depression of the camera?

19. A radiologist uses a gamma ray to treat a tumor that is 3.2 cm beneath a patient's skin. The ray is directed into the skin at an angle of 48°. How far does the ray travel through the patient's body before striking the tumor?

20. The equal sides of an isosceles triangle are 36 cm and the nonequal side is 42 cm. Determine the measure of the interior angles of the triangle.

21. For the drill bit shown in the sketch, find angle θ.

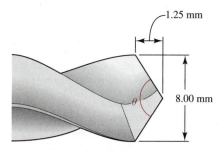

22. Three drill holes are located in an aluminum plate as shown in the sketch. Find the offset distances x and y.

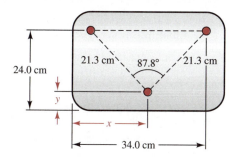

23. For the trapezoidal utility knife blade shown in the sketch, find the distance x.

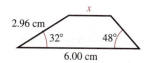

24. For the trapezoidal piece of land shown in the sketch, find angles α and β.

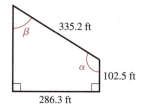

25. For the flat-headed machine screw shown in the sketch, find angles α and β.

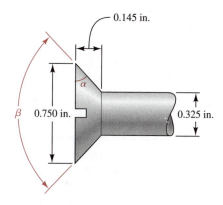

26. Each side of a hexagonal nut is 6 mm, as shown in the sketch. Find the distances x and y.

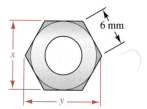

27. For the five intersecting highways sketched in the figure, find the distance between points B and C.

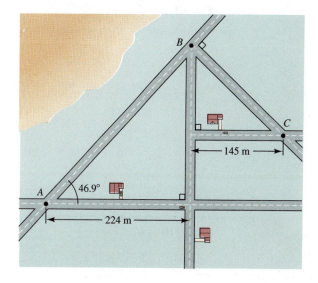

28. For the four-panel Pratt truss shown in the sketch, find the length of each lettered member.

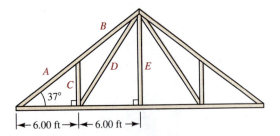

29. An airplane climbs at an angle of 6.5° at a constant speed of 475 mph. How many seconds does it take to ascend from an altitude of 10,500 ft to 12,800 ft?

30. A missile is traveling at a speed of 925 m/s and at an angle of 42.8° from the horizontal. What is its vertical displacement after 1 minute?

31. An observer in a lighthouse determines that the angles of depression to two sailboats directly in line with the lighthouse are 3°36′ and 5°45′. If the observer is 125 ft above sea level, find the distance between the two ships.

32. The horizontal distance from a transit to a smokestack is 126 ft. The angle of elevation to the top of the smokestack is 65°33′ and the angle of depression to the bottom of the smokestack is 3°21′. Find the height of the smokestack.

The equations in Exercises 33–36 describe the simple harmonic motion of an object. Find

(a) the maximum displacement from the equilibrium position,

(b) the frequency, and

(c) the first time ($t > 0$) that the object passes through the equilibrium position.

33. $d = 8 \cos 6\pi t$

34. $d = \frac{1}{2} \cos 30 t$

35. $d = \frac{2}{3} \sin \dfrac{\pi t}{10}$

36. $d = 24 \sin 120 \pi t$

37. The wake from a large ship sets a floating bottle in simple harmonic motion, as shown in the sketch. In 3 seconds the bottle rides from the crest of one wave to the crest of the next wave.

(a) Find the equation that describes the displacement d of the bottle from its equilibrium position if $d = 0$ inches at $t = 0$ seconds.

(b) Find the displacement of the bottle at $t = 4\frac{3}{4}$ s.

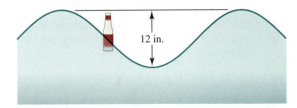

38. A guitar string is lifted up 2 mm and then released at $t = 0$ s. Suppose the note that is struck has a fequency of 440 vibrations per second. Write the equation of simple harmonic motion of a point on the string.

39. A spring with an attached weight is pushed up 20 cm from its equilibrium position and then released at $t = 0$ s. The weight completes one cycle in 0.4 s.

(a) Write the equation of simple harmonic motion and sketch its graph.

(b) Find the displacement of the weight when $t = 6.7$ s.

(c) Find the time when the spring, moving upward for the third time, passes through the equilibrium position.

40. A spring with an attached weight is pulled down 8.5 cm from its equilibrium position and then released at $t = 0$ s. The weight completes one cycle in $\pi/32$ second.

(a) Write the equation of simple harmonic motion and sketch its graph.

(b) Find the displacement of the weight when $t = 1$ s.

(c) Find the time when the spring, moving downward for the second time, passes through the equilibrium position.

Critical Thinking

41. For the cross-belted pulley system shown in the sketch, find the distance x and the angle θ.

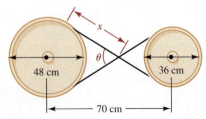

42. A cylindrical oil tank is 36 inches in diameter and 50 inches in length, as shown in the sketch. How many gallons of oil are in the tank if the depth of the oil is 22 inches? [*Hint:* 1 gallon of oil occupies a space equivalent to 231 cubic inches.]

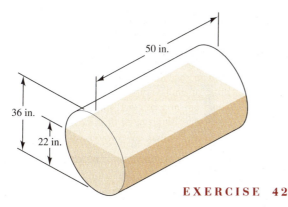

EXERCISE 42

where g is the gravitational constant 32.2 feet per second squared (ft/s^2), and l is the length of the pendulum (in feet). Suppose the length of the pendulum in a grandfather clock is 3.26 feet and the angle θ between the equilibrium position and the position of maximum displacement is 4.00°. Find the equation that describes the displacement d of the pendulum from its equilibrium position if $d = 0$ feet at $t = 0$ seconds.

43. A piece of sheet metal 24 inches wide is bent along its center line to form a V-shaped gutter. Does the gutter have a greater carrying capacity when the angle of the V is 75° or when it is 120°?

44. A painter has two ladders, one of which is twice as long as the other. Each ladder rests on the floor of a room and reaches the same height on the wall. If the shorter ladder makes an angle of 80° with the floor, find the angle between the longer ladder and the floor.

45. A regular octagon is inscribed in a circle. If the radius of the circle is 18 cm, find the length of a side of the octagon.

46. Find the angle between the diagonal of a cube and the diagonal of a face of the cube.

47. Another example of simple harmonic motion is the pendulum on a grandfather clock swinging back and forth through its equilibrium position, as shown in the sketch. In physics it is shown that the frequency f of a pendulum is given by

$$f = \frac{1}{2\pi}\sqrt{\frac{g}{l}},$$

48. Referring to Exercise 47, how many seconds tick off the clock each time the pendulum passes through the equilibrium position?

Calculator Activities

 49. A rain gutter is to be made from a piece of sheet metal 3 feet wide by turning up one-foot strips to make equal angles θ, as shown in the sketch.

(a) Express the distances x and y as trigonometric functions of θ.

(b) Express the trapezoidal cross-sectional area A as a function of θ.

(c) Use a graphing calculator to generate the graph of the function described in part (b). Then use the calculator's trace and zoom features to estimate the angle θ (in radians) that yields the maximum cross-sectional area and, hence, the greatest carrying capacity. Round the answer to four significant digits and convert this angle to the nearest degree.

(d) State the greatest cross-sectional area that can be formed.

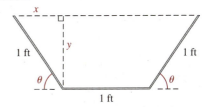

50. A length of rigid copper pipe, when held horizontally, touches the walls of a 6-ft corridor and a 10-ft corridor at points *A*, *B*, and *C*, as shown in the sketch.

(a) Express the lengths *x* and *y* as trigonometric functions of θ.

(b) Express the total length *L* of the rigid copper pipe as a function of θ.

(c) Use a graphing calculator to generate the graph of the function described in part (b). Then use the calculator's trace and zoom features to estimate the shortest length of pipe that touches the walls of the corridors at points *A*, *B*, and *C*. Round the answer to three significant digits.

(d) The length in part (c) actually represents the longest pipe that can be held horizontally and moved from one corridor to the other without bending the pipe or denting the walls. If the height of the ceiling in the corridors is 12 ft, determine the longest pipe that can be moved from one corridor to the other without bending the pipe. Round the answer to three significant digits.

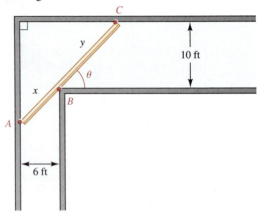

3.2 Law of Sines

In Section 3.1, we discussed the procedure for solving right triangles. We now turn our attention to methods for solving oblique triangles. An **oblique triangle** is a triangle that contains no right angle. In order to solve an oblique triangle, we need to be given either

1. two angles and any one side [AAS (angle-angle-side) or ASA (angle-side-angle)],
2. two sides and an angle opposite one of them (SSA),
3. two sides and their included angle (SAS), or
4. three sides (SSS).

In this section we derive a formula, called the *law of sines*, which enables us to solve an oblique triangle for cases 1 and 2. In Section 3.3 we derive a formula, called the *law of cosines*, which enables us to solve an oblique triangle for cases 3 and 4.

◆ Deriving the Law of Sines

For the oblique triangle in Figure 3.13, we use *a* as the side opposite angle *A*, *b* as the side opposite angle *B*, and *c* as the side opposite angle *C*. By drawing altitude *h* from angle *C* to side *c*, as illustrated in Figure 3.14, we separate the oblique triangle into two right triangles.

Referring to Figure 3.14 and using the right triangle on the left, we have

$$\sin A = \frac{h}{b} \quad \text{or} \quad h = b \sin A.$$

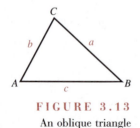

FIGURE 3.13
An oblique triangle

SECTION 3.2 Law of Sines 179

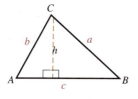

FIGURE 3.14
An oblique triangle with altitude h drawn to side c

Referring to Figure 3.14 and using the right triangle on the right, we have

$$\sin B = \frac{h}{a} \quad \text{or} \quad h = a \sin B.$$

Equating these values of h gives us

$$a \sin B = b \sin A$$

$$\frac{a}{\sin A} = \frac{b}{\sin B} \qquad \text{Divide both sides by } \sin A \sin B$$

Similarly, by drawing the altitude from angle B to side b, we can also show that

$$\frac{a}{\sin A} = \frac{c}{\sin C}.$$

In summary, we conclude that *in a given triangle the ratio of the length of a side to the sine of the angle opposite that side has the same constant value.* We refer to this fact as the **law of sines**.

 Law of Sines

> If A, B, and C are the angles of a triangle and a, b, and c are, respectively, the sides opposite these angles, then
>
> $$\frac{a}{\sin A} = \frac{b}{\sin B} = \frac{c}{\sin C}.$$

Note: We have developed the law of sines for an oblique triangle with angle C acute ($C < 90°$). The law is also valid when C is obtuse ($90° < C < 180°$) or when C is a right angle ($C = 90°$).

◆ **Applying the Law of Sines**

We can use the law of sines to find the missing parts of an oblique triangle whenever one of the ratios

$$\frac{a}{\sin A}, \quad \frac{b}{\sin B}, \quad \text{or} \quad \frac{c}{\sin C}$$

is known and an additional side or angle is given. Hence, the law of sines is of particular use when we are given either

1. two angles and any one side (AAS or ASA) or
2. two sides and an angle opposite one of them (SSA).

EXAMPLE 1 Solve the oblique triangle in Figure 3.15. Round x and y to three significant digits.

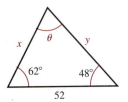

FIGURE 3.15

SOLUTION In this triangle, we are given two angles and the side between them (ASA). To solve the oblique triangle means to find the missing parts, namely, angle θ and sides x and y. Since the sum of the interior angles in a triangle is 180°, we have

$$\theta = 180° - (62° + 48°) = 70°.$$

Now we apply the law of sines by matching each side with the sine of the angle opposite that side:

$$\frac{52}{\sin 70°} = \frac{x}{\sin 48°} = \frac{y}{\sin 62°}$$

Hence,

$$\frac{52}{\sin 70°} = \frac{x}{\sin 48°} \quad \text{and} \quad \frac{52}{\sin 70°} = \frac{y}{\sin 62°}$$

$$x = \frac{52 \sin 48°}{\sin 70°} \qquad\qquad y = \frac{52 \sin 62°}{\sin 70°}$$

$$x \approx 41.1 \qquad\qquad y \approx 48.9 \quad\blacklozenge$$

PROBLEM 1 In any triangle, the longest side is opposite the largest angle and the shortest side is opposite the smallest angle. It is good practice to use this geometric fact as a rough check of the solutions of an oblique triangle. Show that this fact holds for the oblique triangle in Figure 3.15. $\blacklozenge$

In the next example, we solve an oblique triangle in which two sides and the angle opposite the larger one are known (SSA). To avoid *rounding errors* (errors that develop from using approximate values), we carry along a few extra significant digits through the calculating process and then round the final answers to the desired accuracy. For a discussion of rounding errors, see Appendix A.

EXAMPLE 2 Solve the oblique triangle shown in Figure 3.16. Round α and β to the nearest tenth of a degree and round x to three significant digits.

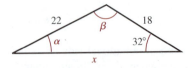

FIGURE 3.16

SOLUTION To solve the oblique triangle, we must find angles α and β and side x. We apply the law of sines by matching each side with the sine of the angle opposite that side:

$$\frac{22}{\sin 32°} = \frac{18}{\sin \alpha} = \frac{x}{\sin \beta}.$$

Hence,

$$\frac{22}{\sin 32°} = \frac{18}{\sin \alpha}$$

$$\sin \alpha = \frac{18 \sin 32°}{22}$$

Since the side opposite α is less than the side opposite 32° (that is, 18 < 22), we know that $\alpha < 32°$. Using the inverse sine function, in conjunction with a calculator set in degree mode, we obtain

$$\alpha = \sin^{-1}\left(\frac{18 \sin 32°}{22}\right) \approx 25.694°.$$

Since the sum of the interior angles in a triangle is 180°, we have

$$\beta \approx 180° - (32° + 25.694°) = 122.306°.$$

Finally, returning to the law of sines with $\beta \approx 122.306°$, we find x as follows:

$$\frac{22}{\sin 32°} \approx \frac{x}{\sin 122.306°}$$

$$x \approx \frac{22 \sin 122.306°}{\sin 32°} \approx 35.089.$$

Rounding to the desired accuracy, we have $\alpha \approx 25.7°$, $\beta \approx 122.3°$, and $x \approx 35.1$. ◆

Referring to Example 2, we cannot apply the Pythagorean theorem to find side x. The Pythagorean theorem holds for right triangles only and cannot be applied to an oblique triangle.

PROBLEM 2 Solve the oblique triangle shown in Figure 3.17. Round x and y to three significant digits. ◆

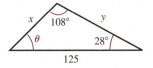

FIGURE 3.17

The Ambiguous Case

For the oblique triangle in Example 2, we were given two sides and the angle opposite the *larger* of these sides. This condition determines a unique triangle. However, if the given angle is acute and the side opposite this angle is *smaller* than the other side, then, as illustrated in Figure 3.18, we can construct *two* oblique triangles (provided the data allows us to construct any triangle). Because of the ambiguity, we refer to this situation as the **ambiguous case.** *The ambiguous case occurs* only when the given angle is acute and the side opposite this angle is smaller than the other given side.

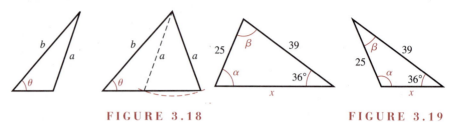

FIGURE 3.18
Two constructions are possible whenever $\theta < 90°$ and $a < b$

FIGURE 3.19

EXAMPLE 3 Solve the oblique triangles shown in Figure 3.19. Round α and β to the nearest tenth of a degree and round x to three significant digits.

SOLUTION To solve these oblique triangles, we must find angles α and β and side x. Applying the law of sines by matching each side with the sine of the angle opposite that side, we have

$$\frac{25}{\sin 36°} = \frac{39}{\sin \alpha} = \frac{x}{\sin \beta}.$$

Hence,

$$\frac{25}{\sin 36°} = \frac{39}{\sin \alpha}$$

$$\sin \alpha = \frac{39 \sin 36°}{25}$$

Since the sine function is positive in both quadrants I and II, we cannot tell from the sine of angle α whether angle α is acute or obtuse. Thus, we first find the reference angle α' associated with α. Using the inverse sine function, in conjunction with a calculator set in degree mode, we obtain

$$\alpha' = \sin^{-1}\left(\frac{39 \sin 36°}{25}\right) \approx 66.483°.$$

Now, placing a reference angle of 66.483° in quadrants I and II, as shown in Figure 3.20, gives us

$\alpha \approx 66.483°$ or $\alpha \approx 180° - 66.483° = 113.517°.$

Use for part (a), since α is acute.   Use for part (b), since α is obtuse.

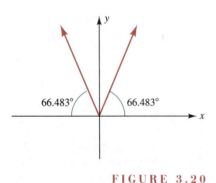

FIGURE 3.20
A reference angle of 66.483° in quadrants I and II

(a) Using $\alpha \approx 66.483°$ and the fact that the sum of the interior angles in a triangle is 180°, we have

SECTION 3.2 Law of Sines

$$\beta \approx 180° - (36° + 66.483°) = 77.517°.$$

Finally, returning to the law of sines with $\beta \approx 77.517°$, we find x as follows:

$$\frac{25}{\sin 36°} \approx \frac{x}{\sin 77.517°}$$

$$x \approx \frac{25 \sin 77.517°}{\sin 36°} \approx 41.527$$

Rounding to the desired accuracy, we have $\alpha \approx 66.5°$, $\beta \approx 77.5°$, and $x \approx 41.5$.

(b) Using $\alpha = 113.517°$ and the fact that the sum of the interior angles in a triangle is 180°, we have

$$\beta \approx 180° - (36° + 113.517°) = 30.483°.$$

Finally, returning to the law of sines with $\beta \approx 30.483°$, we find x as follows:

$$\frac{25}{\sin 36°} \approx \frac{x}{\sin 30.483°}$$

$$x \approx \frac{25 \sin 30.483°}{\sin 36°} \approx 21.576$$

Rounding to the desired accuracy, we have $\alpha \approx 113.5°$, $\beta \approx 30.5°$, and $x \approx 21.6$. ◆

PROBLEM 3 Given that β is obtuse, solve the oblique triangle in Figure 3.21. Round α and β to the nearest tenth of a degree and round x to three significant digits. ◆

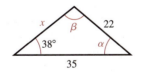

FIGURE 3.21

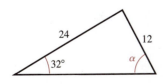

FIGURE 3.22
A triangle with erroneous given information

In order for us to solve an oblique triangle, the data we are given must be adequate for the construction of a triangle. For example, consider the triangle in Figure 3.22. Using the law of sines to solve this oblique triangle for angle α, we obtain

$$\frac{12}{\sin 32°} = \frac{24}{\sin \alpha}$$

$$\sin \alpha = \frac{24 \sin 32°}{12} \approx 1.0598$$

However, 1.0598 is not in the range of the sine function. Hence, this last equation has no solution. To illustrate what is happening, we draw the altitude h as shown in

184 CHAPTER 3 Applications of the Trigonometric Functions

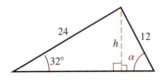

FIGURE 3.23
For this triangle to exist, we must have $12 > h$.

Figure 3.23. For this triangle to exist, we must have $12 > h$. Using the right triangle on the left, we find

$$\sin 32° = \frac{h}{24}$$

$$h = 24 \sin 32° \approx 12.7$$

Since $12 < h$, we conclude that the data supplied is erroneous and no such triangle exists.

The following summarizes the possibilities when we are given two sides and an acute angle opposite one of them.

 Constructions for SSA

Given sides a and b and an acute angle α opposite side a, four possibilities exist for construction of a triangle.

1. If $a \geq b$, one triangle exists.
2. If $a < b$, two triangles exist whenever $a > h$.
3. If $a < b$, no triangle exists whenever $a < h$.
4. If $a < b$, one right triangle exists whenever $a = h$.

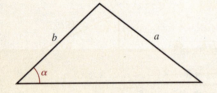 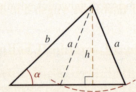 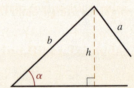 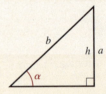

◆ **Application: Surveying**

The law of sines is used in several applications. Our next example illustrates the procedure a surveyor might use to determine a distance that is inaccessible by direct measurement.

EXAMPLE 4 From the surveyor's notes shown in Figure 3.24, determine the width of the river.

FIGURE 3.24

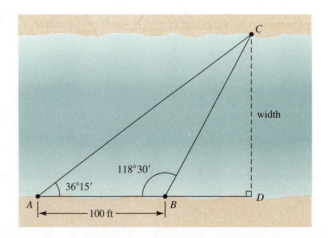

SECTION 3.2   Law of Sines

SOLUTION   Since the sum of the interior angles in a triangle is 180°, we have

$$\text{angle } ACB = 180° - (36°15' + 118°30') = 25°15'$$

Working with oblique triangle ABC, we can determine the length AC from the law of sines:

$$\frac{AC}{\sin 118°30'} = \frac{100}{\sin 25°15'}$$

$$AC = \frac{100 \sin 118°30'}{\sin 25°15'}$$

Now, we can obtain the width of the river CD from right triangle ADC, as follows:

$$\sin 36°15' = \frac{\text{opp}}{\text{hyp}} = \frac{CD}{AC}$$

$$CD = AC \sin 36°15' \qquad \text{Multiply both sides by } AC$$

$$CD = \frac{100 \sin 118°30'}{\sin 25°15'} \sin 36°15' \qquad \text{Substitute for } AC$$

$$CD \approx 121.8 \text{ ft} \qquad \text{Evaluate}$$

Thus the width of the river is approximately 121.8 feet. ◆

PROBLEM 4   Referring to Figure 3.24, use the law of sines on right triangle ADC to find the width of the river CD. You should obtain the same value for CD as in Example 4. ◆

Exercises 3.2

  Basic Skills

In Exercises 1–6, solve each oblique triangle. Round x and y to three significant digits.

1.   2.

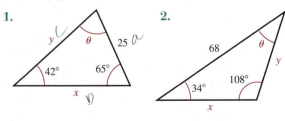

3.   4.

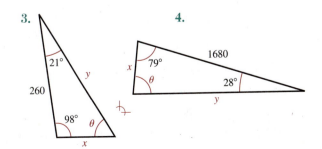

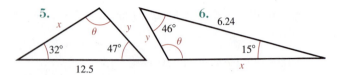

In Exercises 7–12, solve each oblique triangle. Round α and β to the nearest tenth of a degree and round x to three significant digits.

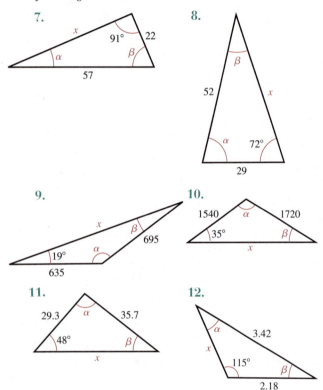

In Exercises 13–24, three parts of an oblique triangle are given. With the understanding that side a is opposite angle α, side b is opposite angle β, and side c is opposite angle θ, solve the oblique triangle with these three given parts, rounding angles to the nearest tenth of a degree and sides to three significant digits. If the data allows for the construction of two triangles, find both solutions. If the data does not allow for the construction of a triangle, state that no such triangle exists.

13. $c = 55.4$, $\alpha = 57°$, $\beta = 86°$
14. $c = 2.74$, $\alpha = 123°$, $\theta = 28°$
15. $b = 35$, $c = 22$, $\beta = 62°$
16. $a = 0.189$, $c = 0.297$, $\theta = 59°$
17. $a = 680$, $c = 990$, $\alpha = 38°$
18. $b = 3.91$, $c = 9.82$, $\beta = 6°$
19. $b = 10.4$, $c = 9.75$, $\theta = 67°$
20. $a = 3.84$, $b = 4.67$, $\alpha = 52°$
21. $a = 280$, $\alpha = 126°$, $\beta = 58°$
22. $a = 151$, $b = 126$, $\beta = 108°$
23. $a = 11$, $b = 16$, $\alpha = 56°$
24. $a = 3.64$, $c = 1.94$, $\theta = 42°$

In Exercises 25–40, solve each applied problem. Round angles to the nearest tenth of a degree and all other measurements to three significant digits.

25. Two angles of a triangle are 62°15′ and 68°30′. If the shortest side is 34.8 m, find the length of the longest side.

26. Two angles of a triangle are 42°20′ and 58°10′. If the longest side is 234 ft, find the length of the shortest side.

27. From the surveyor's notes given in the sketch, find the length x of the pond.

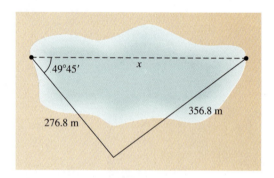

28. For the pulley system shown in the figure, find the distance x between the pulleys.

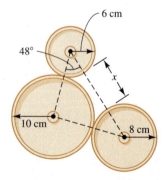

29. A slide at a playground is 22.5 ft long and is inclined 46.8° from the horizontal. The access ladder that reaches to the top of the slide is 18.2 ft long, as shown in the sketch. What acute angle θ does the ladder make with the horizontal?

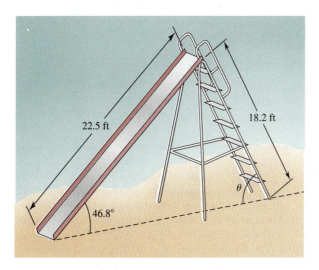

EXERCISE 29

30. A piston and rod assembly has the dimensions shown in the sketch. Find two values of angle β when $\alpha = 18°$.

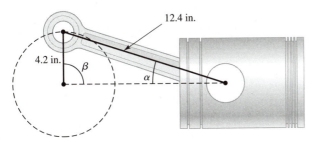

31. Find the amount of fencing needed to enclose the triangular piece of land shown in the figure given that angle C is (a) acute or (b) obtuse.

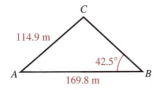

32. Three drill holes are located in a metal plate, as shown in the sketch. Determine the distance x given that angle θ is (a) acute or (b) obtuse.

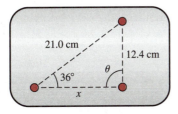

33. The angle of elevation from one edge of a ravine to the top of a tree on the opposite edge is 32.3°. From a point 100 ft back from the edge of the ravine, the angle of elevation to the top of the same tree is 22.9°, as sketched in the figure. Find (a) the width of the ravine and (b) the height of the tree.

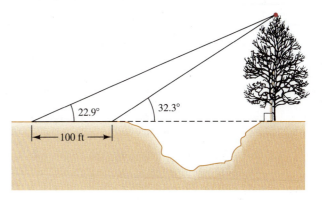

34. For the roof truss shown in the sketch, find the following measurements:

(a) α (b) β (c) BD
(d) θ (e) BC (f) CD

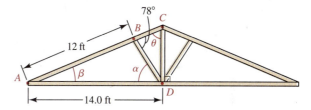

35. The length of a property line from point A to point B is impossible to measure directly. To find the distance AB indirectly, a surveyor selects a point C and finds the distances $BC = 865$ ft, $AC = 520$ ft, and angle $CAB = 38°20'$. What is the length of property line AB?

36. A person in a hot-air balloon 2500 ft above the ground notes that the angle of depression of an object on the ground is 32.6°. After ascending vertically for 15 minutes, the person finds that the angle of depression to the same object is 61.5°. What is the rate of ascent (in mph) of the balloon?

37. A flagpole stands vertically on a hill that is inclined at an angle of 7° to the horizontal. Find the height of the pole if the angle of elevation of the sun is 55°, and the pole casts a shadow of length 26 ft directly (a) down the hill or (b) up the hill.

38. When the angle of elevation of the sun is 66°20′, a leaning telephone pole on level ground casts a shadow of length 10.6 ft. Find the length of the pole if it leans 12°30′ from the vertical (a) toward the sun or (b) away from sun.

39. The angles of elevation to the top of a tower are measured from two points on level ground that are 100 meters apart and are found to be 46.7° and 68.3°. Find the height of the tower if the two points are in the same vertical plane with the tower and are (a) on the same side of the tower or (b) on opposite sides of the tower.

40. The angles of depression to point A at the floor of a canyon are measured from two level points B and C on opposite sides of the canyon at the rim and found to be 59.5° and 71.2°. Find the depth of the canyon if the three points are in the same vertical plane and the distance from B to C is 750 ft.

Critical Thinking

41. Suppose the interior angles of a triangle are A, B, and C, and the sides opposite these angles are a, b, and c, respectively. Furthermore, suppose $b < a < c$. What does this imply about the measures of angles A, B, and C? Explain.

42. Given the oblique triangle in the sketch, find the fallacy in the following argument:

$\dfrac{20}{\sin 30°} = \dfrac{20\sqrt{3}}{\sin \alpha}$	**Apply law of sines**
$\sin \alpha = \sqrt{3} \sin 30°$	**Solve for $\sin \alpha$**
$\sin \alpha = \dfrac{\sqrt{3}}{2}$	**Evaluate $\sin 30°$ and multiply**
$\alpha = \sin^{-1}\left(\dfrac{\sqrt{3}}{2}\right) = 60°$	**Solve for α**

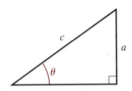

43. Apply the law of sines to the right triangle in the sketch and solve for $\sin \theta$. What familiar formula does this represent?

44. In Chapter 4 we will prove the double-angle formula for sine, $\sin 2\theta = 2 \sin \theta \cos \theta$. Apply the law of sines to an oblique triangle in which one angle is twice as large as another, as shown in the sketch. Then apply this double-angle formula and express $\cos \theta$ in terms of a and b.

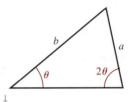

45. For the triangles in the sketch, use the law of sines to help show that

$$h = \frac{a \sin(\alpha - \beta)}{\cos \alpha \cos \beta}.$$

46. For the triangles in the figure, use the law of sines to help show that

$$h = \frac{a \sin \alpha \sin \beta}{\sin(\alpha - \beta)}.$$

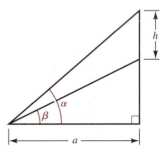

SECTION 3.3 Law of Cosines

 Calculator Activities

Suppose the interior angles of a triangle are A, B, and C, and the sides opposite these angles are a, b, and c, respectively. Mollweide's Formula (after the German mathematician Karl Mollweide, 1774–1825) states that these six parts of a triangle are related by the formula

$$\frac{a-b}{c} = \frac{\sin\frac{A-B}{2}}{\cos\frac{C}{2}}.$$

We can use this formula to check the solutions of a triangle by comparing the ratio on the left-hand side to the ratio on the right-hand side. In Exercises 47–50, use a calculator to compare these ratios and state whether the given information appears to be the solution of a triangle.

47. $A = 34.4°$, $B = 115.7°$, $C = 29.9°$, $a = 3.61$, $b = 5.76$, $c = 3.19$

48. $A = 52.0°$, $B = 65.8°$, $C = 62.2°$, $a = 75.2$, $b = 87.0$, $c = 84.4$

49. $A = 19.7°$, $B = 53.7°$, $C = 106.6°$, $a = 2.508$, $b = 5.001$, $c = 7.134$

50. $A = 35.3°$, $B = 124.8°$, $C = 19.9°$, $a = 7005$, $b = 10{,}367$, $c = 5468$

In physics, a formula that mirrors the law of sines provides a means of calculating the direction taken by a refracted beam of light when it passes from one medium to a second medium. Referred to as Snell's Law, it is given by

$$\frac{v_1}{\sin\alpha} = \frac{v_2}{\sin\beta},$$

where v_1 is the speed of light in the first medium, v_2 is the speed of light in the second medium, and angles α and β are the corresponding angles of refraction, as shown in the sketch. Use this formula in conjunction with a calculator to answer the questions in Exercises 51 and 52.

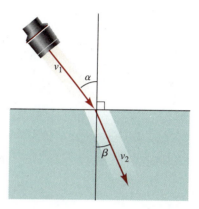

51. Given that the speed of light in air is $v_1 = 2.998 \times 10^{10}$ cm/s, $\alpha = 32.83°$, and $\beta = 21.45°$, find v_2, the speed of light in the second medium. Record the answer in scientific notation.

52. Given that the speed of light in air is $v_1 = 2.998 \times 10^{10}$ cm/s, the speed of light in water is $v_2 = 2.254 \times 10^{10}$ cm/s, and $\alpha = 41.34°$, find angle β.

3.3 Law of Cosines

As we stated in Section 3.2, in order to solve an oblique triangle, we need to be given either

1. two angles and any one side (AAS or ASA),
2. two sides and an angle opposite one of them (SSA),
3. two sides and their included angle (SAS), or
4. three sides (SSS).

Cases 1 and 2 are solved by the law of sines (Section 3.2). However, cases 3 and 4 cannot be solved by the law of sines, since none of the ratios $a/\sin A$, $b/\sin B$, or $c/\sin C$ is known. For cases 3 and 4, we develop a new formula called the law of cosines.

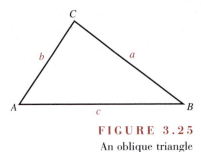

FIGURE 3.25
An oblique triangle

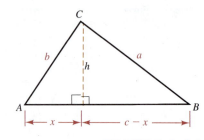

FIGURE 3.26
An oblique triangle with altitude h drawn to side c

◆ Deriving the Law of Cosines

Consider the oblique triangle in Figure 3.25. We use a as the side opposite angle A, b as the side opposite angle B, and c as the side opposite angle C. By drawing altitude h from angle C to side c, as illustrated in Figure 3.26, we separate the oblique triangle into two right triangles and separate side c into two parts, x and $c - x$.

Referring to Figure 3.26, and using the right triangle on the left, we have

$$\cos A = \frac{x}{b} \quad \text{or} \quad x = b \cos A.$$

Also, by the Pythagorean theorem,

$$b^2 = h^2 + x^2 \quad \text{or} \quad h^2 = b^2 - x^2.$$

Referring to Figure 3.26, and using the right triangle on the right, we have

$$a^2 = h^2 + (c - x)^2 \quad \text{or} \quad h^2 = a^2 - (c - x)^2.$$

Equating these values of h^2, we obtain

$$a^2 - (c - x)^2 = b^2 - x^2$$
$$a^2 = b^2 - x^2 + (c - x)^2 \qquad \text{Add } (c - x)^2 \text{ to both sides}$$
$$a^2 = b^2 - x^2 + c^2 - 2cx + x^2 \qquad \text{Expand}$$
$$a^2 = b^2 + c^2 - 2cx \qquad \text{Simplify}$$
$$a^2 = b^2 + c^2 - 2bc \cos A \qquad \text{Replace } x \text{ with } b \cos A$$

Similarly, by drawing the altitudes from angle B to side b and from angle A to side a, we can show that

$$b^2 = a^2 + c^2 - 2ac \cos B \quad \text{and} \quad c^2 = a^2 + b^2 - 2ab \cos C.$$

In summary, we conclude that *the square of the length of any side of a triangle is the sum of the squares of the other two sides minus twice the product of those two*

SECTION 3.3 Law of Cosines

sides and the cosine of the angle between them. We refer to this fact as the **law of cosines.**

Law of Cosines

If A, B, and C are the angles of a triangle and a, b, and c are, respectively, the sides opposite these angles, then

$$a^2 = b^2 + c^2 - 2bc \cos A,$$
$$b^2 = a^2 + c^2 - 2ac \cos B,$$

and

$$c^2 = a^2 + b^2 - 2ab \cos C.$$

Note: We have developed the law of cosines for an oblique triangle with acute angles. The law is also valid when one of the angles is obtuse or when one of the angles is a right angle. Observe that if $C = 90°$, as shown in Figure 3.27, we obtain

$$c^2 = a^2 + b^2 - 2ab \cos 90°$$
$$c^2 = a^2 + b^2 - 2ab(0)$$
$$c^2 = a^2 + b^2 \quad \text{Pythagorean theorem}$$

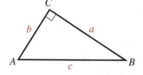

FIGURE 3.27
When $C = 90°$, the law of cosines yields the Pythagorean theorem, $c^2 = a^2 + b^2$.

◆ **Applying the Law of Cosines**

In the next example we use the law of cosines to solve an oblique triangle in which two sides and the included angle are given (case 3, SAS). When using the law of cosines, we can avoid rounding errors (Section 3.2) by carrying along a few extra significant digits through the calculating process and then rounding the final answers to the desired accuracy.

EXAMPLE 1 Solve the oblique triangle shown in Figure 3.28. Round α and β to the nearest tenth of a degree and round x to three significant digits.

SOLUTION In this oblique triangle, we are given two sides and an angle between them (SAS). To solve the oblique triangle, we must find angles α and β and side x. To find side x, we apply the law of cosines:

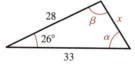

FIGURE 3.28

$$x^2 = 28^2 + 33^2 - 2(28)(33) \cos 26°$$
$$x = \sqrt{28^2 + 33^2 - 2(28)(33) \cos 26°} \approx 14.561$$

Since we now know the length of side x and the angle opposite that side (26°), we can use the law of sines and write

$$\frac{14.561}{\sin 26°} \approx \frac{28}{\sin \alpha} = \frac{33}{\sin \beta}.$$

Hence,

$$\sin \alpha \approx \frac{28 \sin 26°}{14.561} \quad \text{and} \quad \sin \beta \approx \frac{33 \sin 26°}{14.561}.$$

Remember that *we cannot tell from the sine of an angle whether the angle is acute or obtuse*. Thus, in determining the values of α and β, it is best to find the smaller of these angles first (since the smaller angle must be acute). Since $28 < 33$, we begin by finding angle α (the angle opposite 28). Using the inverse sine function, in conjunction with a calculator, we have

$$\alpha \approx \sin^{-1}\left(\frac{28 \sin 26°}{14.561}\right) \approx 57.454°.$$

Now, since the sum of the interior angles of a triangle must be 180°, we conclude

$$\beta \approx 180° - (26° + 57.454°) = 96.546°.$$

Rounding to the desired accuracy, we have $\alpha \approx 57.5°$, $\beta \approx 96.5°$, and $x \approx 14.6$.

◆

PROBLEM 1 From the second part of the law of sines in Example 1, we have

$$\sin \beta = \frac{33 \sin 26°}{14.6}$$

Show how to determine β from this equation.

◆

Since the cosine function is positive in quadrant I and negative in quadrant II, *we can determine from the cosine of the angle whether that angle is acute or obtuse,* that is,

$$\cos \theta > 0 \text{ implies } \theta \text{ acute} \quad \text{and} \quad \cos \theta < 0 \text{ implies } \theta \text{ obtuse.}$$

Thus, to solve an oblique triangle in which three sides are given (case 4, SSS), we first apply the law of cosines to find the angle opposite the *largest* side. We can then find the other two angles of the triangle—which must be acute—by using the law of sines. The procedure is illustrated in the next example.

EXAMPLE 2 Solve the oblique triangle shown in Figure 3.29. Round α, β, and θ to the nearest tenth of a degree.

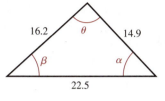

FIGURE 3.29

SOLUTION To solve this oblique triangle, we must find angles α, β, and θ. Since θ is opposite the largest side, we apply the law of cosines and write

$$22.5^2 = \underbrace{16.2^2 + 14.9^2}_{\text{Sides adjacent to } \theta} - 2(16.2)(14.9) \cos \theta$$

$$\underbrace{22.5^2}_{\text{Side opposite } \theta}$$

We now solve for $\cos \theta$ as follows:

$$22.5^2 = 16.2^2 + 14.9^2 - 2(16.2)(14.9) \cos \theta$$

$$22.5^2 - 16.2^2 - 14.9^2 = -2(16.2)(14.9) \cos \theta \qquad \text{Subtract } 16.2^2 \text{ and } 14.9^2 \text{ from both sides}$$

$$\cos \theta = \frac{22.5^2 - 16.2^2 - 14.9^2}{-2(16.2)(14.9)} \qquad \text{Divide both sides by } -2(16.2)(14.9)$$

If $\cos \theta > 0$, then θ is acute, and if $\cos \theta < 0$, then θ is obtuse. Using the inverse cosine function, in conjunction with a calculator set in degree mode, we find that θ is obtuse:

$$\theta = \cos^{-1}\left(\frac{22.5^2 - 16.2^2 - 14.9^2}{-2(16.2)(14.9)}\right) \approx 92.588°$$

Since we now know angle θ and the length of the side opposite θ ($= 22.5$), we can use the law of sines and write

$$\frac{22.5}{\sin 92.588°} \approx \frac{16.2}{\sin \alpha} = \frac{14.9}{\sin \beta}.$$

Hence,

$$\sin \alpha \approx \frac{16.2 \sin 92.588°}{22.5} \quad \text{and} \quad \sin \beta \approx \frac{14.9 \sin 92.588°}{22.5}.$$

Since both α and β must be acute angles (why?), we use the inverse sine function, in conjunction with a calculator set in degree mode, to obtain

$$\alpha \approx \sin^{-1}\left(\frac{16.2 \sin 92.588°}{22.5}\right) \approx 45.994°$$

and

$$\beta \approx \sin^{-1}\left(\frac{14.9 \sin 92.588°}{22.5}\right) \approx 41.418°.$$

Rounding to the desired accuracy, we have $\alpha \approx 46.0°$, $\beta \approx 41.4°$, and $\theta \approx 92.6°$. Note that the sum of these angles is $180.0°$.

In Example 3, we switched from the law of cosines to the simpler law of sines once we had found an additional piece of the triangle. As an alternative approach, we could have continued with the law of cosines, writing

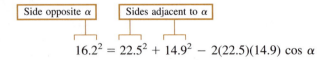

$$16.2^2 = 22.5^2 + 14.9^2 - 2(22.5)(14.9) \cos \alpha$$

and

$$14.9^2 = 16.2^2 + 22.5^2 - 2(16.2)(22.5) \cos \beta$$

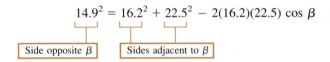

PROBLEM 2 Solve the equations

$$16.2^2 = 22.5^2 + 14.9^2 - 2(22.5)(14.9) \cos \alpha$$

and

$$14.9^2 = 16.2^2 + 22.5^2 - 2(16.2)(22.5) \cos \beta$$

for α and β, respectively. The results should agree with those in Example 2. ◆

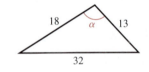

FIGURE 3.30
A triangle with erroneous given information

As we discussed in Section 3.2, in order for us to solve an oblique triangle, the data we are given must be adequate for the construction of a triangle. For example, consider the triangle in Figure 3.30. To find angle α, we apply the law of cosines and write

$$32^2 = 18^2 + 13^2 - 2(18)(13) \cos \alpha.$$

Hence,

$$\cos \alpha = \frac{32^2 - 18^2 - 13^2}{-2(18)(13)} \approx -1.1346.$$

However, -1.1346 is not in the range of the cosine function. Hence, this last equation has no solution. In order to construct a triangle, *the sum of the lengths of any two sides must be greater than the length of the third side*. For the triangle in Figure 3.30, this is not the case:

$$18 + 13 < 32.$$

Therefore, we conclude the data supplied is erroneous, and no such triangle exists.

Area Formulas for Triangles

Consider the oblique triangle in Figure 3.31 in which two sides, *a* and *b*, and the angle between them, θ, are known. From elementary geometry we know that the area *A* of this triangle is half the product of base *b* and height *h*, that is,

$$A = \frac{bh}{2}.$$

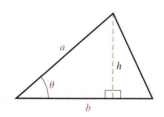

For the right triangle on the left in Figure 3.31, we have

$$\sin \theta = \frac{h}{a} \quad \text{or} \quad h = a \sin \theta.$$

Replacing *h* with $a \sin \theta$ in the basic area formula $A = bh/2$ gives us

$$A = \frac{ab \sin \theta}{2}.$$

FIGURE 3.31
An oblique triangle in which two sides, *a* and *b*, and the angle between them, θ, are given.

Area of a Triangle

> The area *A* of an oblique triangle in which two sides, *a* and *b*, and the angle between them, θ, are known is given by
>
> $$A = \frac{ab \sin \theta}{2}.$$

Note: We have developed this area formula for an oblique triangle in which θ is acute. The formula is also valid when θ is obtuse and when θ is a right angle.

EXAMPLE 3 Find the area of the triangle in Figure 3.29 in Example 2.

SOLUTION In Example 2, we showed that the angle between 16.2 and 14.9 is $\theta \approx 92.588°$. Thus, the area of this triangle is

$$A = \frac{ab \sin \theta}{2} = \frac{(16.2)(14.9) \sin 92.588°}{2} \approx 121 \text{ square units.} \quad \blacklozenge$$

PROBLEM 3 Find the area of the triangle in Figure 3.29 by using

(a) the sides 16.2 and 22.5 and the angle between them.

(b) the sides 22.5 and 14.9 and the angle between them.

Do you obtain the same answer as we did in Example 3? $\quad \blacklozenge$

Around 100 A.D., Hero of Alexandria (Heron) derived a formula for the area of a triangle in which all three sides are known. The formula is now called **Hero's formula.** The proof of this formula is outlined in Exercise 52 of this section.

Hero's Formula

If a, b, and c are the sides of a triangle, then the area A of the triangle is given by

$$A = \sqrt{s(s-a)(s-b)(s-c)}$$

where s is half the perimeter (semiperimeter) of the triangle, that is, $s = \frac{1}{2}(a + b + c)$.

EXAMPLE 4 Use Hero's formula to find the area of the triangle in Figure 3.29.

SOLUTION Since the sides of the triangle in Figure 3.29 are 16.2, 14.9, and 22.5, the semiperimeter s is

$$s = \tfrac{1}{2}(16.2 + 14.9 + 22.5) = 26.8.$$

Thus, by Hero's formula, the area A of this triangle is

$$\begin{aligned} A &= \sqrt{s(s-a)(s-b)(s-c)} \\ &= \sqrt{26.8(26.8 - 16.2)(26.8 - 14.9)(26.8 - 22.5)} \\ &= \sqrt{26.8(10.6)(11.9)(4.3)} \approx 121 \text{ square units,} \end{aligned}$$

which agrees with the answer we obtained in Example 3. ◆

PROBLEM 4 Use Hero's formula to find the area of a right triangle with sides 3, 4, and 5. Check the answer by using the basic area formula $A = bh/2$. ◆

◆ Application: Playing Golf

The law of cosines is useful for several applications. Our next example illustrates the procedure a golfer could use to determine the distance to the hole after the initial drive.

EXAMPLE 5 On a golf course, the distance from a tee to a hole is 355 yards. Suppose a golfer drives the ball 190 yd down the right side of the fairway at an angle of 20° from the center of the fairway. How far is the ball from the hole?

SOLUTION A sketch of the information is shown in Figure 3.32. Note that we are given two sides of a triangle (355 yd and 190 yd) and the angle between them (20°). Thus, to find the distance x to the hole, we apply the law of cosines:

SECTION 3.3 Law of Cosines

FIGURE 3.32
An oblique triangle in which two sides and the included angle are known

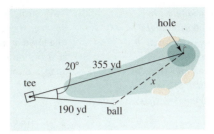

$$x^2 = 355^2 + 190^2 - 2(355)(190)\cos 20°$$
$$x = \sqrt{355^2 + 190^2 - 2(355)(190)\cos 20°} \approx 188.04 \text{ yd}$$

Hence, the ball lies approximately 188 yards from the hole.

PROBLEM 5 Referring to Example 5, suppose the second shot heads directly toward the hole but falls 40 yards short. How far is the ball from the tee after the second shot?

Exercises 3.3

Basic Skills

In Exercises 1–10,

(a) solve each oblique triangle, and round α, β, and θ to the nearest tenth of a degree and x to three significant digits.

(b) determine the area of each triangle, and round the answer to three significant digits.

1.

2.

3.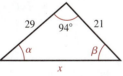

4.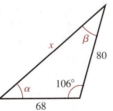

5.

6.

7.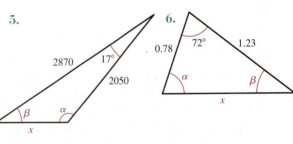

8.

9.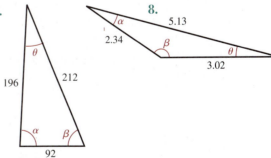

10.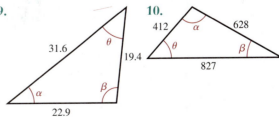

In Exercises 11–24, three parts of an oblique triangle are given. With the understanding that side a is opposite angle α, side b is opposite angle β, and side c is opposite angle θ, solve the oblique triangle with these three given parts, rounding angles to the nearest tenth of a degree and sides to three significant digits. If the data does not allow for the construction of a triangle, state that no such triangle exists.

11. $a = 19$, $b = 21$, $\theta = 42°$
12. $a = 470$, $c = 622$, $\beta = 18°$
13. $b = 3.26$, $c = 5.11$, $\alpha = 112°$
14. $a = 49.2$, $b = 8.3$, $\theta = 142°$
15. $a = 325$, $c = 625$, $\beta = 48°$
16. $b = 3.26$, $c = 5.11$, $\alpha = 112°$
17. $a = 24$, $b = 19$, $c = 10$
18. $a = 0.56$, $b = 1.22$, $c = 0.87$
19. $a = 971$, $b = 1120$, $c = 793$
20. $a = 4200$, $b = 2100$, $c = 5300$
21. $a = 3.67$, $b = 2.27$, $c = 1.76$
22. $a = 12.5$, $b = 11.3$, $c = 19.1$
23. $a = 17$, $b = 32$, $c = 10$
24. $a = 1.72$, $b = 0.98$, $c = 0.62$

In Exercises 25–42, solve each applied problem. Round angles to the nearest tenth of a degree and all other measurements to three significant digits.

25. The sides of a triangle are 7 cm, 10 cm, and 12 cm. Find the measure of the largest angle.

26. The sides of a triangle are 7 ft, 8 ft, and 10 ft. Find the measure of the smallest angle.

27. Determine the length x of the brace for the traffic light shown in the figure.

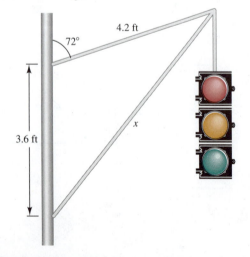

28. Two straight roads intersect with an angle of 64.5° between them. From the intersection point, it is 22.6 miles to a town at point A on one road and 15.8 miles to a town at B on the other road, as shown in the sketch. How far apart are the two towns?

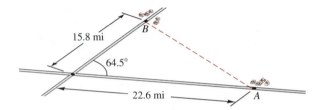

29. The distance between two points A and B, which are on opposite sides of a building, cannot be measured directly. To find the distance AB, a surveyor selects a point C and finds the distances AC and BC and angle ACB to be 132.4 ft, 156.9 ft, and 72°33′, respectively. What is the distance AB?

30. A triangular building lot has frontage of 237.5 ft on one street and frontage of 345.6 ft on another street. If the third side of the lot is 302.6 ft, find (a) the acute angle between the streets and (b) the area of the lot.

31. Two cyclists leave from the same point at the same time and travel along straight courses, which form an angle of 68° with each other. Both cyclists travel at a constant rate of speed—one at 18 mph and the other at 24 mph. What is the distance between the cyclists after 10 minutes of travel?

32. A baseball diamond is a square with 90-ft base paths. The pitcher's mound is 60 ft from home plate and lies directly between home plate and second base. How far is the pitcher's mound from first base?

33. Five drill holes are spaced equally around a circle whose radius is 8 cm, as shown in the figure. Find the distance between the centers of (a) holes A and B and (b) holes A and D.

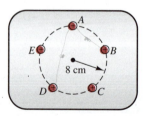

34. For the roof truss shown in the sketch, find the following measurements:
(a) BC (b) α (c) β
(d) θ (e) BD (f) CD

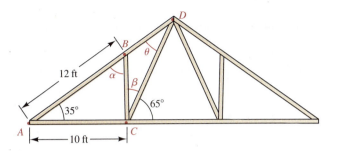

EXERCISE 34

35. Two sides of a parallelogram are 62 m and 86 m, and the acute angle between the sides is 40°.
    (a) Find the length of each diagonal of the parallelogram.
    (b) Find the area of the parallelogram.

36. The diagonals of a parallelogram are 8 cm and 12 cm and the obtuse angle between them is 130°.
    (a) Find the length of each side of the parallelogram.
    (b) Find the area of the parallelogram.

37. A 120-ft amateur radio tower stands vertically on a hill that is inclined at an angle of 9° to the horizontal. Find the length of a guy wire that is attached to the top of the tower and is anchored 85 ft from the tower (a) directly down the hill and (b) directly up the hill.

38. Four airports are located at points A, B, C, and D so that B is 220 mi directly east of A, C is 360 mi directly north of B, and D is 430 mi directly northwest of C. Find the distance between airports (a) at points B and D and (b) at points A and D.

39. From the survey notes shown in the figure, find the length of the pond from A to B.

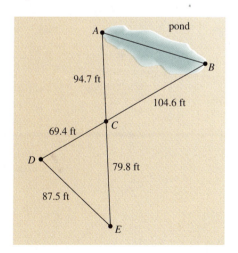

40. From the survey notes shown in the sketch, find the distance from D to E.

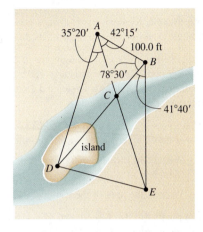

In Exercises 41 and 42, determine the number of acres in each tract of land. [Hint: 43,560 square feet = 1 acre.]

41.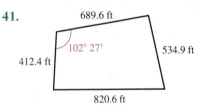

42.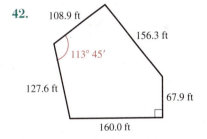

Critical Thinking

43. If a triangle does exist, is it always possible to solve the triangle when given any three of its six parts? Explain.

44. Suppose the sides of a triangle are a, b, and c, and the angle θ opposite side c is nearly 180°. Apply the law of cosines to this oblique triangle, and then give a geometric interpretation of the result.

45. The points $A(2, 3)$, $B(-2, 6)$, and $C(3, -6)$ are the vertices of an oblique triangle in the coordinate plane.

    (a) What is the measure of the largest interior angle of the triangle?

    (b) What is the area of the triangle?

46. The points $A(5, 0)$, $B(0, 0)$, and $C(x, y)$ are the vertices of an oblique triangle in the coordinate plane with point C in quadrant I, angle $ABC = 60°$, and side $BC = 8$ units. Use the law of cosines to find AC, then use the distance formula (Section 1.2) to find the coordinates of point C.

For Exercises 47–50, refer to the oblique triangle shown in the figure.

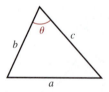

47. Find θ if $a^2 = b^2 + c^2 + bc$.

48. Find $\cos \theta$ if $b = c = 2a$.

49. Find b and c if $\theta = 120°$, $a = 3\sqrt{7}$, and $b = 2c$.

50. Find b and c if $\theta = 60°$, $a = 2\sqrt{7}$, and $b - c = 2$.

51. In a parallelogram, suppose the adjacent sides have lengths a and b and the diagonals have lengths x and y. Express the sum of the squares of the diagonals in terms of a and b.

52. The area A of the triangle in the figure is given by

$$A = \tfrac{1}{2}ab \sin \theta.$$

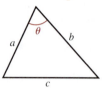

To prove Hero's formula, complete the following steps:

(a) Square both sides of the equation $A = \tfrac{1}{2}ab \sin \theta$ and show that

$$A^2 = \tfrac{1}{4}a^2b^2(1 + \cos \theta)(1 - \cos \theta).$$

(b) Solve the law of cosines, $c^2 = a^2 + b^2 - 2ab \cos \theta$, for $\cos \theta$. Then substitute for $\cos \theta$ in part (a) and show that

$$A^2 = \tfrac{1}{16}[(a + b)^2 - c^2][c^2 - (a - b)^2].$$

(c) Factor the right-hand side of the equation in part (b) and show that

$$A^2 = \tfrac{1}{16} 2s(2s - 2c)(2s - 2b)(2s - 2a),$$

where s is half the perimeter (semiperimeter) of the triangle.

(d) Take the square root of each side of the equation in part (c) to obtain

$$A = \sqrt{s(s - a)(s - b)(s - c)},$$

which is Hero's formula.

Calculator Activities

53. In triangle ABC, line segment $\overline{CD}$ bisects angle C and splits side c into two parts, x and y, as shown in the sketch.

    Given that $a = 4.27$, $b = 6.12$, and $c = 8.38$, solve the triangle as follows:

    (a) Use the law of cosines to help find angles A, B, and ACB, rounding each answer to the nearest hundredth of a degree.

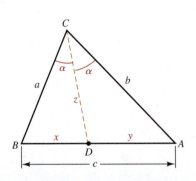

(b) Use the results of part (a) and the law of sines to help find the distances *x*, *y*, and *z*, rounding each answer to three significant digits.

(c) Find and compare the ratios $\dfrac{x}{y}$ and $\dfrac{a}{b}$, rounding each answer to two significant digits. What do you observe?

(d) Prove the observation you noted in part (c).

(e) Compute $\sqrt{ab - xy}$, rounding the answer to three significant digits. Compare this value to the value of *z* from part (b). What do you observe?

(f) Prove the observation you noted in part (e).

54. Using the observations gathered in Exercise 53, find the distances *x*, *y*, and *z* without using the law of sines or cosines. Round each answer to three significant digits.

(a)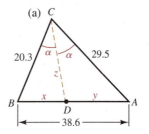

(b)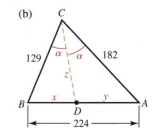

3.4 Vectors

Many quantities cannot be described by *magnitude* alone. For example, if we state that a 100-lb force acts at the end of a diving board, we have not fully described the action taking place on the board. We need to know in what *direction* the force acts. As illustrated in Figure 3.33, a 100-lb force that acts upward has a different effect on the board than a 100-lb force that acts downward.

FIGURE 3.33
The action that takes place on a diving board is influenced by both the magnitude and the direction of the force that is applied.

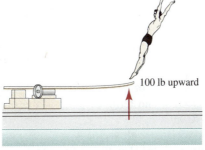
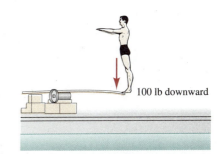

Quantities that are described by both magnitude and direction are called **vector quantities**. Examples of vector quantities are force, velocity, electrical current, displacement, and so on. In this section, we introduce vector operations and show some applications of vectors.

◆ Vectors in Standard Position

In mathematics, we represent vectors with arrows. The length of the arrow (drawn to scale) is the *magnitude* of the vector, and the arrow's orientation in the plane tells us its *direction*. In this text, we use a boldface letter to denote a vector and absolute value bars around the boldface letter to denote the magnitude. Figure 3.34 illustrates vector **A** with magnitude |**A**| and vector **B** with magnitude |**B**|. Note that these vectors have the same length (same magnitude) and are parallel (same direction). We say that vectors with the same magnitude and same direction are **equal**. Hence, we may write

$$\mathbf{A} = \mathbf{B}.$$

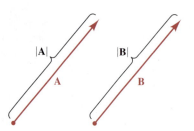

FIGURE 3.34
Vector **A** with magnitude |**A**| and vector **B** with magnitude |**B**|

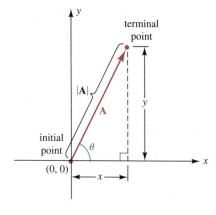

FIGURE 3.35
Vector **A** in standard position

The arrowhead of a vector represents its **terminal point,** and the opposite end of the vector represents its **initial point.** In order to have some sense of direction, we usually place a vector in the coordinate plane with its initial point at the origin $(0, 0)$. The terminal point of the vector then corresponds to some ordered pair (x, y) in the coordinate plane, as illustrated in Figure 3.35. This is called **standard position** of a vector, and the coordinates x and y are called the **components** of the vector in standard position. The notation

$$\mathbf{A} = \langle x, y \rangle$$

is called the **component form** of the vector. We refer to the vector with both initial point and terminal point at the origin as the **zero vector** and denote it by $\mathbf{0} = \langle 0, 0 \rangle$.

Associated with vector $\mathbf{A} = \langle x, y \rangle$ with $\mathbf{A} \neq \mathbf{0}$ is a **direction angle** θ, as shown in Figure 3.35. Usually, angle θ is measured counterclockwise from the positive x-axis, and $0° \leq \theta < 360°$. From our discussion of trigonometry in Chapter 2, we know that x, y, |**A**|, and θ are related as follows.

 Relationships between x, y, |A|, and θ

If |**A**| is the magnitude and θ the direction angle of the nonzero vector $\mathbf{A} = \langle x, y \rangle$, then

1. $|\mathbf{A}| = \sqrt{x^2 + y^2}$
2. $\tan \theta = \dfrac{y}{x}, \quad x \neq 0$
3. $\cos \theta = \dfrac{x}{|\mathbf{A}|}$ or $x = |\mathbf{A}| \cos \theta$
4. $\sin \theta = \dfrac{y}{|\mathbf{A}|}$ or $y = |\mathbf{A}| \sin \theta$

Our first example illustrates the procedure for translating vector **A** into standard position and expressing it in component form $\mathbf{A} = \langle x, y \rangle$. If the component form of a vector is known, then we may find the magnitude and direction angle of the vector with the equations

$$|\mathbf{A}| = \sqrt{x^2 + y^2} \quad \text{and} \quad \tan \theta = \frac{y}{x}.$$

EXAMPLE 1 Given the vector in Figure 3.36,

(a) translate **A** into standard position,

(b) express **A** in component form,

(c) find the magnitude of **A**, and

(d) find the direction angle θ (with $0° \leq \theta < 360°$) for **A**.

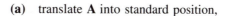

FIGURE 3.36

SOLUTION

(a) To be in standard position, vector **A** must have its initial point at the origin. Therefore, we must adjust its initial point $(-5, 1)$ by moving it five units right and one unit down. To preserve the magnitude and direction of this vector, we must make a similar adjustment to its terminal point. Moving the terminal point $(-2, -3)$ five units right and one unit down, we obtain $(3, -4)$, as illustrated in Figure 3.37.

(b) Referring to Figure 3.37, the x-component of vector **A** is 3 and its y-component is -4. Hence, the component form of vector **A** is given by

$$\mathbf{A} = \langle 3, -4 \rangle.$$

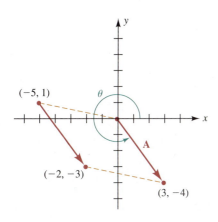

FIGURE 3.37
Vector **A** translated into standard position

(c) The magnitude of $\mathbf{A} = \langle 3, -4 \rangle$ is

$$|\mathbf{A}| = \sqrt{x^2 + y^2} = \sqrt{(3)^2 + (-4)^2} = 5.$$

(d) The direction angle θ for $\mathbf{A} = \langle 3, -4 \rangle$ must satisfy the equation

$$\tan \theta = \frac{y}{x} = \frac{-4}{3}.$$

Note that x is positive and y is negative. Hence, we know that the direction angle θ is a fourth-quadrant angle. The reference angle is

$$\theta' = \tan^{-1} \left| \frac{-4}{3} \right| \approx 53.1°.$$

Thus, we conclude

$$\theta \approx 360° - 53.1° = 306.9°.$$

PROBLEM 1 Repeat Example 1 if vector **A** has initial point $(-2, -3)$ and terminal point $(-5, 1)$. ◆

If the magnitude and direction angle of a vector are known, then we may find the x- and y-components of the vector with the equations

$$x = |\mathbf{A}| \cos \theta \quad \text{and} \quad y = |\mathbf{A}| \sin \theta.$$

EXAMPLE 2 Given that vector **A** has a direction angle 158° and a magnitude 4, find its x- and y-components. Round each answer to three significant digits.

FIGURE 3.38
A vector with magnitude 4 and direction angle 158°

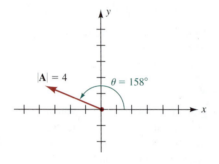

SOLUTION Vector **A** is shown in Figure 3.38. Its x-component is given by

$$x = |\mathbf{A}| \cos \theta = 4 \cos 158° \approx -3.71,$$

and its y-component by

$$y = |\mathbf{A}| \sin \theta = 4 \sin 158° \approx 1.50.$$

Hence,

$$\mathbf{A} \approx \langle -3.71, 1.50 \rangle.$$

◆

PROBLEM 2 Repeat Example 2 given that the direction angle of vector **A** is 212° and its magnitude is 8. ◆

◆ **Vector Operations**

We discuss two fundamental vector operations in this text:

1. scalar multiplication and **2.** vector addition.

We begin with *scalar multiplication*. The **product** of *scalar k* (a real number) and vector $\mathbf{A} = \langle x, y \rangle$ is the vector $k\mathbf{A}$ whose components are k times the corresponding components of vector **A**, as shown in Figure 3.39.

SECTION 3.4 Vectors

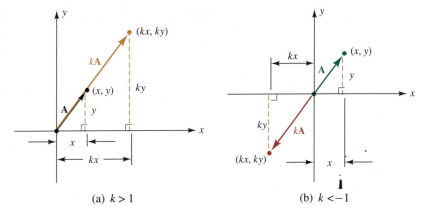

FIGURE 3.39
The components of $k\mathbf{A}$ are k times the corresponding components of $\mathbf{A}$.

(a) $k > 1$ (b) $k < -1$

Scalar Multiplication

The **scalar multiplication** of real number k times vector $\mathbf{A} = \langle x, y \rangle$ is the vector $k\mathbf{A}$ defined by

$$k\mathbf{A} = \langle kx, ky \rangle.$$

In Figure 3.39, we note that the right triangles formed with the x-axis are *similar* (have corresponding angles equal and corresponding sides proportional). Hence, if k is positive, vector $\mathbf{A}$ and vector $k\mathbf{A}$ have the same direction angle (same direction), and if k is negative, vector $\mathbf{A}$ and vector $k\mathbf{A}$ have direction angles that differ by 180° (opposite direction). Also, by similar triangles, the magnitude of vector $k\mathbf{A}$ is $|k|$ times as large as the magnitude of vector $\mathbf{A}$, that is,

$$|k\mathbf{A}| = |k|\,|\mathbf{A}|$$

EXAMPLE 3 Given that $\mathbf{A} = \langle -3, 4 \rangle$, find each vector.

(a) $3\mathbf{A}$ (b) $-\tfrac{2}{3}\mathbf{A}$

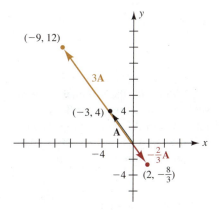

FIGURE 3.40
Relationship between vectors $\mathbf{A}$, $3\mathbf{A}$, and $-\tfrac{2}{3}\mathbf{A}$

SOLUTION

(a) Using scalar multiplication, we have

$$3\mathbf{A} = 3\langle -3, 4 \rangle = \langle 3(-3), 3(4) \rangle = \langle -9, 12 \rangle.$$

Vector $\mathbf{A}$ and vector $3\mathbf{A}$ are sketched in Figure 3.40.

(b) Using scalar multiplication, we have

$$-\tfrac{2}{3}\mathbf{A} = -\tfrac{2}{3}\langle -3, 4 \rangle = \langle -\tfrac{2}{3}(-3), -\tfrac{2}{3}(4) \rangle = \langle 2, -\tfrac{8}{3} \rangle.$$

This vector is also sketched in Figure 3.40. ◆

The **negative** of $\mathbf{A} = \langle x, y \rangle$ is the vector $-\mathbf{A}$ defined by

$$-\mathbf{A} = -1\mathbf{A} = -1\langle x, y \rangle = \langle -x, -y \rangle.$$

PROBLEM 3 Referring to vector $\mathbf{A} = \langle -3, 4 \rangle$ in Example 3, find $-\mathbf{A}$.

The **sum** of two vectors $\mathbf{A} = \langle x_1, y_1 \rangle$ and $\mathbf{B} = \langle x_2, y_2 \rangle$ is the vector $\mathbf{A} + \mathbf{B}$ whose components are the sum of the corresponding components of vectors $\mathbf{A}$ and $\mathbf{B}$, as shown in Figure 3.41.

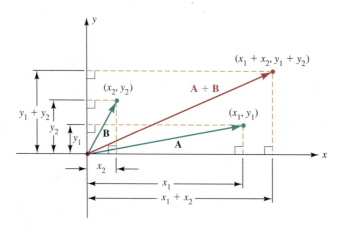

FIGURE 3.41
The components of $\mathbf{A} + \mathbf{B}$ are the sum of the corresponding components of $\mathbf{A}$ and $\mathbf{B}$.

Vector Addition

The **vector addition** of $\mathbf{A} = \langle x_1, y_1 \rangle$ and $\mathbf{B} = \langle x_2, y_2 \rangle$ is the vector $\mathbf{A} + \mathbf{B}$ defined as

$$\mathbf{A} + \mathbf{B} = \langle x_1 + x_2, y_1 + y_2 \rangle.$$

Note: In applied problems, we usually refer to the vector sum $\mathbf{A} + \mathbf{B}$ as the **resultant** of vectors $\mathbf{A}$ and $\mathbf{B}$.

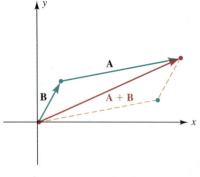

FIGURE 3.42
Vector $\mathbf{A} + \mathbf{B}$ is the diagonal of a parallelogram with adjacent sides $\mathbf{A}$ and $\mathbf{B}$.

Graphically, we may find the sum of vectors $\mathbf{A} = \langle x_1, y_1 \rangle$ and $\mathbf{B} = \langle x_2, y_2 \rangle$ by translating vector $\mathbf{A}$ such that its initial point coincides with the terminal point of vector $\mathbf{B}$, as shown in Figure 3.42 (or, alternatively, by translating vector $\mathbf{B}$ such that its initial point coincides with the terminal point of vector $\mathbf{A}$). Note in Figure 3.42 that the vector $\mathbf{A} + \mathbf{B}$ is the *diagonal of a parallelogram* having $\mathbf{A}$ and $\mathbf{B}$ as adjacent sides. We refer to this observation as the **parallelogram law** for vectors.

EXAMPLE 4 Given that $\mathbf{A} = \langle -2, 3 \rangle$ and $\mathbf{B} = \langle -4, -4 \rangle$, find each vector sum.

(a) $\mathbf{A} + \mathbf{B}$ (b) $-2\mathbf{A} + \frac{1}{4}\mathbf{B}$

SECTION 3.4 Vectors

SOLUTION

(a) Using vector addition, we have

$$\mathbf{A} + \mathbf{B} = \langle -2, 3 \rangle + \langle -4, -4 \rangle = \langle -2 + (-4), 3 + (-4) \rangle = \langle -6, -1 \rangle.$$

A sketch of this vector is shown in Figure 3.43.

(b) Using scalar multiplication first, then vector addition, we have

$$-2\mathbf{A} + \tfrac{1}{4}\mathbf{B} = -2\langle -2, 3 \rangle + \tfrac{1}{4}\langle -4, -4 \rangle = \langle 4, -6 \rangle + \langle -1, -1 \rangle = \langle 3, -7 \rangle$$

This vector is sketched in Figure 3.44. ◆

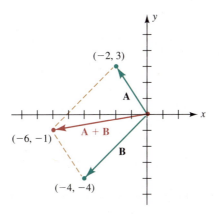

FIGURE 3.43
$\mathbf{A} + \mathbf{B}$ is the diagonal of the parallelogram with sides $\mathbf{A}$ and $\mathbf{B}$.

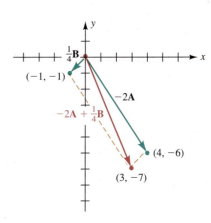

FIGURE 3.44
$-2\mathbf{A} + \tfrac{1}{4}\mathbf{B}$ is the diagonal of the parallelogram with sides $-2\mathbf{A}$ and $\tfrac{1}{4}\mathbf{B}$.

PROBLEM 4 Given vectors $\mathbf{A}$ and $\mathbf{B}$ in Example 4, find $2\mathbf{A} + 3\mathbf{B}$. ◆

To subtract vector $\mathbf{B} = \langle x_2, y_2 \rangle$ from vector $\mathbf{A} = \langle x_1, y_1 \rangle$, we use vector addition to add the *negative* of vector $\mathbf{B}$ to vector $\mathbf{A}$. Hence, we have the following rule for **vector subtraction:**

$$\mathbf{A} - \mathbf{B} = \mathbf{A} + (-\mathbf{B}) = \langle x_1 - x_2, y_1 - y_2 \rangle$$

Since $\mathbf{A} + (-\mathbf{B})$ is the diagonal of a parallelogram formed by sides $\mathbf{A}$ and $-\mathbf{B}$, we may describe $\mathbf{A} - \mathbf{B}$ as the vector from the terminal point of $\mathbf{B}$ to the terminal point of $\mathbf{A}$, as shown in Figure 3.45.

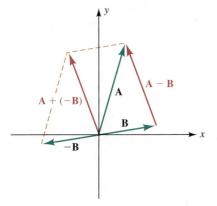

FIGURE 3.45
$\mathbf{A} - \mathbf{B}$ may be described by the vector from the terminal point of $\mathbf{B}$ to the terminal point of $\mathbf{A}$.

EXAMPLE 5 Referring to vectors **A** and **B** in Example 4, find **B** − **A**.

SOLUTION Using vector subtraction, we have

$$\mathbf{B} - \mathbf{A} = \langle -4, -4 \rangle - \langle -2, 3 \rangle = \langle -4 - (-2), -4 - 3 \rangle = \langle -2, -7 \rangle.$$

This vector subtraction is illustrated graphically in Figure 3.46.

FIGURE 3.46
B − **A** may be described by the vector from the terminal point of **A** to the terminal point of **B**.

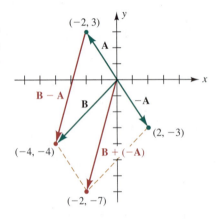

PROBLEM 5 Referring to the vectors **A** and **B** in Example 4, find **A** − **B**.

Many of the properties of scalar multiplication and vector addition are similar to the properties of real numbers. Next, we list nine fundamental properties of scalar multiplication and vector addition.

Fundamental Properties of Scalar Multiplication and Vector Addition

If **A**, **B**, and **C** are vectors, **0** is the zero vector, and c and d are scalars, then

1. **A** + **B** = **B** + **A**   Commutative property of vector addition
2. (**A** + **B**) + **C** = **A** + (**B** + **C**)   Associative property of vector addition
3. **A** + **0** = **A**   Identity property of vector addition
4. **A** + (−**A**) = **0**   Inverse property of vector addition
5. (cd)**A** = $c(d$**A**$)$ Associative property of scalar multiplication
6. 1**A** = **A**   Identity property of scalar multiplication
7. 0**A** = **0**   Multiplicative property of zero
8. $c($**A** + **B**$)$ = c**A** + c**B**   Distributive property of a scalar over vector addition
9. $(c + d)$**A** = c**A** + d**A**   Distributive property of a vector over scalar addition

EXAMPLE 7 A dogsled is pulled with a force of 100 lb at an angle of 30° with the horizontal. (a) Find the force that pulls the sled horizontally and (b) the force that lifts the sled vertically.

SOLUTION We begin by letting **F** be a vector that represents this force and drawing a vector diagram in the coordinate plane, as shown in Figure 3.47. Note that the magnitude of **F** is 100 and the direction angle for **F** is 30°.

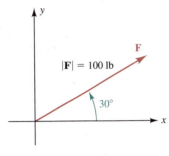

FIGURE 3.47
A vector diagram showing a force of 100 lb applied at an angle of 30°

(a) The force that pulls the sled horizontally is the *x*-component of the vector **F**. Using the fact that $x = |\mathbf{F}| \cos \theta$, we have

$$x = 100 \cos 30° \approx 86.6 \text{ lb}.$$

Hence, the force tending to pull the sled horizontally is 86.6 lb.

(b) The force that lifts the sled vertically is the *y*-component of the vector **F**. Using the fact that $y = |\mathbf{F}| \sin \theta$, we have

$$y = 100 \sin 30° = 50 \text{ lb}.$$

Hence, the force tending to lift the sled vertically is 50 lb. ◆

PROBLEM 7 If the sled in Example 7 weighs 75 lb, determine the magnitude of the force that must be applied at an angle of 30° with the horizontal in order to just lift the sled from the surface. ◆

A **force system** is a group of forces acting on a particular object. For our discussion, a force system is *concurrent* and *coplanar* (that is, the lines of action of all forces pass through a common point and lie in the same plane). A single force that can replace a force system and has the same physical effect upon the object as the system it replaces is called the **resultant force.** To determine the resultant force, we find the vector sum of all the forces in the system.

EXAMPLE 8 For the force system in Figure 3.48, find the magnitude and direction angle of the resultant force.

SOLUTION The resultant force **F** is the vector sum of $\mathbf{F}_1$ and $\mathbf{F}_2$. To find this sum, we express $\mathbf{F}_1$ and $\mathbf{F}_2$ in component form and then apply the rule for vector addition:

$$\mathbf{F} = \mathbf{F}_1 + \mathbf{F}_2$$
$$= \langle 70 \cos 270°, 70 \sin 270° \rangle + \langle 100 \cos 150°, 100 \sin 150° \rangle$$
$$= \langle 70 \cos 270° + 100 \cos 150°, 70 \sin 270° + 100 \sin 150° \rangle$$
$$\approx \langle -86.6, -20 \rangle$$

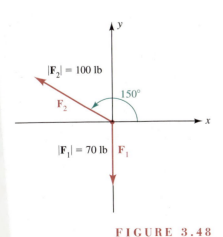

FIGURE 3.48

The resultant force **F** is shown in Figure 3.49. Note that it is the diagonal of the parallelogram with sides $\mathbf{F}_1$ and $\mathbf{F}_2$. The magnitude of **F** is

$$|\mathbf{F}| \approx \sqrt{(-86.6)^2 + (-20)^2} \approx 89 \text{ lb}.$$

Unit Vectors

Any vector whose magnitude is 1 is called a **unit vector.** Two special unit vectors, designated by **i** and **j**, are defined as

$$\mathbf{i} = \langle 1, 0 \rangle \quad \text{and} \quad \mathbf{j} = \langle 0, 1 \rangle.$$

We can use the vectors **i** and **j** to obtain an alternative way of denoting vector $\mathbf{A} = \langle x, y \rangle$ as follows:

$$\mathbf{A} = \langle x, y \rangle = x \langle 1, 0 \rangle + y \langle 0, 1 \rangle$$
$$= x\mathbf{i} + y\mathbf{j}$$

Hence, $\mathbf{A} = \langle x, y \rangle$ and $\mathbf{A} = x\mathbf{i} + y\mathbf{j}$ are simply two different ways of stating that **A** is a vector with components x and y. We can use the unit vectors to describe any vector, such as

$$\langle 2, 3 \rangle = 2\mathbf{i} + 3\mathbf{j} \quad \text{and} \quad \langle 1, -6 \rangle = \mathbf{i} - 6\mathbf{j}$$

or even vectors that have at least one component equal to zero, such as

$$\langle -4, 0 \rangle = -4\mathbf{i} + 0\mathbf{j} = -4\mathbf{i} \quad \text{and} \quad \langle 0, 0 \rangle = 0\mathbf{i} + 0\mathbf{j} = \mathbf{0}.$$

We can also define scalar multiplication, vector addition, and vector subtraction in terms of the unit vectors **i** and **j**:

1. $k(x\mathbf{i} + y\mathbf{j}) = (kx)\mathbf{i} + (ky)\mathbf{j}$
2. $(x_1\mathbf{i} + y_1\mathbf{j}) + (x_2\mathbf{i} + y_2\mathbf{j}) = (x_1 + x_2)\mathbf{i} + (y_1 + y_2)\mathbf{j}$
3. $(x_1\mathbf{i} + y_1\mathbf{j}) - (x_2\mathbf{i} + y_2\mathbf{j}) = (x_1 - x_2)\mathbf{i} + (y_1 - y_2)\mathbf{j}$

EXAMPLE 6 Given that $\mathbf{A} = 2\mathbf{i} - 3\mathbf{j}$ and $\mathbf{B} = \mathbf{i} - 6\mathbf{j}$, find $2\mathbf{A} - \mathbf{B}$.

SOLUTION Using scalar multiplication first, then vector subtraction, we obtain

$$2\mathbf{A} - \mathbf{B} = 2(2\mathbf{i} - 3\mathbf{j}) - (\mathbf{i} - 6\mathbf{j})$$
$$= (4\mathbf{i} - 6\mathbf{j}) - (\mathbf{i} - 6\mathbf{j})$$
$$= (4 - 1)\mathbf{i} + [-6 - (-6)]\mathbf{j}$$
$$= 3\mathbf{i} + 0\mathbf{j} \quad \text{or} \quad \langle 3, 0 \rangle.$$

PROBLEM 6 Given the vectors **A** and **B** in Example 6, find $-\mathbf{A} + 2\mathbf{B}$.

Application: Forces Acting on an Object

Many applied problems in physics and mechanics deal with a *force* or a gr forces acting on a particular object. To analyze such problems, we let a vect resent each force and form a *vector diagram* in the coordinate plane.

Note that both the *x*- and *y*-components of **F** are negative. Hence, we know that the direction angle θ for **F** is between 180° and 270°. The reference angle is

$$\theta' \approx \tan^{-1}\left|\frac{-20}{-86.6}\right| \approx 13.0°.$$

Hence,

$$\theta \approx 180° + 13.0° = 193°.$$

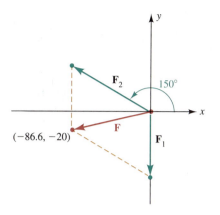

FIGURE 3.49
The resultant force **F** is the diagonal of the parallelogram with sides F_1 and F_2.

If an object is in **equilibrium** (at rest), *the vector sum of all forces acting on the object must equal zero.* For equilibrium to occur in the force system of Example 8, we must add an additional force of the same magnitude as the resultant force **F** but in the opposite direction.

PROBLEM 8 Referring to the force system in Example 8, find an additional force F_3 that produces equilibrium.

◆ **Application: Displacement to a Landing Strip**

The **displacement** between two points is the length and direction of the straight line between the two points. Since displacement is described by both length (magnitude) and direction, it is considered a vector quantity. When displacement is discussed in navigation problems, the direction is usually given with respect to a compass reading, called a **bearing**. A bearing is expressed in degrees east (E) or west (W) of the north (N) or south (S) direction. Referring to Figure 3.50, the direction from point *A* to point *B* is given by the bearing S-30°-E, and the direction from point *A* to point *C* is given by the bearing N-75°-W. The direction North (N) or South (S) is always given as the first reference of a bearing.

FIGURE 3.50
Bearings are measured from the North or South axis.

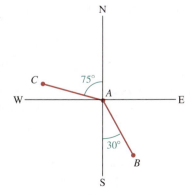

212 CHAPTER 3 Applications of the Trigonometric Functions

EXAMPLE 9 The displacement from an airport to an old landing strip is known to be 200 miles S-85°-W. A small aircraft located 260 miles N-88°-W of the airport has engine trouble and needs to reach the old landing strip. What is the airplane's displacement to the landing strip?

SOLUTION We begin by drawing the given displacements as vectors $\mathbf{D}_1$ and $\mathbf{D}_2$ (see the axes on the right in Figure 3.51).

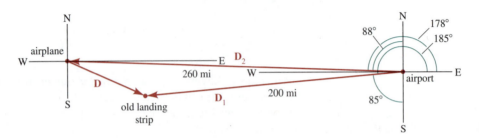

FIGURE 3.51 Graphical display of displacements $\mathbf{D}_1$, $\mathbf{D}_2$, and $\mathbf{D}$

Note in Figure 3.51 that the airplane's displacement $\mathbf{D}$ to the old landing strip is the vector from the terminal point of $\mathbf{D}_2$ to the terminal point of $\mathbf{D}_1$. Hence,

$$\mathbf{D} = \mathbf{D}_1 - \mathbf{D}_2.$$

To find this difference, we express $\mathbf{D}_1$ and $\mathbf{D}_2$ in component form and then apply the rule for vector subtraction:

$$\begin{aligned}\mathbf{D} &= \mathbf{D}_1 - \mathbf{D}_2 \\ &= \langle 200 \cos 185°, 200 \sin 185° \rangle - \langle 260 \cos 178°, 260 \sin 178° \rangle \\ &= \langle 200 \cos 185° - 260 \cos 178°, 200 \sin 185° - 260 \sin 178° \rangle \\ &\approx \langle 60.6, -26.5 \rangle \end{aligned}$$

Hence, the magnitude of $\mathbf{D}$ is

$$|\mathbf{D}| = \sqrt{(60.6)^2 + (-26.5)^2} \approx 66 \text{ mi}.$$

The reference angle associated with $\mathbf{D}$ is

$$\theta' \approx \tan^{-1}\left|\frac{-26.5}{60.6}\right| \approx 23.6°.$$

Measuring from the South axis, we obtain $90° - 23.6° = 66.4°$. Therefore, the airplane should fly 66 miles S-66.4°-E in order to reach the old landing strip. ◆

An alternative approach to solving Example 9 is to use the law of cosines and law of sines. Observe that the vectors in Figure 3.51 form an oblique triangle with sides 200 miles and 260 miles. The angle between these sides is

$180° - (88° + 85°) = 7°$. Since we know two sides and the angle between them (SAS), we can apply the law of cosines to find $|\mathbf{D}|$, and the law of sines to find the angle θ between $\mathbf{D}$ and $\mathbf{D}_2$. Note that the angle between $\mathbf{D}_2$ and the West axis on the right in Figure 3.51 is 2°. Thus, the angle between $\mathbf{D}_2$ and the East axis on the left is also 2°. Hence, the bearing is S-[90° − (θ + 2°)]-E.

PROBLEM 9   Referring to Figure 3.51, use the law of cosines and law of sines to find the airplane's displacement to the landing strip. ◆

Exercises 3.4

  Basic Skills

In Exercises 1–8, translate the vector into standard position and express it in component form.

1.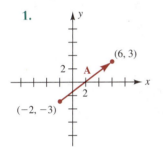

2.

3.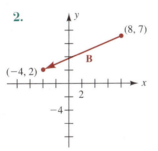

4.

7.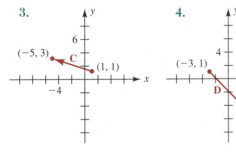

8.

9. Determine the magnitude of vectors **A** and **C** in Exercises 1 and 3.

10. Determine the magnitude of vectors **B** and **D** in Exercises 2 and 4.

5.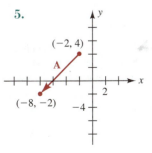

6.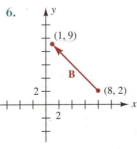

11. Determine the direction angle θ of vectors **A** and **C** in Exercises 5 and 7.

12. Determine the direction angle θ of vectors **B** and **D** in Exercises 6 and 8.

In Exercises 13–20, determine the magnitude and direction angle θ of the vector.

13. $\mathbf{A} = \langle 5, -12 \rangle$
14. $\mathbf{B} = \langle 15, 8 \rangle$
15. $\mathbf{C} = \langle -6, -9 \rangle$
16. $\mathbf{D} = \langle -25, 35 \rangle$
17. $\mathbf{A} = \langle \frac{1}{2}, \frac{1}{4} \rangle$
18. $\mathbf{B} = \langle -\frac{2}{9}, \frac{4}{3} \rangle$
19. $\mathbf{C} = \langle -1, \sqrt{7} \rangle$
20. $\mathbf{D} = \langle -\sqrt{5}, -2 \rangle$

In Exercises 21–28, determine the x- and y-components of a vector whose magnitude and direction angle θ are given.

21. $|\mathbf{A}| = 620, \theta = 45°$
22. $|\mathbf{B}| = 58, \theta = 30°$
23. $|\mathbf{C}| = 84, \theta = 210°$
24. $|\mathbf{F}| = 6\sqrt{2}, \theta = 315°$
25. $|\mathbf{A}| = \sqrt{2}, \theta = 225°$
26. $|\mathbf{B}| = 18, \theta = 120°$
27. $|\mathbf{C}| = \frac{9}{2}, \theta = 270°$
28. $|\mathbf{F}| = 7, \theta = 180°$

In Exercises 29–40, perform the indicated vector operations given that

$$\mathbf{A} = \langle 2, 5 \rangle, \quad \mathbf{B} = \langle -1, 6 \rangle,$$
$$\mathbf{C} = -8\mathbf{i} - 2\mathbf{j}, \quad \mathbf{D} = 6\mathbf{i} - 3\mathbf{j}.$$

Express each answer in component form.

29. $2\mathbf{A}$
30. $-\mathbf{B}$
31. $-\frac{1}{2}\mathbf{C}$
32. $\frac{8}{3}\mathbf{D}$
33. $\mathbf{A} + \mathbf{B}$
34. $\mathbf{C} + \mathbf{D}$
35. $\mathbf{A} - \mathbf{C}$
36. $\mathbf{D} - \mathbf{B}$
37. $4\mathbf{A} - 3\mathbf{C}$
38. $-\mathbf{A} + 3\mathbf{B}$
39. $\frac{1}{3}\mathbf{D} + \frac{1}{4}\mathbf{C}$
40. $\frac{9}{2}\mathbf{C} - 9\mathbf{B}$
41. $\mathbf{A} + (\mathbf{C} + \mathbf{D})$
42. $2\mathbf{A} - (\mathbf{B} - \frac{3}{2}\mathbf{C})$
43. $|\mathbf{A}|(\mathbf{D} + 4\mathbf{C})$
44. $|\mathbf{A} + \mathbf{B}|(\mathbf{C} + \mathbf{D})$

In Exercises 45–50, illustrate graphically the indicated operation.

45. the scalar multiplication in Exercise 31
46. the scalar multiplication in Exercise 32
47. the vector sum in Exercise 33
48. the vector sum in Exercise 34
49. the vector difference in Exercise 35
50. the vector difference in Exercise 36

51. A rocket takes off at an angle of 45° with a constant velocity of 1200 mph, as shown in the sketch. (a) At what rate does the rocket rise vertically? (b) At what rate does the rocket move horizontally?

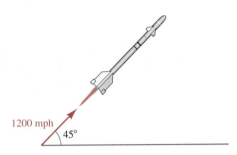

52. A 200-lb force is exerted horizontally to a crate resting on an inclined plane, as shown in the figure. If the angle of inclination of the plane is 30°, find the force that tends to push the crate up the plane.

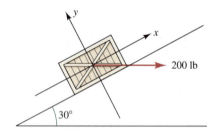

Critical Thinking

53. Express the magnitude of $\mathbf{A} + \mathbf{B}$ in terms of $|\mathbf{A}|$ and $|\mathbf{B}|$, given that vectors $\mathbf{A}$ and $\mathbf{B}$ have (a) the same direction or (b) opposite direction.

54. Suppose $\mathbf{A}$ and $\mathbf{B}$ are two nonzero vectors such that $\mathbf{A} + \mathbf{B} = \mathbf{0}$. What is the relationship between the magnitude and direction of these two vectors?

55. Suppose $\mathbf{A} = \langle x, y \rangle$, $\mathbf{A} \neq \mathbf{0}$. Express the vector $\frac{1}{|\mathbf{A}|}\mathbf{A}$ in component form and then find the magnitude of this vector. What type of vector is $\frac{1}{|\mathbf{A}|}\mathbf{A}$?

56. Using the results of Exercise 55, find a unit vector that has the same direction as each vector.

(a) $\mathbf{A} = \langle -3, 4 \rangle$ (b) $\mathbf{B} = \langle 1, -3 \rangle$
(c) $\mathbf{C} = 5\mathbf{i} + 12\mathbf{j}$ (d) $\mathbf{D} = -8\mathbf{i} - 4\mathbf{j}$

57. Find the scalars c and d such that the given scalar multiplication and vector addition is true.

(a) $c\langle 1, 3 \rangle + d\langle -2, 6 \rangle = \langle 7, -3 \rangle$
(b) $c\langle 2, -5 \rangle + d\langle -1, -1 \rangle = \langle 2, 9 \rangle$

58. The *dot product* of two vectors is a way of multiplying one vector by another. The dot product of two vectors $\mathbf{A} = \langle x_1, y_1 \rangle$ and $\mathbf{B} = \langle x_2, y_2 \rangle$ is denoted $\mathbf{A} \cdot \mathbf{B}$, and

$$\mathbf{A} \cdot \mathbf{B} = x_1 x_2 + y_1 y_2.$$

Note that **A · B** is a *real number,* not a vector. The real number is found by multiplying corresponding components of vectors **A** and **B** and then finding their sum. Find **A · B** for the given vectors.

(a) $\mathbf{A} = \langle -3, 5 \rangle$ and $\mathbf{B} = \langle 4, 2 \rangle$

(b) $\mathbf{A} = 4\mathbf{i} + 8\mathbf{j}$ and $\mathbf{B} = -2\mathbf{i} + \mathbf{j}$

59. Use the definition of the dot product from Exercise 58 and vectors $\mathbf{A} = \langle x_1, y_1 \rangle$, $\mathbf{B} = \langle x_2, y_2 \rangle$, and $\mathbf{C} = \langle x_3, y_3 \rangle$ to show that each statement is true.
(a) $\mathbf{A} \cdot \mathbf{B} = \mathbf{B} \cdot \mathbf{A}$
(b) $\mathbf{A} \cdot (\mathbf{B} + \mathbf{C}) = \mathbf{A} \cdot \mathbf{B} + \mathbf{A} \cdot \mathbf{C}$
(c) $\mathbf{A} \cdot \mathbf{A} = |\mathbf{A}|^2$

60. The figure shows vectors **A, B,** and **A − B,** with angle θ $(0 < \theta \leq 180°)$ between vectors **A** and **B**. By the law of cosines (Section 3.3),

$$|\mathbf{A} - \mathbf{B}|^2 = |\mathbf{A}|^2 + |\mathbf{B}|^2 - 2|\mathbf{A}||\mathbf{B}| \cos \theta.$$

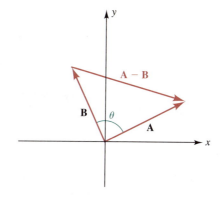

(a) Use the results of Exercise 59 to show that

$$|\mathbf{A} - \mathbf{B}|^2 = (\mathbf{A} - \mathbf{B}) \cdot (\mathbf{A} - \mathbf{B}) =$$
$$|\mathbf{A}|^2 + |\mathbf{B}|^2 - 2\mathbf{A} \cdot \mathbf{B}.$$

(b) Substitute $|\mathbf{A}|^2 + |\mathbf{B}|^2 - 2\mathbf{A} \cdot \mathbf{B}$ for $|\mathbf{A} - \mathbf{B}|^2$ in the law of cosines formula, and then state an alternate formula for **A · B** that involves the angle θ.

(c) What is the value of **A · B** if the angle θ between two vectors **A** and **B** is 90°? 180°?

Calculator Activities

In Exercises 61–64, determine the magnitude and direction angle θ of each vector. Round the magnitude to three significant digits and round the direction angle to the nearest tenth of a degree.

61. $\mathbf{A} = \langle -1.25, 0.76 \rangle$ **62.** $\mathbf{B} = \langle 56.5, -82.7 \rangle$

63. $\mathbf{C} = \langle 0.356, 1.98 \rangle$ **64.** $\mathbf{D} = \langle -123.7, -987.2 \rangle$

In Exercises 65–68, determine the x- and y-components of a vector whose magnitude and direction angle θ are given. Round the x- and y-components to three significant digits.

65. $|\mathbf{A}| = 5.61, \theta = 125.2°$ **66.** $|\mathbf{B}| = 11.2, \theta = 232.4°$

67. $|\mathbf{C}| = 79.6, \theta = 342.6°$ **68.** $|\mathbf{D}| = 0.758, \theta = 21.9°$

For the force systems in Exercises 69–74, find the magnitude and direction angle θ of the resultant force. Round the magnitude to three significant digits and round θ to the nearest tenth of a degree.

69.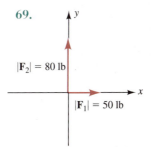

70.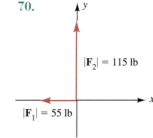

71.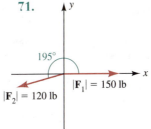

72.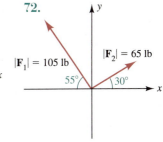

73.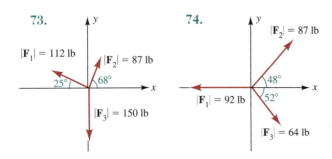

74.

75. For the force system in Exercise 69, find an additional force $\mathbf{F}_3$ that produces equilibrium.

76. For the force system in Exercise 72, find an additional force $\mathbf{F}_3$ that produces equilibrium.

77. A man rows at a rate of 5.2 mph in still water. He wants to row directly across a river, so he keeps the boat moving perpendicular to the flow of the current. The river is 190 ft wide and flows at the rate of 2.3 mph, as shown in the accompanying sketch.

(a) Find the magnitude and direction angle of the resultant velocity of the boat.

(b) Find the distance downstream the boat lands.

(c) In what direction should the rower head the boat in order to land across the river, directly opposite the starting point?

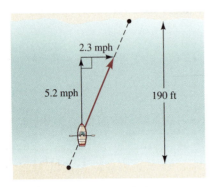

78. Two tugboats are towing an oil barge along a canal. One tugboat exerts a force of 5500 lb on its cable and the other tugboat exerts a force of 4200 lb on its cable, as shown in the figure. The angle between the cables is 17°.

(a) Find the magnitude of the resultant force.

(b) Find the angle between the resultant force and the cable with the force of 5500 lb.

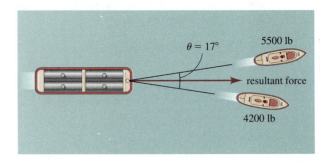

79. A small aircraft leaves airport A and flies 132 miles S-12°-E. It then changes course and flies 214 miles S-69°-E to land at airport B. What is the displacement from airport A to airport B?

80. A sailboat leaves island A and sails 82 miles N-31°-E. It then changes course and sails 107 miles N-78°-E to dock at island B. Find the displacement from island B to island A.

81. The displacement from a harbor on the mainland to a small island is known to be 62.2 miles N-43°30'-E. A sailboat, located 85.3 miles N-37°15'-E of the harbor, learns of an impending tropical storm and heads to the island for safety. What is the sailboat's displacement to the island?

82. The displacement from an observation tower to a fire is known to be 12.7 miles N-8°30'-E. A group of firemen, located 15.2 miles N-12°20'-W of the tower, must reach the fire. What is the displacement from the firemen to the fire?

Chapter 3 Review

Questions for Group Discussion

1. What is meant by *solving a right triangle*? Discuss the procedure for solving a right triangle in which an acute angle and the hypotenuse are known.

2. What is meant by *simple harmonic motion*?

3. What is an *oblique triangle*?

4. State the four cases in which it is possible to solve an oblique triangle. In which of these cases is the *law of sines* applied first? the *law of cosines* applied first?

5. Is it possible to solve an oblique triangle in which only the three angles are given? Explain.

6. Discuss the conditions when the *ambiguous case* may occur in an oblique triangle.

7. Give some examples of oblique triangles with erroneous data that have no solution.

8. When solving an oblique triangle in which all three sides are known, why is it best to find the angle opposite the largest side first?

9. What is a *vector quantity*? Give some examples of vector quantities.

10. When is a vector in *standard position*? How is a vector in standard position specified?

11. What is meant by the *x*- and *y*-components of a vector?

12. Explain the procedure for finding the *magnitude* and *direction angle* of a vector in component form.

13. Illustrate graphically the *scalar multiplication* of real number *k* times vector **A** when (a) $k > 1$, (b) $0 < k < 1$, (c) $k < -1$, and (d) $-1 < k < 0$.

14. Discuss the *parallelogram law* for vectors.

15. Describe graphically the vectors **A** − **B** and **B** − **A**.

16. What is meant by the *resultant force* of a *force system*? Explain the procedure for determining the resultant force.

17. When is a force system in *equilibrium*?

18. The models for simple harmonic motion assume the oscillating motion will repeat itself over and over again. Why is this somewhat unrealistic?

Review Exercises

In Exercises 1–10, (a) solve each oblique triangle and (b) find the area of the triangle.

1.

2.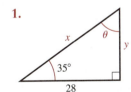

3.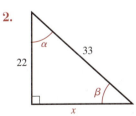

4.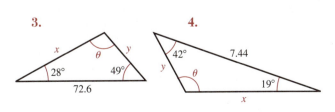

5.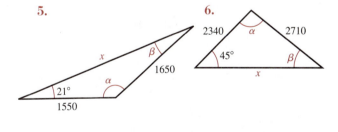

6.

7.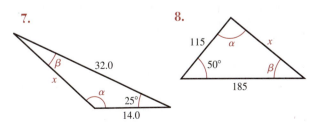

8.

9.

10.

In Exercises 11–22, three parts of an oblique triangle are given. With the understanding that side a is opposite angle α, side b is opposite angle β, and side c is opposite angle θ, solve the oblique triangle with these three given parts. If the data allows for the construction of two triangles, find both solutions. If the data does not allow for the construction of a triangle, state that no such triangle exists.

11. $b = 60.4$, $c = 55.4$, $\beta = 78°$

12. $c = 3.71$, $\alpha = 121°$, $\theta = 33°$

13. $a = 19$, $b = 35$, $c = 22$

14. $a = 0.123$, $b = 0.231$, $\theta = 59°$

15. $a = 610$, $c = 950$, $\alpha = 38°$

16. $b = 5.65$, $c = 4.85$, $\beta = 106°$

17. $b = 12.9$, $c = 9.32$, $\theta = 65°$

18. $a = 456$, $b = 575$, $\alpha = 52°$

19. $a = 2.80$, $b = 6.94$, $\theta = 48°$

20. $a = 152$, $b = 355$, $c = 170$

21. $a = 11$, $b = 16$, $c = 5$

22. $a = 3.64$, $c = 1.94$, $\theta = 42°$

23. Use Hero's formula to find the area A of an equilateral triangle with side a.

24. Given that θ is the angle between two adjacent sides a and b of a parallelogram, express the area of the parallelogram in terms of θ, a, and b.

For the vectors in Exercises 25–28, (a) translate the vector into standard position, (b) express the vector in component form, (c) determine its magnitude, and (d) determine its direction angle.

25. **26.**

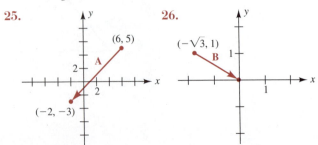

27.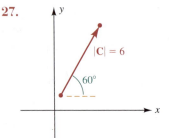

28.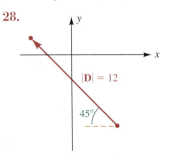

29. Determine the magnitude and direction angle of each vector.

   (a) $\mathbf{A} = \langle 12, -5 \rangle$ (b) $\mathbf{C} = \langle \sqrt{7}, 3 \rangle$

30. Determine the x- and y-components of a vector whose magnitude and direction angle θ are given.

   (a) $|\mathbf{A}| = 72.4$, $\theta = 132.7°$ (b) $|\mathbf{C}| = 643$, $\theta = 250°$

In Exercises 31–38, perform the indicated vector operations given that

$$\mathbf{A} = \langle -3, 4 \rangle, \quad \mathbf{B} = \langle -2, -6 \rangle,$$
$$\mathbf{C} = 3\mathbf{i} - 7\mathbf{j}, \quad \mathbf{D} = 3\mathbf{i} + \mathbf{j}.$$

Express each answer in component form.

31. $-\mathbf{C}$ 32. $\frac{1}{2}\mathbf{B}$

33. $\mathbf{A} + \mathbf{C}$ 34. $\mathbf{D} - \mathbf{B}$

35. $2\mathbf{A} - 5\mathbf{C}$ 36. $-\mathbf{A} + 3(\mathbf{B} - 2\mathbf{C})$

37. $|\mathbf{A}|(\mathbf{D} + 4\mathbf{C})$ 38. $|\mathbf{A} + \mathbf{C}|(\mathbf{C} + \mathbf{D})$

39. Illustrate graphically the vector sum in Exercise 33.

40. Illustrate graphically the vector difference in Exercise 34.

41. As illustrated in the sketch, the length of the shadow of a telephone pole is 48 ft when the angle of elevation of the sun is 32°. What is the height of the pole?

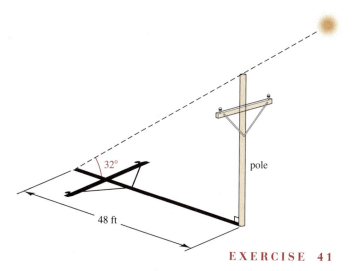

EXERCISE 41

42. For the piece of land shown in the sketch find angles α and β. Round to the nearest tenth of a degree.

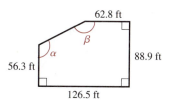

43. A tractor-trailer maintains a speed of 30 mph while climbing a hill that rises at an angle of 8.6°. How many minutes are required for the rig to attain a vertical rise of 2450 ft?

44. An air-traffic controller in a tower notes that two small planes on the ground are directly in line with the tower. The angles of depression to the planes are 9°15′ and 12°30′. If the controller is 150 ft above the ground, find the distance between the two planes.

45. The wake from a small boat sets a fishing bob in simple harmonic motion, as shown in the sketch. In 2 seconds the bob rides from the crest of one wave to the crest of the next wave.

(a) Find the equation that describes the displacement d of the bob from its equilibrium position if $d = 0$ inches at $t = 0$ seconds.

(b) Find the displacement of the bobber at $t = 3\frac{3}{4}$s.

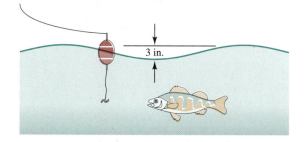

46. The periodic motion of a pendulum is an example of simple harmonic motion. The displacement d of the pendulumn from its vertical position after t seconds can be described by the equation

$$d = a \cos \sqrt{\frac{g}{L}}\, t$$

where $|a|$ is the maximum displacement from the vertical position, g is the gravitational constant 980 cm/s² and L is the length of the pendulum (in centimeters). Suppose a pendulum is 80 cm long and the maximum displacement, 7.0 cm, occurs at $t = 0$ seconds.

(a) Write the equation for the simple harmonic motion.

(b) Sketch the graph of the motion described in part (a).

(c) Find the time when the pendulum passes through the vertical position for the third time.

47. Because of a large oak tree directly on line, a surveyor is unable to measure directly the distance between two points A and B. To determine this distance indirectly, he selects a third point C, which is 75.6 ft from A and 93.6 ft from B. With his transit, he then finds angle BCA to be 72°30′. Determine, to the nearest tenth of a foot, the distance from A to B.

48. Two passenger trains depart from the same terminal at the same time and travel along straight railroads that make an angle of 65° with each other. One of the trains travels at 32 mph and the other at 40 mph. How far apart are the trains at the end of 2 hours 15 minutes?

49. In a baseball game, a right fielder positions himself 290 ft from home plate along a line that is 18° from the foul line, as shown in the sketch. Given that a baseball diamond is a square with 90-ft base paths, find the distance from the right fielder to (a) first base, (b) second base, and (c) third base.

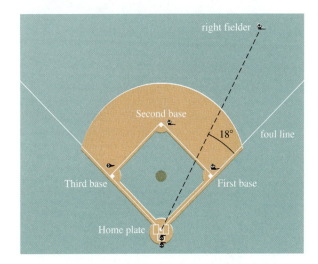

50. A carpenter plans to build a dormer addition to the roof of a Cape Cod cottage, as shown in the figure. Find (a) the length x, (b) angle α, and (c) angle β.

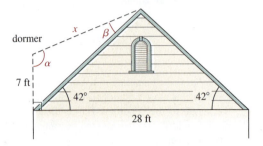

51. As a ship sails into a harbor along a straight course, the captain notices that a buoy off the port side makes an angle of 12° with the path of the ship. Sailing an additional 1200 meters, the captain now observes that the buoy makes an angle of 32° with the path of the ship, as shown in the sketch. What is the closest the ship comes to the buoy?

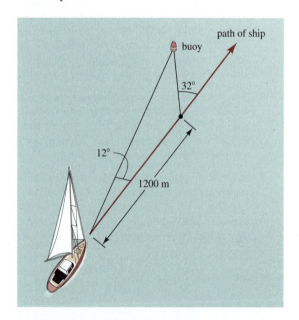

52. The building codes in the city of Boston specify that a single-family dwelling cannot be placed on a lot whose area is less than 15,000 sq ft. If the angle at the street corner of a triangular lot is 73°27′ and the frontage of the lot on one street is 165.6 ft as shown in the sketch, determine the minimum frontage x on the other street so that the lot passes code.

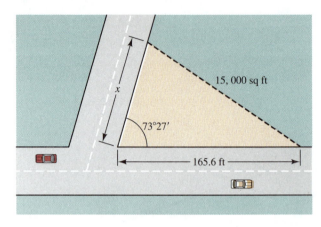

EXERCISE 52

53. A 45-lb force is applied to the handle of a child's wagon at an angle of 34° with the horizontal. Find (a) the force that pulls the wagon horizontally and (b) the force that lifts the wagon vertically.

54. For the force system shown in the sketch, find the magnitude and direction angle of the resultant force.

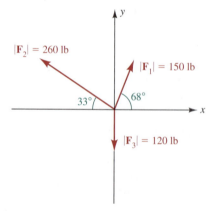

55. For the force system in Exercise 54, find an additional force $\mathbf{F}_4$ that produces equilibrium.

56. Two cranes lift a large fishing boat from the water. One crane exerts a force of 18,500 lb on its cable and the other crane exerts a force of 14,200 lb on its cable. If the angle between the cables is 24°, find the magnitude of the resultant force.

57. A sailboat leaves island A and sails 75 miles S-31°-W. It then changes course and sails 112 miles S-18°-E to dock at island B. What is the displacement from island B to island A?

58. The displacement from a Coast Guard station to a disabled ship is known to be 22.7 miles N-9°40′-W. Find the displacement from a Coast Guard ship, located 25.2 miles N-12°20′-E of the station, to the disabled ship.

59. An aircraft carrier is stationed directly east of an airbase. A jet takes off from the carrier at the speed of 480 mph in the direction N-78°15′-W and, at the same instant, a helicopter takes off from the airbase at the speed of 150 mph. If the jet and the helicopter meet in 30 minutes, find the displacement of the helicopter from the base.

60. Referring to Exercise 59, how far from the airbase is the aircraft carrier?

Cumulative Review Exercises

Chapters 1, 2, & 3

1. Use Hero's formula to find the area A of an isosceles triangle with equal sides a and nonequal side b.

2. Use the law of cosines to show that the sum of the squares of the two diagonals of a parallelogram equals the sum of the squares of the four sides of the parallelogram.

3. Find the exact value of each expression.
   (a) $\sin 210° + \cos(-30°) + \tan 405°$
   (b) $\sin \dfrac{2\pi}{3} \cot \dfrac{17\pi}{6} - \cos \dfrac{3\pi}{4}$
   (c) $\arctan(-\sqrt{3}) + \arccos\left(\cos \dfrac{5\pi}{4}\right)$
   (d) $\tan\left[90° - \sin^{-1} \tfrac{12}{13}\right]$

4. Find the radian measure of acute angles x and y if $\tan(x + 3y) = 1$ and $\cos(x - 2y) = 1$.

5. Given the function f defined by $f(x) = \dfrac{x^2 - 2}{x^2 - 7x + 6}$, state the domain of f using interval notation. Then find $(f \circ f)(2)$, if it is defined.

6. If θ is a fourth-quadrant angle with $\cos \theta = \sqrt{2/3}$, find the exact value of each expression.
   (a) $\sin \theta$ (b) $\tan \theta$ (c) $\sin(\theta + \pi)$ (d) $\cos 2\theta$

7. The power P (in watts) dissipated in a certain electrical circuit is directly proportional to the square of the current i (in amperes) flowing through the circuit. Express P as a function of i if $P = 5$ watts when $i = 20$ milliamperes.

8. Sketch the graph of each function.
   (a) $f(x) = 2 \tan \dfrac{3x}{2}$ (b) $g(x) = \sin(-5\pi x)$
   (c) $F(x) = \sec\left(x - \dfrac{\pi}{3}\right)$ (d) $G(x) = 2 + 3\cos(2x + \pi)$
   (e) $y = \dfrac{\pi}{2} - \arccos x$

9. From a point P, tangent lines are drawn to a circle of circumference 48π. If the angle between the tangent lines is $50°$, determine (a) the distance from point P to the center of the circle and (b) the length of the intercepted arc.

10. State which of the trigonometric functions are even functions and which are odd functions.

11. Find the area of a triangle whose vertices are the points $(8, 8)$, $(-7, -2)$, and $(1, -6)$.

12. Two sides of a triangle are 63.0 m and 86.0 m and the area of the triangle is 2360 square meters. Determine two possible values for the third side of the triangle.

13. Determine the distance from A to B along the centerline of the highway shown in the sketch.

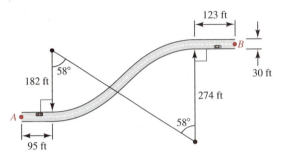

14. The latitude of Boston is $42°20'$. Assuming the earth is a sphere with radius 3960 mi, determine the distance from Boston (a) to the equator and (b) to the North Pole.

15. A tank contains 1000 gallons of water. The spigot is opened and water flows out of the tank at the uniform rate of $\tfrac{2}{3}$ gallons per minute.
    (a) Express the amount A of water (in gallons) in the tank as a function of the time t (in minutes) that the spigot is opened.
    (b) State the domain and range of the function defined in part (a).

16. Given that α and β are the acute angles of a right triangle, a is the side opposite angle α, b is the side opposite angle β, and c is the hypotenuse, express
    (a) a in terms of b and α.   (b) c in terms of b and β.
    (c) α in terms of b and c.   (d) β in terms of a and b.

17. A building lot has the shape of a quadrilateral. The lengths of two adjacent sides of the lot are 250 ft and 354 ft, and the angle between these sides is $69°45'$. The other two sides are, respectively, perpendicular to these adjacent sides. Find (a) the length of the other two sides and (b) the area of the lot.

18. Sketch an angle of radian measure 8 in standard position. Find its reference angle and the smallest positive angle coterminal with 8.

19. Given that $\overline{AD}$ bisects angle A, as shown in the sketch on the next page, find the distance AB.

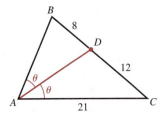

20. Given that $\overline{DE}$ is perpendicular to $\overline{AB}$ as shown in the sketch, find the distance AE.

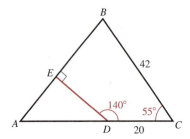

21. Sketch the graph of the equation $y = 1/x^2$. Then use the ideas of shifting and reflecting to help sketch the graphs of the following functions. Label all intercepts and asymptotes.

(a) $f(x) = 2 - \dfrac{1}{x^2}$ (b) $g(x) = \dfrac{1}{(x-1)^2}$

22. Find the x- and y-intercepts for the graph of the equation $y^2 = x(x-2)^2$.

23. For the graph of the equation $y = f(x)$, what is the y-intercept?

24. Given that θ is an obtuse angle, express $\csc \theta$, $\cos \theta$, and $\tan \theta$ in terms of $\sin \theta$.

25. A surveyor finds that the angle of elevation to the top of a radio transmission tower is 32°. Moving directly toward the tower a distance of 150 ft, she finds that the angle of elevation of the tower is now 43°. Assuming the ground is inclined at an angle of 12°, as shown in the sketch, find the height of the tower.

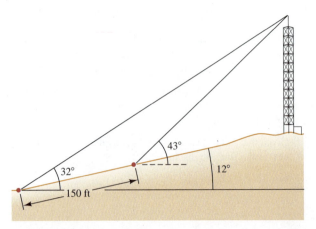

26. The airspeed of a plane heading in the direction N-62°-E is 375 mph. The wind, blowing directly from the south, causes the plane to drift to the north 9°, as shown in the sketch. Determine the velocity of the wind and the ground speed of the plane (its speed relative to the ground).

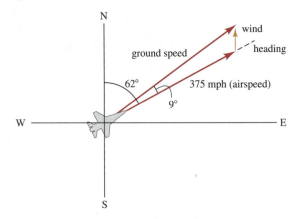

27. Given that $\mathbf{A} = \langle 2, 2 \rangle$ and $\mathbf{B} = \langle -3, 6 \rangle$, find the magnitude and direction angle of each difference.

(a) $\mathbf{A} - \mathbf{B}$ (b) $\mathbf{B} - \mathbf{A}$

28. The structure in the sketch is in equilibrium.

(a) Draw a vector diagram showing the weight W being supported by the cables.

(b) Find the tension (stretching force) in the left and right cables.

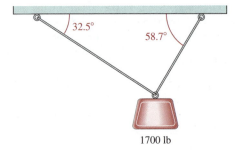

29. Solve each triangle.

(a)

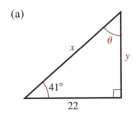

(b)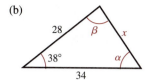

30. Rangers have determined that the population P of deer in a state forest in Massachusetts can be approximated by

$$P = 54 + 29 \sin \frac{\pi}{8} t,$$

where t is the time (in years) with $t = 0$ corresponding to January, 1980. Determine the number of deer in the forest in January, 1992.

CHAPTER 4

Analytic Trigonometry

The number of hours N of daylight on a particular day d of the year in the Boston area may be approximated by

$$N = 12 + 3 \sin\left[\frac{2\pi}{365}(d - 80)\right].$$

Determine the days of the year that have 13 hours 30 minutes of daylight.

(For the solution, see Example 6 in Section 4.2.)

4.1	**Algebraic Manipulations of Trigonometric Expressions**
4.2	**Trigonometric Equations**
4.3	**Sum and Difference Formulas**
4.4	**Multiple-Angle Formulas**
4.5	**Product-to-Sum Formulas and Sum-to-Product Formulas**

4.1 Algebraic Manipulations of Trigonometric Expressions

In this chapter we study *analytic trigonometry,* a branch of mathematics in which algebraic procedures are applied to trigonometry. We apply some basic concepts from algebra to the trigonometric functions from Chapter 2 in order to

1. write trigonometric expressions in simpler form,
2. verify trigonometric identities,
3. solve trigonometric equations, and
4. develop several useful trigonometric formulas.

We begin by discussing the basic algebraic operations of addition, subtraction, multiplication, and division with trigonometric expressions.

◆ **Basic Operations with Trigonometric Expressions**

Because trigonometric expressions, such as sin x, cos x, tan x, csc x, sec x, and cot x, represent real numbers whenever the variable x is a real number for which the trigonometric expression is defined, we can manipulate trigonometric expressions in the same manner that we manipulate algebraic expressions. Hence, to expand

$$(\sin x + \tan x)^2,$$

we think of sin x and tan x as real numbers A and B. Now, since

$$(A + B)^2 = A^2 + 2AB + B^2,$$

we have

$$(\sin x + \tan x)^2 = \sin^2 x + 2 \sin x \tan x + \tan^2 x.$$

Note: In this example, we have followed convention and omitted parentheses when squaring sin x and tan x, that is, we prefer to write $\sin^2 x$ for $(\sin x)^2$ and $\tan^2 x$ for $(\tan x)^2$. We also omit parentheses when writing the product of two or more trigonometric expressions, that is, we prefer sin x tan x to (sin x)(tan x).

EXAMPLE 1 Perform the indicated operations.

(a) $(-2 \sin^3 x)^4$

(b) $3 \cos^2 t (2 \cos^3 t - 5 \cos^4 2t)$

(c) $(2 \tan \theta - 1)(\tan \theta + 3)$

(d) $\dfrac{5}{2 \sec^2 x} - \dfrac{1}{3 \sec x}$

SOLUTION

(a) $(-2 \sin^3 x)^4 = (-2)^4 (\sin^3 x)^4$ Apply $(ab)^n = a^n b^n$

$\qquad\qquad\quad\; = 16 \sin^{12} x$ Apply $(a^m)^n = a^{mn}$

SECTION 4.1 Algebraic Manipulations of Trig Expressions

(b) $3\cos^2 t\,(2\cos^3 t - 5\cos^4 2t)$

$= 3\cos^2 t\,(2\cos^3 t) - 3\cos^2 t\,(5\cos^4 2t)$ **Multiply**

$= 6\cos^5 t - 15\cos^2 t \cos^4 2t$ **Apply $a^m a^n = a^{m+n}$**

Since t is different from $2t$, we cannot add these exponents.

(c) $(2\tan\theta - 1)(\tan\theta + 3)$

$= 2\tan^2\theta + 6\tan\theta - \tan\theta - 3$ **Multiply**

$= 2\tan^2\theta + 5\tan\theta - 3$ **Combine like terms**

(d) $\dfrac{5}{2\sec^2 x} - \dfrac{1}{3\sec x} = \dfrac{3\cdot 5}{3\cdot 2\sec^2 x} - \dfrac{2\sec x \cdot 1}{2\sec x \cdot 3\sec x}$ **Change to equivalent fractions with the same LCD**

The LCD is $6\sec^2 x$

$= \dfrac{15 - 2\sec x}{6\sec^2 x}$ **Subtract fractions** ◆

PROBLEM 1   Repeat Example 1 for $\dfrac{4}{\csc\theta - 1} + \dfrac{3}{1 - \csc\theta}$.

◆ **Factoring Trigonometric Expressions**

As we illustrate in the next example, the techniques of factoring can also be applied to trigonometric expressions.

EXAMPLE 2   Factor each expression completely over the integers.

(a) $3\cos x \sin^2 x - 6\cos x$ (b) $\sin^2\theta - \cos^2\theta$

(c) $6\tan^2 3x - \tan 3x - 2$ (d) $\sec^3 t + 1$

SOLUTION

(a) The expression $3\cos x \sin^2 x - 6\cos x$ contains a common factor of $3\cos x$. Factoring out $3\cos x$, we have

$$3\cos x \sin^2 x - 6\cos x = 3\cos x\,(\sin^2 x - 2).$$

(b) The expression $\sin^2\theta - \cos^2\theta$ is the difference of squares. Thus,

$$\sin^2\theta - \cos^2\theta = (\sin\theta + \cos\theta)(\sin\theta - \cos\theta).$$

(c) The expression $6 \tan^2 3x - \tan 3x - 2$ factors as follows:

$$6 \tan^2 3x - \tan 3x - 2 = (3 \tan 3x - 2)(2 \tan 3x + 1).$$

(d) The expression $\sec^3 t + 1$ is the sum of cubes. Thus,

$$\sec^3 t + 1 = (\sec t + 1)(\sec^2 t - \sec t + 1).$$

PROBLEM 2 Repeat Example 2 for $\csc^2 x - 3 \csc x - 10$.

Simplifying Trigonometric Expressions

The eight fundamental trigonometric identities given in Section 2.2 are frequently used to help simplify trigonometric expressions. For our convenience, these identities are listed here.

Fundamental Trigonometric Identities

1. $\sin x \csc x = 1$
2. $\cos x \sec x = 1$
3. $\tan x \cot x = 1$
4. $\tan x = \dfrac{\sin x}{\cos x}$
5. $\cot x = \dfrac{\cos x}{\sin x}$
6. $\sin^2 x + \cos^2 x = 1$
7. $1 + \tan^2 x = \sec^2 x$
8. $1 + \cot^2 x = \csc^2 x$

EXAMPLE 3 Use algebraic manipulations and the fundamental trigonometric identities to simplify each expression to a single trigonometric function or a constant.

(a) $\cos x \tan^2 x + \cos x$ **(b)** $\dfrac{\cos^2 x}{1 - \sin x} - \sin x$

SOLUTION

(a)
$$\begin{aligned}
\cos x \tan^2 x + \cos x &= \cos x (\tan^2 x + 1) && \text{Factor} \\
&= \cos x \sec^2 x && \text{Apply trig identity 7} \\
&= (\cos x \sec x) \sec x && \text{Rewrite} \\
&= (1) \sec x && \text{Apply trig identity 2} \\
&= \sec x && \text{Simplify}
\end{aligned}$$

(b)
$$\begin{aligned}
\dfrac{\cos^2 x}{1 - \sin x} - \sin x &= \dfrac{\cos^2 x - \sin x (1 - \sin x)}{1 - \sin x} && \text{Subtract fractions} \\
&= \dfrac{\cos^2 x - \sin x + \sin^2 x}{1 - \sin x} && \text{Apply the distributive property}
\end{aligned}$$

SECTION 4.1 Algebraic Manipulations of Trig Expressions

$$= \frac{1 - \sin x}{1 - \sin x} \qquad \text{Apply trig identity 6}$$

$$= 1 \qquad \text{Reduce to lowest terms}$$

PROBLEM 3 Repeat Example 3 for $\sin^2\theta \cot^2\theta + \sin^2\theta$.

The procedures for simplifying the trigonometric expressions in Example 3 are not unique. For instance, another method of simplifying $\cos x \tan^2 x + \cos x$ is to rewrite the expression in terms of sines and cosines and then proceed as follows:

$$\cos x \tan^2 x + \cos x = \cos x \left(\frac{\sin x}{\cos x}\right)^2 + \cos x \qquad \text{Apply trig identity 4}$$

$$= \cos x \frac{\sin^2 x}{\cos^2 x} + \cos x \qquad \text{Apply } (a/b)^n = a^n/b^n$$

$$= \frac{\sin^2 x}{\cos x} + \cos x \qquad \text{Multiply fractions}$$

$$= \frac{\sin^2 x + \cos^2 x}{\cos x} \qquad \text{Add fractions}$$

$$= \frac{1}{\cos x} \qquad \text{Apply trig identity 6}$$

$$= \sec x \qquad \text{Apply trig identity 2}$$

The procedure of rewriting everything in terms of sines and cosines works well when simplifying many trigonometric expressions.

EXAMPLE 4 Use algebraic manipulations and the fundamental trigonometric identities to simplify each expression to a single trigonometric function or a constant.

(a) $\sec^2 x \cot^2 x - \cos^2 x \csc^2 x$
(b) $\dfrac{\cot x - 1}{1 - \tan x}$

SOLUTION

(a) We begin by writing everything in terms of sines and cosines.

$$\sec^2 x \cot^2 x - \cos^2 x \csc^2 x$$

$$= \frac{1}{\cos^2 x} \frac{\cos^2 x}{\sin^2 x} - \cos^2 x \frac{1}{\sin^2 x} \qquad \text{Apply trig identities 2, 5, and 1}$$

$$= \frac{1}{\sin^2 x} - \frac{\cos^2 x}{\sin^2 x} \qquad \text{Multiply fractions}$$

$$= \csc^2 x - \cot^2 x \qquad \text{Apply trig identities 1 and 5}$$

$$= 1 \qquad \text{Apply trig identity 8}$$

(b) We begin by writing everything in terms of sines and cosines.

$$\frac{\cot x - 1}{1 - \tan x} = \frac{\dfrac{\cos x}{\sin x} - 1}{1 - \dfrac{\sin x}{\cos x}} \quad \text{Apply trig identities 5 and 4}$$

$$= \frac{\dfrac{\cos x - \sin x}{\sin x}}{\dfrac{\cos x - \sin x}{\cos x}} \quad \text{Perform the indicated subtractions in the numerator and denominator}$$

$$= \frac{\cos x - \sin x}{\sin x} \cdot \frac{\cos x}{\cos x - \sin x} \quad \text{Rewrite as a product}$$

$$= \frac{\cos x}{\sin x} \quad \text{Multiply fractions}$$

$$= \cot x \quad \text{Apply trig identity 5}$$

PROBLEM 4 Repeat Example 4 for $\sin x + \cos x \cot x$.

◆ **Verifying Trigonometric Identities**

A **trigonometric equation** is a statement declaring that two trigonometric expressions are equal. A trigonometric equation that becomes true when the variable is replaced by every permissible number is called a **trigonometric identity.** To verify that a trigonometric equation is an identity, we can use the fundamental trigonometric identities in conjunction with algebraic manipulations to transform one side of the equation into the other. Although no general method works for verifying every trigonometric identity, we offer the following suggestion.

Suggestion for Verifying a Trigonometric Identity

Reduce the more complicated side of the equation to the simpler side. If no simplication is obvious, express every trigonometric function on the more complicated side of the equation in terms of sines and cosines, and then reduce this expression to the simpler side.

When verifying a trigonometric identity, we are trying to show that the equation is true when the variable is replaced by every permissible number. Thus, it is incorrect to assume the equation is already true and then apply the rules for generating equivalent equations, such as adding an expression or multiplying an expression to both sides.

EXAMPLE 5 Show that each trigonometric equation is an identity.

(a) $\sec x - \sin x \cot x = \tan x \sin x$ **(b)** $2 \sec^2 x - 1 = \dfrac{1 + \sin^2 x}{\cos^2 x}$

SOLUTION

(a) Working with the more complicated left-hand side, we express everything in terms of sines and cosines and then reduce this expression to the simpler right-hand side:

$$\sec x - \sin x \cot x = \tan x \sin x$$

$\dfrac{1}{\cos x} - \sin x \dfrac{\cos x}{\sin x}$	Apply trig identities 2 and 5
$\dfrac{1}{\cos x} - \cos x$	Multiply fractions
$\dfrac{1 - \cos^2 x}{\cos x}$	Subtract fractions
$\dfrac{\sin^2 x}{\cos x}$	Apply trig identity 6
$\dfrac{\sin x}{\cos x} \sin x$	Rewrite
$\tan x \sin x$	Apply trig identity 4

Since we have reduced the left-hand side of this equation to the right-hand side, we conclude that

$$\sec x - \sin x \cot x = \tan x \sin x \quad \text{is a trigonometric identity.}$$

(b) Working with the more complicated right-hand side, we separate this fraction into the sum of two fractions, and then reduce this expression to the simpler left-hand side as follows:

$$2\sec^2 x - 1 = \dfrac{1 + \sin^2 x}{\cos^2 x}$$

$\dfrac{1}{\cos^2 x} + \dfrac{\sin^2 x}{\cos^2 x}$	Separate into two fractions
$\sec^2 x + \tan^2 x$	Apply trig identities 2 and 4
$\sec^2 x + (\sec^2 x - 1)$	Apply trig identity 7
$2\sec^2 x - 1$	Combine like terms

Since we have reduced the right-hand side of this equation to the left-hand side, we conclude that

$$2\sec^2 x - 1 = \frac{1 + \sin^2 x}{\cos^2 x} \quad \text{is a trigonometric identity.} \qquad \blacklozenge$$

PROBLEM 5 Repeat Example 5 for $\tan\theta + \cot\theta = \sec\theta \csc\theta$. $\qquad \blacklozenge$

As illustrated in the next example, sometimes we can verify trigonometric identities by using the *fundamental property of fractions* $\left(\dfrac{a}{b} = \dfrac{ak}{bk} \text{ for all } k, k \neq 0\right)$ or by using the techniques of factoring.

EXAMPLE 6 Show that each equation is a trigonometric identity.

(a) $\dfrac{\sin\theta}{1 + \cos\theta} = \dfrac{1 - \cos\theta}{\sin\theta}$ 　　(b) $\dfrac{\cos^3 x + \sin^3 x}{\cos x + \sin x} = 1 - \sin x \cos x$

SOLUTION

(a) If we multiply the denominator of the left-hand side by $1 - \cos\theta$, we obtain $1 - \cos^2\theta$, which is equivalent to $\sin^2\theta$. Hence, we proceed as follows:

$$\frac{\sin\theta}{1 + \cos\theta} = \frac{1 - \cos\theta}{\sin\theta}$$

$\dfrac{\sin\theta}{1 + \cos\theta} \cdot \dfrac{1 - \cos\theta}{1 - \cos\theta}$	Apply the fundamental property of fractions
$\dfrac{\sin\theta(1 - \cos\theta)}{1 - \cos^2\theta}$	Multiply
$\dfrac{\sin\theta(1 - \cos\theta)}{\sin^2\theta}$	Apply trig identity 6
$\dfrac{1 - \cos\theta}{\sin\theta}$	Reduce

Since we have reduced the left-hand side of this equation to the right-hand side, we conclude that

$$\frac{\sin\theta}{1 + \cos\theta} = \frac{1 - \cos\theta}{\sin\theta} \quad \text{is a trigonometric identity.}$$

(b) Noting that the numerator of the fraction on the left-hand side is the sum of cubes, we proceed as follows:

$$\frac{\cos^3 x + \sin^3 x}{\cos x + \sin x} = 1 - \sin x \cos x$$

$\dfrac{(\cos x + \sin x)(\cos^2 x - \cos x \sin x + \sin^2 x)}{\cos x + \sin x}$	Factor
$\cos^2 x - \cos x \sin x + \sin^2 x$	Reduce
$1 - \cos x \sin x$	Apply trig identity 6

Since we have reduced the left-hand side of this equation to the right-hand side, we conclude that

$$\frac{\cos^3 x + \sin^3 x}{\cos x + \sin x} = 1 - \sin x \cos x \quad \text{is a trigonometric identity.} \quad \blacklozenge$$

PROBLEM 6 Repeat Example 6 for $\sec^4 x - \tan^4 x = \dfrac{1 + \sin^2 x}{\cos^2 x}$. ◆

◆ **Trigonometric Substitution**

In calculus, it is sometimes necessary to rewrite algebraic expressions by making a *trigonometric substitution.* The technique is illustrated in the next example.

EXAMPLE 7 Making the trigonometric substitution $x = 3 \tan \theta$ with $0 < \theta < \pi/2$, express $\sqrt{x^2 + 9}$ as a function of θ in simplified form.

SOLUTION Replacing x with $3 \tan \theta$, we have

$$\sqrt{x^2 + 9} = \sqrt{(3 \tan \theta)^2 + 9}$$

$$= \sqrt{9 \tan^2 \theta + 9} \qquad \text{Apply } (ab)^n = a^n b^n$$

$$= \sqrt{9 (\tan^2 \theta + 1)} \qquad \text{Factor}$$

$$= \sqrt{9 \sec^2 \theta} \qquad \text{Apply trig identity 7}$$

$$= 3 \sec \theta \qquad \text{since } \sec \theta > 0 \text{ for } 0 < \theta < \pi/2 \quad \blacklozenge$$

We can check the results of the trigonometric substitution in Example 7 by using the trigonometric ratios for right triangles (Section 2.2). Since $x = 3 \tan \theta$, we have

$$\tan \theta = \frac{x}{3} = \frac{\text{opp}}{\text{adj}},$$

236   CHAPTER 4   Analytic Trigonometry

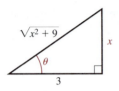

FIGURE 4.1
A right triangle with $\tan \theta = x/3$.

as shown in Figure 4.1. By the Pythagorean theorem, the hypotenuse is

$$\sqrt{x^2 + 9}.$$

Thus,

$$\sec \theta = \frac{\text{hyp}}{\text{adj}} = \frac{\sqrt{x^2 + 9}}{3}.$$

Hence, $\sqrt{x^2 + 9} = 3 \sec \theta$.

PROBLEM 7   Making the trigonometric substitution $x = 2 \sin \theta$ with $0 < \theta < \pi/2$, express $\sqrt{4 - x^2}$ as a function of θ in simplified form. ◆

Exercises 4.1

Basic Skills

In Exercises 1–10, perform the indicated operations.

1. $(3 \cos^2 y \sin y)^3$
2. $(2 \tan^2 x \sin x)(-3 \tan^3 x \sin 3x)$
3. $\dfrac{12 \cos^2 x \cos^{-4} 2x}{3 \cos^{-3} 2x \cos x}$
4. $2 \cos^{-3} y (\cos^3 y - 3 \cos^4 y)$
5. $(3 \tan \theta + 2)^2$
6. $(3 \sin \theta - 2)(\sin \theta + 1)$
7. $\dfrac{8 \tan x \cos 2x}{5 \sin x} \cdot \dfrac{15 \sin^2 x}{\cos 2x}$
8. $\dfrac{3}{2 \sin x} - \dfrac{4}{\sin^2 x}$
9. $\dfrac{1}{1 - \cos x} + \dfrac{1}{\cos x}$
10. $\dfrac{1 + \dfrac{1}{\tan 3y}}{\dfrac{1}{\tan^2 3y}}$

In Exercises 11–16, factor completely over the integers.

11. $3 \tan \theta - 15 \tan^3 \theta$
12. $12 \cos x \cot^2 x - 15 \cos^2 x \cot x$
13. $\cos^2 \theta - \sin^2 \theta$
14. $4 \tan^2 x - 1$
15. $\sin^2 x - 2 \sin x - 3$
16. $2 \csc^2 x - 3 \csc x - 5$

In Exercises 17–20, reduce each fraction to lowest terms.

17. $\dfrac{\sin x + 1}{\sin^2 x - 1}$
18. $\dfrac{\tan^3 x - \tan^2 x - 12 \tan x}{3 + \tan x}$
19. $\dfrac{\sin^4 x - \cos^4 x}{\sin x + \cos x}$
20. $\dfrac{8 - \cot^3 y}{2 - \cot y}$

In Exercises 21–36, simplify each expression to a single trigonometric function or a constant.

21. $\cos x \csc x$
22. $\sin \theta \sec \theta$
23. $\dfrac{\sec x}{\csc x}$
24. $\dfrac{\cot x}{\cos x}$
25. $\csc^2 x \tan^2 x - \tan^2 x$
26. $\sin^2 x + \cot^2 x \sin^2 x$
27. $\cos^2 y (1 - \sec^2 y) + \sin^2 y$
28. $(\csc \theta - \cot \theta)(\csc \theta + \cot \theta)$
29. $(\sin \theta + \cos \theta)^2 + (\sin \theta - \cos \theta)^2$
30. $\dfrac{\sin x \cot x + \cos x}{2 \cot x}$
31. $\dfrac{\tan^2 x + 1}{\tan x + \cot x}$
32. $\dfrac{\sin y + \tan y}{1 + \sec y}$
33. $\dfrac{1 + \csc x}{\sec x} - \cos x$
34. $\dfrac{\sin x}{\sec x + 1} + \dfrac{\sin x}{\sec x - 1}$
35. $(\sec \theta + \tan \theta)^4 (\sec \theta - \tan \theta)^4$
36. $\left(\dfrac{1 + \sin x}{2 \cos^2 x}\right)^3 (1 - \sin x)^3$

In Exercises 37–60, show that each trigonometric equation is an identity.

37. $\sin \theta (\csc \theta - \sin \theta) = \cos^2 \theta$
38. $\cos^2 x (\sec^2 x + \csc^2 x) = \csc^2 x$
39. $(\sin x + \cos x)^2 = 1 + 2 \sin x \cos x$
40. $(\tan \theta + 1)^2 = \sec^2 \theta + 2 \tan \theta$

41. $\cos^2 x - \sin^2 x = 2\cos^2 x - 1$

42. $\cos^2 x - \sin^2 x = 1 - 2\sin^2 x$

43. $\dfrac{1 - \cos^2 y}{\cos y} = \sin y \tan y$

44. $\dfrac{\sec^2 y}{\sec^2 y - 1} = \csc^2 y$

45. $\dfrac{1 + \csc x}{\sec x} = \cos x + \cot x$

46. $\dfrac{\sin^2 x + \sin x + \cos^2 x}{\sin x} = 1 + \csc x$

47. $(1 - \cos \theta)(1 + \sec \theta) = \sin \theta \tan \theta$

48. $(\sin x + \cos x)(\csc x - \sec x) = \cot x - \tan x$

49. $\cot^2 y - \cos^2 y = \cot^2 y \cos^2 y$

50. $\sec^2 \theta + \csc^2 \theta = \sec^2 \theta \csc^2 \theta$

51. $\dfrac{\cos x + \sin x}{\sec x + \csc x} = \sin x \cos x$

52. $\dfrac{\cos x + \sin x}{\cos x - \sin x} = \dfrac{1 + \tan x}{1 - \tan x}$

53. $\dfrac{\cos x}{1 - \sin x} = \dfrac{1 + \sin x}{\cos x}$

54. $\dfrac{1 + \sin x}{1 - \sin x} = (\tan x + \sec x)^2$

55. $\dfrac{1 - \csc^2 x}{1 + \csc x} = \dfrac{\sin x - 1}{\sin x}$

56. $\dfrac{\tan^3 x - 1}{\tan x - 1} = \sec^2 x + \tan x$

57. $\dfrac{\cos^4 y - \sin^4 y}{1 - \tan^4 y} = \cos^4 y$

58. $\dfrac{\sec^4 y - 1}{\tan^2 y} = 2 + \tan^2 y$

59. $2 \tan \theta + \sec^2 \theta = \sin^2 \theta (\sec \theta + \csc \theta)^2$

60. $\cot \theta - \tan \theta = 2 \cos \theta \csc \theta - \sec \theta \csc \theta$

In Exercises 61–68, make the indicated trigonometric substitutions and simplify. Assume $0 < \theta < \pi/2$.

61. For $\sqrt{1 - x^2}$, let $x = \sin \theta$.

62. For $\sqrt{x^2 - 25}$, let $x = 5 \sec \theta$.

63. For $\dfrac{\sqrt{x^2 + 16}}{x}$, let $x = 4 \tan \theta$.

64. For $\dfrac{x}{\sqrt{4 - x^2}}$, let $x = 2 \sin \theta$.

65. For $\dfrac{x^3}{(x^2 - 9)^{3/2}}$, let $x = 3 \sec \theta$.

66. For $\dfrac{x^2}{x^2 + 16}$, let $x = 4 \tan \theta$.

67. For $\dfrac{x}{\sqrt{4 - (x + 1)^2}}$, let $x + 1 = 2 \sin \theta$.

68. For $\sqrt{1 + a^{2x}}$, let $a^x = \tan \theta$.

Critical Thinking

Explain why the algebraic statements in Exercises 69 and 70 are false.

69. $(\sin x + \cos x)^2 = \sin^2 x + \cos^2 x = 1$

70. $\dfrac{\sqrt{\sin^2 x + \cos^2 x}}{\sin x + \cos x} = \dfrac{\sin x + \cos x}{\sin x + \cos x} = 1$

In Exercises 71-74, determine the quadrants associated with x for which each equation is an identity.

71. $\sin x = \sqrt{1 - \cos^2 x}$

72. $\sqrt{1 + \tan^2 x} = -\sec x$

73. $-\sqrt{\csc^2 x - 1} = \cot x$

74. $|\cos x| = \sqrt{1 - \sin^2 x}$

75. Simplify the left side of the equation

$(x \sin x - y \cos x)^2 + (x \cos x + y \sin x)^2 = 25.$

Then describe the graph of this equation.

76. For which trigonometric functions f is the equation $f(-x) = -f(x)$ an identity?

77. Is $\sin (x + y) = \sin x + \sin y$ a trigonometric identity? Select some values of x and y and evaluate each side of the equation, then state your conclusion.

78. Is $\cos (x - y) = \cos x - \cos y$ a trigonometric identity? Select some values of x and y and evaluate each side of the equation, then state your conclusion.

Calculator Activities

 In Exercises 79–86, an equation of the form f(x) = g(x) is stated. Use a graphing calculator or computer with graphing capabilities to generate the graphs of y = f(x) and y = g(x). Compare the graphs, and then state whether the equation appears to be a trigonometric identity.

79. $\sin 2x = 2 \sin x \cos x$

80. $\cos 2x = \cos^2 x - \sin^2 x$

81. $\sin x = \sqrt{1 - \cos^2 x}$

82. $\tan 6x - \tan 2x = \tan 4x$

83. $\tan x = \dfrac{\sin 2x}{1 + \cos 2x}$

84. $\cos 3x \cos 2x - \sin 3x \sin 2x = \cos 5x$

85. $\cos 8x + \cos 2x = 2 \cos 5x \cos 3x$

86. $2 \sin 5x \sin x = \cos 4x - \cos 6x$

4.2 Trigonometric Equations

A trigonometric equation that is true only for some values of the variable but not for others (or is never true for any real number) is called a **conditional trigonometric equation**. In this section, we discuss the procedure for solving conditional trigonometric equations. To solve a conditional trigonometric equation means to find all values of the variable that make the equation true. Unless specifically stated otherwise, the solutions are expressed as real numbers or, equivalently, as angles in radians. We begin with some simple types of equations and then expand to types that are solved by factoring or by the quadratic formula.

◆ Simple Trigonometric Equations

The rules and procedures for solving algebraic equations apply to conditional trigonometric equations as well. For instance, to solve the equation $2 \sin x - 1 = 0$ for x, we begin by generating an equivalent equation:

$$2 \sin x - 1 = 0$$
$$2 \sin x = 1 \quad \text{Add 1 to both sides}$$
$$\sin x = \tfrac{1}{2} \quad \text{Divide both sides by 2}$$

From Table 2.3 in Section 2.3, we know that $\sin(\pi/6) = \tfrac{1}{2}$. Also, we know that the sine function is positive in quadrants I and II. Thus, using a reference angle of $\pi/6$ in the first and second quadrants, as shown in Figure 4.2, we have

$$x = \frac{\pi}{6} \quad \text{and} \quad x = \pi - \frac{\pi}{6} = \frac{5\pi}{6}$$

as values of x in the interval $[0, 2\pi)$ for which $\sin x = \tfrac{1}{2}$. Finally, since the sine

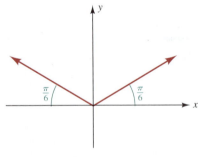

FIGURE 4.2
A reference angle of $\pi/6$ in quadrants I and II

function is periodic with period 2π, we add multiples of 2π to $\pi/6$ and $5\pi/6$ to obtain all real solutions. Thus, the expressions

$$\frac{\pi}{6} + 2\pi n \qquad \text{and} \qquad \frac{5\pi}{6} + 2\pi n,$$

where n is an integer, describe the **general form** of all solutions of the equation $2 \sin x - 1 = 0$. Some of the particular solutions are illustrated graphically in Figure 4.3.

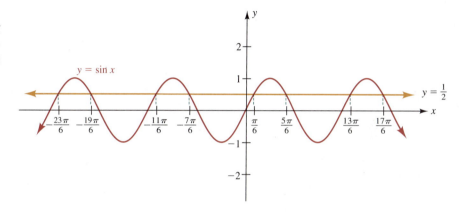

FIGURE 4.3
Graphical display of some solutions of the equation $2 \sin x - 1 = 0$, which is equivalent to $\sin x = \frac{1}{2}$

In a similar manner, we can solve trigonometric equations containing $\cos x$, $\csc x$, or $\sec x$. For equations containing $\tan x$ or $\cot x$, the procedure is slightly different, because the period of the tangent and cotangent functions is π. (The sine, cosine, cosecant, and secant functions have period 2π.) For equations containing $\tan x$ or $\cot x$, we find all solutions in the interval $[0, \pi)$ and then add multiples of π to these solutions to obtain all solutions of the equation. We now summarize the steps for solving a trigonometric equation containing either $\sin x$, $\cos x$, $\csc x$, $\sec x$, $\tan x$, or $\cot x$.

Procedure for Solving a Trigonometric Equation

Equations with $\sin x$, $\cos x$, $\csc x$, or $\sec x$

Step 1: Solve the equation for the trigonometric function.

Step 2: Find all x in the interval $[0, 2\pi)$ that satisfy the equation in Step 1.

Step 3: Add $2\pi n$, where n is an integer, to each solution in step 2. or 360°

Equations with $\tan x$ or $\cot x$

Step 1: Solve the equation for the trigonometric function.

Step 2: Find all x in the interval $[0, \pi)$ that satisfy the equation in Step 1.

Step 3: Add πn, where n is an integer, to each solution in step 2. or 180°

EXAMPLE 1 Find the general form of the solution of each trigonometric equation.

(a) $\sqrt{2}\cos x + 1 = 0$ **(b)** $\tan^2 x - 3 = 0$

SOLUTION

(a) We begin by solving for cos x as follows:

$$\sqrt{2}\cos x + 1 = 0$$
$$\sqrt{2}\cos x = -1 \quad \text{Subtract 1 from both sides}$$
$$\cos x = -\frac{1}{\sqrt{2}} \quad \text{Divide both sides by } \sqrt{2}$$

Now, from Table 2.3 in Section 2.3, we know that $\cos(\pi/4) = 1/\sqrt{2}$. Also, we know that the cosine function is negative in quadrants II and III. Thus, using a reference angle of $\pi/4$ in quadrants II and III, as shown in Figure 4.4, we have

$$x = \pi - \frac{\pi}{4} = \frac{3\pi}{4} \quad \text{and} \quad x = \pi + \frac{\pi}{4} = \frac{5\pi}{4}$$

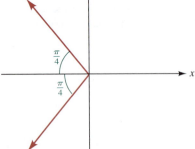

FIGURE 4.4
A reference angle of $\pi/4$ in quadrants II and III

as values of x in the interval $[0, 2\pi)$ for which $\cos x = -1/\sqrt{2}$. Finally, since the cosine function is periodic with period 2π, we add multiples of 2π to $3\pi/4$ and $5\pi/4$ to obtain all real solutions. Thus, the expressions

$$\frac{3\pi}{4} + 2\pi n \quad \text{and} \quad \frac{5\pi}{4} + 2\pi n, \quad \text{where } n \text{ is an integer,}$$

describe the general form of all solutions of the equation $\sqrt{2}\cos x + 1 = 0$.

(b) We begin by solving for tan x as follows:

$$\tan^2 x - 3 = 0$$
$$\tan^2 x = 3 \quad \text{Add 3 to both sides}$$
$$\tan x = \pm\sqrt{3} \quad \text{Take the square root of both sides}$$

From Table 2.3 in Section 2.3, we know that $\tan(\pi/3) = \sqrt{3}$. Also, we know that the tangent function is positive in quadrant I and negative in quadrant II. Thus, using a reference angle of $\pi/3$ in quadrants I and II, as shown in Figure 4.5, we have

$$x = \frac{\pi}{3} \quad \text{and} \quad x = \pi - \frac{\pi}{3} = \frac{2\pi}{3}$$

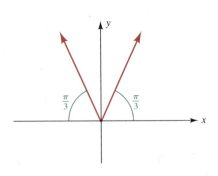

FIGURE 4.5
A reference angle of $\pi/3$ in quadrants I and II

as values of x in the interval $[0, \pi)$ for which $\tan x = \pm\sqrt{3}$. Finally, since

SECTION 4.2 Trigonometric Equations

the tangent function is periodic with period π, we add multiples of π to $\pi/3$ and $2\pi/3$ to obtain all real solutions. Thus, the expressions

$$\frac{\pi}{3} + \pi n \quad \text{and} \quad \frac{2\pi}{3} + \pi n, \quad \text{where } n \text{ is an integer,}$$

describe the general form of all solutions of the equation $\tan^2 x - 3 = 0$. ◆

PROBLEM 1 Repeat Example 1 for $\sqrt{3} \csc x - 2 = 0$. ◆

The method we outlined in Example 1 may be extended to trigonometric equations containing angles of the form $bx + c$, where b and c are constants. We begin by finding all values of $bx + c$ that satisfy the equation, and then solve for x by using ordinary algebraic methods. The procedure is illustrated in the next example.

EXAMPLE 2 Find the general form of the solution of each trigonometric equation.

(a) $2 \cot 3x - 2 = 0$ (b) $4 + \sin\left(2x - \dfrac{\pi}{3}\right) = 5$

SOLUTION

(a) We begin by solving for $\cot 3x$ as follows:

$$2 \cot 3x - 2 = 0$$
$$2 \cot 3x = 2 \qquad \text{Add 2 to both sides}$$
$$\cot 3x = 1 \qquad \text{Divide both sides by 2}$$

From Table 2.3 in Section 2.3, we know that $\cot(\pi/4) = 1$. Hence, in the interval $[0, \pi)$, we have

$$3x = \frac{\pi}{4},$$

and, consequently,

$$3x = \frac{\pi}{4} + \pi n.$$

Now, dividing both sides of this last equation by 3 gives us the solutions

$$x = \frac{\pi}{12} + \frac{\pi}{3} n, \quad \text{where } n \text{ is an integer.}$$

(b) We begin by solving for $\sin\left(2x - \dfrac{\pi}{3}\right)$ as follows:

$$4 + \sin\left(2x - \frac{\pi}{3}\right) = 5$$

$$\sin\left(2x - \frac{\pi}{3}\right) = 1 \qquad \text{Subtract 4 from both sides}$$

From Table 2.2 in Section 2.3, we know that $\sin(\pi/2) = 1$. Hence, in the interval $[0, 2\pi)$, we have

$$2x - \frac{\pi}{3} = \frac{\pi}{2},$$

and, consequently,

$$2x - \frac{\pi}{3} = \frac{\pi}{2} + 2\pi n.$$

Solving this last equation for x gives us

$$2x - \frac{\pi}{3} = \frac{\pi}{2} + 2\pi n$$

$$2x = \frac{5\pi}{6} + 2\pi n \qquad \text{Add } \pi/3 \text{ to both sides}$$

$$x = \frac{5\pi}{12} + \pi n \qquad \text{Divide both sides by 2}$$

Thus, the expression

$$\frac{5\pi}{12} + \pi n, \quad \text{where } n \text{ is an integer,}$$

describes the general form of all the solutions of this equation. ◆

To obtain particular solutions of a trigonometric equation, we substitute integers for n in the general form of the solution. For example, to find the solutions in the interval $[0, 2\pi)$ for the equation in Example 2(a), we let $n = 0, 1, 2, 3, 4, 5$ in the general form of the solution and obtain

$$x = \frac{\pi}{12}, \frac{5\pi}{12}, \frac{3\pi}{4}, \frac{13\pi}{12}, \frac{17\pi}{12}, \text{ and } \frac{7\pi}{4}.$$

Solutions of $2 \cot 3x - 2 = 0$ in the interval $[0, 2\pi)$.

SECTION 4.2 Trigonometric Equations

PROBLEM 2 For the equation in Example 2(b), find the solutions in the interval $[0, 2\pi)$.

◆ Techniques of Solving Trigonometric Equations

Some trigonometric equations can be solved by factoring. This method relies on the *zero product property* [if $ab = 0$, then either $a = 0$ or $b = 0$ (or both a and b are zero)]. That is, we try to write the equation in such a form that a product of trigonometric expressions is zero. The procedure is illustrated in the next example.

EXAMPLE 3 Find the general form of the solution of each trigonometric equation.

(a) $\tan x \sin x - \sin x = 0$ (b) $\sin^2 x + \sin x - 2 = 0$

SOLUTION

(a) Using common-term factoring and the zero product property, we have

$$\tan x \sin x - \sin x = 0$$

$$\sin x (\tan x - 1) = 0$$

$\sin x = 0$ or $\tan x - 1 = 0$

$x = 0, \pi$ $\tan x = 1$

(Solutions in the interval $[0, 2\pi)$)

$x = \dfrac{\pi}{4}$

(Solution in the interval $[0, \pi)$)

Although

$$0 + 2\pi n, \quad \pi + 2\pi n, \quad \text{and} \quad \frac{\pi}{4} + \pi n, \quad \text{where } n \text{ is an integer,}$$

describe all solutions of the original equation, the expressions $0 + 2\pi n$ and $\pi + 2\pi n$ can be combined and written more compactly as πn. Thus, we say that the expressions

$$\pi n \quad \text{and} \quad \frac{\pi}{4} + \pi n, \quad \text{where } n \text{ is an integer,}$$

describe the general form of all solutions of the equation $\tan x \sin x - \sin x = 0$.

(b) Factoring and applying the zero product property, we have

$$\sin^2 x + \sin x - 2 = 0$$

$$(\sin x - 1)(\sin x + 2) = 0$$

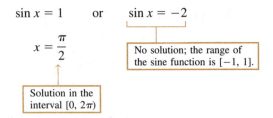

$$\sin x = 1 \quad \text{or} \quad \sin x = -2$$

$$x = \frac{\pi}{2}$$

Solution in the interval $[0, 2\pi)$

No solution; the range of the sine function is $[-1, 1]$.

Thus, the expression

$$\frac{\pi}{2} + 2\pi n, \quad \text{where } n \text{ is an integer,}$$

describes the general form of all solutions of the equation $\sin^2 x + \sin x - 2 = 0$. ◆

PROBLEM 3 Repeat Example 3 for $\sin x \cos x - \cos x = 0$. ◆

To solve other trigonometric equations, we apply the fundamental trigonometric identities in order to write the equation as a product that equals zero. As with algebraic equations, when squaring both sides of a trigonometric equation or multiplying both sides by a variable expression, we must check for *extraneous roots* (roots that don't satisfy the original equation).

EXAMPLE 4 Find the general form of the solution of each trigonometric equation.

(a) $2 \sin x + \cot x = \csc x$ (b) $\sin x + \cos x = 1$

SOLUTION

(a) We begin by applying the fundamental trigonometric identities:

$$2 \sin x + \cot x = \csc x$$

$$2 \sin x + \frac{\cos x}{\sin x} = \frac{1}{\sin x} \quad \textcolor{red}{\text{Apply trig identities}}$$

$$2 \sin^2 x + \cos x = 1 \quad \textcolor{red}{\text{Multiply both sides by } \sin x, \sin x \neq 0}$$

$$2(1 - \cos^2 x) + \cos x = 1 \quad \textcolor{red}{\text{Apply a trig identity}}$$

$$2 \cos^2 x - \cos x - 1 = 0 \quad \textcolor{red}{\text{Collect like terms}}$$

Having written the equation in terms of a single trigonometric function, we use factoring and the zero product property to solve this equation:

$$2 \cos^2 x - \cos x - 1 = 0$$

$$(2 \cos x + 1)(\cos x - 1) = 0$$

$$\cos x = -\tfrac{1}{2} \quad \text{or} \quad \cos x = 1$$

$$x = \frac{2\pi}{3}, \frac{4\pi}{3} \qquad x = 0$$

Possible solutions in the interval $[0, 2\pi)$

To arrive at these apparent solutions, we multiplied both sides of an equation by a variable expression, namely, $\sin x$. Thus, we must check for extraneous roots. Since $\sin 0 = 0$, we must discard $x = 0$ as a root. Thus, on the interval $[0, 2\pi)$, the only solutions are $2\pi/3$ and $4\pi/3$. Hence, the expressions

$$\frac{2\pi}{3} + 2\pi n \quad \text{and} \quad \frac{4\pi}{3} + 2\pi n, \quad \text{where } n \text{ is an integer,}$$

describe the general form of all solutions of the equation $2 \sin x + \cot x = \csc x$.

(b) Squaring both sides of this equation leads to the expressions $\sin^2 x$ and $\cos^2 x$, which are related by the trigonometric identity $\sin^2 x + \cos^2 x = 1$. Thus,

$$\sin x + \cos x = 1$$

$\sin^2 x + 2 \sin x \cos x + \cos^2 x = 1$	**Square both sides**
$1 + 2 \sin x \cos x = 1$	**Apply $\sin^2 x + \cos^2 x = 1$**
$2 \sin x \cos x = 0$	**Subtract 1 from both sides**
$\sin x \cos x = 0$	**Divide both sides by 2**

$$\sin x = 0 \qquad \text{or} \qquad \cos x = 0$$

$$x = 0, \pi \qquad\qquad x = \frac{\pi}{2}, \frac{3\pi}{2}$$

Possible solutions in the interval $[0, 2\pi)$.

To arrive at these apparent solutions, we squared both sides of an equation. Thus, we must check for extraneous roots. Of these four possible solutions, we find that $x = \pi$ and $x = 3\pi/2$ are extraneous. Thus, in the interval $[0, 2\pi)$, the only solutions are 0 and $\pi/2$. Hence, the expressions

$$2\pi n \quad \text{and} \quad \frac{\pi}{2} + 2\pi n, \quad \text{where } n \text{ is an integer,}$$

describe the general form of all solutions of the equation $\sin x + \cos x = 1$. ◆

PROBLEM 4 Repeat Example 4 for $\sin x - \cos x = 1$. ◆

Using a Calculator to Solve Trigonometric Equations

In our preceding discussion, we chose equations in which the solutions unfolded from special values, such as $\pi/6$, $\pi/4$, $\pi/3$, $\pi/2$, and so on. For equations in which these special values do not occur, we use the concept of the reference angle in conjunction with the inverse trigonometric keys on a calculator. We obtain the reference angle by entering the absolute value of the number into a calculator. The procedure is illustrated in the next example.

EXAMPLE 5 For each equation, approximate to four significant digits the solutions in the interval $[0, 2\pi)$.

(a) $5 \sin \theta + 2 = 0$ (b) $\cos^2 \theta - 2 \cos \theta = 2$

SOLUTION

(a) Solving for $\sin \theta$, we have

$$\sin \theta = -\tfrac{2}{5} = -0.4.$$

The reference angle θ' associated with this equation is the positive acute angle whose sine is $|-0.4|$. Using the inverse sine function in conjunction with a calculator set in radian mode, we obtain

$$\theta' = \sin^{-1}|-0.4| \approx 0.4115.$$

Remembering that the sine function is negative in quadrants III and IV, we place a reference angle of 0.4115 rad in the third and fourth quadrants, as shown in Figure 4.6. Hence, we obtain

$$\theta \approx \pi + 0.4115 \approx 3.553 \quad \text{and} \quad \theta \approx 2\pi - 0.4115 \approx 5.872,$$

as values of θ in the interval $[0, 2\pi)$ for which $5 \sin \theta + 2 = 0$.

(b) Writing this quadratic type equation in standard form, we obtain

$$\cos^2 \theta - 2 \cos \theta - 2 = 0.$$

Since this equation is not factorable over the integers, we apply the quadratic formula with $a = 1$, $b = -2$, and $c = -2$, and solve for $\cos \theta$ as follows:

$$\cos \theta = \frac{2 \pm \sqrt{(-2)^2 - 4(1)(-2)}}{2(1)} = \frac{2 \pm \sqrt{12}}{2} = 1 \pm \sqrt{3}.$$

Remember that the range of the cosine function is $[-1, 1]$. Hence, the equation

$$\cos \theta = 1 + \sqrt{3} \approx 2.732,$$

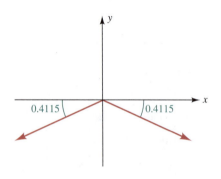

FIGURE 4.6
A reference angle of 0.4115 radians in quadrants III and IV

SECTION 4.2 Trigonometric Equations

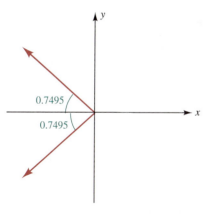

FIGURE 4.7
A reference angle of 0.7495 radians in quadrants II and III

has no solution. The reference angle θ' associated with the equation

$$\cos\theta = 1 - \sqrt{3} \approx -0.73205,$$

is the positive acute angle whose cosine is $|-0.73205|$. Using the inverse cosine function in conjunction with a calculator set in radian mode, we obtain

$$\theta' = \cos^{-1}|-0.73205| \approx 0.7495.$$

Remembering that the cosine function is negative in quadrants II and III, we place a reference angle of 0.7495 rad in quadrants II and III, as shown in Figure 4.7. Hence, we obtain

$$\theta \approx \pi - 0.7495 \approx 2.392 \quad \text{and} \quad \theta \approx \pi + 0.7495 \approx 3.891.$$

as values of θ in the interval $[0, 2\pi)$ for which $\cos^2\theta - 2\cos\theta = 2$. ◆

PROBLEM 5 Repeat Example 5 for $\tan^2 x - 4 = 0$. ◆

◆ **Application: Daylight Hours in the Northern Hemisphere**

At two times during the year, the sun crosses the plane of the earth's equator, making night and day equal in length all over the earth. These times are called the *vernal equinox* (about March 21, the 80th day of the year) and the *autumnal equinox* (about September 21). For a given latitude in the Northern Hemisphere, the number of hours of daylight oscillates in a periodic fashion about the equilibrium position of 12 hours, reaching its maximum value at the *summer solstice* (about June 21) and its minimum value at the *winter solstice* (about December 21), as shown in Figure 4.8. Hence, in the Northern Hemisphere, the number of hours N of daylight for a given latitude on a particular day d of the year may be approximated by

$$N = 12 + k \sin\left[\frac{2\pi}{365}(d - 80)\right],$$

where k is a constant that depends on the given latitude.

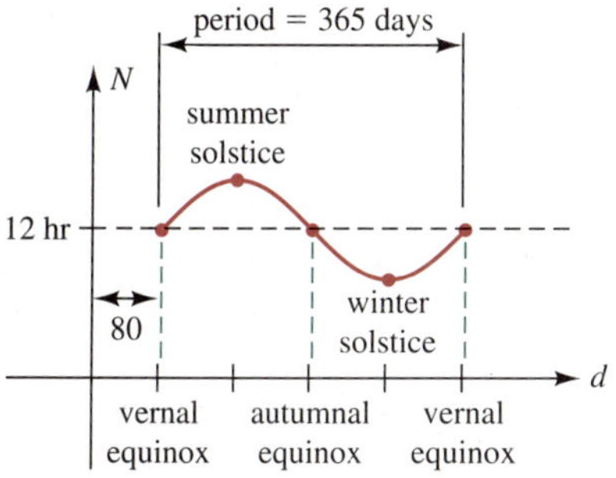

FIGURE 4.8
Daylight hours at a given latitude in the Northern Hemisphere

EXAMPLE 6 The number of hours N of daylight on a particular day d of the year in the Boston area may be approximated by

$$N = 12 + 3\sin\left[\frac{2\pi}{365}(d - 80)\right].$$

Determine the days of the year that have 13 hours 30 minutes of daylight.

SOLUTION To determine the days of the year that have 13 hours 30 minutes of daylight, we must solve the equation

$$13\tfrac{1}{2} = 12 + 3\sin\left[\frac{2\pi}{365}(d-80)\right],$$

for d. To do this, we proceed as follows:

$$13\tfrac{1}{2} = 12 + 3\sin\left[\frac{2\pi}{365}(d-80)\right]$$

$$\frac{3}{2} = 3\sin\left[\frac{2\pi}{365}(d-80)\right] \qquad \text{\color{red}Subtract 12 from both sides}$$

$$\frac{1}{2} = \sin\left[\frac{2\pi}{365}(d-80)\right] \qquad \text{\color{red}Divide both sides by 3}$$

From Table 2.3 in Section 2.3, we know that $\sin(\pi/6) = \tfrac{1}{2}$. Also, we know that the sine function is positive in quadrants I and II. Hence,

$$\frac{2\pi}{365}(d-80) = \frac{\pi}{6} \quad \text{and} \quad \frac{2\pi}{365}(d-80) = \frac{5\pi}{6}$$

$$d - 80 = \frac{365}{12} \qquad\qquad d - 80 = \frac{1825}{12} \qquad \text{\color{red}Divide both sides by $2\pi/365$}$$

$$d \approx 110 \qquad\qquad\qquad d \approx 232 \qquad \text{\color{red}Add 80 to both sides}$$

Thus, 13 hours 30 minutes of daylight occurs on day 110 (April 20th) and on day 232 (August 20th). ◆

PROBLEM 6 Referring to Example 6, determine the days of the year that have 10 hours 30 minutes of daylight. ◆

Exercises 4.2

Basic Skills

In Exercises 1–16, find the general form of the solution of each equation, if it exists.

1. $\sin x = 1$
2. $\cos x = -1$
3. $\tan x = 0$
4. $\sin x = -\tfrac{1}{2}$
5. $\cos x = \dfrac{\sqrt{3}}{2}$
6. $\tan x = -1$
7. $\csc x = \sqrt{2}$
8. $\sec x = -2$
9. $\sqrt{3}\cot x = 1$
10. $2\sin x = -\sqrt{3}$
11. $2\cos x - 1 = 0$
12. $\sqrt{3}\tan x + 3 = 0$
13. $\sin^2 x = \tfrac{1}{4}$
14. $\tan^2 x - 1 = 0$
15. $4\cos^2 x + 3 = 0$
16. $2\sin^2 x - 1 = 0$

In Exercises 17–32, find

(a) the general form of the solution of each equation, if it exists, and

(b) all solutions that exist in the interval $[0, 2\pi)$.

17. $\sin 2x = 1$
18. $\cot 3x = 1$
19. $\csc \dfrac{x}{2} + 1 = 0$
20. $2\cos 3x + 1 = 0$
21. $\sqrt{2}\cos 4x + 1 = 0$
22. $\tan \dfrac{2x}{3} - \sqrt{3} = 0$
23. $4\cos^2 2x - 1 = 0$
24. $2\sin^2 5x + 2 = 0$
25. $\tan\left(2x - \dfrac{\pi}{6}\right) = 1$
26. $\sin\left(3x + \dfrac{\pi}{4}\right) = 0$

27. $4 \sec(x + \pi) = 8$

28. $\sqrt{3} \tan\left(4x - \dfrac{\pi}{3}\right) = 1$

29. $2 \sin 3\left(x + \dfrac{\pi}{4}\right) + 1 = 0$

30. $\sqrt{3} - \cot 2(x - \pi) = 0$

31. $\tan^2\left(\dfrac{x - \pi}{2}\right) = 3$

32. $4 \cos^2\left(2x + \dfrac{\pi}{6}\right) - 3 = 0$

In Exercises 33–60, find the general form of the solution of each equation, if it exists.

33. $(\sin x + 1)(2 \sin x - 1) = 0$

34. $(\cos x - 1)(\cos x + 1) = 0$

35. $(\cot x - 1)(2 \sin^2 x - 1) = 0$

36. $(2 \cos^2 x - 1)(\csc x - 1) = 0$

37. $\cos^2 x - \cos x = 0$

38. $2 \sin^2 x + \sin x = 0$

39. $\tan x - 2 \tan x \cos x = 0$

40. $\tan x \csc x - \csc x = 0$

41. $\cos x \csc^2 x - 2 \cos x = 0$

42. $\sin x \cot^2 x = 3 \sin x$

43. $2 \cos^2 x - \cos x = 1$

44. $\csc^2 x - 3 \csc x + 2 = 0$

45. $2 \sec^3 x + \sec^2 x - 8 \sec x - 4 = 0$

46. $2 \sin^3 x - \sin^2 x + 6 \sin x = 3$

47. $2 \cos^2 x - \sin x - 1 = 0$

48. $2 \sin^2 x - 3 \cos x - 3 = 0$

49. $\tan^2 x + 3 \sec x = -3$

50. $\csc^2 x - \cot x = 1$

51. $4 \tan^2 x = 3 \sec^2 x$

52. $2(1 - 2 \cos^2 x) = 1 - 2 \sin^2 x$

53. $\sec x + \tan x = \cos x$

54. $2 \sin x - 3 \cot x = 3 \csc x$

55. $\cot x - \cos x = 0$

56. $\csc x = \sin x - \cot x$

57. $\sin x + 1 = \cos x$

58. $\cot x + 1 = \csc x$

59. $\tan x = \sec x + 1$

60. $\cot x = 3 \tan x$

61. The displacement d (in centimeters) of an oscillating spring from its equilibrium position is a function of time t (in seconds) and is given by

$$d = 12 \cos 2t.$$

Determine the times in the interval $[0, 2\pi)$ for which $d = -6$ cm.

62. The voltage v (in volts) in an electrical circuit is a function of time t (in seconds) and is given by

$$v = 20 \sin 4\pi t.$$

Determine the times in the interval $[0, 1]$ for which $v = 10$ volts.

63. Rangers have determined that the deer population in a state forest in Vermont can be approximated by

$$P = 186 + 84\sqrt{2} \sin \dfrac{\pi}{8} t,$$

where t is the time (in years) with $t = 0$ corresponding to 1980. Determine the years between 1980 and 2000 when the deer population is 102.

64. The number of hours N of daylight on a particular day d in the Boston area may be approximated by

$$N = 12 + 3 \sin\left[\dfrac{2\pi}{365}(d - 80)\right].$$

Determine the day of the year that has 9 hours of daylight.

Critical Thinking

65. Suppose that x_0 is a solution of the trigonometric equation $\sin bx = k$. State some other solutions of this equation.

66. Does the trigonometric equation $\tan x = k$ have an infinite number of solutions for all real numbers k? Does $\sin x = k$? Does $\cos x = k$? Explain.

67. The graphs of $y = \sin x$ and $y = k$ for $-1 < k < 1$ intersect twice in the interval $[0, 2\pi)$. Hence, the equation $\sin x = k$ must have two solutions in the interval $[0, 2\pi)$. How many solutions does the equation $\sin nx = k$, , where n is an integer and $-1 < k < 1$, have in the interval $[0, 2\pi)$? Explain.

68. A *chord* of a circle is a line segment whose endpoints lie on the circle. In the accompanying sketch, $\overline{AB}$ is a chord.

   (a) Show that the length of the chord $\overline{AB}$ (denoted AB) with central angle θ is given by
   $$AB = 2r \sin \frac{\theta}{2}.$$

   (b) Find θ (in degrees) if $r = 6$ cm and $AB = 6\sqrt{3}$ cm.

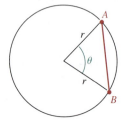

69. Use the graphs of $y = \sin 2x$ and $y = \cos x$ to help solve the equation $\sin 2x - \cos x = 0$.

70. Find the general form of the solution of each equation.
   (a) $(16^{\cos x})^{\tan x} = 4$ (b) $3^{\sec x} \cdot 9^{-\sin^2 x} = 9^{\cos^2 x}$

71. Given that $\sin^2 x = 1$, find the fallacy in the following argument:

$\sin^2 x = 1$	
$\dfrac{\sin^2 x}{\cos^2 x} = \dfrac{1}{\cos^2 x}$	Divide both sides by $\cos^2 x$
$\tan^2 x = \sec^2 x$	Apply trig identities
$\tan^2 x + 1 = \sec^2 x + 1$	Add 1 to both sides
$\sec^2 x = \sec^2 x + 1$	Apply trig identity
$0 = 1$	Subtract $\sec^2 x$ from both sides

72. Sketch the graphs of the given pair of equations on the same coordinate axes, and then find the points of intersection of the two curves.
   (a) $y = 5 \sin x + 1$ and $y = \sin x - 1$
   (b) $y = \cos 2x - 1$ and $y = \sin 2x$

Calculator Activities

In Exercises 73–88, approximate to four significant digits the solutions in the interval $[0, 2\pi)$.

73. $\sin x = \frac{2}{3}$
74. $\cos x = -0.25$
75. $\cot x + 5 = 0$
76. $2 \csc x - 7 = 0$
77. $\sec^2 x - 9 = 0$
78. $4 \tan^2 x - 3 = 0$
79. $5 \sin 3x - 4 = 0$
80. $\cos (2x - 1) = 0.7$
81. $(\cos x + 3)(4 \sin x - 1) = 0$
82. $(2 \sec x - 1)(\sec x + 4) = 0$
83. $\tan^2 x - \tan x = 6$
84. $2 \sin^2 x - 3 \sin x - 1 = 0$
85. $4 \cos^2 x - 7 \sin x + 2 = 0$
86. $2 \tan^2 x - 5 \sec x - 1 = 0$
87. $2 \sin x - \cos x = 0$
88. $\cot x = \csc x + 2$

89. The graph of $y = 3 \tan^2 x + 5 \tan x - 2$ for $-\pi/2 < x < \pi/2$ is shown in the sketch. Approximate the

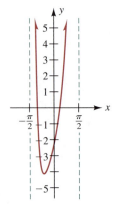

x-intercepts, rounding each answer to three significant digits.

90. The graph of $y = 2 \cos^2 x + 5 \sin x$ for $0 \leq x \leq 2\pi$ is shown in the sketch. Approximate the x-intercepts, rounding each answer to three significant digits.

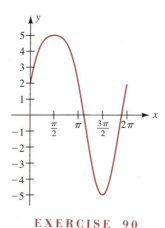

EXERCISE 90

📱 *In Exercises 91–94, an equation of the form $f(x) = 0$ is given. Use a graphing calculator to generate the graph of the function $y = f(x)$ in the interval $[0, 2\pi]$. Then use the calculator's trace and zoom features to estimate the solutions of the equation in the interval $[0, 2\pi]$. Record each answer to three significant digits.*

91. $2\cos^2 x - \cos x - 2 = 0$

92. $3\sin^2 2x - 2\cos 2x - 1 = 0$

93. $2\tan^3 2x - \tan 2x - 1 = 0$

94. $\sin^4 x - \tan x + 2 = 0$

95. The number of hours N of daylight on a particular day d in Fairbanks, Alaska, may be approximated by

$$N = 12 + 11\sin\left[\frac{2\pi}{365}(d - 80)\right].$$

Determine the days of the year that have 20 hours of daylight.

96. The voltage v (in volts) in an electrical circuit is a function of time t (in milliseconds) and is given by

$$v = 12.5\sin(2.4t - 4.5).$$

Approximate to three significant digits the times in the interval $[0, 10]$ for which $v = 9.2$ volts.

4.3 Sum and Difference Formulas

In this section, we derive formulas for the trigonometric functions of $u \pm v$, where u and v are real numbers or angles, measured in either degrees or radians. We refer to these formulas as the *sum and difference formulas* and use them to help prove other trigonometric identities and solve some trigonometric equations. The sum and difference formulas are especially useful when working with the trigonometric functions in calculus.

◆ Sum and Difference Formulas for Cosine

We begin by developing a formula for $\cos(u - v)$. Consider a unit circle in which u and v are angles in standard position, as shown in Figure 4.9. Note that angles u and v intersect the unit circle at the points $A(\cos u, \sin u)$ and $B(\cos v, \sin v)$, respectively, and that the angle between $\overline{OA}$ and $\overline{OB}$ is $u - v$. Figure 4.10 shows angle $u - v$ with a line segment joining points A and B. In Figure 4.11, we redraw angle $u - v$ in standard position and relabel the points on the unit circle $A'(\cos(u - v), \sin(u - v))$ and $B'(1, 0)$, respectively.

Now, line segments AB and $A'B'$ in Figures 4.10 and 4.11 must have the same length. By the distance formula,

$$\begin{aligned}AB &= \sqrt{(\cos u - \cos v)^2 + (\sin u - \sin v)^2} \\ &= \sqrt{(\cos^2 u - 2\cos u \cos v + \cos^2 v) + (\sin^2 u - 2\sin u \sin v + \sin^2 v)} \\ &= \sqrt{2 - 2\cos u \cos v - 2\sin u \sin v} \quad \text{Apply } \sin^2\theta + \cos^2\theta = 1\end{aligned}$$

and

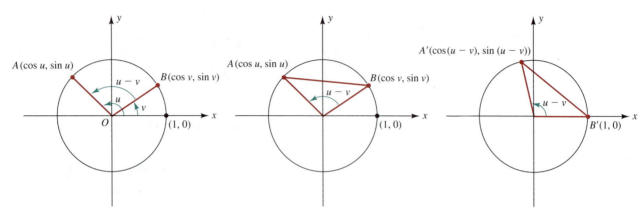

FIGURE 4.9
A unit circle with angles u and v in standard position

FIGURE 4.10
Angle $u - v$ with a line segment joining points A and B

FIGURE 4.11
Angle $u - v$ redrawn in standard position

$$A'B' = \sqrt{[\cos(u-v) - 1]^2 + [\sin(u-v) - 0]^2}$$
$$= \sqrt{[\cos^2(u-v) - 2\cos(u-v) + 1] + \sin^2(u-v)}$$
$$= \sqrt{2 - 2\cos(u-v)} \qquad \textbf{\textcolor{red}{Apply } } \sin^2\theta + \cos^2\theta = 1$$

Equating these lengths, we obtain

$$\sqrt{2 - 2\cos(u-v)} = \sqrt{2 - 2\cos u \cos v - 2\sin u \sin v}$$

$2 - 2\cos(u-v) = 2 - 2\cos u \cos v - 2\sin u \sin v$ **Square both sides**

$-2\cos(u-v) = -2\cos u \cos v - 2\sin u \sin v$ **Subtract 2 from both sides**

$\cos(u-v) = \cos u \cos v + \sin u \sin v$ **Divide both sides by -2**

We refer to this last equation as the *difference formula for cosine*. Although we have developed this formula for positive angles u and v, the formula is also valid for negative angles and for real numbers.

Recall from Section 2.4 that the sine function is odd and the cosine function is even, that is,

$$\sin(-v) = -\sin v \qquad \text{and} \qquad \cos(-v) = \cos v.$$

Using these facts, we can obtain a *sum formula for cosine* as follows:

$\cos(u + v) = \cos[u - (-v)]$ **Rewrite as a difference**

$\qquad\qquad\quad = \cos u \cos(-v) + \sin u \sin(-v)$ **Apply the difference formula for cosine**

$\qquad\qquad\quad = \cos u \cos v - \sin u \sin v$ **Simplify**

Sum and Difference Formulas for Cosine

1. $\cos(u + v) = \cos u \cos v - \sin u \sin v$
2. $\cos(u - v) = \cos u \cos v + \sin u \sin v$

EXAMPLE 1 Using the sum and difference formulas for cosine, simplify each expression.

(a) $\cos\left(\dfrac{\pi}{2} - x\right)$ (b) $\cos 3x \cos 2x - \sin 3x \sin 2x$

SOLUTION

(a) Applying the difference formula for cosine with $u = \pi/2$ and $v = x$, we have

$$\cos\left(\frac{\pi}{2} - x\right) = \cos\frac{\pi}{2} \cos x + \sin\frac{\pi}{2} \sin x$$
$$= (0)\cos x + (1)\sin x \qquad \text{Evaluate}$$
$$= \sin x \qquad \text{Simplify}$$

(b) Applying the sum formula for cosine from right to left with $u = 3x$ and $v = 2x$, we have

$$\cos 3x \cos 2x - \sin 3x \sin 2x = \cos(3x + 2x) = \cos 5x.$$

PROBLEM 1 Repeat Example 1 for $\cos 3x \cos 2x + \sin 3x \sin 2x$.

Sum and Difference Formulas for Sine

Recall from Section 2.2 that the sine and cosine are cofunctions and that a trigonometric function of any acute angle θ is the same as the cofunction of its complementary angle $90° - \theta$. In Example 1(a), we have shown that this relationship holds for any arbitrary angle or real number, that is,

$$\sin x = \cos\left(\frac{\pi}{2} - x\right) \quad \text{for all } x.$$

Now, if we replace x with $\left(\dfrac{\pi}{2} - u\right)$, we obtain

$$\sin\left(\frac{\pi}{2} - u\right) = \cos\left[\frac{\pi}{2} - \left(\frac{\pi}{2} - u\right)\right] = \cos u.$$

Hence, we can also state

$$\cos x = \sin\left(\frac{\pi}{2} - x\right) \quad \text{for all } x.$$

By using these cofunction relationships, we can derive a formula for sin $(u + v)$ as follows:

$$\sin(u+v) = \cos\left[\frac{\pi}{2} - (u+v)\right] \quad \text{Use cofunction relationship}$$

$$= \cos\left[\left(\frac{\pi}{2} - u\right) - v\right] \quad \text{Rewrite}$$

$$= \cos\left(\frac{\pi}{2} - u\right)\cos v + \sin\left(\frac{\pi}{2} - u\right)\sin v \quad \text{Apply the difference formula for cosine}$$

$$= \sin u \cos v + \cos u \sin v \quad \text{Use cofunction relationship}$$

We refer to

$$\sin(u+v) = \sin u \cos v + \cos u \sin v$$

as the *sum formula for sine*.

We can obtain the *difference formula for sine* as follows:

$$\sin(u-v) = \sin[u + (-v)] \quad \text{Rewrite as a sum}$$

$$= \sin u \cos(-v) + \cos u \sin(-v) \quad \text{Apply the sum formula for sine}$$

$$= \sin u \cos v - \cos u \sin v \quad \text{Simplify}$$

Sum and Difference Formulas for Sine

1. $\sin(u+v) = \sin u \cos v + \cos u \sin v$
2. $\sin(u-v) = \sin u \cos v - \cos u \sin v$

EXAMPLE 2 Using the sum and difference formulas for sine, simplify each expression.

(a) $\sin\left(x + \dfrac{3\pi}{2}\right)$
(b) $\sin\left(x + \dfrac{\pi}{6}\right)\cos x - \cos\left(x + \dfrac{\pi}{6}\right)\sin x$

SOLUTION

(a) Applying the sum formula for sine with $u = x$ and $v = 3\pi/2$, we have

$$\sin\left(x + \frac{3\pi}{2}\right) = \sin x \cos\frac{3\pi}{2} + \cos x \sin\frac{3\pi}{2}$$

$$= \sin x\,(0) + \cos x\,(-1) \quad \text{Evaluate}$$

$$= -\cos x \quad \text{Simplify}$$

(b) Applying the difference formula for sine from right to left with $u = x + \dfrac{\pi}{6}$ and $v = x$, we have

$$\sin\left(x + \frac{\pi}{6}\right)\cos x - \cos\left(x + \frac{\pi}{6}\right)\sin x = \sin\left[\left(x + \frac{\pi}{6}\right) - x\right]$$

$$= \sin\frac{\pi}{6} \qquad \textcolor{red}{\text{Simplify}}$$

$$= \tfrac{1}{2} \qquad \textcolor{red}{\text{Evaluate}} \quad \blacklozenge$$

PROBLEM 2 As an alternate method for simplifying Example 2(b), expand $\sin\left(x + \dfrac{\pi}{6}\right)$ and $\cos\left(x + \dfrac{\pi}{6}\right)$ by using the sum formulas for sine and cosine, respectively. Then perform the indicated operations. $\blacklozenge$

♦ **Sum and Difference Formulas for Tangent**

To derive the *sum formula for tangent*, we proceed as follows:

$$\tan(u+v) = \frac{\sin(u+v)}{\cos(u+v)} \qquad \textcolor{red}{\text{Apply } \tan x = \frac{\sin x}{\cos x}}$$

$$= \frac{\sin u \cos v + \cos u \sin v}{\cos u \cos v - \sin u \sin v} \qquad \textcolor{red}{\text{Apply sum formulas for sine and cosine}}$$

$$= \frac{\dfrac{\sin u \cos v + \cos u \sin v}{\cos u \cos v}}{\dfrac{\cos u \cos v - \sin u \sin v}{\cos u \cos v}} \qquad \textcolor{red}{\text{Divide both numerator and denominator by } \cos u \cos v}$$

$$= \frac{\dfrac{\sin u}{\cos u} + \dfrac{\sin v}{\cos v}}{1 - \dfrac{\sin u \sin v}{\cos u \cos v}} \qquad \textcolor{red}{\text{Split and reduce fractions}}$$

$$= \frac{\tan u + \tan v}{1 - \tan u \tan v} \qquad \textcolor{red}{\text{Apply } \frac{\sin x}{\cos x} = \tan x}$$

Recall that the tangent function is an odd function, that is,

$$\tan(-v) = -\tan v.$$

We use this fact to obtain the *difference formula for tangent*:

$$\tan(u - v) = \tan[u + (-v)] \qquad \text{Rewrite as a sum}$$

$$= \frac{\tan u + \tan(-v)}{1 - \tan u \tan(-v)} \qquad \text{Apply the sum formula for tangent}$$

$$= \frac{\tan u - \tan v}{1 + \tan u \tan v} \qquad \text{Simplify}$$

Sum and Difference Formulas for Tangent

1. $\tan(u + v) = \dfrac{\tan u + \tan v}{1 - \tan u \tan v}$
2. $\tan(u - v) = \dfrac{\tan u - \tan v}{1 + \tan u \tan v}$

EXAMPLE 3 Using the sum and difference formulas for tangent, simplify each expression.

(a) $\tan(\pi + x)$ (b) $\dfrac{\tan 87° - \tan 27°}{1 + \tan 87° \tan 27°}$

SOLUTION

(a) Applying the sum formula for tangent with $u = \pi$ and $v = x$, we have

$$\tan(\pi + x) = \frac{\tan \pi + \tan x}{1 - \tan \pi \tan x} = \frac{0 + \tan x}{1 - (0)\tan x} = \tan x.$$

(b) Applying the difference formula for tangent from right to left with $u = 87°$ and $v = 27°$, we have

$$\frac{\tan 87° - \tan 27°}{1 + \tan 87° \tan 27°} = \tan(87° - 27°) = \tan 60° = \sqrt{3}. \qquad \blacklozenge$$

PROBLEM 3 Use a calculator to find the approximate value of $\dfrac{\tan 87° - \tan 27°}{1 + \tan 87° \tan 27°}$. Compare your answer with the exact value from Example 3(b). $\blacklozenge$

◆ **Applying the Sum and Difference Formulas**

The next four examples illustrate a variety of uses of sum and difference formulas. We begin by showing how to evaluate a trigonometric function whose angle is the sum (or difference) of the special angles 30°, 45°, and 60°.

EXAMPLE 4 Find the exact value of cos 75°.

SOLUTION We can think of 75° as the sum of the special angles 30° and 45°. Hence,

$$\cos 75° = \cos(30° + 45°) \qquad \text{Rewrite with special angles}$$
$$= \cos 30° \cos 45° - \sin 30° \sin 45° \qquad \text{Apply the sum formula for cosine}$$
$$= \frac{\sqrt{3}}{2} \cdot \frac{1}{\sqrt{2}} - \frac{1}{2} \cdot \frac{1}{\sqrt{2}} \qquad \text{Use Table 2.3 to evaluate}$$
$$= \frac{\sqrt{3} - 1}{2\sqrt{2}} \text{ or } \frac{\sqrt{6} - \sqrt{2}}{4} \qquad \text{Simplify} \qquad \blacklozenge$$

PROBLEM 4 Use a calculator to find the approximate value of cos 75°. Compare your answer with the exact value in Example 4. ◆

If a trigonometric function of u and a trigonometric function of v are given and the quadrants containing u and v are known, then the values of the trigonometric functions of $u \pm v$ can be determined. This procedure is shown in the next example.

EXAMPLE 5 Given that u and v are second-quadrant angles with $\sin u = \frac{3}{5}$ and $\tan v = -\frac{1}{4}$, find the exact value of $\tan(u + v)$.

SOLUTION We can use the sum formula for tangent to evaluate $\tan(u + v)$, provided we know the values of $\tan u$ and $\tan v$. We are given that $\tan v = -\frac{1}{4}$. From the trigonometric identity $\sin^2 u + \cos^2 u = 1$, we have

$$\cos u = \pm\sqrt{1 - \sin^2 u}.$$

However, since it is given that u is a second-quadrant angle, we know that $\cos u$ must be negative. Thus,

$$\cos u = -\sqrt{1 - \sin^2 u} = -\sqrt{1 - (\tfrac{3}{5})^2} = -\sqrt{\tfrac{16}{25}} = -\tfrac{4}{5}.$$

Therefore,

$$\tan u = \frac{\sin u}{\cos u} = \frac{\tfrac{3}{5}}{-\tfrac{4}{5}} = -\frac{3}{4}$$

Hence,

$$\tan(u + v) = \frac{\tan u + \tan v}{1 - \tan u \tan v} = \frac{(-\tfrac{3}{4}) + (-\tfrac{1}{4})}{1 - (-\tfrac{3}{4})(-\tfrac{1}{4})} = -\frac{16}{13}. \qquad \blacklozenge$$

PROBLEM 5 Use the information in Example 5 to find the exact value of $\sin(u - v)$. ◆

As illustrated in the next example, sum and difference formulas are helpful in verifying certain trigonometric identities.

EXAMPLE 6 Show that the given trigonometric equation is an identity:

$$\frac{\cos x}{\sec 3x} + \frac{\sin x}{\csc 3x} = \cos 2x$$

SOLUTION Working with the left-hand side, we proceed as follows:

$$\frac{\cos x}{\sec 3x} + \frac{\sin x}{\csc 3x} = \cos 2x$$

$\dfrac{\cos x}{1/\cos 3x} + \dfrac{\sin x}{1/\sin 3x}$	Apply fundamental trig identities
$\cos 3x \cos x + \sin 3x \sin x$	Divide
$\cos (3x - x)$	Apply difference formula for cosine
$\cos 2x$	Simplify

Since we have reduced the left-hand side of this equation to the right-hand side, we conclude that

$$\frac{\cos x}{\sec 3x} + \frac{\sin x}{\csc 3x} = \cos 2x \quad \text{is a trigonometric identity.} \quad \blacklozenge$$

PROBLEM 6 Show that the following trigonometric equation is an identity:

$$\sin 3x \cot x + \cos 3x = \sin 4x \csc x \quad \blacklozenge$$

Finally, we illustrate the use of sum and difference formulas to help solve a trigonometric equation.

EXAMPLE 7 Find the general form of the solution of the equation

$$\sin 3x \cos x = 1 + \cos 3x \sin x.$$

SOLUTION To solve this equation, we proceed as follows:

$\sin 3x \cos x = 1 + \cos 3x \sin x$	
$\sin 3x \cos x - \cos 3x \sin x = 1$	Subtract $\cos 3x \sin x$ from both sides
$\sin (3x - x) = 1$	Apply the difference formula for sine
$\sin 2x = 1$	Simplify

SECTION 4.3 Sum and Difference Formulas 259

Solving for x, we obtain

$$2x = \frac{\pi}{2} + 2\pi n$$

$$x = \frac{\pi}{4} + \pi n, \quad \text{where } n \text{ is an integer.}$$

PROBLEM 7 Find the general form of the solution of the equation

$$\sin 3x \cos x = 1 - \cos 3x \sin x.$$

The Function $f(t) = A \sin bt + B \cos bt$

In physics and many branches of engineering, we may encounter functions of the form

$$f(t) = A \sin bt + B \cos bt,$$

where A, B, and b are real numbers and t is time (in seconds). When working with this type of function, we usually replace the expression $A \sin bt + B \cos bt$ with

$$a \sin (bt + c),$$

where a is the distance from the origin to a point P having coordinates (A, B) and c is the measure of the angle in standard position with terminal side $\overline{OP}$, as shown in Figure 4.12. To show that these two expressions are equivalent, refer to Figure 4.12 and observe that

$$\cos c = \frac{A}{a} \quad \text{and} \quad \sin c = \frac{B}{a}$$

$$A = a \cos c \qquad B = a \sin c$$

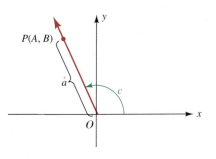

FIGURE 4.12

The point $P(A, B)$ determines distance a and angle c

Thus,

$$\begin{aligned}
A \sin bt + B \cos bt &= (a \cos c) \sin bt + (a \sin c) \cos bt \\
&= a \sin bt \cos c + a \cos bt \sin c & \textbf{Rewrite} \\
&= a(\sin bt \cos c + \cos bt \sin c) & \textbf{Factor} \\
&= a \sin (bt + c) & \textbf{Apply the sum formula for sine}
\end{aligned}$$

When a function f is written in the form

$$f(t) = a \sin (bt + c),$$

we can determine the amplitude, period, and phase shift of f quite easily. Recall from Section 2.4 that

$$\text{Amplitude: } |a| \qquad \text{Period: } \frac{2\pi}{|b|} \qquad \text{Phase shift: } \left|\frac{c}{b}\right|$$

EXAMPLE 8 The current i in an electrical circuit is a function of time t and is defined by

$$i(t) = \sqrt{3}\sin 4t + \cos 4t,$$

where i is in amps and t is in seconds.

(a) Express this function in the form $i(t) = a\sin(bt + c)$.

(b) Determine the amplitude, period, and phase shift of i, and sketch its graph.

(c) Use both forms of $i(t)$ to find $i(\pi/24)$, the current in the circuit at $\pi/24$ second.

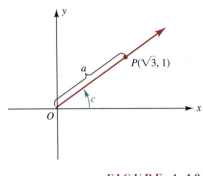

FIGURE 4.13
The values of a and c can be determined from the point $P(\sqrt{3}, 1)$.

SOLUTION

(a) For $i(t) = \sqrt{3}\sin 4t + \cos 4t$, we have $A = \sqrt{3}$ and $B = 1$. Plotting the point $P(\sqrt{3}, 1)$, as shown in Figure 4.13, we find

$$a = \sqrt{A^2 + B^2} = \sqrt{(\sqrt{3})^2 + (1)^2} = \sqrt{4} = 2$$

and

$$\tan c = \frac{B}{A} = \frac{1}{\sqrt{3}}$$

Since $0 < c \leq \pi/2$, we have

$$c = \tan^{-1}\frac{1}{\sqrt{3}} = \frac{\pi}{6}.$$

Hence,

$$i(t) = \sqrt{3}\sin 4t + \cos 4t = 2\sin\left(4t + \frac{\pi}{6}\right).$$

(b) For $i(t) = 2\sin\left(4t + \frac{\pi}{6}\right)$, we have $a = 2$, $b = 4$, and $c = \pi/6$. Thus,

$$\text{Amplitude: } |a| = 2 \qquad \text{Period: } \frac{2\pi}{|b|} = \frac{2\pi}{4} = \frac{\pi}{2}$$

$$\text{Phase shift: } \left|\frac{c}{b}\right| = \frac{\pi/6}{4} = \frac{\pi}{24} \text{ unit to the left}$$

SECTION 4.3 Sum and Difference Formulas

The graph of $i(t) = \sqrt{3} \sin 4t + \cos 4t = 2 \sin\left(4t + \frac{\pi}{6}\right)$ is shown in Figure 4.14.

(c) Using $i(t) = \sqrt{3} \sin 4t + \cos 4t$, we find

$$i\left(\frac{\pi}{24}\right) = \sqrt{3} \sin \frac{\pi}{6} + \cos \frac{\pi}{6} = \sqrt{3}\left(\frac{1}{2}\right) + \frac{\sqrt{3}}{2} = \sqrt{3} \text{ amps.}$$

Using $i(t) = 2 \sin\left(4t + \frac{\pi}{6}\right)$, we find

$$i\left(\frac{\pi}{24}\right) = 2 \sin\left(\frac{\pi}{6} + \frac{\pi}{6}\right) = 2 \sin\left(\frac{\pi}{3}\right) = 2\left(\frac{\sqrt{3}}{2}\right) = \sqrt{3} \text{ amps.}$$

As expected, both forms of $i(t)$ yield the same answer.

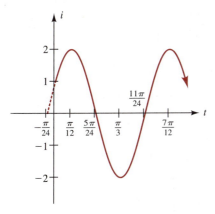

FIGURE 4.14
Graph of
$i(t) = \sqrt{3} \sin 4t + \cos 4t = 2 \sin\left(4t + \frac{\pi}{6}\right)$

PROBLEM 8 Referring to Example 8, find all times t in the interval $[0, \pi/2]$ for which $i(t) = 1$ amp. Use the graph in Figure 4.14 to check your answers.

Exercises 4.3

Basic Skills

In Exercises 1–16, use sum and difference formulas to simplify each expression.

1. $\cos\left(\frac{\pi}{2} + x\right)$

2. $\cos(\pi - x)$

3. $\sin(\pi - x)$

4. $\sin\left(\frac{3\pi}{2} + x\right)$

5. $\tan(2\pi + x)$

6. $\tan(\pi - x)$

7. $\sin\left(\frac{\pi}{6} - x\right) + \cos\left(x - \frac{2\pi}{3}\right)$

8. $\sin\left(\frac{\pi}{4} + x\right) - \sin\left(\frac{\pi}{4} - x\right)$

9. $\sin 35° \cos 55° + \cos 35° \sin 55°$

10. $\cos 230° \cos 50° + \sin 230° \sin 50°$

11. $\cos 3x \cos 5x - \sin 3x \sin 5x$

12. $\sin 5x \cos 6x - \cos 5x \sin 6x$

13. $\sin\left(\frac{\pi}{4} - x\right) \cos \frac{\pi}{4} + \cos\left(\frac{\pi}{4} - x\right) \sin \frac{\pi}{4}$

14. $\cos\left(\frac{\pi}{6} - x\right) \cos x - \sin\left(\frac{\pi}{6} - x\right) \sin x$

15. $\dfrac{\tan 205° + \tan 20°}{1 - \tan 205° \tan 20°}$

16. $\dfrac{\tan\left(\frac{\pi}{3} + x\right) - \tan x}{1 + \tan\left(\frac{\pi}{3} + x\right) \tan x}$

For the trigonometric functions in Exercises 17–22, express each angle as the sum or difference of 30°, 45°, or 60°. Then find the exact value of the trigonometric function by applying a sum or difference formula.

17. $\sin 75°$

18. $\sin 15°$

19. $\cos 15°$

20. $\cos 105°$

21. $\tan 105°$

22. $\tan 15°$

In Exercises 23–30, find the exact value of the indicated trigonometric function from the given information.

23. $\tan x = 4$; find $\tan\left(x + \frac{\pi}{4}\right)$

24. $\tan x = -2$; find $\tan\left(x - \frac{\pi}{4}\right)$

25. $\sin x = \frac{4}{5}$, x in quadrant I; find $\sin\left(x - \frac{\pi}{6}\right)$

26. $\cos x = \frac{12}{13}$, x in quadrant I; find $\cos\left(x + \frac{2\pi}{3}\right)$

27. $\sin x = \frac{3}{5}$, $\cos y = -\frac{5}{13}$, x and y in quadrant II; find $\cos(x+y)$

28. $\cos x = \frac{1}{3}$, $\tan y = -\sqrt{2}$, x and y in quadrant IV; find $\tan(x-y)$

29. $\cot x = \frac{3}{4}$, $\tan y = 3$, x in quadrant I and y in quadrant III; find $\sin(x+y)$

30. $\sec x = 4$, $\sec y = 2$, x and y in quadrant I; find $\cos(x-y)$

In Exercises 31–42, show that each trigonometric equation is an identity.

31. $\sin 2x \cot x - \cos 2x = 1$

32. $\cos 2x - \sin 2x \tan 3x = \cos 5x \sec 3x$

33. $\dfrac{\cos x}{\csc 4x} + \dfrac{\cos 4x}{\csc x} = \sin 5x$

34. $\dfrac{\sin x}{\sec 2x} - \dfrac{\cos x}{\csc 2x} = -\sin x$

35. $\dfrac{1}{\cot 2x} + \dfrac{1}{\tan 4x} = \csc 4x$

36. $\tan 6x - \tan 3x = \dfrac{\tan 3x}{\cos 6x}$

37. $\dfrac{1 + \tan x}{\tan\left(x + \frac{\pi}{4}\right)} = 1 - \tan x$

38. $\dfrac{\sin(x+y)}{\cos x \cos y} = \tan x + \tan y$

39. $\sin(x+y)\sin(x-y) = \sin^2 x - \sin^2 y$

40. $\cos(x+y)\cos(x-y) = \cos^2 x - \sin^2 y$

41. $\sec(x+y) = \dfrac{\sec x \sec y}{1 - \tan x \tan y}$

42. $\csc(x-y) = \dfrac{\csc x \csc y}{\cot y - \cot x}$

In Exercises 43–54, find the general form of the solution of each equation.

43. $\cos 3x \cos x + \sin 3x \sin x = 0$

44. $\cos 2x \cos x = 1 + \sin 2x \sin x$

45. $\sqrt{2}\sin 3x \cos 2x = 1 + \sqrt{2}\cos 3x \sin 2x$

46. $2\sin x \cos 2x + 2\cos x \sin 2x = 1$

47. $\dfrac{\tan 2x + \tan x}{1 - \tan 2x \tan x} = 1$

48. $1 + \tan x \tan 3x = \tan x - \tan 3x$

49. $\tan 2x \cos x = \sin x$ 50. $\cot 3x + \tan x = 0$

51. $\cos x + \sin\left(\frac{\pi}{2} - x\right) = 1$

52. $2\tan x + \tan(\pi - x) = \sqrt{3}$

53. $\cos\left(x + \frac{\pi}{4}\right) + \cos\left(x - \frac{\pi}{4}\right) = 1$

54. $\sin\left(x + \frac{\pi}{6}\right) - \sin\left(x - \frac{\pi}{6}\right) = \frac{1}{2}$

In Exercises 55–58,

(a) express each function f in the form $f(x) = a\sin(bx + c)$, and

(b) determine the amplitude, period, and phase shift of f.

55. $f(x) = 3\sin 2x + 3\cos 2x$

56. $f(x) = 2\sin 3x - 2\cos 3x$

57. $f(x) = -\sqrt{3}\sin 5x + \cos 5x$

58. $f(x) = -\sin 4x - \sqrt{3}\cos 4x$

Critical Thinking

59. In general, $\sin(u+v) \neq \sin u + \sin v$.

    (a) Use $u = \pi/3$ and $v = \pi/6$ to illustrate this fact.

    (b) Are there any real numbers u and v for which $\sin(u+v) = \sin u + \sin v$? If so, give an example.

60. In general, $\tan(u-v) \neq \tan u - \tan v$.

    (a) Use $u = \pi/3$ and $v = \pi/6$ to illustrate this fact.

    (b) Are there any real numbers u and v for which $\tan(u-v) = \tan u - \tan v$? If so, give an example.

61. Use sum and difference formulas to find the exact value of each expression.

    (a) $\sin(\arctan 2 + \arctan 3)$

    (b) $\cos(\cos^{-1}\frac{1}{4} - \tan^{-1}\frac{1}{2})$

62. Sketch the graph of each function.
   (a) $f(x) = \cos 5x \cos 3x - \sin 5x \sin 3x$
   (b) $f(x) = \sin 5x \cos 3x - \cos 5x \sin 3x$

63. Suppose A, B, and C are the interior angles of an acute triangle with $\sin A = a$ and $\sin B = b$. Express $\sin C$ in terms of a and b. [*Hint:* The sum of the interior angles of a triangle is 180°.]

64. Given the right triangle in the sketch, express each function in terms of a, b, and c.
   (a) $\tan \alpha$
   (b) $\tan \beta$
   (c) $\tan \theta$; use the fact that $\theta = \beta - \alpha$ and apply the difference formula for tangent

65. For functions such as $f(t) = A \sin bt + B \cos bt$, we may also replace the expression $A \sin bt + B \cos bt$ with $a \cos (bt - c)$, where a is the distance from the origin to a point P having coordinates (B, A) and c is the measure of the angle in standard position with terminal side $\overline{OP}$, as shown in the sketch. Show that these two expressions are equivalent.

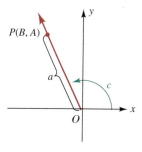

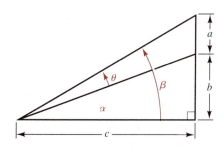

66. Given that $f(x) = \sin x$, show that the difference quotient

$$\frac{f(x + \Delta x) - f(x)}{\Delta x} =$$

$$\sin x \left(\frac{\cos \Delta x - 1}{\Delta x} \right) + \cos x \left(\frac{\sin \Delta x}{\Delta x} \right).$$

◆ Calculator Activities

In Exercises 67–72, assume that α is a fourth-quadrant angle and β is a second-quadrant angle with $\cos \alpha = 0.5299$ and $\sin \beta = 0.5592$. Use a calculator to evaluate to three significant digits each trigonometric expression by

(a) *determining the values of α and β and then directly evaluating the given expression and*

(b) *applying the appropriate sum or difference formula.*

67. $\sin (\alpha + \beta)$ 68. $\sin (\alpha - \beta)$
69. $\cos (\alpha - \beta)$ 70. $\cos (\alpha + \beta)$
71. $\tan (\alpha + \beta)$ 72. $\tan (\alpha - \beta)$

73. Express each function f in the form $f(x) = a \sin (bx + c)$, rounding a and c to three significant digits.
   (a) $f(x) = 12.6 \sin 5.2x - 10.3 \cos 5.2x$
   (b) $f(x) = -1.35 \sin 3.7x + 2.88 \cos 3.7x$

74. Use a graphing calculator or a computer with graphing capabilities to generate the graphs of both forms of the function f in each of the Exercises 73(a) and 73(b). Are the graphs identical?

75. The current i (in amps) in an electrical circuit is a function of time t (in seconds) and is given by

$$i(t) = 5 \sin 6\pi t + 12 \cos 6\pi t.$$

   (a) Determine a and c such that $i(t) = a \sin (6\pi t + c)$.
   (b) Find the times t between 0 and 1 second for which $i = 8.5$ amps.

76. The motion of a weight attached to an oscillating spring is given by

$$d = 24 \sin 8t - 7 \cos 8t,$$

where d is the displacement (in centimeters) from its equilibrium position at time t (in seconds).

   (a) Determine a and c such that $d = a \sin (8t + c)$.
   (b) Find the times t between 0 and 1 second for which $d = -9.5$ cm.

4.4 Multiple-Angle Formulas

Sum and difference formulas (Section 4.3) may be used to develop several other trigonometric formulas. In this section, we derive formulas for the trigonometric functions of *ku*, where *k* is a constant and *u* is a real number or an angle, measured in either degrees or radians. We refer to these formulas as *multiple-angle formulas* and use them to help prove other trigonometric identities and solve some trigonometric equations. Multiple-angle formulas are especially useful for working with the trigonometric functions in calculus.

◆ Double-Angle Formulas

The sum formula for sine, given in Section 4.3, states that

$$\sin(u + v) = \sin u \cos v + \cos u \sin v.$$

If we replace v with u in this formula, then we obtain

$$\sin(u + u) = \sin u \cos u + \cos u \sin u$$

$$\sin 2u = 2 \sin u \cos u \qquad \textbf{Simplify both sides}$$

We refer to this last equation as the *double-angle formula for sine*. In a similar manner, we can use the sum formula for cosine and tangent to derive *double-angle formulas for cosine and tangent*.

Double-Angle Formulas

1. $\sin 2u = 2 \sin u \cos u$
2. $\cos 2u = \cos^2 u - \sin^2 u$
3. $\tan 2u = \dfrac{2 \tan u}{1 - \tan^2 u}$

EXAMPLE 1 Using double-angle formulas, simplify each expression.

(a) $6 \sin 5x \cos 5x$ (b) $\cos^2 \dfrac{x}{2} - \sin^2 \dfrac{x}{2}$

SOLUTION

(a) Applying the double-angle formula for sine from right to left with $u = 5x$, we have

$$6 \sin 5x \cos 5x = 3(2 \sin 5x \cos 5x) \qquad \textbf{Rewrite}$$
$$= 3 \sin 2(5x) \qquad \textbf{Apply double-angle formula}$$
$$= 3 \sin 10x \qquad \textbf{Simplify}$$

(b) Applying the double-angle formula for cosine from right to left with $u = x/2$, we have

$$\cos^2 \frac{x}{2} - \sin^2 \frac{x}{2} = \cos 2\left(\frac{x}{2}\right) = \cos x.$$

◆

PROBLEM 1 Repeat Example 1 for $\dfrac{2 \tan 3x}{1 - \tan^2 3x}$.

◆ **Applying the Double-Angle Formulas**

Double-angle formulas have a variety of uses. In the next example, we use these formulas to help verify a trigonometric identity.

EXAMPLE 2 Show that the following trigonometric equation is an identity:

$$\csc 2x - \cot 2x = \tan x$$

SOLUTION Working with the left-hand side, we proceed as follows:

$$\csc 2x - \cot 2x = \tan x$$

$\dfrac{1}{\sin 2x} - \dfrac{\cos 2x}{\sin 2x}$	Apply fundamental trig identities
$\dfrac{1 - \cos 2x}{\sin 2x}$	Subtract fractions
$\dfrac{1 - (\cos^2 x - \sin^2 x)}{2 \sin x \cos x}$	Apply double-angle formulas
$\dfrac{(1 - \cos^2 x) + \sin^2 x}{2 \sin x \cos x}$	Rewrite
$\dfrac{2 \sin^2 x}{2 \sin x \cos x}$	Replace $1 - \cos^2 x$ with $\sin^2 x$ and simplify
$\dfrac{\sin x}{\cos x}$	Reduce
$\tan x$	Apply fundamental trig identity

◆

PROBLEM 2 Repeat Example 2 for $2 \cos x \csc 2x = \csc x$.

◆

To help solve certain trigonometric equations in which the angles have the ratio 2 to 1, we apply double-angle formulas to obtain trigonometric functions with the same angle.

EXAMPLE 3 Find the general form of the solution of the equation

$$\cos 2x = 2 \sin^2 x.$$

SOLUTION To solve this equation, we must first express the trigonometric functions with the same angle. If we apply the double-angle formula for cosine, we obtain an equation in which all angles are x. Thus, we proceed as follows:

$\cos 2x = 2 \sin^2 x$
$\cos^2 x - \sin^2 x = 2 \sin^2 x$ Apply double-angle formula for cosine
$(1 - \sin^2 x) - \sin^2 x = 2 \sin^2 x$ Replace $\cos^2 x$ with $1 - \sin^2 x$
$1 = 4 \sin^2 x$ Add $2 \sin^2 x$ to both sides
$\sin^2 x = \frac{1}{4}$ Divide both sides by 4
$\sin x = \pm \frac{1}{2}$ Solve for $\sin x$

Consequently, for $\sin x = \frac{1}{2}$, we have

$$x = \frac{\pi}{6} + 2\pi n \quad \text{or} \quad x = \frac{5\pi}{6} + 2\pi n$$

and for $\sin x = -\frac{1}{2}$, we have

$$x = \frac{7\pi}{6} + 2\pi n \quad \text{or} \quad x = \frac{11\pi}{6} + 2\pi n.$$

These general forms of the solutions may be written more compactly as

$$x = \frac{\pi}{6} + \pi n \quad \text{or} \quad x = \frac{5\pi}{6} + \pi n,$$

where n is an integer. ◆

PROBLEM 3 Repeat Example 3 for $\cos x + \sin 2x = 0$. ◆

By using double-angle formulas in conjunction with the sum formulas from Section 4.3, we can develop other multiple-angle formulas. In the next example, we develop a triple-angle formula for sine.

EXAMPLE 4 Derive a formula for $\sin 3x$ in terms of $\sin x$.

SOLUTION Writing $3x$ as $2x + x$, we proceed as follows:

$$\begin{aligned}
\sin 3x &= \sin(2x + x) \\
&= \sin 2x \cos x + \cos 2x \sin x & &\text{Apply sum formula} \\
&= (2 \sin x \cos x) \cos x + (\cos^2 x - \sin^2 x) \sin x & &\text{Apply double-angle formulas} \\
&= 2 \sin x \cos^2 x + \cos^2 x \sin x - \sin^3 x & &\text{Multiply} \\
&= 2 \sin x (1 - \sin^2 x) + (1 - \sin^2 x) \sin x - \sin^3 x & &\text{Replace } \cos^2 x \text{ with } 1 - \sin^2 x \\
&= 2 \sin x - 2 \sin^3 x + \sin x - \sin^3 x - \sin^3 x & &\text{Multiply} \\
&= 3 \sin x - 4 \sin^3 x & &\text{Collect like terms}
\end{aligned}$$

Hence, the *triple-angle formula for sine* is

$$\sin 3x = 3 \sin x - 4 \sin^3 x. \qquad \blacklozenge$$

PROBLEM 4 Referring to Example 4, show that the triple-angle formula for sine is true when x is replaced by $\pi/6$. $\qquad \blacklozenge$

◆ Power Reduction Formulas

Two alternate forms for the double-angle formula for cosine can be obtained as follows:

$$\begin{aligned}
\cos 2u &= \cos^2 u - \sin^2 u & \cos 2u &= \cos^2 u - \sin^2 u \\
&= (1 - \sin^2 u) - \sin^2 u & &= \cos^2 u - (1 - \cos^2 u) \\
&= 1 - 2 \sin^2 u & &= 2 \cos^2 u - 1
\end{aligned}$$

$$\boxed{\cos 2u = 1 - 2 \sin^2 u \qquad \cos 2u = 2 \cos^2 u - 1}$$

From these two equations, we now derive the *power reduction formulas* for sine, cosine, and tangent. Solving the equation $\cos 2u = 1 - 2 \sin^2 u$ for $\sin^2 u$ and the equation $\cos 2u = 2 \cos^2 u - 1$ for $\cos^2 u$, we obtain

$$\sin^2 u = \frac{1 - \cos 2u}{2} \quad \text{and} \quad \cos^2 u = \frac{1 + \cos 2u}{2}.$$

Since $\tan^2 u = \dfrac{\sin^2 u}{\cos^2 u}$, we also have

$$\tan^2 u = \frac{1 - \cos 2u}{1 + \cos 2u}.$$

Power Reduction Formulas

1. $\sin^2 u = \dfrac{1 - \cos 2u}{2}$ or $\tfrac{1}{2}(1 - \cos 2u)$

2. $\cos^2 u = \dfrac{1 + \cos 2u}{2}$ or $\tfrac{1}{2}(1 + \cos 2u)$

3. $\tan^2 u = \dfrac{1 - \cos 2u}{1 + \cos 2u}$

Power reduction formulas are used in calculus to express even powers of sine and cosine as first powers of the cosine function. The procedure is illustrated in the next example.

EXAMPLE 5 Express $\sin^4 x$ in terms of first powers of the cosine function.

SOLUTION

$\sin^4 x = \sin^2 x \sin^2 x$ — Rewrite with exponents of 2

$= \tfrac{1}{2}(1 - \cos 2x) \tfrac{1}{2}(1 - \cos 2x)$ — Apply power reduction formula for sine

$= \tfrac{1}{4}(1 - 2 \cos 2x + \cos^2 2x)$ — Multiply

$= \tfrac{1}{4}[1 - 2 \cos 2x + \tfrac{1}{2}(1 + \cos 4x)]$ — Apply power reduction formula for cosine

When applying a power reduction formula, be sure to double *the angle.*

$= \tfrac{1}{4}(1 - 2 \cos 2x + \tfrac{1}{2} + \tfrac{1}{2} \cos 4x)$ — Distribute $\tfrac{1}{2}$

$= \tfrac{1}{4}(\tfrac{3}{2} - 2 \cos 2x + \tfrac{1}{2} \cos 4x)$ — Combine like terms

$= \tfrac{1}{8}(3 - 4 \cos 2x + \cos 4x)$ — Factor out $\tfrac{1}{2}$

PROBLEM 5 Repeat Example 5 for $\cos^4 x$.

◆ Half-Angle Formulas

Alternate forms of the power reduction formulas are obtained by replacing u with $v/2$ and then taking the square root of both sides of the equation. We refer to the equations that are formed by this procedure as *half-angle formulas*.

Half-Angle Formulas

1. $\sin \dfrac{v}{2} = \pm\sqrt{\dfrac{1 - \cos v}{2}}$

2. $\cos \dfrac{v}{2} = \pm\sqrt{\dfrac{1 + \cos v}{2}}$

3. $\tan \dfrac{v}{2} = \pm\sqrt{\dfrac{1 - \cos v}{1 + \cos v}}$

SECTION 4.4 Multiple-Angle Formulas

Note: Whether we select the positive or negative square root in a half-angle formula depends on the quadrant in which $v/2$ lies.

EXAMPLE 6 Find the exact value of $\sin \dfrac{\pi}{12}$.

SOLUTION We can think of $\pi/12$ as a first-quadrant angle that is half of $\pi/6$. Since the sine of a first-quadrant angle is positive, we use the positive square root in the half-angle formula for sine and proceed as follows:

$$\sin \frac{\pi}{12} = \sin \frac{\pi/6}{2} = \sqrt{\frac{1 - \cos(\pi/6)}{2}}$$

$$= \sqrt{\frac{1 - \sqrt{3}/2}{2}} \qquad \text{Evaluate } \cos(\pi/6)$$

$$= \frac{\sqrt{2 - \sqrt{3}}}{2} \qquad \text{Simplify} \qquad \blacklozenge$$

PROBLEM 6 Use a calculator to find the approximate value of $\sin \dfrac{\pi}{12}$. Compare your answer with its exact value. $\blacklozenge$

EXAMPLE 7 Given that θ is a fourth-quadrant angle with $\cos \theta = 0.62$, find the exact value of $\cos \dfrac{\theta}{2}$.

SOLUTION If θ is a fourth-quadrant angle, then

$$\frac{3\pi}{2} < \theta < 2\pi$$

$$\frac{3\pi}{4} < \frac{\theta}{2} < \pi \qquad \text{Divide each member by 2}$$

Therefore, $\theta/2$ is a second-quadrant angle. Since the cosine of a second-quadrant angle is negative, we use the negative square root in the half-angle formula for cosine and proceed as follows:

$$\cos \frac{\theta}{2} = -\sqrt{\frac{1 + \cos \theta}{2}} = -\sqrt{\frac{1 + 0.62}{2}} = -\sqrt{0.81} = -0.9. \qquad \blacklozenge$$

PROBLEM 7 Given that θ is a third-quadrant angle with $\cos \theta = -0.28$, find the exact value of $\sin \dfrac{\theta}{2}$. $\blacklozenge$

270 CHAPTER 4 Analytic Trigonometry

Alternate forms of the half-angle formula for tangent may be obtained as follows:

$$\tan \frac{v}{2} = \frac{\sin(v/2)}{\cos(v/2)}$$

$$= \frac{2\sin(v/2)\cos(v/2)}{2\cos^2(v/2)} \quad \text{\textcolor{darkred}{Multiply numerator and denominator by } } 2\cos(v/2)$$

$$= \frac{\sin v}{2\left(\dfrac{1+\cos v}{2}\right)} \quad \text{\textcolor{darkred}{Apply double-angle and power reduction formulas}}$$

$$= \frac{\sin v}{1+\cos v} \quad \text{\textcolor{darkred}{Simplify}}$$

$$= \frac{1-\cos v}{\sin v} \quad \text{\textcolor{darkred}{See Example 6(a) in Section 4.1}}$$

$$\boxed{\tan \frac{v}{2} = \frac{\sin v}{1+\cos v} \qquad \tan \frac{v}{2} = \frac{1-\cos v}{\sin v}}$$

The advantage of these alternate forms is that they contain no radical and have no ambiguity of sign.

EXAMPLE 8 Show that the following trigonometric equation is an identity:

$$\tan \frac{x}{2} + \cot \frac{x}{2} = 2 \csc x$$

SOLUTION

$$\tan \frac{x}{2} + \cot \frac{x}{2} = 2 \csc x$$

$$\tan \frac{x}{2} + \frac{1}{\tan \dfrac{x}{2}} \qquad \text{\textcolor{darkred}{Apply fundamental trig identity}}$$

$$\frac{\sin x}{1+\cos x} + \frac{1+\cos x}{\sin x} \qquad \text{\textcolor{darkred}{Apply half-angle formula}}$$

$$\frac{\sin^2 x + 1 + 2\cos x + \cos^2 x}{(1+\cos x)\sin x} \qquad \text{\textcolor{darkred}{Add fractions}}$$

SECTION 4.4 Multiple-Angle Formulas

$$\frac{2 + 2\cos x}{(1 + \cos x)\sin x}$$ Replace $\sin^2 x + \cos^2 x$ with 1

$$\frac{2(1 + \cos x)}{(1 + \cos x)\sin x}$$ Factor

$$\frac{2}{\sin x}$$ Reduce

$$2\csc x$$ Apply fundamental trig identity

Since we have reduced the left-hand side of this equation to the right-hand side, we conclude that

$$\tan\frac{x}{2} + \cot\frac{x}{2} = 2\csc x \quad \text{is a trigonometric identity.}$$ ◆

PROBLEM 8   Rework Example 8 by applying the formula $\tan\dfrac{v}{2} = \dfrac{1 - \cos v}{\sin v}$. ◆

◆ **Application: Design of an A-Frame Structure**

We conclude this section with an applied problem concerning an A-frame structure (a structure in the form of an inverted V).

EXAMPLE 9   An architect is to design an A-frame house with rafters of length 20 feet.

(a) Express the cross-sectional area A of the house as a trigonometric function of the angle θ between the rafters.

(b) Determine the largest possible cross-sectional area.

SOLUTION

(a) The cross-sectional area A of the house is the area of an isosceles triangle. Recall from elementary geometry that if we construct an altitude to the base of an isosceles triangle, the vertex angle θ and the base are both cut in half, as shown in Figure 4.15. Working with either of the right triangles in Figure 4.15, we find that

$$\sin\frac{\theta}{2} = \frac{x}{20} \quad \text{and} \quad \cos\frac{\theta}{2} = \frac{h}{20}$$

$$x = 20\sin\frac{\theta}{2} \quad\quad\quad\quad\quad h = 20\cos\frac{\theta}{2}$$

272 CHAPTER 4 Analytic Trigonometry

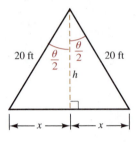

FIGURE 4.15
The altitude to the base of an isosceles triangle cuts the vertex angle and base in half.

Thus, using the fact that the area of a triangle is half the product of its base and height, we have

$$A = xh = \left(20 \sin \frac{\theta}{2}\right)\left(20 \cos \frac{\theta}{2}\right) = 400 \sin \frac{\theta}{2} \cos \frac{\theta}{2}$$

$$= 200\left(2 \sin \frac{\theta}{2} \cos \frac{\theta}{2}\right) \quad \text{Rewrite}$$

$$= 200 \sin \theta \quad \text{Apply double-angle formula}$$

(b) Note that angle θ must be between 0° and 180°. Since $\sin \theta$ reaches a maximum value of 1 when $\theta = 90°$, we conclude that the largest cross-sectional area is 200 square feet, and this occurs when $\theta = 90°$. ◆

PROBLEM 9 Referring to Example 9, find an acute angle θ (to the nearest tenth of a degree) associated with a cross-sectional area of 170 square feet. ◆

Exercises 4.4

Basic Skills

In Exercises 1–16, use double-angle formulas to simplify each expression.

1. $10 \sin 3x \cos 3x$
2. $2 \sin \frac{x}{2} \cos \frac{x}{2}$
3. $\cos^2 2x - \sin^2 2x$
4. $\dfrac{6 \tan 2x}{1 - \tan^2 2x}$
5. $\sec 6x \, (\sin^2 3x - \cos^2 3x)$
6. $2 \sin 4x \, (\cos^2 2x - \sin^2 2x)$
7. $\frac{1}{2} \cot 3x \, (1 - \tan^2 3x)$
8. $\frac{1}{2} \sec 4x \csc 4x$
9. $1 - 2 \sin^2 \dfrac{x}{2}$
10. $2 \cos^2 5x - 1$
11. $\dfrac{\sin 2x}{2 \sin x}$
12. $2 \cos x - \sin 2x \csc x$
13. $(\sin x + \cos x)^2 - \sin 2x$
14. $\cos^4 2x - \sin^4 2x$
15. $\dfrac{2}{1 + \cos 6x}$
16. $\dfrac{2}{\cot 3x - \tan 3x}$

In Exercises 17–28, use double-angle formulas to help show that each trigonometric equation is an identity.

17. $2 \cos x \csc 2x = \csc x$
18. $\sec^2 x \cos 2x + 2 \tan^2 x = \sec^2 x$
19. $2 \sin 2x - 4 \sin 2x \sin^2 x = \sin 4x$
20. $2 \tan x - \sin 2x = 2 \tan x \sin^2 x$
21. $\dfrac{\tan 2x}{2 \tan x} = \dfrac{1}{2 - \sec^2 x}$
22. $\dfrac{2 \tan 3x}{1 + \tan^2 3x} = \sin 6x$
23. $\dfrac{\sin 6x}{1 + \cos 6x} = \tan 3x$
24. $\dfrac{1 + \cos 2x}{1 - \cos 2x} = \cot^2 x$
25. $\dfrac{2 \cos x}{\csc x - 2 \sin x} = \tan 2x$
26. $\dfrac{2 \cos x - \sec x}{2 \sin x} = \cot 2x$
27. $\dfrac{1 + \cot x}{\cot x - 1} = \dfrac{1 + \sin 2x}{\cos 2x}$
28. $2 \sin x + \sin 2x = \dfrac{2 \sin^3 x}{1 - \cos x}$

In Exercises 29–38, use double-angle formulas to help find the general form of the solution of each equation.

29. $2 \sin x = \sin 2x$
30. $\sin 2x - \cos x = 0$
31. $\sin x + \cos 2x = 1$
32. $\cos 2x + 3 \sin x = 2$
33. $2 \sin^2 2x = 2 + \cos 4x$
34. $2 \cos^2 3x + \cos 6x = 1$
35. $\tan 2x - 2 \sin x = 0$
36. $\sin 2x = \cot x$
37. $\tan 2x + \tan x = 0$
38. $\tan 4x - \tan 2x = 0$

In Exercises 39–42, derive a formula for the indicated trigonometric expression.

39. $\cos 3x$ in terms of $\cos x$. **40.** $\cos 4x$ in terms of $\cos x$.
41. $\sin 5x$ in terms of $\sin x$. **42.** $\cos 5x$ in terms of $\cos x$.

In Exercises 43–46, use power reduction formulas to express the even powers of sine and cosine in terms of first powers of the cosine function.

43. $\cos^2 3x$ **44.** $\sin^2 x \cos^2 x$
45. $\sin^4 x \cos^2 x$ **46.** $\sin^6 x$

In Exercises 47–50, find the exact value of each trigonometric function by applying a half-angle formula.

47. $\cos 15°$ **48.** $\tan 15°$
49. $\sin 22.5°$ **50.** $\cos 22.5°$

In Exercises 51–54, find the exact value of the indicated trigonometric function from the given information.

51. Given $\sin x = \frac{3}{5}$, x in quadrant I, find

(a) $\sin 2x$ (b) $\cos 2x$ (c) $\sin \frac{x}{2}$ (d) $\cos \frac{x}{2}$

52. Given $\cos x = -\frac{12}{13}$, x in quadrant III, find

(a) $\cos 2x$ (b) $\tan 2x$ (c) $\cos \frac{x}{2}$ (d) $\tan \frac{x}{2}$

53. Given $\tan x = -3$, x in quadrant IV, find

(a) $\tan 2x$ (b) $\sec 2x$ (c) $\sin \frac{x}{2}$ (d) $\cot \frac{x}{2}$

54. Given $\csc x = 1.25$, x in quadrant II, find

(a) $\sin 2x$ (b) $\cot 2x$ (c) $\cos \frac{x}{2}$ (d) $\csc \frac{x}{2}$

In Exercises 55–60, use half-angle formulas to help show that each trigonometric equation is an identity.

55. $2\cos^2 \frac{x}{2} - \cos x = 1$

56. $2\sin^2 \frac{x}{2} \sec x + 1 = \sec x$

57. $\csc^2 \frac{x}{2} - 1 = \frac{1 + \cos x}{1 - \cos x}$

58. $\tan \frac{x}{2} = \csc x - \cot x$

59. $\dfrac{2 \tan \frac{x}{2}}{1 + \tan^2 \frac{x}{2}} = \sin x$

60. $\cot \frac{x}{2} - \tan \frac{x}{2} = 2 \cot x$

In Exercises 61–66, use half-angle formulas to help find the general form of the solution of each equation.

61. $\sin^2 \frac{x}{2} - \cos x + 1 = 0$ **62.** $2 \cos^2 \frac{x}{2} = 2 - \cos x$

63. $\sin \frac{x}{2} - \sin x = 0$ **64.** $\cos x - \cos \frac{x}{2} + 1 = 0$

65. $\tan \frac{x}{2} + \sin 2x = \csc x$ **66.** $\tan \frac{x}{2} + 1 = \cos x$

67. The cross section of a water trough is in the shape of an isosceles triangle with dimensions as shown in the sketch.

(a) Express the volume V of the trough as a trigonometric function of the angle θ between its sides.

(b) Determine the largest possible volume of the trough.

(c) Find θ (to the nearest tenth of a degree) if the volume is 30 cubic feet.

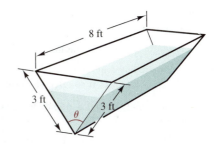

68. A rectangle is inscribed in a semicircle of radius 1 meter, as shown in the sketch.

(a) Express the area A of the rectangle as a trigonometric function of angle θ.

(b) What value of θ yields the rectangle with the maximum area? What is the maximum area?

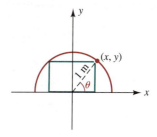

69. The height of one building is three times the height of another building. When standing midway between the two buildings, an observer notes that the angle of elevation to the top of the taller building is twice the angle of elevation to the top of the other building. Assuming the buildings are on level ground, find these angles of elevation.

70. A house has an amateur radio tower fixed on top. The house is 15 ft high and the tower is 25 ft high, as shown in the sketch. Using a transit, a person observes that the angle of elevation to the top of the roof is the same as the angle from the top of the roof to the top of the tower. How far away from the house is the transit?

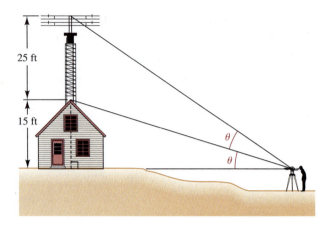

EXERCISE 70

 Critical Thinking

71. Find the amplitude and period of each function.
    (a) $y = 6 \sin 4x \cos 4x$ (b) $y = \sin x \cos x$

72. In general, $\sin 2u \ne 2 \sin u$.
    (a) Use $u = \pi/6$ to illustrate this fact.
    (b) Are there any real numbers u for which $\sin 2u = 2 \sin u$? If so, give an example.

73. Use double-angle formulas to find the exact value of each expression.
    (a) $\sin [2 (\arctan 3)]$ (b) $\cos [2 (\cos^{-1} \frac{1}{4})]$

74. Use half-angle formulas to find the exact value of each expression.
    (a) $\tan [\frac{1}{2} (\arcsin \frac{2}{3})]$ (b) $\sin [\frac{1}{2} (\tan^{-1} 2)]$

75. A *rhombus* is a parallelogram all of whose sides have the same length. The diagonals of a rhombus are perpendicular to one another and bisect the interior angles, as shown in the sketch. Use this fact to show that the area A of a rhombus is

$$A = a^2 \sin \theta,$$

where a is the length of one side of the rhombus and θ is the measure of one of its interior angles.

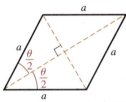

EXERCISE 75

76. Refer to the right triangle in the sketch.
    (a) Use the double-angle formula for sine to find $\sin 2\alpha$ and $\sin 2\beta$. What do you observe? Explain.
    (b) Use the double-angle formula for cosine to find $\cos 2\alpha$ and $\cos 2\beta$. What do you observe? Explain.

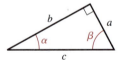

 Calculator Activities

77. Given that $\cos 10° \approx 0.9848$, use double-angle formulas to find $\cos 20°$, $\sin 20°$, and $\tan 20°$. Check your answers by using the $\boxed{\text{SIN}}$, $\boxed{\text{COS}}$, and $\boxed{\text{TAN}}$ keys on your calculator.

78. Given that $\cos 10° \approx 0.9848$, use half-angle formulas to find $\cos 5°$, $\sin 5°$, and $\tan 5°$. Check you answers by using the $\boxed{\text{SIN}}$, $\boxed{\text{COS}}$, and $\boxed{\text{TAN}}$ keys on your calculator.

A function and its graph are shown in Exercises 79 and 80. Approximate the x-intercepts of the graph, rounding each answer to three significant digits.

79. $y = \cos x - 2 \sin 2x$ for $0 \leq x \leq 2\pi$

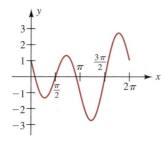

80. $y = 2 \cos x + \cos 2x$ for $0 \leq x \leq 2\pi$

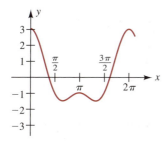

In Exercises 81–84, an equation of the form $f(x) = 0$ is given.

(a) Use a graphing calculator to generate the graph of the function $y = f(x)$ in the interval $[0, 2\pi)$. Then use the calculator's trace and zoom features to estimate the solutions of each equation in the interval $[0, 2\pi)$. Record each answer to three significant digits.

(b) Use double-angle formulas to help find the solutions in the interval $[0, 2\pi)$. Compare your answers to the estimated values from part (a). Explain any discrepancies.

81. $2 \cos^2 x - \cos 2x - 2 \sin 2x = 0$

82. $2 \sin^2 2x + 2 \cos 4x - 1 = 0$

83. $\tan 2x + 3 \cot x = 0$ **84.** $3 \tan 2x - \tan 4x = 0$

In Exercises 85–88, an equation of the form $f(x) = g(x)$ is stated. Use a graphing calculator or computer with graphing capabilities to generate the graphs of $y = f(x)$ and $y = g(x)$. Compare the graphs and then state whether the equation appears to be a trigonometric identity.

85. $\sin 4x = 8 \cos^3 x \sin x - 4 \cos x \sin x$

86. $\cos 6x = 32 \cos^6 x - 48 \cos^4 x + 18 \cos^2 x - 1$

87. $\sin 6x = 32 \cos^5 x \sin x - 32 \cos^3 x \sin x + 6 \cos x \sin x$

88. $\tan 3x = \dfrac{3 \tan x - \tan^3 x}{1 - 3 \tan^2 x}$

4.5 Product-to-Sum Formulas and Sum-to-Product Formulas

In this section, we discuss trigonometric formulas that allow us to write the *product* of certain trigonometric functions as a *sum*, and the *sum* of certain trigonometric functions as a *product*. We refer to these formulas as the *product-to-sum formulas* and *sum-to-product formulas* and use them to help prove other trigonometric identities and solve some trigonometric equations. We begin by deriving the product-to-sum formulas.

◆ Deriving the Product-to-Sum Formulas

Recall (from Section 4.3) the sum and difference formulas for cosine and sine:

1. $\cos(u + v) = \cos u \cos v - \sin u \sin v$
2. $\cos(u - v) = \cos u \cos v + \sin u \sin v$
3. $\sin(u + v) = \sin u \cos v + \cos u \sin v$
4. $\sin(u - v) = \sin u \cos v - \cos u \sin v$

If we add the respective members of the second equation to the first, we obtain

$$\cos(u+v) = \cos u \cos v - \sin u \sin v$$
$$\cos(u-v) = \cos u \cos v + \sin u \sin v$$

$$\cos(u+v) + \cos(u-v) = 2\cos u \cos v$$

Now, multiplying both sides of this equation by $\frac{1}{2}$ gives us

$$\cos u \cos v = \tfrac{1}{2}[\cos(u+v) + \cos(u-v)].$$

Note that this formula expresses the *product* $\cos u \cos v$ as a *sum*. We refer to this formula as a *product-to-sum formula*.

In a similar manner, by subtracting the respective members of the first equation from the second, adding the respective members of the fourth equation to the third, and subtracting the respective members of the fourth equation from the third, we obtain three other *product-to-sum formulas*.

Product-to-Sum Formulas

1. $\cos u \cos v = \tfrac{1}{2}[\cos(u+v) + \cos(u-v)]$
2. $\sin u \sin v = \tfrac{1}{2}[\cos(u-v) - \cos(u+v)]$
3. $\sin u \cos v = \tfrac{1}{2}[\sin(u+v) + \sin(u-v)]$
4. $\cos u \sin v = \tfrac{1}{2}[\sin(u+v) - \sin(u-v)]$

EXAMPLE 1 Express the product as a sum or difference.

(a) $2\sin 5x \sin x$ (b) $\sin 3x \cos 2x$

SOLUTION

(a) Using product-to-sum formula 2 with $u = 5x$ and $v = x$, we have

$$2\sin 5x \sin x = 2\{\tfrac{1}{2}[\cos(5x-x) - \cos(5x+x)]\}$$
$$= \cos 4x - \cos 6x.$$

(b) Using product-to-sum formula 3 with $u = 3x$ and $v = 2x$, we have

$$\sin 3x \cos 2x = \tfrac{1}{2}[\sin(3x+2x) + \sin(3x-2x)]$$
$$= \tfrac{1}{2}\sin 5x + \tfrac{1}{2}\sin x.$$ ◆

PROBLEM 1 Rework Example 1(b) by using product-to-sum formula 4 with $u = 2x$ and $v = 3x$. Explain why the answer is the same as that obtained in the example. ◆

◆ Deriving the Sum-to-Product Formulas

If we let

$$u + v = a \quad \text{and} \quad u - v = b,$$

then

$$(u + v) + (u - v) = a + b \quad \text{and} \quad (u + v) - (u - v) = a - b$$
$$2u = a + b \qquad\qquad\qquad 2v = a - b$$
$$u = \frac{a + b}{2} \qquad\qquad\qquad v = \frac{a - b}{2}$$

Making the substitutions

$$u + v = a, \quad u - v = b, \quad u = \frac{a + b}{2}, \quad \text{and} \quad v = \frac{a - b}{2}$$

into product-to-sum formula 1, we obtain

$$\cos \frac{a + b}{2} \cos \frac{a - b}{2} = \tfrac{1}{2}[\cos a + \cos b].$$

Now, multiplying both sides of this equation by 2 gives us

$$\cos a + \cos b = 2 \cos \frac{a + b}{2} \cos \frac{a - b}{2}.$$

This formula expresses the *sum* cos a + cos b as a *product*, and we refer to this formula as a *sum-to-product formula*. Similarly, by making these substitutions into the other product-to-sum formulas, we obtain three other *sum-to-product formulas*.

Sum-to-Product Formulas

1. $\cos a + \cos b = 2 \cos \dfrac{a + b}{2} \cos \dfrac{a - b}{2}$
2. $\cos a - \cos b = -2 \sin \dfrac{a + b}{2} \sin \dfrac{a - b}{2}$
3. $\sin a + \sin b = 2 \sin \dfrac{a + b}{2} \cos \dfrac{a - b}{2}$
4. $\sin a - \sin b = 2 \cos \dfrac{a + b}{2} \sin \dfrac{a - b}{2}$

EXAMPLE 2 Express the sum or difference as a product.

(a) $\cos 8x + \cos 2x$ (b) $\sin x - \sin 4x$

SOLUTION

(a) Using sum-to-product formula 1 with $a = 8x$ and $b = 2x$, we have

$$\cos 8x + \cos 2x = 2 \cos \frac{8x + 2x}{2} \cos \frac{8x - 2x}{2}$$

$$= 2 \cos 5x \cos 3x.$$

(b) Using sum-to-product formula 4 with $a = x$ and $b = 4x$, we have

$$\sin x - \sin 4x = 2 \cos \frac{x + 4x}{2} \sin \frac{x - 4x}{2}$$

$$= 2 \cos \frac{5x}{2} \sin \left(-\frac{3x}{2}\right)$$

$$= -2 \cos \frac{5x}{2} \sin \frac{3x}{2} \qquad \text{Use } \sin(-u) = -\sin u$$

◆

PROBLEM 2 Express $\sin 5x + \sin 7x$ as a product. ◆

◆ **Applying the Formulas**

The next four examples illustrate a variety of uses of product-to-sum and sum-to-product formulas. In the next example, we apply a product-to-sum formula to help find the exact value of the product of two trigonometric functions.

EXAMPLE 3 Use a product-to-sum formula to find the exact value of $\cos 75° \cos 15°$.

SOLUTION Using product-to-sum formula 1 with $u = 75°$ and $v = 15°$, we have

$$\cos 75° \cos 15° = \tfrac{1}{2} [\cos (75° + 15°) + \cos (75° - 15°)]$$

$$= \tfrac{1}{2} (\cos 90° + \cos 60°)$$

$$= \tfrac{1}{2} (0 + \tfrac{1}{2}) \qquad \text{Evaluate}$$

$$= \tfrac{1}{4} \qquad \text{Simplify}$$

◆

PROBLEM 3 Use a sum-to-product formula to find the exact value of $\cos 75° - \cos 15°$. ◆

The product-to-sum formulas and sum-to-product formulas are also helpful in verifying certain trigonometric identities.

EXAMPLE 4 Show that the following equation is an identity:

$$2 \cos^2 x \sin x - 2 \sin^3 x = \sin 3x - \sin x$$

SOLUTION Working with the left-hand side, we have

$$2\cos^2 x \sin x - 2\sin^3 x = \sin 3x - \sin x$$

$2\sin x(\cos^2 x - \sin^2 x)$	Factor
$2\sin x \cos 2x$	Apply double-angle formula
$2\{\tfrac{1}{2}[\sin(x+2x) + \sin(x-2x)]\}$	Apply product-to-sum formula
$\sin 3x + \sin(-x)$	Simplify
$\sin 3x - \sin x$	Use $\sin(-u) = -\sin u$

Since we have reduced the left-hand side of this equation to the right-hand side, we conclude that

$$2\cos^2 x \sin x - 2\sin^3 x = \sin 3x - \sin x \quad \text{is a trigonometric identity.} \blacklozenge$$

PROBLEM 4 Rework Example 4 by working with the right-hand side of the equation. ◆

We can find the solutions to certain trigonometric equations with the aid of product-to-sum formulas and sum-to-product formulas.

EXAMPLE 5 Find the general form of the solution of the equation

$$\sin 2x + \sin 4x = \sin 3x.$$

SOLUTION

$\sin 2x + \sin 4x = \sin 3x$	
$2\sin 3x \cos(-x) = \sin 3x$	Apply sum-to-product formula
$2\sin 3x \cos x = \sin 3x$	Use $\cos(-u) = \cos u$
$2\sin 3x \cos x - \sin 3x = 0$	Subtract $\sin 3x$ from both sides
$\sin 3x(2\cos x - 1) = 0$	Factor
$\sin 3x = 0 \quad \text{or} \quad 2\cos x - 1 = 0$	Apply zero product property
$3x = \pi n \qquad\qquad \cos x = \tfrac{1}{2}$	
$x = \dfrac{\pi n}{3} \qquad\qquad x = \dfrac{\pi}{3} + 2\pi n,\ \dfrac{5\pi}{3} + 2\pi n$	

Thus, the expression

$$\frac{\pi n}{3}, \quad \text{where } n \text{ is an integer,}$$

describes the general form of all solutions of this equation. ◆

PROBLEM 5 Referring to Example 5, find the six solutions in the interval $[0, 2\pi)$. ◆

We may use sum-to-product formulas to help find the *x*-intercepts of the graph of an equation that involves the sum of sine or cosine functions with different angles.

EXAMPLE 6 Find the *x*-intercepts of the graph of the given equation in the interval $[0, 2\pi)$:

$$y = \cos 3x + \cos x$$

SOLUTION We may find the *x*-intercepts by letting $y = 0$ and solving for x as follows:

$$\cos 3x + \cos x = 0$$
$$2 \cos 2x \cos x = 0 \qquad \textbf{Apply sum-to-product formula}$$
$$2 \cos 2x = 0 \quad \text{or} \quad \cos x = 0 \qquad \textbf{Apply zero product property}$$
$$2x = \frac{\pi}{2} + \pi n \qquad x = \frac{\pi}{2} + \pi n$$
$$x = \frac{\pi}{4} + \frac{\pi}{2} n$$

Thus, in the interval $[0, 2\pi)$, the *x*-intercepts are

$$\frac{\pi}{4}, \frac{3\pi}{4}, \frac{5\pi}{4}, \frac{7\pi}{4} \quad \text{and} \quad \frac{\pi}{2}, \frac{3\pi}{2}.$$ ◆

We can sketch the graph of $y = \cos 3x + \cos x$ (Example 6) by plotting some additional points using a technique called **addition of ordinates** (*y*-values). We begin by selecting *x*-values for which $\cos 3x$ (the function with the shorter period) are 0 or ± 1, namely,

$$x = \frac{\pi}{6} n, \quad \text{where } n \text{ is an integer.}$$

These *x*-values allow us to compute the sum of $\cos 3x$ and $\cos x$ quite easily. Table 4.1 summarizes the computations.

SECTION 4.5 Product-to-Sum and Sum-to-Product Formulas

Table 4.1
Table of values for $y = \cos 3x + \cos x$ in the interval $[0, 2\pi]$

x	0	$\dfrac{\pi}{6}$	$\dfrac{\pi}{3}$	$\dfrac{\pi}{2}$	$\dfrac{2\pi}{3}$	$\dfrac{5\pi}{6}$	π	$\dfrac{7\pi}{6}$	$\dfrac{4\pi}{3}$	$\dfrac{3\pi}{2}$	$\dfrac{5\pi}{3}$	$\dfrac{11\pi}{6}$	2π
$\cos 3x$	1	0	-1	0	1	0	-1	0	1	0	-1	0	1
$\cos x$	1	$\dfrac{\sqrt{3}}{2}$	$\dfrac{1}{2}$	0	$-\dfrac{1}{2}$	$-\dfrac{\sqrt{3}}{2}$	-1	$-\dfrac{\sqrt{3}}{2}$	$-\dfrac{1}{2}$	0	$\dfrac{1}{2}$	$\dfrac{\sqrt{3}}{2}$	1
y	2	$\dfrac{\sqrt{3}}{2}$	$-\dfrac{1}{2}$	0	$\dfrac{1}{2}$	$-\dfrac{\sqrt{3}}{2}$	-2	$-\dfrac{\sqrt{3}}{2}$	$\dfrac{1}{2}$	0	$-\dfrac{1}{2}$	$\dfrac{\sqrt{3}}{2}$	2

We plot the x-intercepts from Example 6 and the points (x, y) given in Table 4.1 and then connect these points to form the graph of $y = \cos 3x + \cos x$ in the interval $[0, 2\pi)$. The graph is shown in Figure 4.16.

FIGURE 4.16
Graph of $y = \cos 3x + \cos x$

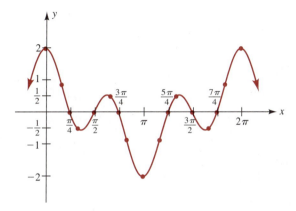

Note: If you have access to a graphing calculator or a computer with graphing capabilities, you may wish to verify the sketch that is shown in Figure 4.16.

PROBLEM 6 In calculus, it is shown that the x-coordinates of the relative maximum and minimum points for the graph of $y = \cos 3x + \cos x$ are the solutions of the equation

$$3 \sin 3x + \sin x = 0.$$

Use the triple-angle formula for sine (see Example 4 in Section 4.4) to help estimate the relative maximum and minimum points in the interval $[0, 2\pi)$. ◆

◆ Summary of Trigonometric Identities

We conclude this chapter by listing many of the important trigonometric identities that we have discussed in Chapters 2 and 4.

Fundamental Trigonometric Identities

1. $\sin x \csc x = 1$
2. $\cos x \sec x = 1$
3. $\tan x \cot x = 1$
4. $\tan x = \dfrac{\sin x}{\cos x}$
5. $\cot x = \dfrac{\cos x}{\sin x}$
6. $\sin^2 x + \cos^2 x = 1$
7. $1 + \tan^2 x = \sec^2 x$
8. $1 + \cot^2 x = \csc^2 x$

Even-Odd Identities

1. $\sin(-x) = -\sin x$
2. $\cos(-x) = \cos x$
3. $\tan(-x) = -\tan x$
4. $\csc(-x) = -\csc x$
5. $\sec(-x) = \sec x$
6. $\cot(-x) = -\cot x$

Cofunction Identities

1. $\sin\left(\dfrac{\pi}{2} - x\right) = \cos x$
2. $\cos\left(\dfrac{\pi}{2} - x\right) = \sin x$
3. $\tan\left(\dfrac{\pi}{2} - x\right) = \cot x$
4. $\cot\left(\dfrac{\pi}{2} - x\right) = \tan x$
5. $\sec\left(\dfrac{\pi}{2} - x\right) = \csc x$
6. $\csc\left(\dfrac{\pi}{2} - x\right) = \sec x$

Sum and Difference Formulas

1. $\cos(u + v) = \cos u \cos v - \sin u \sin v$
2. $\cos(u - v) = \cos u \cos v + \sin u \sin v$
3. $\sin(u + v) = \sin u \cos v + \cos u \sin v$
4. $\sin(u - v) = \sin u \cos v - \cos u \sin v$
5. $\tan(u + v) = \dfrac{\tan u + \tan v}{1 - \tan u \tan v}$
6. $\tan(u - v) = \dfrac{\tan u - \tan v}{1 + \tan u \tan v}$

Double-Angle Formulas

1. $\sin 2u = 2 \sin u \cos u$
2. $\cos 2u = \cos^2 u - \sin^2 u$
   $= 2\cos^2 u - 1$
   $= 1 - 2\sin^2 u$
3. $\tan 2u = \dfrac{2 \tan u}{1 - \tan^2 u}$

Power Reduction Formulas

1. $\sin^2 u = \dfrac{1 - \cos 2u}{2}$
2. $\cos^2 u = \dfrac{1 + \cos 2u}{2}$
3. $\tan^2 u = \dfrac{1 - \cos 2u}{1 + \cos 2u}$

Half-Angle Formulas

1. $\sin\dfrac{v}{2} = \pm\sqrt{\dfrac{1 - \cos v}{2}}$
2. $\cos\dfrac{v}{2} = \pm\sqrt{\dfrac{1 + \cos v}{2}}$
3. $\tan\dfrac{v}{2} = \pm\sqrt{\dfrac{1 - \cos v}{1 + \cos v}}$
   $= \dfrac{\sin v}{1 + \cos v}$
   $= \dfrac{1 - \cos v}{\sin v}$

Note: Selecting the positive or negative square root in a half-angle formula depends on the quadrant in which $v/2$ lies.

Product-to-Sum Formulas

1. $\cos u \cos v = \frac{1}{2}[\cos(u+v) + \cos(u-v)]$
2. $\sin u \sin v = \frac{1}{2}[\cos(u-v) - \cos(u+v)]$
3. $\sin u \cos v = \frac{1}{2}[\sin(u+v) + \sin(u-v)]$
4. $\cos u \sin v = \frac{1}{2}[\sin(u+v) - \sin(u-v)]$

Sum-to-Product Formulas

1. $\cos a + \cos b = 2 \cos \frac{a+b}{2} \cos \frac{a-b}{2}$
2. $\cos a - \cos b = -2 \sin \frac{a+b}{2} \sin \frac{a-b}{2}$
3. $\sin a + \sin b = 2 \sin \frac{a+b}{2} \cos \frac{a-b}{2}$
4. $\sin a - \sin b = 2 \cos \frac{a+b}{2} \sin \frac{a-b}{2}$

Exercises 4.5

Basic Skills

In Exercises 1–8, express each product as a sum or difference with positive multiples of x.

1. $2 \sin 3x \sin x$
2. $2 \sin 5x \sin 4x$
3. $2 \cos 2x \cos 6x$
4. $2 \cos 4x \cos x$
5. $\cos 3x \sin 4x$
6. $\cos \frac{x}{2} \sin \frac{3x}{2}$
7. $6 \sin 5x \cos(-2x)$
8. $7 \sin(-3x) \cos 7x$

In Exercises 9–16, express each sum or difference as a product with positive multiples of x.

9. $\cos 4x + \cos 2x$
10. $\cos x + \cos 3x$
11. $\sin 5x + \sin(-3x)$
12. $\sin 4x + \sin 5x$
13. $\cos \frac{2x}{3} - \cos \frac{4x}{3}$
14. $\cos(-2x) - \cos 7x$
15. $\frac{1}{2}(\sin 3x - \sin 6x)$
16. $5 \sin 9x - 5 \sin 4x$

In Exercises 17–24, use product-to-sum or sum-to-product formulas to find the exact value of each trigonometric expression.

17. $\sin 75° \sin 15°$
18. $\cos 75° \sin 15°$
19. $\cos 15° \sin 105°$
20. $\cos 75° \cos 105°$
21. $\cos 195° + \cos 105°$
22. $\sin 165° + \sin 105°$
23. $\sin 165° - \sin 75°$
24. $\cos 15° - \cos 195°$

In Exercises 25–36, show that each trigonometric equation is an identity.

25. $2 \cos 4x \sin x + \sin 3x = \sin 5x$
26. $2 \sin 2x \cos x - \sin x = \sin 3x$
27. $\frac{1}{2} \sec x (\sin x + \sin 3x) = \sin 2x$
28. $\cos 3x + 3 \cos x = 4 \cos^3 x$
29. $2 \sin x \cos 3x + \sin 2x = \sin 4x$
30. $4 \sin 6x \cos^2 2x - 2 \sin 6x = \sin 10x + \sin 2x$
31. $\frac{\sin 7x + \sin 3x}{\cos 3x + \cos 7x} = \tan 5x$
32. $\frac{\sin 3x - \sin 5x}{\cos 5x - \cos 3x} = \cot 4x$
33. $\frac{\cos x - \cos 3x}{\sin^2 x} = 4 \cos x$
34. $\frac{\sin 3x + \sin x}{\sin x \cos x} = 4 \cos x$
35. $\sin 12x + \sin 4x - \sin 8x = 4 \cos 6x \cos 4x \sin 2x$
36. $\cos 10x - \cos 6x + \cos 4x = 1 - 4 \sin 5x \sin 2x \cos 3x$

In Exercises 37–48, find the general form of the solution of each trigonometric equation.

37. $\sin 3x + \sin x = 0$
38. $\cos 3x + \cos 7x = 0$
39. $\cos 6x = \cos 2x$
40. $\sin x = \sin 5x$
41. $\cos 2x - \cos 4x - 2 \sin 3x = 0$
42. $\sin x - \cos 2x = \sin 3x$
43. $\sin 2x + \sin 4x = \sin 3x$
44. $\cos 5x + \cos x = 2 \cos 2x$

45. $\dfrac{\cos 5x - \cos x}{2 \sin 3x} = 1$

46. $\dfrac{\cos 4x}{\sin 5x - \sin 3x} = -1$

47. $\dfrac{\sin 2x}{\cos 3x} + 2 \sin x = 0$

48. $\dfrac{\cos 3x + \cos x}{2 \cos^3 x - 2 \cos x} = 2$

In Exercises 49–52,

(a) use sum-to-product formulas to help find the x-intercepts of the graph of f, and

(b) plot some additional points by using addition of ordinates, and then sketch the graph of f.

49. $f(x) = \cos 2x + \cos x$

50. $f(x) = \cos x - \cos 2x$

51. $f(x) = \sin 4x - \sin 2x$

52. $f(x) = \sin 3x + \sin 2x$

53. In calculus, it is shown that the x-coordinates of the relative maximum and minimum points for the graph of $f(x) = \cos 2x + \cos x$ (Exercise 49) are the solutions of the equation $2 \sin 2x + \sin x = 0$. Find the coordinates of the relative maximum and minimum points in the interval $[0, 2\pi)$.

54. In calculus, it is shown that the x-coordinates of the relative maximum and minimum points for the graph of $f(x) = \cos x - \cos 2x$ (Exercise 50) are the solutions of the equation $2 \sin 2x - \sin x = 0$. Find the coordinates of the relative maximum and minimum points in the interval $[0, 2\pi)$.

Critical Thinking

55. Express $\sin 190° + \sin 70°$ as the sine of an acute angle.

56. Express $2 \cos 130° \cos 140°$ as the cosine of an acute angle.

57. Given that x and y are supplementary angles, use sum-to-product formulas to evaluate each expression. What do you observe? Explain.

    (a) $\cos x + \cos y$ (b) $\sin x - \sin y$

58. Given that x and y are supplementary angles, use sum-to-product formulas to write each expression as trigonometric functions of x only. What do you observe? Explain.

    (a) $\cos x - \cos y$ (b) $\sin x + \sin y$

59. The graph of $y = \cos 3x + \cos x$ (Figure 4.16) completes one cycle from 0 to 2π. Hence, the period of $y = \cos 3x + \cos x$ is 2π. Is the period of $y = \cos nx + \cos x$, where n is an integer, always 2π? Explain.

60. Given that $f(x) = \cos x$, use a sum-to-product formula to show that the difference quotient

$$\dfrac{f(x + \Delta x) - f(x)}{\Delta x} = -\dfrac{\sin (\Delta x/2)}{\Delta x/2} \sin \left(x + \dfrac{\Delta x}{2}\right).$$

Calculator Activities

61. Rewrite each product as a sum. Then use a calculator to evaluate both the original expression and the sum. The values should be the same.

    (a) $2 \cos 56.3° \cos 25.7°$

    (b) $2 \sin 147° \, 46' \sin 112° \, 52'$

    (c) $2 \sin \dfrac{3\pi}{10} \cos \dfrac{\pi}{5}$

    (d) $2 \cos 3.02 \sin 2.31$

62. Rewrite each sum as a product. Then use a calculator to evaluate both the original expression and the product. The values should be the same.

    (a) $\cos 36.9° + \cos 26.2°$

    (b) $\cos 227° \, 53' - \cos 115° \, 41'$

    (c) $\sin \dfrac{3\pi}{8} + \sin \dfrac{\pi}{4}$

    (d) $\sin 5.33 - \sin 3.29$

63. Consider the function f defined by $f(x) = \cos 4x - \cos 2x$.

    (a) Use a graphing calculator to generate the graph of this function in the interval $[0, 2\pi)$, and then state the period of this function.

    (b) Estimate the x-intercepts in the interval $[0, 2\pi)$ by tracing to these points. Round each intercept to three significant digits.

    (c) Estimate the coordinates of the relative maximum and minimum points in the interval $[0, 2\pi)$ by tracing to these points. Round each coordinate to three significant digits.

(d) Use a sum-to-product formula to help determine the exact value of each x-intercept in the interval $[0, 2\pi)$. Compare your answers to the estimated values in part (b).

64. In calculus, it is shown that the x-coordinates of the relative maximum and minimum points for the graph of $f(x) = \cos 4x - \cos 2x$ (Exercise 63) are the solutions of the equation $2 \sin 2x - 4 \sin 4x = 0$. Solve this equation for x and then state the coordinates of the relative maximum and minimum points in the interval $[0, 2\pi)$. Compare your answers to Exercise 63(c).

65. Rework Exercise 63 for the function f defined by $f(x) = \sin 3x - \sin x$.

66. In calculus, it is shown that the x-coordinates of the relative maximum and minimum points for the graph of $f(x) = \sin 3x - \sin x$ (Exercise 65) are the solutions of the equation $3 \cos 3x - \cos x = 0$. Use the triple-angle formula for cosine (see Exercise 39 in Section 4.4) to help solve this equation for x, and then state the coordinates of the relative maximum and minimum points in the interval $[0, 2\pi)$. Compare your answers to Exercise 65(c).

Chapter 4 Review

Questions for Group Discussion

1. What is a *trigonometric identity*? Discuss the procedure for determining whether a trigonometric equation is an identity.

2. Explain why the equation $\cos x = \sqrt{1 - \sin^2 x}$ is *not* a trigonometric identity. Give another example of a trigonometric equation that is true for several values of the variable but is not an identity.

3. Discuss the procedure for solving a *conditional trigonometric equation*.

4. What is meant by the *general form* of the solution of a trigonometric equation? How can particular solutions be developed from the general form?

5. Discuss the procedure for using a calculator to solve a trigonometric equation. Illustrate with an example.

6. When is it necessary to check a trigonometric equation for extraneous roots?

7. Complete the *sum and difference formulas:*
   (a) $\sin(x + y) =$ (b) $\cos(x + y) =$ (c) $\tan(x + y) =$
   (d) $\sin(x - y) =$ (e) $\cos(x - y) =$ (f) $\tan(x - y) =$

8. Discuss $\tan(x + y)$ if $\sin x = 0$ and if $\cos x = 0$.

9. Complete the *double-angle formulas* and discuss any restrictions on x:
   (a) $\sin 2x =$ (b) $\cos 2x =$ (c) $\tan 2x =$

10. Use the double-angle formula for cosine to derive the *power reduction formulas*.

11. Complete the *half-angle formulas* and discuss any restrictions on x:
    (a) $\sin \dfrac{x}{2} =$ (b) $\cos \dfrac{x}{2} =$ (c) $\tan \dfrac{x}{2} =$

12. How is it possible to determine whether to select the positive or negative square root in a half-angle formula?

13. Complete the *product-to-sum formulas:*
    (a) $\cos x \cos y =$ (b) $\sin x \sin y =$ (c) $\sin x \cos y =$

14. Discuss how *sum-to-product formulas* can be used to help find the x-intercepts of the graph of $y = \cos ax + \cos bx$, where a and b are positive integers.

15. Explain the procedure for writing $A \sin bt + B \cos bt$ in the form $a \sin(bt + c)$.

16. For the function f defined by $f(t) = A \sin bt + B \cos bt$, what is the advantage of replacing the expression $A \sin bt + B \cos bt$ with $a \sin(bt + c)$?

Review Exercises

In Exercises 1–14, write each expression as a single trigonometric function or a constant.

1. $\tan 2x \cos 2x$
2. $\dfrac{\csc x}{\sec x}$
3. $\sec^2 \theta - \tan^2 \theta$
4. $\sqrt{1 - \sin^2 x}$, $0 \le x \le \pi/2$
5. $\cos 5x \cos 4x - \sin 5x \sin 4x$
6. $\sin(\theta - y)\cos y + \cos(\theta - y)\sin y$
7. $\sin\left(\dfrac{\pi}{2} + x\right)$
8. $\tan(\pi + x)$
9. $2 \sin 3x \cos 3x$
10. $\cos^2 3y - \sin^2 3y$
11. $\dfrac{2 \tan 4x}{1 - \tan^2 4x}$
12. $\dfrac{\tan 5x - \tan 3x}{1 + \tan 5x \tan 3x}$
13. $\sqrt{\dfrac{1 - \cos 10x}{2}}$
14. $\dfrac{\sin 4x}{1 + \cos 4x}$

In Exercises 15–18, express each product as a sum and each sum as a product.

15. $2 \cos x \sin 5x$
16. $\cos 4y \cos 3y$
17. $\cos 3x - \cos 2x$
18. $\sin 8x + \sin 4x$

In Exercises 19–22, find the exact value of each trigonometric expression.

19. $\sin \dfrac{\pi}{8}$
20. $\tan 75°$
21. $\sin 75° \cos 15°$
22. $\cos \dfrac{5\pi}{12} + \cos \dfrac{7\pi}{12}$

In Exercises 23–26, find the exact value of the indicated trigonometric function from the given information.

23. Given $\sin x = \tfrac{3}{5}$, $0 < x < \pi/2$, find
    (a) $\sin\left(x + \dfrac{\pi}{4}\right)$
    (b) $\cos\left(x - \dfrac{\pi}{6}\right)$
    (c) $\sin 2x$
    (d) $\cos \dfrac{x}{2}$

24. Given $\tan x = -2$, $\pi/2 < x < \pi$, find
    (a) $\tan\left(x - \dfrac{\pi}{4}\right)$
    (b) $\cos\left(x + \dfrac{\pi}{3}\right)$
    (c) $\cos 2x$
    (d) $\tan \dfrac{x}{2}$

25. Given $\sin x = -\tfrac{2}{3}$, $\cos y = -\tfrac{12}{13}$, x and y in quadrant III, find
    (a) $\cos(x - y)$
    (b) $\sin(x + y)$
    (c) $\tan 2y$
    (d) $\sin \dfrac{x}{2}$

26. Given $\sec x = 3$, $\csc y = -2$, x and y in quadrant IV, find
    (a) $\sin(x - y)$
    (b) $\tan(x + y)$
    (c) $\sin 2y$
    (d) $\cos \dfrac{x}{2}$

In Exercises 27 and 28, make the indicated trigonometric substitutions and simplify. Assume $0 < \theta < \pi/2$.

27. For $\sqrt{x^2 + 9}$, let $x = 3 \tan \theta$.
28. For $\dfrac{\sqrt{16 - x^2}}{x}$, let $x = 4 \sin \theta$.

In Exercises 29 and 30, use power reduction formulas to express the even powers of the sine and cosine in terms of first powers of the cosine function. Then use a product-to-sum formula to eliminate products of trigonometric functions.

29. $\sin^2 x \cos^4 x$
30. $\cos^6 x$

In Exercises 31–60, show that each trigonometric equation is an identity.

31. $1 - 2 \sin^2 x = 2 \cos^2 x - 1$
32. $\tan x \csc x \cos x = 1$
33. $\dfrac{1 + \tan^2 x}{\csc^2 x} = \tan^2 x$
34. $\dfrac{\cos x}{\csc^2 x - 1} = \sin x \tan x$
35. $\csc x + \cot x = \dfrac{\sin x}{1 - \cos x}$
36. $\cot x + \tan x = \cot x \sec^2 x$
37. $\dfrac{1 + \sin x}{1 - \sin x} = \dfrac{\csc x + 1}{\csc x - 1}$
38. $\dfrac{1 + \sec x}{\csc x} = \sin x + \tan x$
39. $\dfrac{1}{\cos x + 1} - \dfrac{1}{\cos x - 1} = 2 \csc^2 x$
40. $\dfrac{\sec^3 x + \cos^3 x}{\sec x + \cos x} = \sec^2 x - \sin^2 x$

41. $\dfrac{\cos^2 x}{1 + \sin x} = 1 - \sin x$

42. $1 + \tan x = \dfrac{2 \tan x + \sec^2 x}{1 + \tan x}$

43. $\dfrac{1 + \sin^2 x - \cos^2 x}{\sin 2x} = \tan x$

44. $\dfrac{1 + \tan^2 x}{1 - \tan^2 x} = \sec 2x$

45. $2 \csc 4x = \csc 2x \sec 2x$

46. $\tan x \sin 2x = 1 - \cos 2x$

47. $\dfrac{2 \sin x \cos x}{\cos^2 x - \sin^2 x} = \dfrac{2 \tan x}{1 - \tan^2 x}$

48. $\cot 2x + \csc 2x = \cot x$

49. $\dfrac{\sin 2x}{\sin x} - \dfrac{\cos 2x}{\cos x} = \sec x$

50. $\dfrac{\csc x - \sec x}{\csc x + \sec x} = \dfrac{\cos 2x}{1 + \sin 2x}$

51. $\tan\left(x + \dfrac{\pi}{4}\right) = \dfrac{1 + \tan x}{1 - \tan x}$

52. $\sin(x + y) + \sin(x - y) = 2 \sin x \cos y$

53. $\dfrac{\sin 3x}{\sec x} - \dfrac{\cos 3x}{\csc x} = \sin 2x$

54. $\tan 3x + \cot 6x = \csc 6x$

55. $2 \sin^2 \dfrac{x}{2} + \cos x = 1$

56. $\tan \dfrac{x}{2} = \csc x - \cot x$

57. $2 \sin 7x \sin 2x + \cos 9x = \cos 5x$

58. $\dfrac{\cos 2x + \cos 6x}{\cos^2 x - \sin^2 x} = 2 \cos 4x$

59. $\dfrac{\sin 7x + \sin 3x}{\sin 10x} = \dfrac{\sec 5x}{\sec 2x}$

60. $\dfrac{2 \sin 3x \cos 2x - \sin x}{\cos x - 2 \sin 3x \sin 2x} = \tan 5x$

In Exercises 61–86,

(a) find the general form of the solution of each equation, and

(b) find the solutions in the interval $[0, 2\pi)$.

61. $2 \cos x + 1 = 0$
62. $\sqrt{3} \tan x + 1 = 0$
63. $4 \sin^2 x - 3 = 0$
64. $\sec^2 x - 2 = 0$
65. $\cot 2x + 1 = 0$
66. $\sin 3x - 1 = 0$
67. $\sin\left(3x - \dfrac{\pi}{6}\right) = 1$
68. $\tan\left(2x + \dfrac{\pi}{4}\right) = \sqrt{3}$
69. $2 \tan x \sin x - \tan x = 0$
70. $\sin x + \sin x \sin 3x = 0$
71. $\sec^2 x - \sec x = 2$
72. $2 \cos^2 x - 5 \cos x + 3 = 0$
73. $1 + \cos^2 x + \sin x = 0$
74. $\sec^2 x + \tan x = 1$
75. $\csc x = 1 - \cot x$
76. $\sin x + \cos x + 1 = 0$
77. $\sin 2x + \sin x = 0$
78. $\sin^2 x + \cos 2x = 0$
79. $\sin 3x \cos x - \cos 3x \sin x = 1$
80. $\tan 5x \sin 3x + \cos 3x = 0$
81. $\cos^2 \dfrac{x}{2} + \cos x = 2$
82. $2 \tan \dfrac{x}{2} = \csc x$
83. $\sin 5x + \sin 3x = 0$
84. $\cos 3x - \cos x = \sin x$
85. $\cos 4x + 2 \cos^2 2x = 2$
86. $\sin 2x = \sqrt{2} \sin\left(\dfrac{3\pi}{2} - x\right)$

In Exercises 87–96, approximate to four significant digits the solutions that are in the interval $[0, 2\pi)$.

87. $5 \sin x + 2 = 0$
88. $2 \sec x - 5 = 0$
89. $\tan^2 x - 4 = 0$
90. $9 \sin^2 x - 4 = 0$
91. $\cot^2 x - 5 \cot x + 6 = 0$
92. $3 \sin^2 x + 4 \sin x - 4 = 0$
93. $3 \sin x + \sin^2 x = \cos^2 x$
94. $\tan^2 x + 2 = 3 \sec x$
95. $9 \cos^2 x - 9 \cos 2x = 1$
96. $6 \sin\left(\dfrac{\pi}{2} + x\right) + 4 \cos 2x = 5$

97. Express each function f in the form $f(x) = a \sin(bx + c)$. Determine the amplitude, period, and phase shift of f, and sketch its graph.

(a) $f(x) = \sin 3x + \sqrt{3} \cos 3x$

(b) $f(x) = 2 \sin 4\pi x - 2 \cos 4\pi x$

98. Determine the x-intercepts for the graph of each function. Then use addition of ordinates to sketch the graph in the interval $[0, 2\pi)$.

(a) $y = \sin 2x + \sin x$ (b) $y = \cos 4x - \cos x$

99. The voltage v (in volts) in an electrical circuit is a function of time t (in seconds) and is given by

$$v(t) = 24 \sin 8\pi t + 32 \cos 8\pi t.$$

(a) Determine a and c such that $v(t) = a \sin(8\pi t + c)$.

(b) Find the approximate times t between 0 and 1 second for which $v = 10$ volts.

100. A mansarded roof has a cross section in the shape of an isosceles trapezoid with dimensions as shown in the sketch.

(a) Show that the cross-sectional area A of the mansarded roof is given by $A = 16 \sin \theta + 8 \sin 2\theta$.

(b) Using calculus, it can be shown that the value of θ for the maximum cross-sectional area is a solution of the equation $16 \cos \theta + 16 \cos 2\theta = 0$. Find the maximum cross-sectional area of this roof.

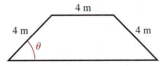

101. In physics it is shown that if a golf ball is driven from a tee with an initial velocity of v ft/s at an angle θ from the horizontal, as shown in the sketch, then the horizontal distance d (in feet) of the drive is given by

$$d = \tfrac{1}{32} v^2 \sin \theta \cos \theta.$$

If $v = 240$ ft/s, find values of θ (to the nearest tenth of a degree) that result in a 250-yard drive.

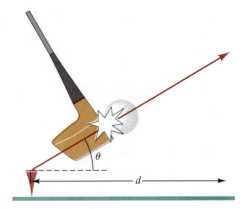

102. The number of hours N of daylight on a particular day d in Quebec, Canada, may be approximated by

$$N = 12 + 4 \sin \left[\frac{2\pi}{365} (d - 80) \right].$$

Determine the days of the year that have 15 hours of daylight.

CHAPTER 5

Complex Numbers

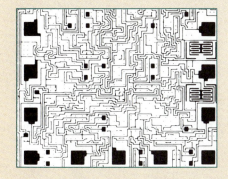

Given an AC circuit in which $R = 6\ \Omega$, $X_L = 5\ \Omega$, and $X_C = 8\ \Omega$, find
(a) the magnitude of the impedance.
(b) the phase angle between the voltage and current.

(For the solution, see Example 5 in Section 5.2)

5.1   **The Complex Number System**
5.2   **Trigonometric Form of Complex Numbers**
5.3   **Powers and Roots of Complex Numbers**

5.1 The Complex Number System

◆ Introductory Comments

One of the basic properties of a real number is that its square is always *nonnegative*. That is, if x is a real number, then

$$x^2 \geq 0.$$

Hence, the principal square root of a *negative* number *cannot* be a real number. For example, $\sqrt{-4}$ and $\sqrt{-9}$ *cannot* be real numbers, since no real number squared is -4 and no real number squared is -9. In this section, we extend the real number system to a larger system that includes the real numbers as well as those numbers whose squares are negative. We refer to this larger system as the **complex number system.**

◆ Pure Imaginary Numbers

In the seventeenth century, the French mathematician René Descartes used the word "imaginary" to describe roots like $\sqrt{-4}$ and $\sqrt{-9}$, and in the eighteenth century a Swiss mathematician, Leonhard Euler, introduced the **imaginary unit i.**

Imaginary Unit i

$$i = \sqrt{-1} \quad \text{where} \quad i^2 = -1$$

Thus, if a is a real number and $a > 0$, then we can find the **principal square root of $-a$** as follows:

$$\sqrt{-a} = \sqrt{-1 \cdot a} = \sqrt{-1} \cdot \sqrt{a} = i\sqrt{a}.$$

Principal Square Root of $-a$

If a is a real number and $a > 0$, then

$$\sqrt{-a} = i\sqrt{a}.$$

Thus, for  $\sqrt{-4}$ and $\sqrt{-9}$, we have

$$\sqrt{-4} = i\sqrt{4} = 2i \quad \text{and} \quad \sqrt{-9} = i\sqrt{9} = 3i.$$

Note that the squares of $2i$ and $3i$ are -4 and -9, respectively:

$$(2i)^2 = 2^2 i^2 = 4(-1) = -4 \quad \text{and} \quad (3i)^2 = 3^2 i^2 = 9(-1) = -9.$$

Numbers such as 2*i* and 3*i*, whose squares are negative, are called *pure imaginary numbers*.

Pure Imaginary Number

> Any number of the form *bi*, where *i* is the imaginary unit and *b* is a real number such that $b \neq 0$, is a **pure imaginary number.**

We can add, subtract, multiply, and divide pure imaginary numbers by applying the commutative, associative, and distributive properties as we did with the real numbers. Some illustrations:

1. $\sqrt{-4} + \sqrt{-9} = 2i + 3i = (2 + 3)i = 5i$
2. $\sqrt{-4} - \sqrt{-9} = 2i - 3i = (2 - 3)i = -i$
3. $\sqrt{-4} \cdot \sqrt{-9} = (2i)(3i) = (2 \cdot 3)(i \cdot i) = 6i^2 = 6(-1) = -6$
4. $\dfrac{\sqrt{-4}}{\sqrt{-9}} = \dfrac{2i}{3i} = \dfrac{2}{3}$

From the preceding illustrations, note that the sum and difference of 2*i* and 3*i* are pure imaginary numbers, while the product and quotient of 2*i* and 3*i* are real numbers.

Caution The property of radicals that states $\sqrt{a}\sqrt{b} = \sqrt{ab}$ does not apply when both *a* and *b* are negative numbers. For example, to write

$$\sqrt{-4} \cdot \sqrt{-9} = \sqrt{(-4)(-9)} = \sqrt{36} = 6 \text{ is WRONG!}$$

When working with negative radicands, be sure to apply $\sqrt{-a} = i\sqrt{a}$ before using any of the properties of radicals from basic algebra.

Note that for the product $(\sqrt{-1})(-\sqrt{-1})$, we have

$$(\sqrt{-1})(-\sqrt{-1}) = (i)(-i) = -i^2 = -(-1) = 1.$$

Since the product $(i)(-i)$ equals 1, we know that *i* and $-i$ are *reciprocals* (or multiplicative inverses) of each other. Thus,

$$\dfrac{1}{i} = -i \quad \text{and} \quad \dfrac{1}{-i} = i$$

EXAMPLE 1 Simplify, and express each answer in the form *b* or *bi*, where *b* is a real number.

(a) $2\sqrt{-3} + \sqrt{-27}$

(b) $3\sqrt{5} \cdot \sqrt{\dfrac{-16}{5}}$

(c) $\sqrt{-2}(\sqrt{-18} + \sqrt{-8})$

(d) $\dfrac{8}{-\sqrt{-25}}$

SOLUTION

(a) $2\sqrt{-3} + \sqrt{-27} = 2i\sqrt{3} + 3i\sqrt{3} = 5i\sqrt{3}$

(b) $3\sqrt{5} \cdot \sqrt{\dfrac{-16}{5}} = 3\sqrt{5} \cdot \sqrt{\dfrac{16}{5}}\,i = 3\sqrt{5} \cdot \dfrac{4}{\sqrt{5}}i = 3 \cdot 4i = 12i$

(c) $\sqrt{-2}(\sqrt{-18} + \sqrt{-8}) = i\sqrt{2}(3i\sqrt{2} + 2i\sqrt{2})$
$= 6i^2 + 4i^2 = 10i^2 = -10$

(d) $\dfrac{8}{-\sqrt{-25}} = \dfrac{8}{-5i} = \dfrac{8}{5} \cdot \dfrac{1}{-i} = \dfrac{8}{5}i$

The reciprocal of $-i$ is i.

PROBLEM 1 Repeat Example 1 for each expression.

(a) $3\sqrt{-80} - \tfrac{1}{3}\sqrt{-45}$

(b) $\dfrac{9}{\sqrt{-36}}$

◆ Powers of *i*

When we raise *i* to successive positive integer powers, an interesting pattern develops:

$i^1 = i$ $\qquad i^2 = -1$

$i^3 = i^2 i = (-1)i = -i$ $\qquad i^4 = i^2 i^2 = (-1)(-1) = 1$

$i^5 = i^4 i = 1(i) = i$ $\qquad i^6 = i^4 i^2 = 1(-1) = -1$

$i^7 = i^4 i^3 = 1(-i) = -i$ $\qquad i^8 = i^4 i^4 = 1(1) = 1$

$i^9 = i^8 i = 1(i) = i$ $\qquad i^{10} = i^8 i^2 = 1(-1) = -1$

$i^{11} = i^8 i^3 = 1(-i) = -i$ $\qquad i^{12} = i^8 i^4 = 1(1) = 1$

Notice the values repeat in cycles of four according to the pattern $i, -1, -i, 1$. For higher powers of *i*, we use the fact that $i^4 = 1$ and apply the laws of exponents from basic algebra.

EXAMPLE 2 Simplify, and express the answer in the form b or bi, where b is a real number.

(a) $3i^{100}$ (b) $\dfrac{1}{2i^{13}}$

SECTION 5.1 The Complex Number System

SOLUTION

(a) $3i^{100} = 3(i^4)^{25} = 3(1)^{25} = 3$

(b) $\dfrac{1}{2i^{13}} = \dfrac{1}{2(i^4)^3 i} = \dfrac{1}{2(1)^3 i} = \dfrac{1}{2i} = \dfrac{1}{2}(-i) = -\dfrac{1}{2}i$

The reciprocal of i is $-i$.

PROBLEM 2 Repeat Example 2 for each expression.

(a) i^{22} (b) $\dfrac{1}{i^{51}}$

◆ Operations with Complex Numbers

The *product* of a real number a ($a \neq 0$) and a pure imaginary bi is the pure imaginary number $(ab)i$. However, the *sum* of a real number and a pure imaginary number is neither a real number nor a pure imaginary number. In the nineteenth century, the German mathematician Carl Friedrich Gauss denoted the sum of a real number a and a pure imaginary number bi by writing the complex number $a + bi$.

Complex Number

> Any number of the form
>
> $$a + bi,$$
>
> where a and b are real numbers and i is the imaginary unit, is a **complex number.**

In the complex number $a + bi$, the number a is the *real part* and bi is the *imaginary part* of the number. If $b = 0$, then $a + bi$ becomes $a + 0i = a$. A complex number of the form $a + 0i$ is a real number. If $a = 0$, then $a + bi$ becomes $0 + bi = bi$. A complex number of the form $0 + bi$ is a *pure imaginary number.* If $b \neq 0$, then $a + bi$ is referred to simply as an **imaginary number.** Some examples of imaginary numbers are

$$2 + 3i \qquad \pi - 6i \qquad 5i \qquad i\sqrt{2}$$

The set of imaginary numbers and the set of real numbers are the two major subsets of the set of complex numbers. The relationship between the various sets of numbers is shown in Figure 5.1. All the numbers shown in Figure 5.1 are part of the *complex number system.*

FIGURE 5.1
Some subsets of the complex numbers.

Complex Numbers			
Imaginary Numbers		**Real Numbers**	
$2 + 3i$ $\pi - 6i$		**Rational Numbers**	**Irrational Numbers**
	$1 + i\sqrt{2}$	$\frac{3}{8}$ $-\frac{1}{2}$	π $\sqrt{2}$
Pure Imaginary Numbers		**Integers**	$\sqrt[3]{7}$
$5i$ $-12i$		0 32	
$\frac{3}{7}i$	i	-7	$0.020020002...$
	$-i\sqrt{2}$	1 -18	
$1 + i$			$\sqrt[4]{11}$
	$-3 + \frac{3}{4}i$	0.75 $2.333...$	$\sqrt{3}$

The commutative, associative, and distributive properties are valid for complex numbers and are used when adding, subtracting, multiplying, and dividing complex numbers. For the *sum* of the complex numbers $a + bi$ and $c + di$, we have

$$(a + bi) + (c + di) = (a + c) + (bi + di)$$
$$= (a + c) + (b + d)i$$

where $(a+c)$ is the Sum of the real parts and $(b+d)i$ is the Sum of the imaginary parts.

Addition of Complex Numbers

If $a + bi$ and $c + di$ are complex numbers, then

$$(a + bi) + (c + di) = (a + c) + (b + d)i.$$

The *additive identity* for any complex number $a + bi$ is $0 + 0i$, since

$$(a + bi) + (0 + 0i) = (a + 0) + (b + 0)i = a + bi.$$

The *additive inverse* for the complex number $c + di$ is $-c - di$, since

$$(c + di) + (-c - di) = (c - c) + (d - d)i = 0 + 0i.$$

Thus, for the *difference* $(a + bi) - (c + di)$, we add $a + bi$ with the additive inverse of $c + di$ as follows:

$$(a + bi) - (c + di) = (a + bi) + (-c - di)$$
$$= (a - c) + (b - d)i$$

where $(a-c)$ is the Difference of the real parts and $(b-d)i$ is the Difference of the imaginary parts.

Subtraction of Complex Numbers

If $a + bi$ and $c + di$ are complex numbers, then

$$(a + bi) - (c + di) = (a - c) + (b - d)i.$$

EXAMPLE 3 Perform the indicated operations, and express each answer in the form $a + bi$.

(a) $(6 - 3i) + (-4 + 7i)$ (b) $(3 + \sqrt{-25}) - (3 - \sqrt{-36})$

SOLUTION

(a) $(6 - 3i) + (-4 + 7i) = (6 - 4) + (-3 + 7)i = 2 + 4i$

(b) $(3 + \sqrt{-25}) - (3 - \sqrt{-36}) = (3 + 5i) - (3 - 6i)$
$= (3 - 3) + [5 - (-6)]i = 0 + 11i$ or $11i$ ◆

PROBLEM 3 Repeat Example 3 for $(2 + 3i) - (5 - 4i) + (-3 + i)$. ◆

For the product $(a + bi)(c + di)$, we apply the distributive property and multiply each part of the first complex number by each part of the second complex number. The procedure is similar to the way we multiply two binomials. Hence,

$$(a + bi)(c + di) = ac + ad\,i + bc\,i + bd\,i^2$$
$$= ac + (ad + bc)i + bd(-1)$$
$$= (ac - bd) + (ad + bc)i$$

Multiplication of Complex Numbers

If $a + bi$ and $c + di$ are complex numbers, then

$$(a + bi)(c + di) = (ac - bd) + (ad + bc)i.$$

EXAMPLE 4 Perform the indicated operations, and express each answer in the form $a + bi$.

(a) $(3 + 4i)(5 - i)$ (b) $(1 - \sqrt{-4})^2$

SOLUTION

(a) $(3 + 4i)(5 - i) = 15 - 3i + 20i - 4i^2$
$= 15 + 17i - 4(-1)$
$= 19 + 17i$

(b) $(1 - \sqrt{-4})^2 = (1 - 2i)^2 = (1 - 2i)(1 - 2i)$
$= 1 - 2i - 2i + 4i^2$
$= 1 - 4i + 4(-1)$
$= -3 - 4i$ ◆

PROBLEM 4 Repeat Example 4 for $(1 + 3i)(2 + 5i)$. ◆

To find the quotient $\dfrac{a + bi}{c + di}$, we multiply numerator and denominator by the **complex conjugate** of the denominator, namely $c - di$. The *product* of complex conjugates is

$$(c + di)(c - di) = c^2 - cdi + cdi - d^2i^2 = \underbrace{c^2 + d^2}_{\text{A \textit{nonnegative} real number}}.$$

The fact that the product of complex conjugates is a nonnegative real number enables us to express the quotient $\dfrac{a + bi}{c + di}$ as a complex number:

$$\dfrac{a + bi}{c + di} = \dfrac{(a + bi)\underbrace{(c - di)}_{\text{Complex conjugate of } c + di}}{(c + di)(c - di)} = \dfrac{(ac + bd) + (bc - ad)i}{c^2 + d^2}$$

$$= \dfrac{ac + bd}{c^2 + d^2} + \dfrac{bc - ad}{c^2 + d^2}i$$

Division of Complex Numbers

If $a + bi$ and $c + di$ are complex numbers, then

$$\dfrac{a + bi}{c + di} = \dfrac{ac + bd}{c^2 + d^2} + \dfrac{bc - ad}{c^2 + d^2}i$$

EXAMPLE 5 Perform the indicated operations and express each answer in the form $a + bi$.

(a) $\dfrac{1}{2 + i}$ (b) $\dfrac{3 + 5i}{1 - 5i}$

SOLUTION

(a) $\dfrac{1}{2 + i} = \dfrac{1\underbrace{(2 - i)}_{\text{Complex conjugate of } 2 + i}}{(2 + i)(2 - i)} = \dfrac{2 - i}{4 - i^2} = \dfrac{2 - i}{5} = \dfrac{2}{5} - \dfrac{1}{5}i$

(b) $\dfrac{3 + 5i}{1 - 5i} = \dfrac{(3 + 5i)\underbrace{(1 + 5i)}_{\text{Complex conjugate of } 1 - 5i}}{(1 - 5i)(1 + 5i)} = \dfrac{3 + 15i + 5i + 25i^2}{1 - 25i^2}$

$= \dfrac{-22 + 20i}{26}$

$= \dfrac{-22}{26} + \dfrac{20}{26}i = -\dfrac{11}{13} + \dfrac{10}{13}i$

PROBLEM 5 Repeat Example 5 for $\dfrac{1 + \sqrt{-9}}{1 - \sqrt{-9}}$.

Exercises 5.1

Basic Skills

In Exercises 1–20, simplify each expression and write the answer in the form b or bi, where b is a real number.

1. $\sqrt{-16} + \sqrt{-25}$
2. $\sqrt{-36} - \sqrt{-49}$
3. $\sqrt{3} \cdot \sqrt{-12} \cdot \sqrt{-9}$
4. $\dfrac{\sqrt{-125} \cdot \sqrt{36}}{\sqrt{-5}}$
5. $5\sqrt{-5} + \sqrt{-20} - \frac{4}{3}\sqrt{-45}$
6. $2\sqrt{-3} - 2\sqrt{-27} + 6\sqrt{-48}$
7. $\sqrt{-3}(\sqrt{-12} - \sqrt{-27})$
8. $\sqrt{2}(\sqrt{-18} - \sqrt{-50})$
9. $\dfrac{5\sqrt{18}}{10\sqrt{-2}}$
10. $\dfrac{24}{2(\sqrt{-6})^2}$
11. $8\sqrt{7} \cdot \sqrt{\dfrac{36}{-7}}$
12. $6 \cdot \sqrt{\dfrac{-4}{5}} \cdot \sqrt{5}$
13. $5i^{26}$
14. $13i^{36}$
15. $2i^{-23}$
16. $\dfrac{1}{6i^{37}}$
17. $i + i^2 + i^3 + i^4$
18. $i^{-1} - i^{-2} + i^{-3} - i^{-4}$
19. $(3i)^3(2i)^2$
20. $(2i)^4(-3i)^3$

In Exercises 21-42, perform the indicated operations and express each answer in the form a + bi.

21. $(9 - 3i) + (6 + 2i) + 4i$
22. $(12 + 6i) - (4 - 3i) + (1 + i)$
23. $(3 + 2\sqrt{-36}) - (5 - 3\sqrt{-49})$
24. $(8 - 4\sqrt{-9}) + (6 + 5\sqrt{-16})$
25. $8i(2 - 3i)$
26. $\sqrt{-25}(3\sqrt{-4} - 8)$
27. $(3 - \sqrt{-4})(4 + 3\sqrt{-9})$
28. $(8 - 2i)(1 - i)$
29. $\dfrac{1 - 2i}{1 + 2i}$
30. $\dfrac{4 - i}{2 - 3i}$
31. $\dfrac{\sqrt{-4}}{2 - \sqrt{-6}}$
32. $\dfrac{9}{1 + 2\sqrt{-3}}$
33. $\dfrac{4 + 2i}{i}$
34. $\dfrac{1 - \sqrt{-16}}{3i}$
35. $(-1 + 2\sqrt{-25})^2$
36. $(-4 - 5i)^2$
37. $(3 - 2i)^3$
38. $\dfrac{1}{(2 + \sqrt{-9})^3}$
39. $\dfrac{\sqrt{-9}(3 + \sqrt{-4})^2}{1 + \sqrt{-1}}$
40. $\dfrac{(2 - i)^2(1 + i)}{3 - 2i}$
41. $\dfrac{i^5 - i^2}{(2 + i)^2}$
42. $\dfrac{(3i^2 - i)^2}{2i^3 + i^{10}}$

Critical Thinking

43. Show that $1 + 0i$ is the *multiplicative identity* for the complex number $a + bi$.

44. Find the *multiplicative inverse*, or reciprocal, of the complex number $a + bi$.

45. Evaluate the polynomial $x^2 - 2x + 2$ when (a) $x = 1 + i$ and (b) $x = 1 - i$.

46. Evaluate the polynomial $x^2 - 6x + 13$ when (a) $x = 3 - 2i$ and (b) $x = 3 + 2i$.

47. Evaluate the polynomial $x^3 - 3x^2 + x + 5$ when (a) $x = 2 - i$ and (b) $x = 2 + i$.

48. Evaluate the polynomial $x^3 - 4x^2 + 14x - 20$ when (a) $x = 1 + 3i$ and (b) $x = 1 - 3i$.

49. Show that $\dfrac{\sqrt{2} + i\sqrt{2}}{2}$ is a square root of i.

50. Show that $\dfrac{\sqrt{3} + i}{2}$ is a cube root of i.

Calculator Activities

51. When working with the *square root key* () on your calculator, you must enter nonnegative inputs. (A negative input causes an error message to appear in the display window.) Using your calculator in conjunction with the definition of the principal square root of $-a$, write each expression in the form bi, with b rounded to three significant digits.

(a) $\sqrt{-18}$ (b) $\sqrt{-506}$
(c) $\sqrt{-153.6}$ (d) $\sqrt{-0.8765}$

52. Evaluate

$$\frac{-b \pm \sqrt{b^2 - 4ac}}{2a}$$

for the given values of a, b, and c. Express each answer in the form $a + bi$ with a and b rounded to three significant digits.

(a) $a = 2, b = 6, c = 7$
(b) $a = 13, b = -5, c = 10$

53. The magnitude of the current (in amperes, A) flowing through the series circuit shown in the sketch may be found by expressing

$$\frac{200}{150 + (188.5 - 76.4)i}$$

in the complex number form $a + bi$, and then computing $\sqrt{a^2 + b^2}$, where a is the real part and b the coefficient of the imaginary part of this complex number. Determine the magnitude of the current in this circuit.

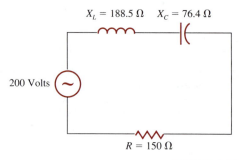

EXERCISE 53

54. The magnitude of the total impedance (in ohms, Ω) for the parallel circuit shown in the sketch may be found by expressing

$$\frac{1}{\dfrac{1}{150} + \dfrac{1}{188.5i} - \dfrac{1}{76.4i}}$$

in the complex number form $a + bi$, and then computing $\sqrt{a^2 + b^2}$, where a is the real part and b the coefficient of the imaginary part of this complex number. Determine the magnitude of the total impedance for this circuit.

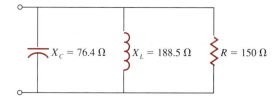

5.2 Trigonometric Form of Complex Numbers

In Section 5.1 we introduced the complex number $a + bi$, where a and b are real numbers, and $i = \sqrt{-1}$. We refer to $a + bi$ as a **complex number in standard form**. In this section we introduce the *trigonometric form* of a complex number. Expressing complex numbers in trigonometric form makes the operations of multiplication and division easier to perform. In order to express a complex number in trigonometric form, we must first discuss the graphical representation of a complex number.

◆ Graphical Representation of a Complex Number

For a graphical representation of a complex number, we assign $a + bi$ the ordered pair (a, b) in the coordinate plane. When the coordinate plane is used in this man-

SECTION 5.2 Trigonometric Form of Complex Numbers

ner, we call it the **complex plane.** In the complex plane, the horizontal axis is referred to as the **real axis** and the vertical axis as the **imaginary axis.**

EXAMPLE 1 Plot each complex number in the complex plane.

(a) $3 + 4i$ (b) $1 - i$ (c) $2i$ (d) -3

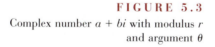

SOLUTION

(a) The complex number $3 + 4i$ is assigned the point $(3, 4)$ in the complex plane (see Figure 5.2).

(b) The complex number $1 - i$ is assigned the point $(1, -1)$ in the complex plane (see Figure 5.2).

(c) Since $2i = 0 + 2i$, the pure imaginary number $2i$ is assigned the point $(0, 2)$ in the complex plane (see Figure 5.2).

(d) Since $-3 = -3 + 0i$, the real number -3 is assigned the point $(-3, 0)$ in the complex plane (see Figure 5.2). ◆

FIGURE 5.2
Some complex numbers plotted in the complex plane

Recall from Section 5.1 that the complex numbers $a + bi$ and $a - bi$ are called *complex conjugates.*

PROBLEM 1 Plot the complex conjugate of each complex number in Example 1. ◆

FIGURE 5.3
Complex number $a + bi$ with modulus r and argument θ

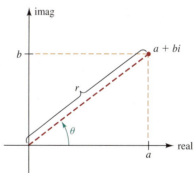

◆ **Changing between Trigonometric and Standard Form**

In the complex plane, the distance r from the origin $(0, 0)$ to the complex number $a + bi$ is called the **modulus** (or **absolute value**) of the complex number. The angle θ, measured counterclockwise from the positive real axis as shown in Figure 5.3, is called an **argument** of the complex number. Normally, angle θ is measured in radians with $0 \leq \theta < 2\pi$. However, any angle coterminal with θ may also be considered an argument of the complex number. From our discussion of trigonometry in Chapter 2, we know that a, b, r, and θ are related as follows.

> **Relationships between a, b, r, and θ**
>
> If r is the modulus and θ is the argument of a nonzero complex number $a + bi$, then
>
> 1. $r = \sqrt{a^2 + b^2}$
> 2. $\tan \theta = \dfrac{b}{a}, \quad a \neq 0$
> 3. $\cos \theta = \dfrac{a}{r}$ or $a = r \cos \theta$
> 4. $\sin \theta = \dfrac{b}{r}$ or $b = r \sin \theta$

If we substitute $r \cos \theta$ for a and $r \sin \theta$ for b in the complex number $a + bi$, we obtain

$$a + bi = (r \cos \theta) + (r \sin \theta)i = r(\cos \theta + i \sin \theta).$$

We refer to

$$r(\cos \theta + i \sin \theta)$$

as the **trigonometric form** of a complex number. To change from standard form to trigonometric form, we apply the equations

$$r = \sqrt{a^2 + b^2} \quad \text{and} \quad \tan \theta = \frac{b}{a}.$$

EXAMPLE 2 Express each complex number in trigonometric form with $0 \leq \theta < 2\pi$.

(a) $3 + 3i$ (b) $4 - 2i$

SOLUTION

(a) For $3 + 3i$, we have $a = 3$ and $b = 3$. Thus, the modulus is

$$r = \sqrt{a^2 + b^2} = \sqrt{(3)^2 + (3)^2} = \sqrt{18} = 3\sqrt{2}.$$

Since $3 + 3i$ lies in quadrant I, the argument θ must be a first-quadrant angle that satisfies the equation

$$\tan \theta = \frac{b}{a} = \frac{3}{3} = 1.$$

Hence,

$$\theta = \tan^{-1}(1) = \frac{\pi}{4},$$

as shown in Figure 5.4. Thus,

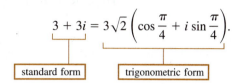

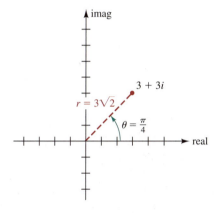

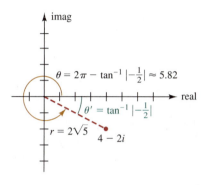

FIGURE 5.4
For the complex number $3 + 3i$, the modulus is $3\sqrt{2}$ and the argument is $\pi/4$.

FIGURE 5.5
For the complex number $4 - 2i$, the modulus is $2\sqrt{5}$ and the argument is $2\pi - \tan^{-1}\left|-\tfrac{1}{2}\right| \approx 5.82$.

(b) For $4 - 2i$, we have $a = 4$ and $b = -2$. Thus, the modulus is

$$r = \sqrt{a^2 + b^2} = \sqrt{(4)^2 + (-2)^2} = \sqrt{20} = 2\sqrt{5}.$$

Since $4 - 2i$ lies in quadrant IV, the argument θ must be a fourth-quadrant angle that satisfies the equation

$$\tan\theta = \frac{b}{a} = \frac{-2}{4} = -\frac{1}{2}.$$

The reference angle θ' associated with θ is the positive acute angle whose tangent is $\left|-\tfrac{1}{2}\right|$. Hence,

$$\theta = 2\pi - \tan^{-1}\left|-\tfrac{1}{2}\right| \approx 5.82,$$

as shown in Figure 5.5. Thus,

$$4 - 2i \approx 2\sqrt{5}\,(\cos 5.82 + i\sin 5.82).$$

     standard form   trigonometric form

◆

PROBLEM 2   Repeat Example 2 for $-\sqrt{3} + i$. ◆

To check the results of Example 2, we reverse the procedure and change trigonometric form to standard form.

EXAMPLE 3 Change the complex number $3\sqrt{2}\left(\cos\dfrac{\pi}{4} + i\sin\dfrac{\pi}{4}\right)$ to standard form.

SOLUTION To express this complex number in standard form, we proceed as follows:

$$3\sqrt{2}\left(\cos\dfrac{\pi}{4} + i\sin\dfrac{\pi}{4}\right) = 3\sqrt{2}\left[\left(\dfrac{1}{\sqrt{2}}\right) + i\left(\dfrac{1}{\sqrt{2}}\right)\right] \quad \text{Evaluate the trigonometric functions}$$

$$= 3\sqrt{2}\left(\dfrac{1}{\sqrt{2}}\right) + 3\sqrt{2}\left(\dfrac{1}{\sqrt{2}}\right)i \quad \text{Multiply}$$

$$= 3 + 3i \quad \text{Simplify}$$

This result agrees with Example 2(a). ◆

PROBLEM 3 Using a calculator set in radian mode, check the result of Example 2(b) by changing $2\sqrt{5}\,(\cos 5.82 + i\sin 5.82)$ to standard form. ◆

◆ Multiplying and Dividing Complex Numbers

In Section 5.1 we found the sum, difference, product, and quotient of complex numbers expressed in standard form. You may wish to review these operations before reading further.

Consider the complex numbers z_1 and z_2 defined as

$$z_1 = r_1\,(\cos\theta_1 + i\sin\theta_1) \quad \text{and} \quad z_2 = r_2\,(\cos\theta_2 + i\sin\theta_2).$$

The product of z_1 and z_2 is

$$z_1 z_2 = [r_1\,(\cos\theta_1 + i\sin\theta_1)][r_2\,(\cos\theta_2 + i\sin\theta_2)]$$

$$= r_1 r_2 (\cos\theta_1 + i\sin\theta_1)(\cos\theta_2 + i\sin\theta_2) \quad \text{Rewrite the product}$$

$$= r_1 r_2 \,[(\cos\theta_1\cos\theta_2 + i\cos\theta_1\sin\theta_2 + i\sin\theta_1\cos\theta_2 - \sin\theta_1\sin\theta_2)] \quad \text{Multiply}$$

$$= r_1 r_2 \,[(\cos\theta_1\cos\theta_2 - \sin\theta_1\sin\theta_2) + i(\sin\theta_1\cos\theta_2 + \cos\theta_1\sin\theta_2)] \quad \text{Rearrange the terms}$$

Apply sum formula for cosine (Section 4.3)

Apply sum formula for sine (Section 4.3)

$$= r_1 r_2 \,[\cos(\theta_1 + \theta_2) + i\sin(\theta_1 + \theta_2)]$$

Product of moduli of z_1 and z_2

Sum of arguments of z_1 and z_2

Observe that *the modulus of the product z_1z_2 is the product of the moduli of z_1 and z_2, and an argument of z_1z_2 is the sum of the arguments of z_1 and z_2.*

The quotient of z_1 and z_2 is

$$\frac{z_1}{z_2} = \frac{r_1(\cos\theta_1 + i\sin\theta_1)}{r_2(\cos\theta_2 + i\sin\theta_2)}$$

$$= \frac{r_1(\cos\theta_1 + i\sin\theta_1)}{r_2(\cos\theta_2 + i\sin\theta_2)} \cdot \frac{(\cos\theta_2 - i\sin\theta_2)}{(\cos\theta_2 - i\sin\theta_2)} \qquad \text{Introduce the conjugate}$$

$$= \frac{r_1}{r_2}\left[\frac{\cos\theta_1\cos\theta_2 - i\cos\theta_1\sin\theta_2 + i\sin\theta_1\cos\theta_2 + \sin\theta_1\sin\theta_2}{\cos^2\theta_2 + \sin^2\theta_2}\right] \qquad \text{Multiply}$$

$$= \frac{r_1}{r_2}[\underbrace{(\cos\theta_1\cos\theta_2 + \sin\theta_1\sin\theta_2)}_{\text{Apply difference formula for cosine (Section 4.3)}} + i\underbrace{(\sin\theta_1\cos\theta_2 - \cos\theta_1\sin\theta_2)}_{\text{Apply difference formula for sine (Section 4.3)}}] \qquad \text{Simplify and rewrite}$$

$$= \frac{r_1}{r_2}[\underbrace{\cos(\theta_1 - \theta_2)}_{\substack{\text{Quotient of moduli}\\\text{of } z_1 \text{ and } z_2}} \quad + \quad \underbrace{i\sin(\theta_1 - \theta_2)}_{\substack{\text{Difference of arguments}\\\text{of } z_1 \text{ and } z_2}}]$$

Observe that *the modulus of the quotient z_1/z_2 is the quotient of the moduli of z_1 and z_2, and an argument of z_1/z_2 is the difference of the arguments of z_1 and z_2.*

◆ Rules for Multiplying and Dividing Complex Numbers in Trigonometric Form

If $z_1 = r_1(\cos\theta_1 + i\sin\theta_1)$ and $z_2 = r_2(\cos\theta_2 + i\sin\theta_2)$, then

1. $z_1z_2 = r_1r_2[\cos(\theta_1 + \theta_2) + i\sin(\theta_1 + \theta_2)]$

and

2. $\dfrac{z_1}{z_2} = \dfrac{r_1}{r_2}[\cos(\theta_1 - \theta_2) + i\sin(\theta_1 - \theta_2)], \quad z_2 \neq 0.$

EXAMPLE 4 Given that $z_1 = 12\left(\cos\dfrac{\pi}{3} + i\sin\dfrac{\pi}{3}\right)$ and $z_2 = 3\left(\cos\dfrac{7\pi}{6} + i\sin\dfrac{7\pi}{6}\right)$, perform the indicated operations and record the results in standard form.

(a) z_1z_2 (b) $\dfrac{z_1}{z_2}$.

SOLUTION

(a) $z_1 z_2 = \left[12\left(\cos\frac{\pi}{3} + i\sin\frac{\pi}{3}\right)\right]\left[3\left(\cos\frac{7\pi}{6} + i\sin\frac{7\pi}{6}\right)\right]$

$= 12 \cdot 3\left[\cos\left(\frac{\pi}{3} + \frac{7\pi}{6}\right) + i\sin\left(\frac{\pi}{3} + \frac{7\pi}{6}\right)\right]$ **Multiply moduli and add arguments**

$= 36\left[\cos\frac{3\pi}{2} + i\sin\frac{3\pi}{2}\right]$ **Simplify**

$= 36[0 + i(-1)] = -36i$ **Change to standard form**

(b) $\dfrac{z_1}{z_2} = \dfrac{12\left(\cos\dfrac{\pi}{3} + i\sin\dfrac{\pi}{3}\right)}{3\left(\cos\dfrac{7\pi}{6} + i\sin\dfrac{7\pi}{6}\right)}$

$= 4\left[\cos\left(\frac{\pi}{3} - \frac{7\pi}{6}\right) + i\sin\left(\frac{\pi}{3} - \frac{7\pi}{6}\right)\right]$ **Divide moduli and subtract arguments**

$= 4\left[\cos\left(-\frac{5\pi}{6}\right) + i\sin\left(-\frac{5\pi}{6}\right)\right]$ **Simplify**

$= 4\left[-\frac{\sqrt{3}}{2} + i\left(-\frac{1}{2}\right)\right] = -2\sqrt{3} - 2i$ **Change to standard form** ◆

We can check the results of Example 4 by first changing z_1 and z_2 to standard form and then performing the operations of multiplication and division by using the methods discussed in Section 5.1:

$z_1 = 12\left(\cos\frac{\pi}{3} + i\sin\frac{\pi}{3}\right)$ and $z_2 = 3\left(\cos\frac{7\pi}{6} + i\sin\frac{7\pi}{6}\right)$

$= 12\left[\left(\frac{1}{2}\right) + i\left(\frac{\sqrt{3}}{2}\right)\right]$ $= 3\left[\left(-\frac{\sqrt{3}}{2}\right) + i\left(-\frac{1}{2}\right)\right]$

$= 6 + 6\sqrt{3}\, i$ $= -\dfrac{3\sqrt{3}}{2} - \dfrac{3}{2}i$

Standard forms of z_1 and z_2

Multiplying the standard forms of z_1 and z_2, we obtain

$$z_1 z_2 = (6 + 6\sqrt{3}\, i)\left(-\frac{3\sqrt{3}}{2} - \frac{3}{2}i\right)$$

$$= -9\sqrt{3} - 9i - 27i + 9\sqrt{3} = -36i.$$

This agrees with the answer in Example 4(a).

PROBLEM 4 Check the result of Example 4(b) by dividing $z_1 = 6 + 6\sqrt{3}\,i$ by $z_2 = -\dfrac{3\sqrt{3}}{2} - \dfrac{3}{2}i.$

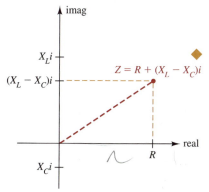

FIGURE 5.6

Representation of impedance Z and its components R, X_L, and X_C with $X_L > X_C$

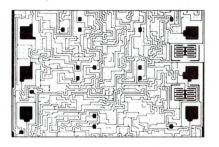

Application: Alternating Current (AC) Circuits

An electrical circuit in which the voltage source is sinusoidal (a sine wave) is called an *alternating current* (AC) *circuit*. The **impedance** of an AC circuit is the opposition to the flow of electrons in the circuit. The unit of measure for impedance is the ohm (Ω).

Impedance (Z) consists of three components: **resistance** *(R)*, **inductive reactance** *(X_L)*, and **capacitive reactance** *(X_C)*. In a purely resistive circuit, the voltage and current are *in phase*, that is, they both reach their peak values at the same time. In a purely inductive circuit, the voltage *leads* the current by 90°, that is, the voltage reaches its peak $\frac{1}{4}$ cycle sooner than the current. In a purely capacitive circuit, the voltage *lags* the current by 90°, that is, the current reaches its peak $\frac{1}{4}$ cycle sooner than the voltage.

The phase relationships of the three components make the complex plane ideally suited for describing the impedance Z of an AC circuit. Resistance R is represented by the positive real axis, inductive reactance X_L is represented by the positive imaginary axis, and capacitive reactance X_C is represented by the negative imaginary axis, as shown in Figure 5.6. Mathematically, we have

$$Z = R + (X_L - X_C)i$$

EXAMPLE 5 Given an AC circuit in which $R = 6\Omega$, $X_L = 5\Omega$, and $X_C = 8\Omega$, find (a) the magnitude of the impedance and (b) the phase angle between the voltage and current.

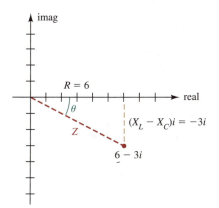

FIGURE 5.7

Representation of the impedance $Z = 6 - 3i$

SOLUTION

(a) The impedance Z is given by

$$\begin{aligned} Z &= R + (X_L - X_C)i \\ &= 6 + (5 - 8)i \\ &= 6 - 3i \end{aligned}$$

and is shown in Figure 5.7. The magnitude (or modulus) of Z, written $|Z|$, is

$$\begin{aligned} |Z| &= \sqrt{R^2 + (X_L - X_C)^2} \\ &= \sqrt{6^2 + (-3)^2} = \sqrt{45} = 3\sqrt{5}\ \Omega \approx 6.71\Omega \end{aligned}$$

(b) The phase angle is the angle θ measured from R to Z, as shown in Figure 5.7. Using the fact that

$$\tan\theta = \frac{X_L - X_C}{R} = \frac{-3}{6} = -\frac{1}{2},$$

we have

$$\theta = \tan^{-1}\left(-\frac{1}{2}\right) \approx -26.6°.$$

The negative value indicates that the voltage lags the current by 26.6°. ◆

PROBLEM 5 Repeat Example 5 given that $R = 6\Omega$, $X_L = 5\Omega$, and $X_C = 3\Omega$. ◆

Exercises 5.2

Basic Skills

For each complex number in Exercises 1–20,
(a) plot the number in the complex plane,
(b) find its modulus,
(c) find its argument, and
(d) express the complex number in trigonometric form with $0 \leq \theta < 2\pi$.

1. $1 + i$
2. $-4 + 4i$
3. $\frac{5}{2} - \frac{5}{2}i$
4. $-\sqrt{5} - \sqrt{5}i$
5. $-1 - \sqrt{3}i$
6. $\sqrt{3} + i$
7. $-2\sqrt{3} - 2i$
8. $7 - 7\sqrt{3}i$
9. $-4i$
10. $\sqrt{2}i$
11. π
12. $-\frac{8}{3}$
13. $3 + 4i$
14. $-12 + 5i$
15. $7 - i$
16. $-2 - 3i$
17. $-\sqrt{7} + 3i$
18. $5 + \sqrt{11}i$
19. $-\frac{1}{2} - \frac{2}{3}i$
20. $\frac{5}{4} - \sqrt{2}i$

In Exercises 21–30, express each complex number in standard form.

21. $12\left(\cos\frac{\pi}{6} + i\sin\frac{\pi}{6}\right)$
22. $5\left(\cos\frac{5\pi}{6} + i\sin\frac{5\pi}{6}\right)$
23. $\frac{1}{2}\left(\cos\frac{2\pi}{3} + i\sin\frac{2\pi}{3}\right)$
24. $4\left(\cos\frac{5\pi}{3} + i\sin\frac{5\pi}{3}\right)$
25. $\sqrt{2}\left(\cos\frac{5\pi}{4} + i\sin\frac{5\pi}{4}\right)$
26. $6\left(\cos\frac{\pi}{4} + i\sin\frac{\pi}{4}\right)$
27. $9\left(\cos\frac{\pi}{2} + i\sin\frac{\pi}{2}\right)$
28. $3(\cos 270° + i\sin 270°)$
29. $\frac{2}{3}(\cos 180° + i\sin 180°)$
30. $8(\cos 8\pi + i\sin 8\pi)$

In Exercises 31–42, perform the indicated operations and record the results (a) in trigonometric form with $0 \leq \theta < 2\pi$ and (b) in standard form.

31. $\left[2\left(\cos\frac{\pi}{3} + i\sin\frac{\pi}{3}\right)\right]\left[9\left(\cos\frac{7\pi}{6} + i\sin\frac{7\pi}{6}\right)\right]$
32. $\left[4\left(\cos\frac{3\pi}{4} + i\sin\frac{3\pi}{4}\right)\right]\left[5\left(\cos\frac{\pi}{4} + i\sin\frac{\pi}{4}\right)\right]$
33. $\left[\frac{2}{3}\left(\cos\frac{\pi}{2} + i\sin\frac{\pi}{2}\right)\right]\left[6\left(\cos\frac{3\pi}{4} + i\sin\frac{3\pi}{4}\right)\right]$
34. $\left[12\left(\cos\frac{11\pi}{9} + i\sin\frac{11\pi}{9}\right)\right]\left[\frac{5}{4}\left(\cos\frac{17\pi}{18} + i\sin\frac{17\pi}{18}\right)\right]$
35. $\left[5\left(\cos\frac{17\pi}{12} + i\sin\frac{17\pi}{12}\right)\right][\sqrt{2} + \sqrt{2}i]$
36. $[\sqrt{3}(\cos 150° + i\sin 150°)][\cos 240° + i\sin 240°]$

37. $\dfrac{8\left(\cos\dfrac{4\pi}{3} + i\sin\dfrac{4\pi}{3}\right)}{2\left(\cos\dfrac{5\pi}{6} + i\sin\dfrac{5\pi}{6}\right)}$

38. $\dfrac{12\left(\cos\dfrac{5\pi}{4} + i\sin\dfrac{5\pi}{4}\right)}{16\left(\cos\dfrac{\pi}{4} + i\sin\dfrac{\pi}{4}\right)}$

42. $\dfrac{\sqrt{2}\left(\cos\dfrac{\pi}{2} + i\sin\dfrac{\pi}{2}\right)}{-1 - i}$

39. $\dfrac{16\left(\cos\dfrac{19\pi}{20} + i\sin\dfrac{19\pi}{20}\right)}{\dfrac{8}{3}\left(\cos\dfrac{\pi}{5} + i\sin\dfrac{\pi}{5}\right)}$

40. $\dfrac{3\left(\cos\dfrac{\pi}{4} + i\sin\dfrac{\pi}{4}\right)}{15\left(\cos\dfrac{5\pi}{12} + i\sin\dfrac{5\pi}{12}\right)}$

41. $\dfrac{2 - 2\sqrt{3}\,i}{4\left(\cos\dfrac{\pi}{6} + i\sin\dfrac{\pi}{6}\right)}$

In Exercises 43–46, the values of R, X_L, and X_C in an AC circuit are given. Find (a) the magnitude of the impedance Z and (b) the phase angle θ (in degrees) between the voltage and current.

43. $R = 5\Omega$, $X_L = 7\Omega$, and $X_C = 2\Omega$

44. $R = 9\Omega$, $X_L = 16\Omega$, and $X_C = 25\Omega$

45. $R = 4\sqrt{3}\;\Omega$, $X_L = 4\Omega$, and $X_C = 8\Omega$

46. $R = 2\Omega$, $X_L = 4\sqrt{3}\;\Omega$, and $X_C = 2\sqrt{3}\;\Omega$

Critical Thinking

47. If the product of two complex numbers in trigonometric form equals 1, then the complex numbers are said to be *multiplicative inverses* (or *reciprocals*) of each other. Find the multiplicative inverse of $r(\cos\theta + i\sin\theta)$.

48. If the product of two complex numbers in trigonometric form equals r^2, then the complex numbers are said to be *complex conjugates* of each other. Find the complex conjugate of $r(\cos\theta + i\sin\theta)$.

49. Find the product $(1 + i\sqrt{3})(1 + i)$ by

    (i) multiplying as you would two binomials, and

    (ii) changing each factor to trigonometric form, and then using the technique discussed in this section.

    Equate the real and imaginary parts of your two answers to find the exact values of each trigonometric function:

    (a) $\cos 105°$ (b) $\sin 105°$

50. Find the product $(\cos\theta + i\sin\theta)(\cos\theta + i\sin\theta)$ by

    (i) multiplying as you would two binomials.

    (ii) using the technique discussed in this section.

    Equate the real and imaginary parts of your two answers. What two familiar trigonometric identities evolve?

Calculator Activities

In Exercises 51–54, express each complex number in trigonometric form with $0 \le \theta < 2\pi$. Round the modulus and argument to three significant digits.

51. $2.65 + 7.91i$

52. $-23 + 18i$

53. $0.65 - 1.23i$

54. $-567.1 - 129.1i$

In Exercises 55–58, express each complex number in the standard form $a + bi$, with a and b rounded to three significant digits.

55. $3\sqrt{5}\,(\cos 1.107 + i\sin 1.107)$

56. $\sqrt{34}\,(\cos 3.682 + i\sin 3.682)$

57. $14\,(\cos 2 + i\sin 2)$

58. $33\,(\cos 175° + i\sin 175°)$

In Exercises 59–64, perform the indicated operations and record the answer in the standard form $a + bi$, with a and b rounded to three significant digits.

59. $[1.67\,(\cos 2.36 + i\sin 2.36)][2.45\,(\cos 1.61 + i\sin 1.61)]$

60. $[1 - 3i][9\,(\cos 3 + i\sin 3)]$

61. $\dfrac{7\,(\cos 2 + i\sin 2)}{4\,(\cos 6 + i\sin 6)}$

62. $\dfrac{\cos 42.1° + i\sin 42.1°}{\cos 73.8° + i\sin 73.8°}$

63. $\dfrac{(-2.34 + 1.66i)(\cos 1.89 + i\sin 1.89)}{\cos 1.34 + i\sin 1.34}$

64. $\dfrac{(0.56 + 1.02i)(\cos 2.46 + i\sin 2.46)}{1.25i}$

In an AC circuit, the relationship between the voltage V (in volts), current I (in amps), and impedance Z (in ohms) is given by V = IZ. Use this formula to answer Exercises 65 and 66.

65. Find the magnitude and phase angle of the voltage V in an AC circuit in which $R = 4.2\ \Omega$, $X_L = 10.1\ \Omega$, $X_C = 13.2\ \Omega$, and $I = 8.3\ (\cos 1.25 + i \sin 1.25)$ amps.

66. Find the magnitude and phase angle of the current I in an AC circuit in which $R = 5.2\ \Omega$, $X_L = 15.3\ \Omega$, $X_C = 3.9\ \Omega$, and $V = 65.1\ (\cos 1.38 + i \sin 1.38)$ volts.

5.3 Powers and Roots of Complex Numbers

In this section, we discuss the procedures for raising a complex number to an integer power and for finding the *n*th roots of a complex number. If a complex number is expressed in the trigonometric form $r(\cos \theta + i \sin \theta)$, then we may find the powers and roots of that complex number by using De Moivre's theorem. We begin by considering positive integer powers of complex numbers.

◆ Positive Integer Powers of Complex Numbers

Consider the complex number

$$z = r(\cos \theta + i \sin \theta).$$

To raise z to a positive integer power, we apply the rule for multiplying complex numbers in trigonometric form (Section 5.2):

Second power:

$$z^2 = z \cdot z = r(\cos \theta + i \sin \theta)\, r(\cos \theta + i \sin \theta)$$
$$= r^2(\cos 2\theta + i \sin 2\theta) \qquad \text{\textbf{Multiply moduli and add arguments}}$$

Third power:

$$z^3 = z \cdot z^2 = r(\cos \theta + i \sin \theta)\, r^2(\cos 2\theta + i \sin 2\theta)$$
$$= r^3(\cos 3\theta + i \sin 3\theta) \qquad \text{\textbf{Multiply moduli and add arguments}}$$

Fourth power:

$$z^4 = z \cdot z^3 = r(\cos \theta + i \sin \theta)\, r^3(\cos 3\theta + i \sin 3\theta)$$
$$= r^4(\cos 4\theta + i \sin 4\theta) \qquad \text{\textbf{Multiply moduli and add arguments}}$$

Do you see the pattern? It appears that if $z = r(\cos \theta + i \sin \theta)$ and n is a positive integer, then the modulus of z^n is r^n and an argument of z^n is $n\theta$. This important fact concerning powers of complex numbers is attributed to mathematician Abraham De Moivre and appropriately called *De Moivre's theorem*. A formal proof of De Moivre's theorem can be made by using mathematical induction (see Exercise 51).

De Moivre's Theorem

If $z = r(\cos\theta + i\sin\theta)$ and n is a positive integer, then

$$z^n = r^n(\cos n\theta + i \sin n\theta).$$

EXAMPLE 1 Find the indicated power of the given complex number and express the answer in the standard form $a + bi$.

(a) $\left[2\left(\cos\dfrac{5\pi}{12} + i\sin\dfrac{5\pi}{12}\right)\right]^4$ (b) $(-1+i)^{12}$

SOLUTION

(a) By De Moivre's theorem, we have

$$\left[2\left(\cos\dfrac{5\pi}{12} + i\sin\dfrac{5\pi}{12}\right)\right]^4 = 2^4\left[\cos(4)\dfrac{5\pi}{12} + i\sin(4)\dfrac{5\pi}{12}\right]$$

$$= 16\left(\cos\dfrac{5\pi}{3} + i\sin\dfrac{5\pi}{3}\right)$$

$$= 16\left[\dfrac{1}{2} + i\left(-\dfrac{\sqrt{3}}{2}\right)\right]$$

$$= 8 - 8\sqrt{3}\,i$$

(b) The complex number must be in trigonometric form before we can apply De Moivre's theorem. For $-1 + i$ the modulus is

$$r = \sqrt{a^2 + b^2} = \sqrt{(-1)^2 + (1)^2} = \sqrt{2}.$$

Since $-1 + i$ lies in quadrant II, the argument θ must be a second-quadrant angle that satisfies the equation

$$\tan\theta = \dfrac{b}{a} = \dfrac{1}{-1} = -1.$$

Thus, $\theta = \dfrac{3\pi}{4}$ and

$$-1 + i = \sqrt{2}\left(\cos\dfrac{3\pi}{4} + i\sin\dfrac{3\pi}{4}\right).$$

Now, applying De Moivre's theorem, we have

$$(-1+i)^{12} = \left[\sqrt{2}\left(\cos\frac{3\pi}{4} + i\sin\frac{3\pi}{4}\right)\right]^{12}$$

$$= (\sqrt{2})^{12}\left[\cos(12)\frac{3\pi}{4} + i\sin(12)\frac{3\pi}{4}\right]$$

$$= 64(\cos 9\pi + i\sin 9\pi)$$

$$= -64 + 0i \quad \text{or} \quad -64 \qquad \blacklozenge$$

PROBLEM 1 Repeat Example 1 for $(\sqrt{3} + i)^9$. $\qquad \blacklozenge$

◆ **Zero and Negative Integer Exponents**

How should we define z^0 or z^{-n}, where n is a positive integer? Certainly, we would like De Moivre's theorem to hold for zero and negative integer exponents as well. If $z = r(\cos\theta + i\sin\theta)$ and De Moivre's theorem is valid for the zero exponent, then

$$z^0 = r^0[\cos(0)\theta + i\sin(0)\theta]$$

$$= 1(\cos 0 + i\sin 0) \qquad \textbf{Simplify}$$

$$= 1[(1) + i(0)] \qquad \textbf{Evaluate trigonometric functions}$$

$$= 1 \qquad \textbf{Multiply}$$

If $z = r(\cos\theta + i\sin\theta)$ and De Moivre's theorem is valid for a negative integer exponent, then

$$z^{-n} = r^{-n}[\cos(-n\theta) + i\sin(-n\theta)].$$

Since the cosine function is even and the sine function is odd, we have

$$\cos(-n\theta) = \cos n\theta \quad \text{and} \quad \sin(-n\theta) = -\sin n\theta.$$

Thus

$$z^{-n} = r^{-n}(\cos n\theta - i\sin n\theta)$$

$$= \frac{\cos n\theta - i\sin n\theta}{r^n} \cdot \frac{\cos n\theta + i\sin n\theta}{\cos n\theta + i\sin n\theta} \qquad \textbf{Introduce the conjugate of the numerator}$$

$$= \frac{\cos^2 n\theta + \sin^2 n\theta}{r^n(\cos n\theta + i\sin n\theta)} \qquad \textbf{Multiply}$$

$$= \frac{1}{r^n(\cos n\theta + i\sin n\theta)} \qquad \textbf{Apply trig identity}$$

$$= \frac{1}{z^n} \qquad \textbf{Substitute}$$

Hence, we make the following definitions, which are consistent with the definitions of the zero and negative integer exponents of real numbers.

Definitions of z^0 and z^{-n}

For all complex numbers z, with $z \neq 0$, and any integer n,

$$z^0 = 1 \quad \text{and} \quad z^{-n} = \frac{1}{z^n}.$$

With this understanding of the zero and negative integer exponents, we extend De Moivre's theorem to include any integral power.

Extension of De Moivre's Theorem

If $z = r(\cos \theta + i \sin \theta)$ and k is any integer, then

$$z^k = r^k (\cos k\theta + i \sin k\theta).$$

EXAMPLE 2 Find the indicated power of the given complex number.

(a) $\left[2 \left(\cos \frac{5\pi}{12} + i \sin \frac{5\pi}{12} \right) \right]^0$

(b) $(-1 + i)^{-12}$

SOLUTION

(a) By definition $z^0 = 1$, provided $z \neq 0$. Therefore, we conclude

$$\left[2 \left(\cos \frac{5\pi}{12} + i \sin \frac{5\pi}{12} \right) \right]^0 = 1.$$

Also, by De Moivre's theorem, we have

$$\left[2 \left(\cos \frac{5\pi}{12} + i \sin \frac{5\pi}{12} \right) \right]^0 = 2^0 \left[\cos (0)\frac{5\pi}{12} + i \sin (0)\frac{5\pi}{12} \right]$$

$$= 1 (\cos 0 + i \sin 0)$$

$$= 1[1 + i(0)]$$

$$= 1$$

(b) In Example 1(b), we used De Moivre's theorem to show that

$$(-1 + i)^{12} = \left[\sqrt{2} \left(\cos \frac{3\pi}{4} + i \sin \frac{3\pi}{4} \right) \right]^{12} = -64.$$

Since $z^{-n} = \frac{1}{z^n}$, provided $z \neq 0$, we have

$$(-1 + i)^{-12} = \frac{1}{(-1 + i)^{12}} = -\frac{1}{64}.$$

Also, by De Moivre's theorem, we have

$$(-1+i)^{-12} = \left[\sqrt{2}\left(\cos\frac{3\pi}{4} + i\sin\frac{3\pi}{4}\right)\right]^{-12}$$

$$= (\sqrt{2})^{-12}\left[\cos(-12)\frac{3\pi}{4} + i\sin(-12)\frac{3\pi}{4}\right]$$

$$= \frac{1}{(\sqrt{2})^{12}}[\cos(-9\pi) + i\sin(-9\pi)]$$

$$= \frac{1}{64}[(-1) + i(0)] = -\frac{1}{64}.$$

PROBLEM 2 Repeat Example 2 for $(\sqrt{3} + i)^{-9}$.

◆ *n*th Root of a Complex Number

We define the *n*th root of a complex number as follows.

_n_th Root of _z_

> If z and w are complex numbers, n is a positive integer, and $w^n = z$, then w is an **_n_th root of _z_.**

Now, if w is an nth root of z, and

$$w = s(\cos\alpha + i\sin\alpha) \quad \text{and} \quad z = r(\cos\theta + i\sin\theta),$$

then by the definition of an nth root and De Moivre's theorem, we have

$$w^n = z$$
$$[s(\cos\alpha + i\sin\alpha)]^n = r(\cos\theta + i\sin\theta)$$
$$s^n(\cos n\alpha + i\sin n\alpha) = r(\cos\theta + i\sin\theta)$$

Since two equal complex numbers must have the same modulus, it follows that

$$s^n = r \quad \text{or} \quad s = r^{1/n}.$$

Furthermore, since the sine and cosine functions have a period 2π, the arguments of equal complex numbers must either be equal or differ by an integral multiple of 2π. Thus,

$$n\alpha = \theta + 2\pi k \quad \text{or} \quad \alpha = \frac{\theta + 2\pi k}{n},$$

where k is an integer. For $k = 0, 1, 2, \ldots, n - 1$, we obtain n distinct values of α, which in turn identify n distinct complex roots of z. For $k = n, n + 1, n + 2, \ldots$

SECTION 5.3 Powers and Roots of Complex Numbers

as well as $k = -1, -2, -3, \ldots$, we obtain repetitions of these n distinct complex roots. For instance, if $k = n$, then

$$\alpha = \frac{\theta + 2\pi k}{n} = \frac{\theta + 2\pi n}{n} = \frac{\theta}{n} + 2\pi,$$

which is coterminal with the angle we obtain when $k = 0$. In summary, we state the **nth root formula** for complex numbers.

*n*th Root Formula

If $z = r(\cos \theta + i \sin \theta)$, with $z \neq 0$, and n is a positive integer, then z has exactly n distinct nth roots $w_0, w_1, w_2, \ldots, w_{n-1}$ given by

$$w_k = r^{1/n} \left[\cos\left(\frac{\theta + 2\pi k}{n}\right) + i \sin\left(\frac{\theta + 2\pi k}{n}\right) \right]$$

for $k = 0, 1, 2, \ldots, n - 1$.

EXAMPLE 3 Find the indicated roots of the given complex number.

(a) Square roots of $2\left(\cos\frac{5\pi}{12} + i\sin\frac{5\pi}{12}\right)$.   (b) Cube roots of $-1 + i$.

SOLUTION

(a) By the nth root formula, $2\left(\cos\frac{5\pi}{12} + i\sin\frac{5\pi}{12}\right)$ has two distinct square roots, and they are given by

$$w_k = 2^{1/2} \left\{ \cos\left[\frac{(5\pi/12) + 2\pi k}{2}\right] + i \sin\left[\frac{(5\pi/12) + 2\pi k}{2}\right] \right\}$$

for $k = 0, 1$. Hence, the two square roots are

$$w_0 = 2^{1/2} \left\{ \cos\left[\frac{(5\pi/12) + 2\pi(0)}{2}\right] + i \sin\left[\frac{(5\pi/12) + 2\pi(0)}{2}\right] \right\}$$

$$= 2^{1/2}\left(\cos\frac{5\pi}{24} + i\sin\frac{5\pi}{24}\right) \approx 1.122 + 0.861i$$

and

$$w_1 = 2^{1/2} \left\{ \cos\left[\frac{(5\pi/12) + 2\pi(1)}{2}\right] + i \sin\left[\frac{(5\pi/12) + 2\pi(1)}{2}\right] \right\}$$

$$= 2^{1/2}\left(\cos\frac{29\pi}{24} + i\sin\frac{29\pi}{24}\right) \approx -1.122 - 0.861i.$$

(b) By the nth root formula, $-1 + i = \sqrt{2}\left(\cos\dfrac{3\pi}{4} + i\sin\dfrac{3\pi}{4}\right)$ has three distinct cube roots, and they are given by

$$w_k = (\sqrt{2})^{1/3}\left\{\cos\left[\dfrac{(3\pi/4) + 2\pi k}{3}\right] + i\sin\left[\dfrac{(3\pi/4) + 2\pi k}{3}\right]\right\}$$

for $k = 0, 1, 2$. Hence, the three cube roots are

$$w_0 = (\sqrt{2})^{1/3}\left\{\cos\left[\dfrac{(3\pi/4) + 2\pi(0)}{3}\right] + i\sin\left[\dfrac{(3\pi/4) + 2\pi(0)}{3}\right]\right\}$$

$$= 2^{1/6}\left(\cos\dfrac{\pi}{4} + i\sin\dfrac{\pi}{4}\right) \approx 0.794 + 0.794i,$$

$$w_1 = (\sqrt{2})^{1/3}\left\{\cos\left[\dfrac{(3\pi/4) + 2\pi(1)}{3}\right] + i\sin\left[\dfrac{(3\pi/4) + 2\pi(1)}{3}\right]\right\}$$

$$= 2^{1/6}\left(\cos\dfrac{11\pi}{12} + i\sin\dfrac{11\pi}{12}\right) \approx -1.084 + 0.291i,$$

$$w_2 = (\sqrt{2})^{1/3}\left\{\cos\left[\dfrac{(3\pi/4) + 2\pi(2)}{3}\right] + i\sin\left[\dfrac{(3\pi/4) + 2\pi(2)}{3}\right]\right\}$$

$$= 2^{1/6}\left(\cos\dfrac{19\pi}{12} + i\sin\dfrac{19\pi}{12}\right) \approx 0.291 - 1.084i. \qquad \blacklozenge$$

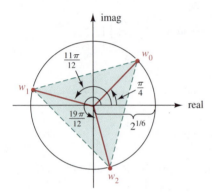

FIGURE 5.8
The cube roots of $-1 + i$ represent the vertices of an equilateral triangle in the complex plane.

Figure 5.8 shows the three cube roots of $-1 + i$ in the complex plane. Note that all roots are spaced equally from each other on a circle with center at the origin and radius $2^{1/6}$. Hence, we can think of w_0, w_1, and w_2 as the vertices of an equilateral triangle. In general, the nth roots of a complex number always represent the vertices of a *regular n-sided polygon* in the complex plane whenever $n \geq 3$. It is good practice to use this fact to check the solutions whenever we find the roots of a complex number.

PROBLEM 3 Find the fourth roots of $\sqrt{3} + i$. Show that these roots represent the vertices of a square in the complex plane. $\qquad \blacklozenge$

◆ **Applying the *n*th Root Formula to Polynomial Equations**

In more advanced algebra courses, it is shown that every polynomial function P of degree $n \geq 1$ has exactly n zeros. Consequently, every nth-degree polynomial equation has n solutions in the complex number system. For example, consider the polynomial equation

$$x^4 + 64 = 0.$$

Since this is a fourth-degree equation, it must have four solutions in the complex number system. One method of finding these four solutions is to use factoring and the zero product property:

$$x^4 + 64 = 0$$

$$(x^4 + 16x^2 + 64) - 16x^2 = 0 \quad \text{Add and subtract } 16x^2 \text{ and group terms as shown}$$

$$(x^2 + 8)^2 - 16x^2 = 0 \quad \text{Factor the perfect square trinomial}$$

$$[(x^2 + 8) - 4x][(x^2 + 8) + 4x] = 0 \quad \text{Factor the difference of squares}$$

$$(x^2 - 4x + 8)(x^2 + 4x + 8) = 0 \quad \text{Rewrite}$$

$$x^2 - 4x + 8 = 0 \quad \text{or} \quad x^2 + 4x + 8 = 0 \quad \text{Apply the zero product property}$$

$$x = 2 \pm 2i \quad \text{or} \quad x = -2 \pm 2i \quad \text{Apply the quadratic formula}$$

Hence, the four solutions of the equation $x^4 + 64 = 0$ are $2 + 2i$, $2 - 2i$, $-2 + 2i$, and $-2 - 2i$. As illustrated in the next example, we may also use the *n*th root formula to find the solutions of the equation $x^4 + 64 = 0$.

EXAMPLE 4 Use the *n*th root formula to find the solutions of the equation $x^4 + 64 = 0$.

SOLUTION Subtracting 64 from both sides of the equation, we obtain

$$x^4 = -64.$$

The solutions of this equation are the four fourth roots of -64. By the *n*th root formula, the four fourth roots of

$$-64 = -64 + 0i = 64 (\cos \pi + i \sin \pi)$$

are given by

$$w_k = 64^{1/4} \left[\cos \left(\frac{\pi + 2\pi k}{4} \right) + i \sin \left(\frac{\pi + 2\pi k}{4} \right) \right]$$

for $k = 0, 1, 2, 3$. Hence, the four fourth roots are

$$w_0 = 64^{1/4} \left\{ \cos \left[\frac{\pi + 2\pi(0)}{4} \right] + i \sin \left[\frac{\pi + 2\pi(0)}{4} \right] \right\}$$

$$= 2\sqrt{2} \left(\cos \frac{\pi}{4} + i \sin \frac{\pi}{4} \right) = 2 + 2i,$$

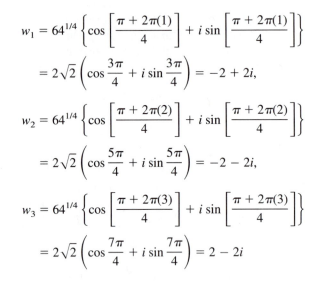

$$w_1 = 64^{1/4}\left\{\cos\left[\frac{\pi + 2\pi(1)}{4}\right] + i\sin\left[\frac{\pi + 2\pi(1)}{4}\right]\right\}$$

$$= 2\sqrt{2}\left(\cos\frac{3\pi}{4} + i\sin\frac{3\pi}{4}\right) = -2 + 2i,$$

$$w_2 = 64^{1/4}\left\{\cos\left[\frac{\pi + 2\pi(2)}{4}\right] + i\sin\left[\frac{\pi + 2\pi(2)}{4}\right]\right\}$$

$$= 2\sqrt{2}\left(\cos\frac{5\pi}{4} + i\sin\frac{5\pi}{4}\right) = -2 - 2i,$$

$$w_3 = 64^{1/4}\left\{\cos\left[\frac{\pi + 2\pi(3)}{4}\right] + i\sin\left[\frac{\pi + 2\pi(3)}{4}\right]\right\}$$

$$= 2\sqrt{2}\left(\cos\frac{7\pi}{4} + i\sin\frac{7\pi}{4}\right) = 2 - 2i$$

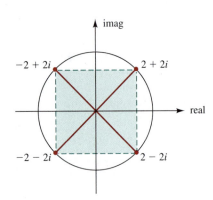

FIGURE 5.9
The solutions of $x^4 + 64 = 0$ represent the vertices of a square in the complex plane.

Thus, the four solutions of the equation $x^4 + 64 = 0$ are $2 \pm 2i$ and $-2 \pm 2i$. As shown in Figure 5.9, these roots form the vertices of a square in the complex plane. ◆

PROBLEM 4 Solve the equation $x^3 + 1 = 0$ by using the nth root formula. ◆

Exercises 5.3

Basic Skills

In Exercises 1–20, find the indicated power of each complex number. Record the results (a) in trigonometric form with $0 \leq \theta < 2\pi$ and (b) in standard form.

1. $\left[4\left(\cos\frac{\pi}{6} + i\sin\frac{\pi}{6}\right)\right]^3$

2. $\left[\sqrt{3}\left(\cos\frac{\pi}{4} + i\sin\frac{\pi}{4}\right)\right]^4$

3. $\left[\sqrt{2}\left(\cos\frac{15\pi}{8} + i\sin\frac{15\pi}{8}\right)\right]^6$

4. $\left[\frac{1}{2}\left(\cos\frac{11\pi}{12} + i\sin\frac{11\pi}{12}\right)\right]^3$

5. $[4\sqrt{5}(\cos 2.137 + i\sin 2.137)]^0$

6. $[9(\cos 240° + i\sin 240°)]^2$

7. $\left[\frac{2}{3}\left(\cos\frac{11\pi}{9} + i\sin\frac{11\pi}{9}\right)\right]^{-3}$

8. $\left(\cos\frac{5\pi}{3} + i\sin\frac{5\pi}{3}\right)^{-5}$

9. $(\cos 210° + i\sin 210°)^{-7}$

10. $\left[5\left(\cos\frac{17\pi}{18} + i\sin\frac{17\pi}{18}\right)\right]^{-3}$

11. $(1 + i)^8$

12. $(-2 + 2i)^4$

13. $(-1 - \sqrt{3}\,i)^5$

14. $(\sqrt{3} - i)^{10}$

15. $(0 - \sqrt{2}\,i)^{14}$

16. $(0 + 3i)^6$

17. $(3 - 3i)^{-2}$

18. $(1 - \sqrt{3}\,i)^{-4}$

19. $(-2\sqrt{3} - 2i)^{-3}$

20. $(9 - 3i)^0$

In Exercises 21–34, find the indicated roots of each complex number. Record the results (a) in trigonometric form with $0 \leq \theta < 2\pi$ and (b) in standard form.

21. Square roots of $49\left(\cos\frac{5\pi}{3} + i\sin\frac{5\pi}{3}\right)$

22. Square roots of $18\left(\cos\frac{2\pi}{3} + i\sin\frac{2\pi}{3}\right)$

23. Cube roots of $8\left(\cos\frac{\pi}{2} + i\sin\frac{\pi}{2}\right)$

24. Fourth roots of $16\left(\cos\dfrac{4\pi}{3} + i\sin\dfrac{4\pi}{3}\right)$

25. Fourth roots of 25 (cos 240° + i sin 240°)

26. Cube roots of 64 (cos 180° + i sin 180°)

27. Square roots of i

28. Square roots of -4

29. Cube roots of 1

30. Cube roots of $-8i$

31. Fourth roots of -16

32. Sixth roots of 1

33. Square roots of $-4 + 4\sqrt{3}\,i$

34. Square roots of $9 - 9\sqrt{3}\,i$

In Exercises 35–38, plot the indicated solutions in the complex plane. Describe the regular polygon that has these points as its vertices.

35. Solutions to Exercise 23
36. Solutions to Exercise 24
37. Solutions to Exercise 31
38. Solutions to Exercise 32

In Exercises 39–46, use the nth root formula to find all the solutions of the given equation. Write each answer in standard form, a + bi.

39. $x^3 + 8 = 0$
40. $x^3 - 1 = 0$
41. $x^4 - 81 = 0$
42. $x^6 + 729 = 0$
43. $x^3 - i = 0$
44. $x^3 + 64i = 0$
45. $x^2 - 1 = \sqrt{3}\,i$
46. $x^4 + 1 = \sqrt{3}\,i$

Critical Thinking

47. What is the relationship between the solutions of the equation $x^8 = 1$ and a regular octagon? Explain.

48. What is the relationship between the zeros of the function $f(x) = x^{10} + 1$ and a regular decagon? Explain.

49. One of the cube roots of a complex number is $5\left(\cos\dfrac{3\pi}{8} + i\sin\dfrac{3\pi}{8}\right)$. Determine the other two cube roots. [*Hint:* Consider the graphical representation of the roots in the complex plane.]

50. One of the sixth roots of a complex number is $2\sqrt{3} + 2i$. Determine the other five sixth roots. [*Hint:* Consider the graphical representation of the roots in the complex plane.]

51. The following two-step procedure, called *proof by mathematical induction*, is used to show that a mathematical statement is true for all positive integers n:
    Step 1. Show that the statement is true for $n = 1$.
    Step 2. Assume that the statement is true for an arbitrary positive integer k, and show, by using this assumption, that the statement is true when $n = k + 1$.

    Use proof by mathematical induction to prove DeMoivre's theorem.

52. Many objects in nature have a spiral pattern similar to the seashell shown in the sketch. Show that the nonnegative integer powers of $z = 1 + i$, when plotted in the same complex plane, form a spiral pattern similar to that of the seashell.

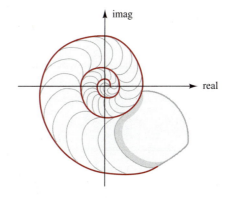

Calculator Activities

In Exercises 53–60, find the indicated power of each complex number. Record the answer in standard form, a + bi, with a and b rounded to three significant digits.

53. $[1.2\,(\cos 5 + i\sin 5)]^5$
54. $[3.3\,(\cos 70° + i\sin 70°)]^6$

318 CHAPTER 5 Complex Numbers

55. $[2(\cos 1 + i \sin 1)]^{-3}$

56. $[2.3(\cos 1.24 + i \sin 1.24)]^{-4}$

57. $(1 - 2i)^3$

58. $(1.56 + 2.23i)^5$

59. $(-12.2 - 15.7i)^{-4}$

60. $(-14 + 8i)^{-3}$

In Exercises 61–68, find the indicated roots of each complex number. Record each answer in standard form, $a + bi$, with a and b rounded to three significant digits.

61. Cube roots of $18\left(\cos \dfrac{3\pi}{4} + i \sin \dfrac{3\pi}{4}\right)$

62. Fourth roots of $54(\cos 200° + i \sin 200°)$

63. Square roots of $5 - 12i$

64. Square roots of $-15 - 8i$

65. Sixth roots of i

66. Fifth roots of -32

67. Fourth roots of $6.2 + 3.4i$

68. Cube roots of $-12.4 + 17.3i$

Chapter 5 Review

Questions for Group Discussion

1. What is an *imaginary number*? How does it differ from a *pure imaginary number*?

2. What is a *complex number*? What are the two major subsets of the set of complex numbers? List some other subsets of the set of complex numbers.

3. State the procedures for adding and subtracting complex numbers? Illustrate with an example.

4. Is the product of two pure imaginary numbers always a real number? Explain.

5. What type of number is the product of *complex conjugates*? How is a complex conjugate used in finding the quotient of two complex numbers? Illustrate with an example.

6. What is the reciprocal of *i*? Does every complex number have a reciprocal? Explain.

7. Explain the pattern that develops when *i* is raised to successive positive integer powers?

8. In the *complex plane*, what are the horizontal and vertical axes called?

9. Explain the procedure for finding the *modulus* and *argument* of a complex number in standard form.

10. What is the *trigonometric form* of a complex number? Explain the procedure for multiplying and dividing complex numbers in trigonometric form.

11. State *DeMoivre's Theorem* and illustrate its use with an example.

12. Discuss the geometrical significance of the *n*th roots of a complex number when plotted in the complex plane.

Review Exercises

In Exercises 1–8, simplify each expression and write the answer in the form b or bi, where b is a real number.

1. $4\sqrt{-45} + 3\sqrt{-80} - \sqrt{-20}$

2. $\sqrt{-24} \cdot \sqrt{6} \cdot \sqrt{-16}$

3. $\sqrt{-6}(\sqrt{-54} - \sqrt{-24})$

4. $\dfrac{10}{-5(\sqrt{-2})^3}$

5. $(2i)^6$

6. $5i^{-33}$

7. $\dfrac{1}{3i^{17}}$ 8. $i - i^2 + i^3 - i^4$

In Exercises 9 – 16, perform the indicated operations and express the answer in standard form $a + bi$.

9. $(9 + i) + (3 - 2i)$
10. $(4 - 3\sqrt{-9}) - (9 + 3\sqrt{-25})$
11. $(3 - \sqrt{-25})(4 + \sqrt{-49})$
12. $(2 - 3i)^2$
13. $\dfrac{3 + 4i}{3 - 4i}$
14. $\dfrac{\sqrt{-9}}{5 - \sqrt{-1}}$
15. $\dfrac{1}{(1 - \sqrt{-25})^2}$
16. $\dfrac{(1 - i)^2}{3 + 4i}$

In Exercises 17–24,
(a) plot each complex number in the complex plane.
(b) find the modulus of each complex number.
(c) find the argument of each complex number.
(d) express each complex number in trigonometric form with $0 \le \theta < 2\pi$.

17. $6 - 6i$
18. $-\sqrt{2} + \sqrt{2}\,i$
19. $-5\sqrt{3} - 5i$
20. $\tfrac{5}{4} + \tfrac{5}{4}\sqrt{3}\,i$
21. $1 + 2i$
22. $-3 - 4i$
23. -2
24. $\sqrt{-9}$

In Exercises 25–34, perform the indicated operations and record the results in a) trigonometric form with $0 \le \theta < 2\pi$ and b) standard form.

25. $\left[3\left(\cos\dfrac{\pi}{3} + i\sin\dfrac{\pi}{3}\right)\right]\left[6\left(\cos\dfrac{\pi}{4} + i\sin\dfrac{\pi}{4}\right)\right]$

26. $\left[5\left(\cos\dfrac{3\pi}{4} + i\sin\dfrac{3\pi}{4}\right)\right]\left[12\left(\cos\dfrac{17\pi}{12} + i\sin\dfrac{17\pi}{12}\right)\right]$

27. $\dfrac{\cos 81° + i\sin 81°}{\cos 141° + i\sin 141°}$

28. $\dfrac{12(\cos 4\pi/3 + i\sin 4\pi/3)}{4 + 4i}$

29. $\left[2\left(\cos\dfrac{17\pi}{12} + i\sin\dfrac{17\pi}{12}\right)\right]^8$

30. $\left[\dfrac{1}{3}(\cos 36° + i\sin 36°)\right]^{-5}$

31. $(1 + i)^{-10}$
32. $(\sqrt{2} - i\sqrt{2})(\sqrt{3} + i)^6$

33. $\dfrac{[\sqrt{2}\,(\cos \pi/2 + i\sin \pi/2)]^4}{(-1 - i)^5}$

34. $\dfrac{(8 + 6i)^0}{(1 - \sqrt{3}\,i)^3(1 + i)^{-4}}$

In Exercises 35–40, find the indicated roots of each complex number. Record the results in a) trigonometric form with $0 \le \theta < 2\pi$ and b) standard form $a + bi$.

35. Square roots of $\dfrac{1}{4}\left(\cos\dfrac{2\pi}{3} + i\sin\dfrac{2\pi}{3}\right)$

36. Cube roots of $27\left(\cos\dfrac{3\pi}{4} + i\sin\dfrac{3\pi}{4}\right)$

37. Fourth roots of 1   38. Fifth roots of i
39. Cube roots of $4\sqrt{2} - 4\sqrt{2}\,i$
40. Fourth roots of $-2 - 2\sqrt{3}\,i$

41. One of the fourth roots of a complex number is $\sqrt{3} - i$. Determine the other three fourth roots.

42. Find *all* the solutions of the given equations.
    (a) $x^3 - 27 = 0$
    (b) $x^5 + 1 = -i$

In Exercises 43 and 44, use the quadratic formula to solve each quadratic equation, writing the answers in trigonometric form with $0 \le \theta < 2\pi$.

43. $x^2 - 2x + 2 = 0$ 44. $x^2 - 6x + 13 = 0$

45. Suppose $r(\cos\theta + i\sin\theta) = a + bi$. Express each of the following complex numbers in standard form.
    (a) $r[\cos(\theta + 2\pi) + i\sin(\theta + 2\pi)]$
    (b) $r[\cos(\theta + \pi) + i\sin(\theta + \pi)]$
    (c) $r[\cos(-\theta) + i\sin(-\theta)]$
    (d) $r^2[\cos(2\theta) + i\sin(2\theta)]$

46. The notations $\bar{z}$ and $\bar{w}$ are often used to represent the complex conjugates of the complex numbers z and w, respectively. Using $z = a + bi$ and $w = c + di$ show that each of the following statements is true.
    (a) $\bar{z} + \bar{w} = \overline{z + w}$, where $\overline{z + w}$ is the complex conjugate of the sum $z + w$.
    (b) $\bar{z}\,\bar{w} = \overline{zw}$, where $\overline{zw}$ is the complex conjugate of the product zw.

For the AC circuits in Exercises 47 and 48, find

(a) *the magnitude of the impedance.*

(b) *the phase angle (in degrees) between the voltage and current.*

47.

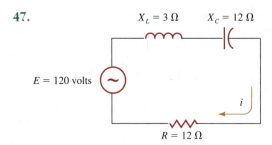

48.

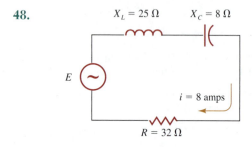

49. The current I in amperes (amps) flowing through the AC circuit shown in Exercise 47 may be found by dividing the voltage E by the impedance Z. Find the current in this circuit.

50. The voltage E in volts (v) flowing through the AC circuit shown in Exercise 48 may be found by multiplying the current I by the impedance Z. Find the voltage in this circuit.

CHAPTER 6

Conic Sections

Earth is closest to the sun (at its perihelion) in January, when the distance between them is approximately 9.14×10^7 miles, and is furthest from the sun (aphelion) in July, when the distance between them is approximately 9.44×10^7 miles.

(a) Determine the polar equation of the elliptical orbit of the earth if the sun (focus) is at the pole, and its corresponding directrix is perpendicular to the polar axis and to the right of the pole.

(b) State the eccentricity of the orbit.

(For the solution, see Example 8 in Section 6.5.)

- 6.1 **The Parabola**
- 6.2 **The Ellipse**
- 6.3 **The Hyperbola and the Common Definition of the Conic Sections**
- 6.4 **Rotation of Axes**
- 6.5 **Polar Equations**
- 6.6 **Parametric Equations**

6.1 The Parabola

◆ **Introductory Comments**

Early Greek mathematicians noted that when a double cone is sliced with a plane, a special family of curves is formed. As illustrated in Figure 6.1, this family of curves has four main members: *circle, ellipse, parabola,* and *hyperbola.* Collectively, this family of curves is called the **conic sections.**

FIGURE 6.1
The conic sections

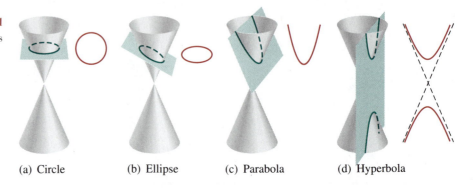

(a) Circle   (b) Ellipse   (c) Parabola   (d) Hyperbola

When a plane intersects a double cone, it is also possible to obtain a single point, one line, or a pair of intersecting lines. Do you see how? These special cases are called the **degenerate conic sections.**

An equation of the form

$$Ax^2 + Bxy + Cy^2 + Dx + Ey + F = 0$$

where A, B, C, D, E, and F are real numbers, is called a **general quadratic equation in two unknowns.** As we will see, if the graph of this type of equation exists, it is either a conic section or a degenerate conic section.

In the first three sections of this chapter, we study the geometric properties of each conic section and the distinguishing characteristics of their equations. In this section we discuss the parabola, in Section 6.2 the ellipse (and the circle, which is a special form of an ellipse), and in Section 6.3 the hyperbola. In these sections, we limit our discussion to general quadratic equations in two unknowns in which $B = 0$. By choosing $B = 0$, we keep the axes of symmetry of the conic section parallel to the xy-axes. In Section 6.4, we consider general quadratic equations in two unknowns in which $B \neq 0$. As we will see, the presence of an xy-term rotates the conic section so that its axes are not parallel to the xy-axes. The last two sections of this chapter discuss the polar and parametric equations of the conic sections.

Equations of Parabolas

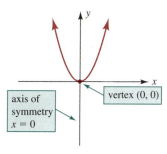

FIGURE 6.2
The graph of $y = x^2$ is a parabola with vertex $(0, 0)$ and axis of symmetry $x = 0$.

Recall from Section 1.3 that the graph of the equation $y = x^2$ is a cupped-shaped curve that is symmetric with respect to the y-axis. We refer to this curve as a **parabola** and the *vertical line $x = 0$* (the y-axis) as its **axis of symmetry.** The point at which the parabola intersects its axis of symmetry is called the **vertex** of the parabola (see Figure 6.2).

By the vertical stretch and shrink rules (Section 1.4), we know that the graph of $y = ax^2$ is also a parabola, but with a *wider* opening than $y = x^2$ if $0 < a < 1$, or a *narrower* opening if $a > 1$, as shown in Figure 6.3. By the x-axis reflection rule (Section 1.4) the parabola opens *upward* if $a > 0$, or *downward* if $a < 0$ (see Figure 6.4). In summary, the value of a in the equation $y = ax^2$ determines the width and direction of the opening of the parabola.

Interchanging x and y in the equation $y = ax^2$ gives us $x = ay^2$, whose graph is also a parabola with its vertex at the origin. However, its axis of symmetry is the *horizontal line $y = 0$* (the x-axis) instead of the vertical line $x = 0$. If $a > 0$ in the equation $x = ay^2$, the parabola opens *to the right* and if $a < 0$, the parabola opens *to the left,* as shown in Figure 6.5.

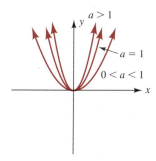

FIGURE 6.3
Graph of $y = ax^2$ for $0 < a < 1$, $a = 1$, and $a > 1$.

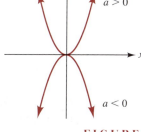

FIGURE 6.4
Graph of $y = ax^2$ for $a > 0$ and $a < 0$.

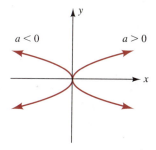

FIGURE 6.5
Graph of $x = ay^2$ for $a > 0$ and $a < 0$.

If h and k are real numbers and we replace x with $(x - h)$ and y with $(y - k)$ in the equations $y = ax^2$ and $x = ay^2$, we obtain

$$y - k = a(x - h)^2 \quad \text{and} \quad x - h = a(y - k)^2$$
$$y = a(x - h)^2 + k \quad\quad\quad\quad x = a(y - k)^2 + h,$$

respectively. By the vertical and horizontal shift rules (Section 1.4), we can state that the graph of $y = a(x - h)^2 + k$ is the same as the graph of $y = ax^2$ but with the vertex shifted to the point (h, k) and axis of symmetry the line $x = h$ (see Figure 6.6). The graph of $x = a(y - k)^2 + h$ is also a parabola with vertex (h, k) but with axis of symmetry $y = k$ (see Figure 6.7). We refer to $y = a(x - h)^2 + k$ and $x = a(y - k)^2 + h$ as the *algebraic definitions of a parabola* or the **equations of a parabola in standard form.**

324 CHAPTER 6 Conic Sections

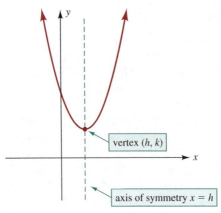

FIGURE 6.6
Graph of $y = a(x - h)^2 + k$ with $a > 0$, $h > 0$, and $k > 0$.

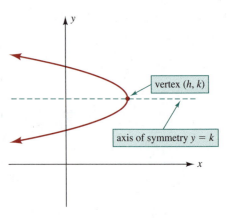

FIGURE 6.7
Graph of $x = a(y - k)^2 + h$ with $a < 0$, $h > 0$, and $k > 0$.

Equations of a Parabola in Standard Form

> The **equation of a parabola in standard form** with vertex (h, k) is
> $$y = a(x - h)^2 + k$$
> if the axis of symmetry of the parabola is *vertical*, or
> $$x = a(y - k)^2 + h$$
> if the axis of symmetry of the parabola is *horizontal*.
>
> In these equations, a is a nonzero constant that determines the width and direction of the opening of the parabola.

We can determine the equation of a parabola if we know the vertex of the parabola, the axis of symmetry, and the coordinates of just one other point on the parabola.

EXAMPLE 1 Determine the equation of a parabola in standard form with vertex $(2, 3)$ and vertical axis of symmetry if the parabola passes through the point $(4, 1)$.

SOLUTION The parabola with the given characteristics is shown in Figure 6.8. Since the vertex is $(2, 3)$ and the axis of symmetry is the *vertical line* $x = 2$, the equation of this parabola must have the form

$$y = a(x - 2)^2 + 3.$$

Since the parabola *opens downward*, we know that $a < 0$. To determine the value of a, we use the fact that the parabola passes through the point $(4, 1)$.

FIGURE 6.8
A parabola passing through $(4, 1)$ with vertex $(2, 3)$ and vertical axis of symmetry.

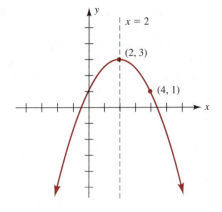

$$y = a(x - 2)^2 + 3$$
$$1 = a(4 - 2)^2 + 3 \quad \text{Replace } x \text{ with 4 and } y \text{ with 1}$$
$$1 = 4a + 3 \quad \text{Solve for } a$$
$$-2 = 4a$$
$$a = -\frac{1}{2}.$$

Thus, the required equation is $y = -\frac{1}{2}(x - 2)^2 + 3$. ◆

PROBLEM 1 Rework Example 1 if the axis of symmetry is horizontal instead of vertical. ◆

If we expand $(x - h)^2$ in the equation $y = a(x - h)^2 + k$, distribute the a, and collect all terms on one side of the equation, we obtain a general quadratic equation in two unknowns of the form

$$Ax^2 + Dx + Ey + F = 0,$$

where A, D, E, and F are real numbers with $A \neq 0$ and $E \neq 0$. Similarly, if we expand $(y - k)^2$ in the equation $x = a(y - k)^2 + h$, distribute the a, and collect all terms on one side of the equation, we obtain a general quadratic equation in two unknowns of the form

$$Cy^2 + Dx + Ey + F = 0,$$

where C, D, E, and F are real numbers with $C \neq 0$ and $D \neq 0$. We refer to

$$\boxed{Ax^2 + Dx + Ey + F = 0}$$

and

$$\boxed{Cy^2 + Dx + Ey + F = 0}$$

as the **equations of a parabola in general form** with vertical and horizontal axes of symmetry, respectively. These general quadratic equations in two unknowns are characterized by the presence of either an x^2 or y^2 term, but never both.

If the equation of a parabola is given in general form, we can sketch its graph by converting the equation to standard form and observing the coordinates of the vertex (h, k). To accomplish this task, it may be necessary to use the algebraic process of *completing the square*.

EXAMPLE 2 Sketch the graph of the equation $y^2 - 2x - 4y + 10 = 0$.

SOLUTION The y^2 term indicates that this is the equation of a parabola in general form with a horizontal axis of symmetry. To sketch its graph, we solve for x and then complete the square as follows:

$y^2 - 2x - 4y + 10 = 0$

$2x = y^2 - 4y + 10$ **Add $2x$ to both sides**

$x = \frac{1}{2}y^2 - 2y + 5$ **Divide both sides by 2**

$x = \left(\frac{1}{2}y^2 - 2y\ \ \ \right) + 5$ **Regroup**

$x = \frac{1}{2}(y^2 - 4y\ \ \ \ ) + 5$ **Factor out $\frac{1}{2}$**

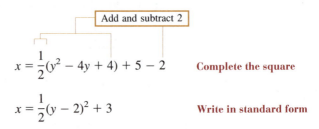

$x = \frac{1}{2}(y^2 - 4y + 4) + 5 - 2$ **Complete the square**

$x = \frac{1}{2}(y - 2)^2 + 3$ **Write in standard form**

Thus the *vertex* has coordinates $(3, 2)$ and, since $a = \frac{1}{2}$ ($a > 0$), the parabola opens to the right from the point $(3, 2)$. Hence, the graph has no y-intercept. We find the x-intercept by letting $y = 0$ in the equation $x = \frac{1}{2}y^2 - 2y + 5$:

$x = \frac{1}{2}(0)^2 - 2(0) + 5 = 5$ **x-intercept**

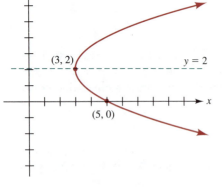

FIGURE 6.9
The graph of $y^2 - 2x - 4y + 10 = 0$ is a parabola that opens to the right with vertex $(3, 2)$.

The graph of the equation $y^2 - 2x - 4y + 10 = 0$ is shown in Figure 6.9. We draw the bottom half of the parabola by connecting the vertex $(3, 2)$ and x-intercept $(5, 0)$. We then use the fact that the parabola is symmetric with respect to the line $y = 2$ to draw the upper half of the curve. ◆

PROBLEM 2 Repeat Example 2 for the equation $x^2 - 2y - 4x + 10 = 0$. ◆

Geometric Definition of a Parabola

Geometrically, we define a parabola as follows.

Geometric Definition of a Parabola

> A **parabola** is the set of all points in a plane that are equidistant from a fixed line and a fixed point not on the line. The fixed point is called the **focus** of the parabola and the fixed line is called the **directrix** of the parabola.

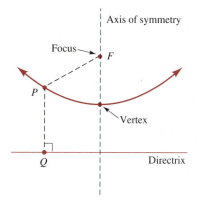

FIGURE 6.10
A parabola is the set of all points P such that $PQ = PF$.

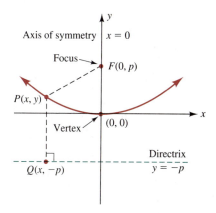

FIGURE 6.11
The vertex of a parabola is the halfway point between the focus and directrix.

We now show that our geometric definition of a parabola is consistent with our algebraic definition. Applying our geometric definition, a parabola is the set of all points P such that

$$PQ = PF,$$

as shown in Figure 6.10. For convenience, let's place the parabola that is shown in Figure 6.10 in the xy-coordinate system so that it opens upward with vertex at the origin and axis of symmetry $x = 0$ (see Figure 6.11). The vertex, like every other point on the parabola, is equidistant from the focus and the directrix. Hence, if the focus F has coordinates $(0, p)$, then the equation of the directrix must be $y = -p$. Referring to Figure 6.11, we have

$$PQ = PF$$
$$\sqrt{(x-x)^2 + (y+p)^2} = \sqrt{(x-0)^2 + (y-p)^2} \quad \text{Apply distance formula}$$
$$(x-x)^2 + (y+p)^2 = (x-0)^2 + (y-p)^2 \quad \text{Square both sides}$$
$$y^2 + 2py + p^2 = x^2 + y^2 - 2py + p^2 \quad \text{Expand both sides}$$
$$4py = x^2 \quad \text{Solve for } y$$
$$y = \frac{1}{4p}x^2$$

From our algebraic definition of a parabola, we recognize

$$y = \frac{1}{4p}x^2 = a(x-0)^2 + 0$$

as the equation of a parabola in standard form with vertex $(0, 0)$ and vertical axis of symmetry $x = 0$. In this equation, $a = 1/(4p)$ is the nonzero constant that determines the width and direction of the opening of the parabola. Hence, we may rewrite the equation of a parabola in standard form with a vertical axis of symmetry as

$$y = \frac{1}{4p}(x-h)^2 + k$$

where (h, k) is the vertex and $|p|$ is the distance from the focus to the vertex. If $p > 0$, the focus is *above* the vertex and the parabola opens upward, and, if $p < 0$, the focus is *below* the vertex and the parabola opens downward.

In a similar manner, we can show that if a parabola is placed in the xy-coordinate system so that it opens to the right with vertex at the origin, then the geometric definition of a parabola yields

$$x = \frac{1}{4p}y^2$$

Hence, we may rewrite the equation of a parabola in standard form with horizontal axis of symmetry as

$$x = \frac{1}{4p}(y-k)^2 + h.$$

where (h, k) is the vertex and $|p|$ is the distance from the focus to the vertex. If $p > 0$, the focus is to the *right* of the vertex and the parabola opens to the right, and, if $p < 0$, the focus is to the *left* of the vertex and the parabola opens left.

EXAMPLE 3 Find the focus and directrix of each parabola.

(a) $y^2 + 12x = 0$ (b) $x^2 + 3x - 6y = 0$

SOLUTION

(a) Solving for x, we obtain

$$x = -\frac{1}{12}y^2 = -\frac{1}{12}(y-0)^2 + 0,$$

which is the equation of a parabola in standard form with horizontal axis of symmetry, vertex at $(0, 0)$, and

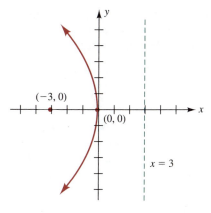

FIGURE 6.12
The focus of the parabola $y^2 + 12x = 0$ is $(-3, 0)$ and its directrix is $x = 3$.

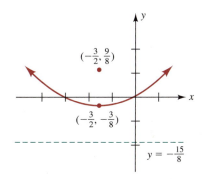

FIGURE 6.13
The focus of the parabola $x^2 + 3x - 6y = 0$ is $\left(-\frac{3}{2}, \frac{9}{8}\right)$ and the directrix is $y = -\frac{15}{8}$.

$$\frac{1}{4p} = -\frac{1}{12}$$

$$-4p = 12$$

$$p = -3$$

Since $p < 0$, the focus is 3 units left of the vertex, and the directrix is 3 units right of the vertex. Hence, the coordinates of the focus are $(0, -3)$ and the directrix is the vertical line $x = 3$, as shown in Figure 6.12.

(b) First, we write the equation in standard form by completing the square as follows:

$$x^2 + 3x - 6y = 0$$

$$y = \frac{1}{6}(x^2 + 3x \qquad) \qquad \text{Solve for } y$$

Add and subtract $\frac{3}{8}$

$$y = \frac{1}{6}\left(x^2 + 3x + \frac{9}{4}\right) - \frac{3}{8} \qquad \text{Complete the square}$$

$$y = \frac{1}{6}\left(x + \frac{3}{2}\right)^2 - \frac{3}{8} \qquad \text{Factor}$$

We recognize this last equation as the equation of a parabola in standard form with vertical axis of symmetry, vertex at $\left(-\frac{3}{2}, -\frac{3}{8}\right)$, and

$$\frac{1}{4p} = \frac{1}{6}$$

$$4p = 6$$

$$p = \frac{3}{2}$$

Since $p > 0$, the focus is $\frac{3}{2}$ units *above* the vertex, and the directrix is $\frac{3}{2}$ units *below* the vertex. Hence, the coordinates of the focus are $\left(-\frac{3}{2}, \frac{9}{8}\right)$, and the directrix is the horizontal line $y = -\frac{15}{8}$, as shown in Figure 6.13. ◆

PROBLEM 3 Find the focus and directrix for the parabola $3y^2 - 8x = 0$. ◆

As illustrated in the next example, if we know the focus and directrix of a parabola, we can determine its equation.

EXAMPLE 4 Find the equation of a parabola in standard form with focus $F(-5, 1)$ and directrix $x = 2$.

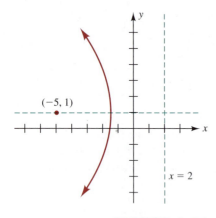

FIGURE 6.14
A parabola with a vertical directrix must have a horizontal axis of symmetry.

SOLUTION Since the directrix is the vertical line $x = 2$, the parabola must have a *horizontal* axis of symmetry that passes through the focus $F(-5, 1)$, as shown in Figure 6.14. Thus, the equation of this parabola has the form

$$x = \frac{1}{4p}(y - k)^2 + h$$

where (h, k) is the vertex and $|p|$ is the distance from the focus to the vertex. The vertex of a parabola is always located halfway between its focus and directrix. Hence, by the midpoint formula (Section 1.2), the coordinates of the vertex are

$$\left(\frac{-5 + 2}{2}, 1\right) = \left(-\frac{3}{2}, 1\right).$$

The distance from the focus to the vertex is

$$|p| = \left|-5 - \left(-\frac{3}{2}\right)\right| = \frac{7}{2}.$$

Since the focus is to the left of the vertex, the parabola opens to the left. Hence, we choose $p = -\frac{7}{2}$. Therefore, the equation of this parabola is

$$x = \frac{1}{4p}(y - k)^2 + h$$

$$x = \frac{1}{4(-\frac{7}{2})}(y - 1)^2 + \left(-\frac{3}{2}\right)$$

$$x = -\frac{1}{14}(y - 1)^2 - \frac{3}{2}$$

◆

PROBLEM 4 Referring to Example 4, express the equation of the parabola in general form with integer coefficients. ◆

◆ **Application-Design of a Flashlight**

A law of physics states that the angle of incidence equals the angle of reflection; that is, the angle at which a light or sound wave hits a surface is equal to the angle at which it reflects from the surface, as shown in Figure 6.15. When this physical law is applied to a parabolic surface, an interesting phenomenon occurs: all light or sound waves that move parallel to the axis of the parabola, after bouncing off the parabola, pass through the focus of the parabola (see Figure 6.16). We refer to this phenomenon as the **reflection property of a parabola**.

Engineers and scientist use the reflection property of a parabola to design telescopes, radar antennas, satellite dishes, solar heating devices, microphones, and other communications equipment. The reflection property of a parabola is also used in the design of searchlights and flashlights. The light rays emanating from a bulb located at the focus of a parabolic surface bounce off the surface and shine paral-

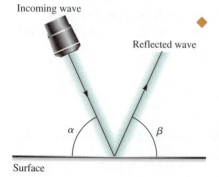

FIGURE 6.15
The angle of incidence α equals the angle of reflection β.

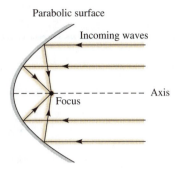

FIGURE 6.16
Light or sound waves entering parallel to the axis reflect to the focus

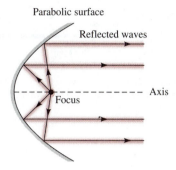

FIGURE 6.17
Light or sound waves emanating from the focus reflect parallel to the axis.

lel to the axis of the parabola, thus keeping the light beam at maximum intensity (see Figure 6.17).

EXAMPLE 5 The open end of a parabolic reflector in a flashlight has a diameter of 3 inches and a depth of 1 inch. How far from the vertex of the parabola should the filament of the bulb be located so that all the light rays emanating from the bulb are parallel to the axis of the parabola?

SOLUTION Let's place the parabolic reflector in the *xy*-coordinate system so that its vertex is at the origin and it opens to the right, as shown in Figure 6.18. Now, if the diameter of the opening of the parabolic reflector is 3 inches and the depth is 1 inch, then the coordinates of point Q in Figure 6.18 are $(1, \frac{3}{2})$. Hence,

$$x = \frac{1}{4p}y^2$$

$$1 = \frac{1}{4p}\left(\frac{3}{2}\right)^2 \quad \text{Substitute } x = 1 \text{ and } y = \frac{3}{2}$$

$$\frac{4}{9} = \frac{1}{4p} \quad \text{Multiply both sides by } \frac{4}{9}$$

$$p = \frac{9}{16} \quad \text{Solve for } p$$

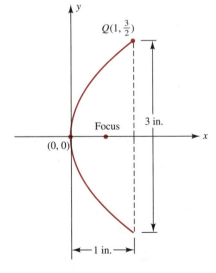

FIGURE 6.18
If the diameter of the parabolic reflector is 3 inches and the depth is 1 inch, then point Q has coordinates $\left(1, \frac{3}{2}\right)$.

Since p represents the distance from the vertex to the focus and light rays emanating from the focus are all parallel to the axis of the parabola, we conclude that the filament of the bulb should be placed $\frac{9}{16}$ inch from the vertex of the parabolic reflector. ◆

PROBLEM 5 Taking the axes as shown in Figure 6.18, determine the equation of the parabola that may be rotated about the *x*-axis to generate this parabolic reflector. ◆

Exercises 6.1

Basic Skills

In Exercises 1–16, sketch the graph of each parabola. Label the vertex and any x- and y-intercept(s).

1. $x = y^2$
2. $x = -3y^2$
3. $y = 2x^2 - 8$
4. $x = 1 - 4y^2$
5. $x = -(y - 1)^2 + 4$
6. $y = 2(x + 1)^2 - 2$
7. $y = 4x - x^2$
8. $4x = y^2 - 4$
9. $x = y^2 - 3y - 4$
10. $y = 6 - x - \frac{1}{3}x^2$
11. $y^2 - 2x - 4y + 2 = 0$
12. $y^2 + 3x + 6y + 3 = 0$
13. $x^2 + 8x - 2y + 16 = 0$
14. $x^2 - 3y - 6x + 15 = 0$
15. $2y^2 + 3x + 16y + 26 = 0$
16. $3y^2 + 4x - 6y + 3 = 0$

Find the focus and directrix of the parabola whose equation is given in the indicated exercise.

17. Exercise 3
18. Exercise 6
19. Exercise 9
20. Exercise 10
21. Exercise 15
22. Exercise 16

In Exercises 23–32, determine the equation of the parabola with the given characteristics.

23. Vertex at the origin, passing through the point (2, 4), and symmetric with respect to the *x*-axis.
24. Vertex at the origin, passing through the point (−1, 3), and symmetric with respect to the *y*-axis.
25. Vertex at (−1, −1), passing through the point (0, 1), and symmetric with respect to the line $x = -1$.
26. Vertex at (2, 1), passing through the origin, and symmetric with respect to the line $y = 1$.
27. Focus at (0, 1) and directrix $x = 2$.
28. Focus at (−4, −2) and directrix $y = 2$.
29. Vertex at (−3, 2) and focus at (−3, −3).
30. Vertex at (9, −12) and focus at (1, −12).
31. Vertex at (3, −1) and directrix $x = -1$.
32. Vertex at (−8, −4) and directrix $y = 2$.

33. The cable that supports the weight of a suspension bridge approximates a parabola with vertex located halfway between the supports, as shown in the sketch. Determine the length *x* of the vertical cable that is 30 meters from the end of the support.

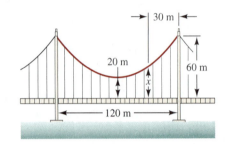

34. The main entrance into a new shopping mall is a parabolic arch with span 12 feet and height 18 feet, as shown in the sketch. Calculate the height *h* of a security camera that is located 4 feet from the center of the arch.

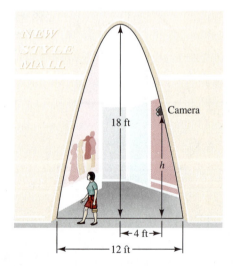

35. A parabolic satellite dish for receiving television signals has its receiver located at the focus, as shown in the sketch. Find the depth *d* of the satellite dish.

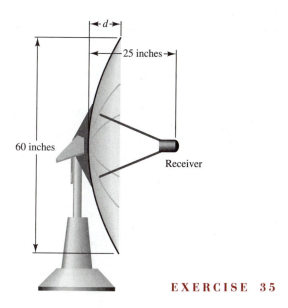

EXERCISE 35

36. A solar collector for a domestic water heater is parabolic in shape, as shown in the sketch. At what distance d from the vertex should the water pipe be located in order to take advantage of the reflection property of a parabola?

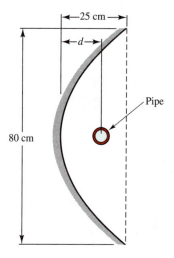

Critical Thinking

37. If a parabola has two distinct x-intercepts a and b, what is the x-coordinate of its vertex?

38. Find all values of k such that the vertex of the graph of $y = x^2 + kx + 9$ lies on the x-axis.

39. If the graph of $x = ay^2 + by + c$ passes through the origin, what is the value of c?

40. Determine the value of a such that the parabola $x = ay^2 - 4y + 3$ has its axis of symmetry along the line $y = 5$.

41. What positive values of p make the opening of the parabola $y = \dfrac{1}{4p}x^2$ narrower than the opening of the parabola $y = x^2$? Wider than $y = x^2$?

42. The chord of a parabola that passes through its focus and is perpendicular to its axis of symmetry is called the **latus rectum** of the parabola. The latus rectum of the parabola $x = \dfrac{1}{4p}y^2$, with $p > 0$, is the line segment AB, as shown in the sketch. What is the length of the latus rectum?

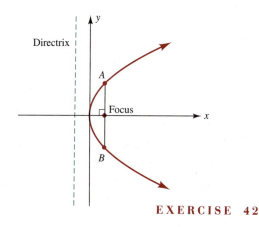

EXERCISE 42

43. The equation of a parabola is $Ax^2 + Ey = 0$. Find the coordinates of its focus in terms of A and E.

44. What point on a parabola is always closest to its focus?

Calculator Activities

In Exercises 45–48, find the coordinates of the vertex of each parabola. Round each coordinate to three significant digits.

45. $12.5x^2 - 18.6x + 2.4y = 10.7$

46. $1.24x^2 + 4.56x = 9.55y$

47. $23.4y^2 - 45.8y - 12.6x = 0$

48. $y^2 - 2.34x - 5.74y = -6.87$

If you have access to a graphing calculator, you can check the results of Exercises 45–48. To display the graph of the parabola $Ax^2 + Dx + Ey + F = 0$ on a graphing calculator, simply solve the equation for y and enter the function that results. To display the graph of the parabola $Cy^2 + Dx + Ey + F = 0$ on a graphing calculator, rewrite the equation as a quadratic in y, $Cy^2 + Ey + (Dx + F) = 0$, and apply the quadratic formula to obtain

$$y = \frac{-E \pm \sqrt{E^2 - 4C(Dx + F)}}{2C}.$$

Then graph the two functions that result. In Exercises 49–52, use a graphing calculator to generate the graph of each parabola in the viewing rectangle. Then estimate the coordinates of the vertex by tracing to this point. Round each coordinate to three significant digits.

49. The parabola in Exercise 45.
50. The parabola in Exercise 46.
51. The parabola in Exercise 47.
52. The parabola in Exercise 48.

In Exercises 53–56, determine the equation in standard form of the parabola with the given characteristics.

53. Vertex at $(2.67, -1.98)$, y-intercept 1.05, and axis of symmetry $x = 2.67$.

54. Vertex at $(-2.43, 1.67)$, horizontal axis of symmetry, and passing through the origin.

55. Focus at $(12.7, -23.1)$ and directrix $y = 28.5$.

56. Vertex at $(-32.5, 21.3)$ and focus at $(-13.1, 21.3)$.

57. In an engineer's design for a headlight in a new car, a parabola is rotated about the x-axis, as shown in the sketch. Determine the equation of the parabola that is used to generate the shape of the headlight. Write the equation in the form $x = ay^2$, with a rounded to three significant digits.

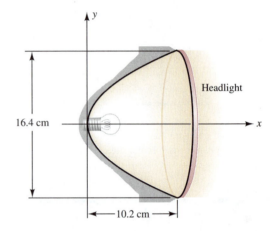

58. A simply supported beam with a concentrated load at midspan deflects in the shape of a parabola, as shown in the sketch. Determine the equation of the parabola, which is called the *elastic curve*. Write the equation in the form $y = ax^2 + bx$ with a and b rounded to three significant digits.

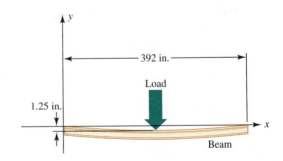

6.2 The Ellipse

◆ Introductory Comments

Like the parabola, the ellipse has a reflection property. This property accounts for the "whispering gallery effect" in elliptically-shaped dome rooms. If a person stand-

FIGURE 6.19
In an elliptical-shaped dome room, the whispers of one person can be heard on the other side of the room.

ing at a special point on one side of the room whispers, all the sound waves bounce off the sides of the dome and return to another special point on the other side of the room, allowing a person standing at that point to hear the whisper (see Figure 6.19). The two special points in the room where this phenomenon occurs are called the *foci* of the ellipse. The whispering gallery effect may be observed in the National Statuary Hall in the Capitol building in Washington, D.C. In this section, we study the geometric properties of the ellipse and the distinguishing characteristics of its equation.

◆ Equations of Circles and Ellipses

Consider the equation

$$\frac{x^2}{a^2} + \frac{y^2}{b^2} = 1$$

with $a > 0$ and $b > 0$. This is a general quadratic equation in two unknowns with $A = 1/a^2$, $C = 1/b^2$, $F = -1$, and $B = D = E = 0$. If $a = b$, then we have

$$\frac{x^2}{a^2} + \frac{y^2}{a^2} = 1 \quad \text{or} \quad x^2 + y^2 = a^2.$$

Recall from Section 1.2 that the graph of $x^2 + y^2 = a^2$ is a circle with center at the origin and radius a. If a and b have different values, then the circle is either flattened or stretched to form an egg-shaped curve called an **ellipse**. The center of the ellipse is at the origin with x-intercepts $\pm a$ and y-intercepts $\pm b$. Figure 6.20 shows the three possibilities for the graph of $(x^2/a^2) + (y^2/b^2) = 1$.

In each of the graphs sketched in Figure 6.20, the horizontal and vertical line segments $\overline{AB}$ and $\overline{CD}$ are the **axes** of the ellipse, and their lengths are $2a$ and $2b$, respectively. We refer to the longer line segment as the **major axis** and the shorter line segment as the **minor axis**. If $a > b$, as in Figure 6.20(b), then $\overline{AB}$ is the major axis and $\overline{CD}$ is the minor axis. However, if $a < b$, as in Figure 6.20(c), then $\overline{CD}$ is the major axis and $\overline{AB}$ is the minor axis. We refer to the endpoints of the major axis as the **vertices** of the ellipse and the midpoint of the major axis and minor axis, as its **center**.

If we replace x with $(x - h)$ and y with $(y - k)$ in the equation $(x^2/a^2) + (y^2/b^2) = 1$, we obtain

$$\frac{(x-h)^2}{a^2} + \frac{(y-k)^2}{b^2} = 1.$$

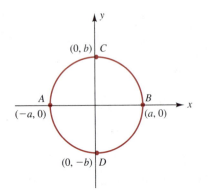 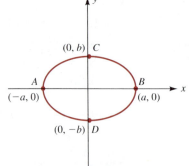 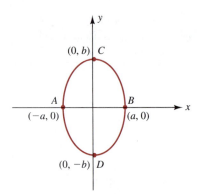

(a) If $a = b$, the graph is a circle with center at the origin.

(b) If $a > b$, the graph is an ellipse, elongated horizontally, with center at the origin.

(c) If $a < b$, the graph is an ellipse, elongated vertically, with center at the origin.

FIGURE 6.20 The three possibilities for the graph of $\dfrac{x^2}{a^2} + \dfrac{y^2}{b^2} = 1$

If a and b are positive real numbers such that $a = b$, we have

$$\frac{(x-h)^2}{a^2} + \frac{(y-k)^2}{a^2} = 1 \quad \text{or} \quad (x-h)^2 + (y-k)^2 = a^2,$$

which is the equation of a circle in standard form with center (h, k) and radius a (see Section 1.2). If a and b have different positive values, we conclude that $(x-h)^2/a^2 + (y-k)^2/b^2 = 1$ is the *equation of an ellipse in standard form* with center (h, k), horizontal axis of length $2a$, and vertical axis of length $2b$. We refer to this equation as the *algebraic definition of an ellipse*.

Equation of an Ellipse in Standard Form

The **equation of an ellipse in standard form** with center (h, k), horizontal axis of length $2a$, and vertical axis of length $2b$, is

$$\frac{(x-h)^2}{a^2} + \frac{(y-k)^2}{b^2} = 1.$$

EXAMPLE 1 Determine the equation of an ellipse in standard form if its vertices are $(3, 3)$ and $(-5, 3)$ and its minor axis has length 4.

SOLUTION The vertices of an ellipse are the endpoints of its major axis. Since the y-coordinates of the vertices are equal, we know the major axis is horizontal. Thus, the minor axis is vertical. Now, if the minor axis is vertical and has a length of 4, we have

$$2b = 4 \text{ or } b = 2.$$

The length of the major axis is the distance between the x-coordinates of the vertices, or $|3 - (-5)| = 8$. Thus,

$$2a = 8 \quad \text{or} \quad a = 4.$$

The midpoint of the major axis is the center of the ellipse. Using the midpoint formula (Section 1.2), we find the coordinates of the center:

$$(h, k) = \left(\frac{3 + (-5)}{2}, \frac{3 + 3}{2}\right) = (-1, 3).$$

Thus, the equation of the ellipse in standard form is

$$\frac{(x - h)^2}{a^2} + \frac{(y - k)^2}{b^2} = 1$$

$$\frac{(x - (-1))^2}{4^2} + \frac{(y - 3)^2}{2^2} = 1 \qquad \text{Substitute}$$

$$\frac{(x + 1)^2}{16} + \frac{(y - 3)^2}{4} = 1 \qquad \text{Simplify}$$

The graph of this equation is shown in Figure 6.21 ◆

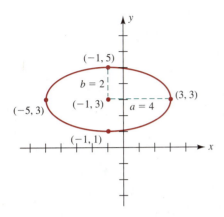

FIGURE 6.21
The equation of an ellipse with vertices $(-5, 3)$ and $(3, 3)$ and minor axis of length 4 is $\frac{(x - 1)^2}{16} + \frac{(y - 3)^2}{4} = 1$.

PROBLEM 1 Determine the y-intercepts of the ellipse in Example 1. ◆

If we take the equation of an ellipse in standard form and multiply both sides of the equation by a^2b^2, we obtain an equation of the form

$$A(x - h)^2 + C(y - k)^2 = N,$$

where $A = b^2$, $C = a^2$, and $N = a^2b^2$. If we expand $(x - h)^2$ and $(y - k)^2$, we obtain

$$A(x^2 - 2hx + h^2) + C(y^2 - 2ky + k^2) = N$$

or $\qquad Ax^2 + Cy^2 - 2Ahx - 2Cky + (Ah^2 + Ck^2 - N) = 0.$

Now, replacing the constants $-2Ah$, $-2Ck$, and $(Ah^2 + Ck^2 - N)$ with D, E, and F, respectively, gives us

$$\boxed{Ax^2 + Cy^2 + Dx + Ey + F = 0.}$$

This is the **equation of an ellipse in general form** with vertical and horizontal axes of symmetry. This general quadratic equation in two unknowns is characterized by the presence of x^2 and y^2 terms that have *different coefficients* but *like signs*. The difference between the general form of the equation of an ellipse and that of a circle is that the coefficients of the x^2 and y^2 terms of an ellipse are different, whereas those of a circle are the same.

EXAMPLE 2 Sketch the graph of the equation.

(a) $4x^2 + y^2 = 9$. (b) $x^2 + 9y^2 + 8x - 36y + 43 = 0$

SOLUTION

(a) We begin by writing the equation in standard form:

$$4x^2 + y^2 = 9$$

$$\frac{4x^2}{9} + \frac{y^2}{9} = 1 \quad \text{Divide both sides by 9}$$

$$\frac{x^2}{\frac{9}{4}} + \frac{y^2}{9} = 1 \quad \text{Invert } \tfrac{4}{9} \text{ and divide}$$

$$\frac{(x-0)^2}{\left(\frac{3}{2}\right)^2} + \frac{(y-0)^2}{3^2} = 1 \quad \text{Write in standard form}$$

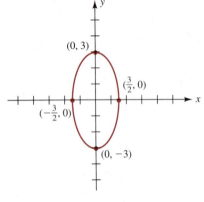

FIGURE 6.22
The graph of $4x^2 + y^2 = 9$ is an ellipse, elongated vertically, with center $(0, 0)$ and vertices $(0, 3)$ and $(0, -3)$.

This last equation tells us that the graph of $4x^2 + y^2 = 9$ is an ellipse with $a = \tfrac{3}{2}$, $b = 3$, and center at the origin. Since $b > a$, the ellipse is elongated vertically with a major axis of length $2b = 6$ and a minor axis of length $2a = 3$. The graph of the equation is shown in Figure 6.22.

(b) We begin by writing the equation in standard form. To do this, we use the process of completing the square and proceed as follows:

$$x^2 + 9y^2 + 8x - 36y + 43 = 0$$

$$(x^2 + 8x \quad) + (9y^2 - 36y \quad) = -43 \quad \text{Regroup}$$

$$(x^2 + 8x \quad) + 9(y^2 - 4y \quad) = -43 \quad \text{Factor out 9}$$

Add 16 to both sides

$$(x^2 + 8x + 16) + 9(y^2 - 4y + 4) = -43 + (16 + 36) \quad \text{Complete the squares}$$

Add 36 to both sides

$$(x + 4)^2 + 9(y - 2)^2 = 9 \quad \text{Factor}$$

$$\frac{(x + 4)^2}{9} + \frac{(y - 2)^2}{1} = 1 \quad \text{Divide both sides by 9}$$

$$\frac{(x - (-4))^2}{3^2} + \frac{(y - 2)^2}{1^2} = 1 \quad \text{Write in standard form}$$

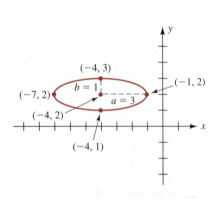

FIGURE 6.23
The graph of $x^2 + 9y^2 + 8x - 36y + 43 = 0$ is an ellipse, elongated horizontally, with center $(-4, 2)$ and vertices $(-7, 2)$ and $(-1, 2)$.

This last equation tells us that the graph of $x^2 + 9y^2 + 8x - 36y + 43 = 0$ is an ellipse with $a = 3$, $b = 1$, and center $(h, k) = (-4, 2)$. Since $a > b$, the ellipse is elongated horizontally with major axis of length $2a = 6$ and minor axis of length $2b = 2$. The graph of the equation is shown in Figure 6.23. ◆

SECTION 6.2 The Ellipse

 An ellipse is not the graph of every equation of the form $Ax^2 + Cy^2 + Dx + Ey + F = 0$, where the x^2 and y^2 terms have different coefficients but like signs. For example, if this equation is written in the form

$$A(x - h)^2 + C(y - k)^2 = N \quad \text{and} \quad N = 0,$$

then the graph of the equation is the single point (h, k). We refer to this single point as a *degenerate ellipse*. Also, if

$$A(x - h)^2 + C(y - k)^2 = N \quad \text{and} \quad N < 0,$$

then this equation has no graph, since no point (x, y) with real coordinates satisfies this equation.

PROBLEM 2 Is $3x^2 + 2y^2 - 6x + 8y + 15 = 0$ the equation of an ellipse? Explain. ◆

◆ **Geometric Definition of an Ellipse**

Geometrically, we define an ellipse as follows:

Geometric Definition of an Ellipse

An **ellipse** is the set of all points in a plane, the sum of whose distances from two fixed points is a constant, which is equal to the length of the major axis of the ellipse. The two fixed points are called the **foci** (plural of *focus*).

We now proceed to show that our geometric definition of an ellipse is consistent with our algebraic definition. Applying our geometric definition, an ellipse is the set of all points P such that

$$PF_1 + PF_2 = \text{length of major axis},$$

as shown in Figure 6.24. For convenience, let's place the ellipse that is shown in Figure 6.24 in the xy-coordinate system so that its center is at the origin, its major axis is horizontal with length $2a$, and its minor axis is vertical with length $2b$ (see Figure 6.25). We have labeled the coordinates of the foci $(-c, 0)$ and $(c, 0)$, where $a > c$.

The constants a, b, and c have a special relationship. Referring to Figure 6.26, the point $(0, b)$ is on the ellipse and is equidistant from the foci. Since the sum of the distances from the foci to the point $(0, b)$ must be $2a$, the distance from each focus to $(0, b)$ is a. Thus, the shaded right triangle in Figure 6.26 has legs b and c and hypotenuse a. Hence, by the Pythagorean theorem, we have the relation

$$a^2 = b^2 + c^2 \quad \text{or} \quad a^2 - c^2 = b^2$$

between a, b, and c in this ellipse.

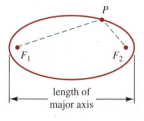

FIGURE 6.24
An ellipse is the set of all points P such that $PF_1 + PF_2 = $ length of major axis.

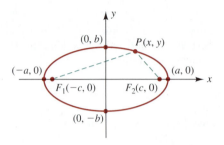

FIGURE 6.25

An ellipse in the xy-coordinate system with center at the origin, major axis of length $2a$, minor axis of length $2b$, and foci with coordinates $(-c, 0)$ and $(c, 0)$.

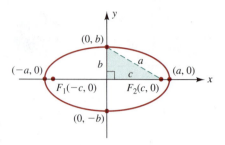

FIGURE 6.26

The constants a, b, and c are related by the Pythagorean theorem.

Now referring to Figure 6.25, we have

$$PF_1 + PF_2 = 2a$$

$$\sqrt{(x+c)^2 + y^2} + \sqrt{(x-c)^2 + y^2} = 2a \qquad \text{Apply the distance formula}$$

$$\sqrt{(x+c)^2 + y^2} = 2a - \sqrt{(x-c)^2 + y^2} \qquad \text{Isolate one of the radicals}$$

$$(x+c)^2 + y^2 = 4a^2 - 4a\sqrt{(x-c)^2 + y^2} + (x-c)^2 + y^2 \qquad \text{Square both sides}$$

$$4a\sqrt{(x-c)^2 + y^2} = 4a^2 + (x-c)^2 + y^2 - (x+c)^2 - y^2 \qquad \text{Isolate the radical}$$

$$a\sqrt{(x-c)^2 + y^2} = a^2 - cx \qquad \text{Simplify the right-hand side, then divide both sides by 4}$$

$$a^2[(x-c)^2 + y^2] = a^4 - 2a^2cx + c^2x^2 \qquad \text{Square both sides}$$

$$(a^2 - c^2)x^2 + a^2y^2 = a^2(a^2 - c^2) \qquad \text{Collect like terms}$$

$$\frac{x^2}{a^2} + \frac{y^2}{a^2 - c^2} = 1 \qquad \text{Divide both sides by } a^2(a^2 - c^2)$$

$$\frac{x^2}{a^2} + \frac{y^2}{b^2} = 1 \qquad \text{Replace } a^2 - c^2 \text{ with } b^2$$

From our algebraic definition of an ellipse, we recognize

$$\frac{x^2}{a^2} + \frac{y^2}{b^2} = 1$$

as the equation of an ellipse in standard form with center $(0, 0)$, horizontal axis of length $2a$, and vertical axis of length $2b$.

The **eccentricity** e of an ellipse is a measure of its "ovalness" and is defined as

$$e = \frac{\text{Distance from center to focus}}{\text{Distance from center to vertex}}$$

SECTION 6.2 The Ellipse

(Do not confuse the eccentricity e with base e in exponential and logarithmic expressions.) The eccentricity e of an ellipse is a number between 0 and 1. If e is close to 1, the ellipse is very long and narrow, and if e is close to 0, the ellipse is nearly circular.

EXAMPLE 3 Find the foci and eccentricity of each ellipse.

(a) $16x^2 + 9y^2 = 144$ (b) $x^2 + 4y^2 - 6x + 5 = 0$

SOLUTION

(a) Dividing both sides by 144, we obtain

$$\frac{x^2}{9} + \frac{y^2}{16} = 1 \quad \text{or} \quad \frac{(x-0)^2}{3^2} + \frac{(y-0)^2}{4^2} = 1,$$

which is the equation of an ellipse in standard form with $a = 3$, $b = 4$, and center at the origin. Since $b > a$, the ellipse is elongated vertically with major axis of length $2b = 8$ and minor axis of length $2a = 6$, as shown in Figure 6.27.

To find the coordinates of the foci F_1 and F_2, we apply the geometric definition of an ellipse. Referring to Figure 6.27, the point $(3, 0)$ is on the ellipse and is equidistant from the foci. Since the sum of the distances from the foci to the point $(3, 0)$ must be $2b = 8$, the distance from each focus to $(3, 0)$ is $b = 4$. Thus, the shaded right triangle in Figure 6.27 has legs c and 3 and hypotenuse 4. Hence, by the Pythagorean theorem, we have

$$4^2 = c^2 + 3^2,$$

which implies $c = \sqrt{7}$. Hence, the coordinates of the foci are

$$F_1(0, \sqrt{7}) \quad \text{and} \quad F_2(0, -\sqrt{7}).$$

FIGURE 6.27
The coordinates of the foci of the ellipse $16x^2 + 9y^2 = 144$ are found by appling the Pythagorean theorem.

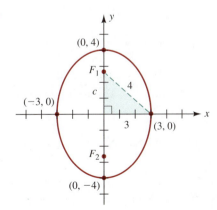

The eccentricity e of this ellipse is

$$e = \frac{\text{Distance from center to focus}}{\text{Distance from center to vertex}} = \frac{\sqrt{7}}{4} \approx 0.66.$$

(b) First, we write the equation in standard form, by completing the square as follows:

$$x^2 + 4y^2 - 6x + 5 = 0$$

$(x^2 - 6x \quad) + 4y^2 = -5$ Regroup

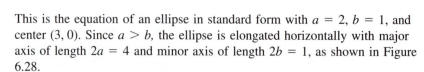

$(x^2 - 6x + 9) + 4y^2 = -5 + 9$ Complete the square

$(x - 3)^2 + 4y^2 = 4$ Factor

$$\frac{(x-3)^2}{2^2} + \frac{(y-0)^2}{1^2} = 1$$ Divide both sides by 4 and write in standard form

This is the equation of an ellipse in standard form with $a = 2$, $b = 1$, and center $(3, 0)$. Since $a > b$, the ellipse is elongated horizontally with major axis of length $2a = 4$ and minor axis of length $2b = 1$, as shown in Figure 6.28.

To find the coordinates of the foci F_1 and F_2, we apply the geometric definition of an ellipse. Referring to Figure 6.28, the point $(3, 1)$ is on the ellipse and is equidistant from the foci. Since the sum of the distances from the foci to the point $(3, 1)$ must be $2a = 4$, the distance from each focus to $(3, 1)$ is $a = 2$. Thus, the shaded right triangle in Figure 6.28 has legs c and 1 and hypotenuse 2. Hence, by the Pythagorean theorem, we have

$$2^2 = c^2 + 1^2,$$

which implies $c = \sqrt{3}$. Hence, the coordinates of the foci are

$$F_1(3 - \sqrt{3}, 0) \quad \text{and} \quad F_2(3 + \sqrt{3}, 0).$$

The eccentricity e of this ellipse is

$$e = \frac{\text{Distance from center to focus}}{\text{Distance from center to vertex}} = \frac{\sqrt{3}}{2} \approx 0.87. \quad \blacklozenge$$

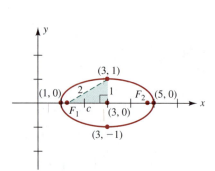

FIGURE 6.28
The coordinates of the foci of the ellipse $x^2 + 4y^2 - 6x + 5 = 0$ are found by applying the Pythagorean theorem.

PROBLEM 3   Find the foci and eccentricity for the ellipse $9x^2 + y^2 - 9 = 0$. $\blacklozenge$

As illustrated in the next example, if we know the foci and eccentricity of an ellipse, we can determine its equation.

EXAMPLE 4
Find the equation of an ellipse in standard form with foci $F_1(2, 3)$ and $F_2(2, -1)$ and eccentricity $\frac{1}{2}$.

SOLUTION The foci of the ellipse are on the vertical line $x = 2$ and the center of the ellipse is located halfway between the foci. Hence, the major axis of the ellipse is vertical with length $2b$ and the center of the ellipse is $(h, k) = (2, 1)$. Now, referring to Figure 6.29,

$$\frac{\text{Distance from center to focus}}{\text{Distance from center to vertex}} = \frac{2}{b} = \frac{1}{2}.$$

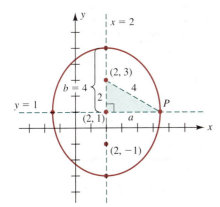

FIGURE 6.29
An ellipse with foci $(2, 3)$ and $(2, -1)$ and eccentricity $1/2$ has center $(2, 1)$ and vertical major axis $2b = 8$.

which implies $b = 4$. Since point P is on the ellipse and is equidistant from the foci, the sum of the distances from the foci to point P must be $2b = 8$. Hence, the distance from each focus to point P is $b = 4$. Now, applying the Pythagorean theorem to the shaded right triangle in Figure 6.29, we obtain

$$4^2 = a^2 + 2^2,$$

which implies $a = 2\sqrt{3}$. Since $(h, k) = (2, 1)$, $a^2 = (2\sqrt{3})^2 = 12$, and $b^2 = (4)^2 = 16$, the equation of the ellipse in standard form is

$$\frac{(x - 2)^2}{12} + \frac{(y - 1)^2}{16} = 1. \blacklozenge$$

PROBLEM 4
Referring to Example 4, express the equation of the ellipse in general form with integer coefficients. $\blacklozenge$

Exercises 6.2

◆ Basic Skills

In Exercises 1–16, sketch the graph of each ellipse. Label the center of the ellipse if it is not at the origin, and label the endpoints of the major and minor axes.

1. $\dfrac{x^2}{16} + \dfrac{y^2}{4} = 1$
2. $\dfrac{x^2}{4} + \dfrac{y^2}{9} = 1$
3. $25x^2 + 4y^2 = 100$
4. $16x^2 + 25y^2 = 400$
5. $x^2 + 16y^2 = 9$
6. $4x^2 + 9y^2 = 16$
7. $3x^2 + 4y^2 - 9 = 0$
8. $x^2 + 5y^2 - 10 = 0$
9. $\dfrac{(x-2)^2}{16} + \dfrac{(y+1)^2}{4} = 1$
10. $\dfrac{x^2}{9} + \dfrac{(y-3)^2}{25} = 1$
11. $9x^2 + 4(y + 2)^2 = 36$
12. $16(x - 4)^2 + (y - 5)^2 = 16$
13. $25x^2 + 4y^2 - 50x + 24y + 45 = 0$
14. $9x^2 + 25y^2 - 36x + 50y = 39$
15. $4x^2 + y^2 + 40x - 4y + 103 = 0$
16. $2x^2 + y^2 + 8x + 8y + 23 = 0$

Find the foci and eccentricity of the ellipse whose equation is given in the indicated exercise.

17. Exercise 3
18. Exercise 4
19. Exercise 9
20. Exercise 12
21. Exercise 13
22. Exercise 16

In Exercises 23–32, determine the equation of the ellipse with the given characteristics.

23. Center at the origin, minor axis of length 2, and horizontal major axis of length 5.
24. Vertices at $(0, \pm 3)$ and minor axis of length 2.

25. Vertices at (2, −6) and (2, 2) and minor axis of length 3.

26. Vertices at (−1, 2) and (5, 2) and endpoints of minor axis at (2, 4) and (2, 0).

27. Vertices at (−2, −3) and (−2, 3) and one focus at (−2, 1).

28. Foci at (−4, −1) and (2, −1) and one of the vertices at (4, −1)

29. Foci at (±2, 0) and major axis of length 6.

30. Foci at (1, ±4) and minor axis of length 6.

31. Foci at (0, 0) and (3, 0) and eccentricity $\frac{1}{2}$.

32. Foci at (1, −3) and (1, 5) and eccentricity $\frac{2}{3}$.

33. A road passes through a tunnel whose cross section is a semiellipse, which is 12 feet high at the center and 30 feet wide, as shown in the sketch. What is the tallest tractor-trailer rig that can fit through the tunnel if the trailor is 10 feet wide?

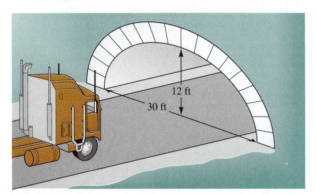

34. A window positioned above an 8-foot-wide glass door is a semiellipse, as shown in the sketch. What is the length of each brace?

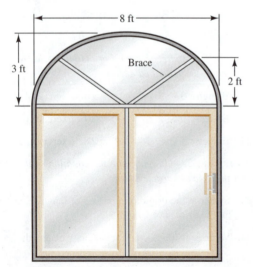

35. An architect's design for an elliptically-shaped ceiling calls for rotating the upper portion of the ellipse $x^2 + 4y^2 = 400$, (for x and y in feet), 180° about its major axis, as shown in the sketch.

    (a) Find the length and width of this ceiling.

    (b) By the reflection property of an ellipse, a whisper emanating from one focus can be heard at the other focus. Where are the whispering and listening positions located?

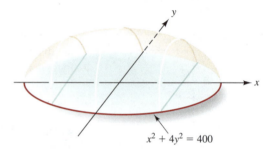

36. A *lithotripter* is a medical device that uses the reflection property of an ellipse. Ultrahigh-frequency (UHF) sound waves emanating from one focus of a lithotripter are reflected to the other focus of the device to break up kidney stones located at this point. Suppose a lithotripter is formed by rotating the portion of an ellipse below its minor axis about its major axis, as shown in the sketch. How far from the kidney stone should the opening of the lithotripter be placed in order to break up the stone?

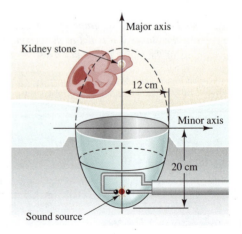

Critical Thinking

37. Suppose you are given two thumbtacks, a piece of string, and a pencil. Using this material, explain how to construct an ellipse.

38. The eccentricity of one ellipse is the constant k_1 and the eccentricity of another ellipse is the constant k_2, where $k_1 > k_2$. Which ellipse is more circular? Explain.

39. Find the equation of the largest circle that can fit inside the ellipse

$$\frac{x^2}{a^2} + \frac{y^2}{b^2} = 1 \quad \text{if } a < b.$$

40. The equation of an ellipse with a horizontal major axis is $Ax^2 + Cy^2 = E$. Find the coordinates of its foci in terms of A, C, and E.

41. The chord of an ellipse that passes through a focus and is perpendicular to its major axis is called a **latus rectum** of the ellipse. For the ellipse $\frac{x^2}{a^2} + \frac{y^2}{b^2} = 1$ with $a > b$, a latus rectum is the line segment AB, as shown in the sketch. Show that the length of the latus rectum is $2b^2/a$.

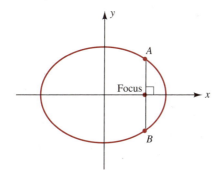

42. A square is inscribed in the ellipse $9x^2 + 16y^2 = 144$, which is sketched in the figure. What is the length of the side of the square?

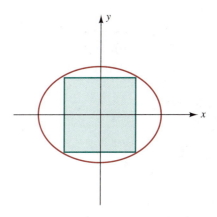

If a and b are positive real numbers, then solving $(x^2/a^2) + (y^2/b^2) = 1$ for y yields $y = \pm(b/a)\sqrt{a^2 - x^2}$. Taken separately, the equations

$$y = (b/a)\sqrt{a^2 - x^2} \quad \text{and} \quad y = -(b/a)\sqrt{a^2 - x^2}$$

define y as a function of x, and the graph of each function is a semiellipse (half of an ellipse) whenever $a \neq b$. In Exercises 43 and 44, state the domain and range of each function, and sketch its graph.

43. $y = \frac{2}{3}\sqrt{9 - x^2}$ 44. $y = -\sqrt{25 - 9x^2}$

Calculator Activities

In Exercises 45–48, find the coordinate of the vertices of each ellipse. Round each coordinate to three significant digits.

45. $23.4x^2 + 19.6y^2 = 87.3$

46. $8.67x^2 + 3.67y^2 = 9.87y$

47. $0.25x^2 + 0.14y^2 - 1.24y - 8.12 = 0$

48. $5.4x^2 + 3.4y^2 - 2.2x - 4.8y - 11.6 = 0$

If you have access to a graphing calculator, you can check the results of Exercises 45–48. To display the graph of the ellipse $Ax^2 + Cy^2 + Dx + Ey + F = 0$ on a graphing calculator, first solve the equation of the ellipse for y, and then enter the two functions that result. To do this, rewrite the equation as a quadratic in y, $Cy^2 + Ey + (Ax^2 + Dx + F) = 0$, and then apply the quadratic formula to obtain

$$y = \frac{-E \pm \sqrt{E^2 - 4C(Ax^2 + Dx + F)}}{2C}$$

In Exercises 49–52, use a graphing calculator to generate the graph of each ellipse in the viewing rectangle. Then estimate the coordinates of the vertices by tracing to these points. Round each coordinate to three significant digits.

49. The ellipse in Exercise 45.
50. The ellipse in Exercise 46.
51. The ellipse in Exercise 47.
52. The ellipse in Exercise 48.

In Exercises 53–56, determine the equation in standard form of the ellipse with the given characteristics.

53. Vertices at $(3.54, -5.67)$ and $(3.54, 1.53)$ and minor axis of length 2.40.
54. Vertices at $(25.5, -31.7)$ and $(-21.1, -31.7)$ and one focus at $(19.5, -31.7)$.
55. Foci at $(\pm 52.3, 0)$ and major axis of length 124.2.
56. Foci at $(1.23, -3.67)$ and $(1.23, 5.33)$ and eccentricity 0.65.
57. A stream 2.3 meters deep passes through an elliptical culvert, as shown in the sketch. What is the width w of the stream?

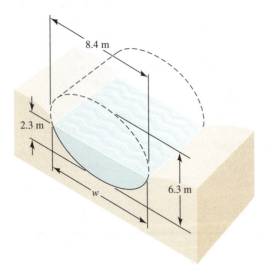

58. In astronomy, the average distance from the earth to the sun is called 1 astronomical unit (AU), and 1AU $\approx$ 9.26×10^7 miles. It is known that Halley's comet travels around the sun in an elliptical orbit with the sun at one focus. The major axis of the ellipse is 36.18 AU and the minor axis is 9.12AU.

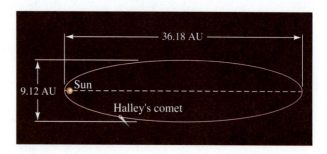

(a) What is the distance between the foci of the ellipse?
(b) What is the eccentricity of Halley's comet?
(c) What is the closest that Halley's comet comes toward the sun in miles?
(d) What is the furthest that Halley's comet moves from the sun in miles?

6.3 The Hyperbola and the Common Definition of the Conic Sections

◆ Introductory Comments

Like the parabola and ellipse, the hyperbola also has a reflection property. A ray of light directed at one focus of a hyperbolic mirror will be reflected so as to pass through its other focus. The reflection property of a hyperbola is used in the design of reflecting telescopes, such as the Hale telescope on Mount Palomar, California. As illustrated in Figure 6.30, light from the moon hits the primary parabolic mirror in the telescope and is reflected back toward the parabola's focus F_1, which is also one of the two foci of the hyperbolic mirror. The light hits the hyperbolic mirror and is reflected toward the hyperbola's second focus F_2, which is also one of the two foci of an elliptical mirror. The light then hits the elliptical mirror and reflects to the ellipses's second focus F_3, which is the eyepiece of the telescope. In this section, we study the geometric properties of a hyperbola and the distinguishing characteristics of its equation.

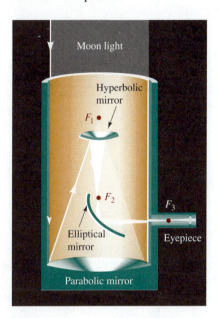

FIGURE 6.30
A reflecting telescope uses the reflection properties of the parabola, ellipse, and hyperbola to bring an image to the eyepiece.

◆ **Equations of Hyperbolas**

Consider the equation

$$\frac{x^2}{a^2} - \frac{y^2}{b^2} = 1$$

with $a > 0$ and $b > 0$. This is a general quadratic equation in two unknowns with $A = 1/a^2$, $C = -1/b^2$, $F = -1$, and $B = D = E = 0$. Since replacing x with $-x$ or y with $-y$ does not change the equation, we know that the graph of this equation is symmetric to both coordinate axes and to the origin. Letting $y = 0$, we find that the x-intercepts are $\pm a$. However, letting $x = 0$, we find that the graph has no y-intercept, since the equation

$$-\frac{y^2}{b^2} = 1 \quad \text{or} \quad y^2 = -b^2$$

has no real solution. If we solve the equation $(x^2/a^2) - (y^2/b^2) = 1$ for y, we obtain

$$y = \pm \frac{b}{a}\sqrt{x^2 - a^2}.$$

Now for y to be a real number, $x^2 - a^2$ must be either positive or zero; that is,

$$x^2 - a^2 \geq 0$$

so

$$x \geq a \quad \text{or} \quad x \leq -a$$

This means that the graph exists only to the right of the x-intercept a and to the left of the x-intercept $-a$. Intuitively, we can see that for very large values of x (either negative or positive),

$$y = \pm \frac{b}{a} \sqrt{x^2 - a^2} \approx \pm \frac{b}{a} \sqrt{x^2} = \pm \frac{b}{a} x,$$

since x^2 is so much larger than a^2. That is, the graph of $(x^2/a^2) - (y^2/b^2) = 1$ approaches the lines $y = \pm (b/a)x$ as $|x|$ increases without bound. We refer to the lines

$$y = \pm \frac{b}{a} x$$

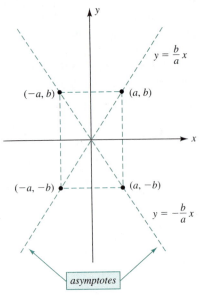

FIGURE 6.31

The asymptotes $y = \pm (b/a)x$ can be constructed from the diagonals of the rectangle with vertices (a, b), $(-a, b)$, $(-a, -b)$, and $(a, -b)$.

as the **asymptotes** of the curve.

The diagonals of a rectangle with vertices (a, b), $(-a, b)$, $(-a, -b)$, and $(a, -b)$ have slopes $\pm b/a$. Extending these diagonals gives us the asymptotes of the curve, as shown in Figure 6.31. Using the x-intercepts and the asymptotes, we can sketch the graph of $(x^2/a^2) - (y^2/b^2) = 1$. The graph, called a **hyperbola,** is shown in Figure 6.32. The two disconnected curves that make up the graph are the **branches** of the hyperbola.

In Figure 6.32, the points $(-a, 0)$ and $(a, 0)$ are the **vertices** of the hyperbola, and the horizontal and vertical line segments $\overline{AB}$ and $\overline{CD}$ are called the **axes** of the hyperbola. Note that the lengths of $\overline{AB}$ and $\overline{CD}$ are $2a$ and $2b$, respectively. The line segment that connects the vertices of the hyperbola is called the **transverse axis** and the other axis of the hyperbola is the **conjugate axis.** In this figure the horizontal line segment $\overline{AB}$ is the transverse axis and the vertical line segment $\overline{CD}$ is the conjugate axis. The midpoint of both the transverse axis and the conjugate axis is the **center** of the hyperbola. The hyperbola in Figure 6.32 has its center at the origin.

The graph of the equation

$$\frac{y^2}{b^2} - \frac{x^2}{a^2} = 1$$

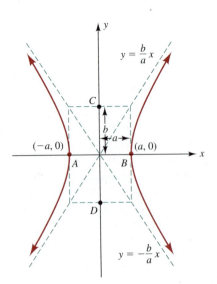

FIGURE 6.32

The graph of $(x^2/a^2) - (y^2/b^2) = 1$ is called a hyperbola. The x-intercepts are $\pm a$ and the asymptotes are $y = \pm(b/a)x$.

with $a > 0$ and $b > 0$ is also a hyperbola. However, the y-intercepts are now $\pm b$ and the graph has no x-intercept. Thus, the vertices of this hyperbola are $(0, b)$ and $(0, -b)$. The transverse axis is vertical with length $2b$ and the conjugate axis is horizontal with length $2a$. The asymptotes are still the two lines $y = \pm(b/a)x$, and the center of the hyperbola remains at the origin. The graph of $(y^2/b^2) - (x^2/a^2) = 1$ is shown in Figure 6.33

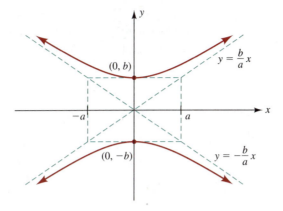

FIGURE 6.33
The graph of $(y^2/b^2) - (x^2/a^2) = 1$ is a hyperbola with y-intercepts $\pm b$ and asymptotes $y = \pm(b/a)x$.

The graphs of the equations

$$\frac{(x-h)^2}{a^2} - \frac{(y-k)^2}{b^2} = 1 \quad \text{and} \quad \frac{(y-k)^2}{b^2} - \frac{(x-h)^2}{a^2} = 1$$

are the same as the graphs of $(x^2/a^2) - (y^2/b^2) = 1$ and $(y^2/b^2) - (x^2/a^2) = 1$, respectively, except the center of the hyperbola is now (h, k). We refer to

$$\frac{(x-h)^2}{a^2} - \frac{(y-k)^2}{b^2} = 1 \quad \text{and} \quad \frac{(y-k)^2}{b^2} - \frac{(x-h)^2}{a^2} = 1$$

with $a > 0$ and $b > 0$ as the *equations of a hyperbola in standard form*, or the *algebraic definitions of a hyperbola*.

Equations of a Hyperbola in Standard Form

The **equation of a hyperbola in standard form** with center (h, k), *horizontal* transverse axis of length $2a$, and vertical conjugate axis of length $2b$ is

$$\frac{(x-h)^2}{a^2} - \frac{(y-k)^2}{b^2} = 1.$$

The **equation of a hyperbola in standard form** with center (h, k), *vertical* transverse axis of length $2b$, and horizontal conjugate axis of length $2a$ is

$$\frac{(y-k)^2}{b^2} - \frac{(x-h)^2}{a^2} = 1.$$

EXAMPLE 1 Determine the equation of a hyperbola in standard form if its asymptotes are $y = \frac{1}{2}x + 1$ and $y = -\frac{1}{2}x + 1$, and its horizontal transverse axis has length 6.

SOLUTION The intersection point of the asymptotes is the center of the hyperbola. Since the lines $y = \frac{1}{2}x + 1$ and $y = -\frac{1}{2}x + 1$ both have y-intercept at 1, we conclude that the center of the hyperbola is

$$(h, k) = (0, 1).$$

Since the horizontal transverse axis has length 6, we have

$$2a = 6 \quad \text{or} \quad a = 3.$$

We can determine b by using the fact that the slopes of the asymptotes are $\pm b/a = \pm 1/2$. Since $a = 3$, we have

$$\pm \frac{b}{3} = \pm \frac{1}{2} \quad \text{or} \quad b = \frac{3}{2}.$$

Now the equation of the hyperbola is

$$\frac{(x - h)^2}{a^2} - \frac{(y - k)^2}{b^2} = 1$$

$$\frac{(x - 0)^2}{3^2} - \frac{(y - 1)^2}{(\frac{3}{2})^2} = 1 \qquad \text{Substitute}$$

$$\frac{x^2}{9} - \frac{4(y - 1)^2}{9} = 1 \qquad \text{Simplify}$$

The graph of this equation is shown in Figure 6.34. ◆

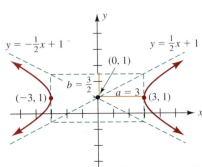

FIGURE 6.34

The equation of the hyperbola with asymptotes $y = \pm \frac{1}{2}x + 1$ and horizontal transverse axis of length 6 is $\frac{x^2}{9} - \frac{4(y-1)^2}{9} = 1$.

PROBLEM 1 Rework Example 1 if the transverse axis is vertical with length 6. ◆

If we take the equations of a hyperbola in standard form and multiply both sides of these equations by a^2b^2, we obtain equations of the forms

$$A(x - h)^2 - C(y - k)^2 = N \quad \text{and} \quad C(y - k)^2 - A(x - h)^2 = N,$$

where $A = b^2$, $C = a^2$, and $N = a^2b^2$.

If we expand $(x - h)^2$ and $(y - k)^2$ in the equation $A(x - h)^2 - C(y - k)^2 = N$, we obtain

$$A(x^2 - 2hx + h^2) - C(y^2 - 2ky + k^2) = N$$
$$Ax^2 - Cy^2 - 2Ahx + 2Cky + (Ah^2 - Ck^2 - N) = 0.$$

Now, replacing the constants $-2Ah$, $2Ck$, and $(Ah^2 - Ck^2 - N)$ with D, E, and F, respectively, gives us

$$\boxed{Ax^2 - Cy^2 + Dx + Ey + F = 0}$$

SECTION 6.3 The Hyperbola and the Common Definition

If we expand $(x - h)^2$ and $(y - k)^2$ in the equation $C(y - k)^2 - A(x - h)^2 = N$, we obtain

$$C(y^2 - 2ky + k^2) - A(x^2 - 2hx + h^2) = N$$
$$Cy^2 - Ax^2 + 2Ahx - 2Cky + (Ck^2 - Ah^2 - N) = 0.$$

Now, replacing the constants $2Ah$, $-2Ck$, and $(Ck^2 - Ah^2 - N)$ with D, E, and F, respectively, gives us

$$Cy^2 - Ax^2 + Dx + Ey + F = 0$$

We refer to

$$Ax^2 - Cy^2 + Dx + Ey + F = 0 \text{ and } Cy^2 - Ax^2 + Dx + Ey + F = 0$$

as the **equations of a hyperbola in general form** with vertical and horizontal axes of symmetry. These general quadratic equations in two unknowns are characterized by the presence of x^2 and y^2 terms that have *different signs*.

EXAMPLE 2 Sketch the graph of the equation.

(a) $4x^2 - 5y^2 = 20$ (b) $4x^2 - 9y^2 + 8x + 36y + 4 = 0$

SOLUTION

(a) We begin by writing the equation in standard form:

$$4x^2 - 5y^2 = 20$$
$$\frac{x^2}{5} - \frac{y^2}{4} = 1 \qquad \text{Divide both sides by 20}$$
$$\frac{(x - 0)^2}{(\sqrt{5})^2} - \frac{(y - 0)^2}{2^2} = 1 \qquad \text{Write in standard form}$$

This last equation tells us that the graph of $4x^2 - 5y^2 = 20$ is a hyperbola with $a = \sqrt{5}$, $b = 2$, center at the origin, and a horizontal transverse axis of length $2a = 2\sqrt{5}$. As an aid in graphing the hyperbola, we may draw a rectangle using the values of a and b, and then extend the diagonals of the rectangle to form the asymptotes of the hyperbola. The graph of the hyperbola is shown in Figure 6.35.

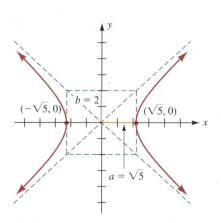

FIGURE 6.35
The graph of $4x^2 - 5y^2 = 20$ is a hyperbola with center at the origin and vertices $(\pm\sqrt{5}, 0)$.

(b) We begin by writing the equation in standard form. To do this, we use the process of completing the square as follows:

$$4x^2 - 9y^2 + 8x + 36y + 4 = 0$$
$$(4x^2 + 8x \quad) + (-9y^2 + 36y \quad) = -4 \qquad \text{Regroup}$$
$$4(x^2 + 2x \quad) - 9(y^2 - 4y \quad) = -4 \qquad \text{Factor out 4 and } -9$$

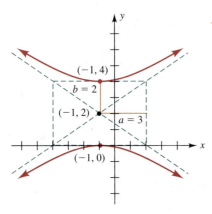

FIGURE 6.36
The graph of $4x^2 - 9y^2 + 8x + 36y + 4 = 0$ is a hyperbola with center $(-1, 2)$ and vertices $(-1, 4)$ and $(-1, 0)$.

$$4(x^2 + 2x + 1) - 9(y^2 - 4y + 4) = -4 + (4 - 36)$$ **Complete the squares** (Add 4 to both sides; Add -36 to both sides)

$$4(x + 1)^2 - 9(y - 2)^2 = -36$$ **Factor**

$$\frac{(y - 2)^2}{4} - \frac{(x + 1)^2}{9} = 1$$ **Divide both sides by -36**

$$\frac{(y - 2)^2}{2^2} - \frac{(x + 1)^2}{3^2} = 1$$ **Write in standard form**

The last equation tells us that the graph of $4x^2 - 9y^2 + 8x + 36y + 4 = 0$ is a hyperbola with $a = 3$, $b = 2$, center $(h, k) = (-1, 2)$, and a vertical transverse axis of length $2b = 4$. The graph of this hyperbola is shown in Figure 6.36.

A hyperbola is not the graph of every equation of the form

$$Ax^2 - Cy^2 + Dx + Ey + F = 0 \quad \text{or} \quad Cy^2 - Ax^2 + Dx + Ey + F = 0$$

where the x^2 and y^2 terms have different signs. For example, when $Ax^2 - Cy^2 + Dx + Ey + F = 0$ is written in the form $A(x - h)^2 - C(y - k)^2 = N$ with $N = 0$, the graph of the equation is a pair of intersecting lines. We refer to those two lines as a *degenerate hyperbola*.

PROBLEM 2 Is $x^2 - 4y^2 - 6x + 8y + 5 = 0$ the equation of a hyperbola? Explain.

◆ Geometric Definition of a Hyperbola

Geometrically, we define a hyperbola as follows:

Geometric Definition of a Hyperbola

A **hyperbola** is the set of all points in a plane, the difference of whose distances from two fixed points is a constant, which is equal to the length of the transverse axis of the hyperbola. The two fixed points are called the **foci** (plural of focus).

For convenience, we place a hyperbola in the *xy*-coordinate system so that its center is at the origin, its transverse axis is horizontal with length $2a$, and its conjugate axis is vertical with length $2b$ (see Figure 6.37). We have labeled the coordinates of the foci $(-c, 0)$ and $(c, 0)$, where $a < c$.

The relationship between the constants a, b, and c in any hyperbola is

$$c^2 = a^2 + b^2$$

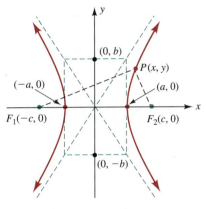
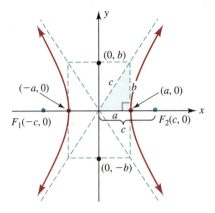

FIGURE 6.37
A hyperbola in the xy-coordinate system with center at the origin, transverse axis of length $2a$, conjugate axis of length $2b$, and foci with coordinates $(-c, 0)$ and $(c, 0)$.

FIGURE 6.38
The constants a, b, and c are related by the Pythagorean theorem.

Figure 6.38 illustrates this relationship. If c is the length of the line segment from the center to the focus, then c is also the length of the diagonal of the rectangle that is formed by a and b. Now, referring to Figure 6.37 and applying our geometric definition of a hyperbola, we have

$$PF_1 - PF_2 = 2a$$
$$\sqrt{(x+c)^2 + y^2} - \sqrt{(x-c)^2 + y^2} = 2a$$

Using the same algebraic procedure as we used for an ellipse in Section 6.2 (isolating the radical, squaring both sides, simplifying), and using the fact that $c^2 = a^2 + b^2$ in any hyperbola, we obtain

$$\frac{x^2}{a^2} - \frac{y^2}{b^2} = 1.$$

Hence, we conclude that our algebraic definition of a hyperbola is consistent with our geometric definition.

The **eccentricity** e of a hyperbola is a measure of the "spread of its wings" and is defined as

$$e = \frac{\text{Distance from center to focus}}{\text{Distance from center to vertex}}$$

The eccentricity e of a hyperbola is always a number greater than 1. The larger the number, the wider the opening of each branch.

EXAMPLE 3 Find the foci and eccentricity of each hyperbola.

(a) $16x^2 - 9y^2 = 144$ (b) $y^2 - 4x^2 - 6y + 5 = 0$

SOLUTION

(a) Dividing both sides by 144, we obtain

$$\frac{x^2}{9} - \frac{y^2}{16} = 1 \quad \text{or} \quad \frac{(x-0)^2}{3^2} - \frac{(y-0)^2}{4^2} = 1,$$

which is the equation of a hyperbola in standard form with center at the origin, horizontal transverse axis of length $2a = 2(3) = 6$ and vertical conjugate axis of length $2b = 2(4) = 8$, as shown in Figure 6.39. If c is the length of the diagonal of the rectangle that is formed by $a = 3$ and $b = 4$, then by the Pythagorean theorem

$$c^2 = 3^2 + 4^2,$$

which implies $c = 5$. Now, $c = 5$ is also the length of the line segment from the center to a focus. Therefore, the coordinates of the foci are

$$F_1(-5, 0) \quad \text{and} \quad F_2(5, 0).$$

The eccentricity e of this hyperbola is

$$e = \frac{\text{Distance from center to focus}}{\text{Distance from center to vertex}} = \frac{5}{3} \approx 1.67$$

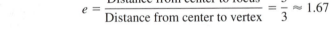

FIGURE 6.39
The coordinates of the foci of the hyperbola $16x^2 - 9y^2 = 144$ are found by applying the Pythagorean theorem.

(b) First, we write the equation in standard form by completing the square as follows:

$$y^2 - 4x^2 - 6y + 5 = 0$$
$$(y^2 - 6y \quad) - 4x^2 = -5 \qquad \text{Regroup}$$

Add 9 to both sides

$$(y^2 - 6y + 9) - 4x^2 = -5 + 9 \qquad \text{Complete the square}$$
$$(y - 3)^2 - 4x^2 = 4 \qquad \text{Factor}$$
$$\frac{(y-3)^2}{2^2} - \frac{(x-0)^2}{1^2} = 1 \qquad \text{Divide both sides by 4 and write in standard form}$$

This is the equation of a hyperbola in standard form with center $(h, k) = (0, 3)$, vertical transverse axis of length $2b = 2(2) = 4$, and horizontal conjugate axis of length $2a = 2(1) = 2$, as shown in Figure 6.40. If c is the length of the diagonal of the rectangle that is formed by $a = 1$ and $b = 2$, then by the Pythagorean theorem

$$c^2 = 1^2 + 2^2,$$

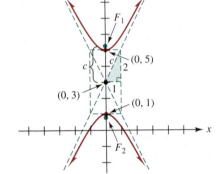

FIGURE 6.40
The coordinates of the foci of the hyperbola $y^2 - 4x^2 - 6y + 5 = 0$ are found by applying the Pythagorean theorem.

which implies $c = \sqrt{5}$. Now, $c = \sqrt{5}$ is also the length of the line segment from the center to a focus. Therefore, the coordinates of the foci are

$$F_1(0, 3 + \sqrt{5}) \quad \text{and} \quad F_2(0, 3 - \sqrt{5})$$

The eccentricity e of this hyperbola is

$$e = \frac{\text{Distance from center to focus}}{\text{Distance from center to vertex}} = \frac{\sqrt{5}}{2} \approx 1.12.$$

PROBLEM 3 Find the foci and eccentricity for the hyperbola $9x^2 - y^2 - 9 = 0$.

As illustrated in the next example, if we know the foci and eccentricity of a hyperbola, we can determine its equation.

EXAMPLE 4 Find the equation of the hyperbola in standard form with foci $F_1(2, 5)$ and $F_2(2, -3)$ and eccentricity 2.

SOLUTION The foci of the hyperbola are on the vertical line $x = 2$ and the center of the hyperbola is located halfway between the foci. Hence, the transverse axis of the hyperbola is vertical with length $2b$, and the center of the hyperbola is $(h, k) = (2, 1)$. Now referring to Figure 6.41,

$$\frac{\text{Distance from center to focus}}{\text{Distance from center to vertex}} = \frac{4}{b} = 2,$$

which implies $b = 2$. Hence, the coordinates of the vertices of the hyperbola are $(2, 3)$ and $(2, -1)$. Since $c = 4$ is the length of the line segment from the center to the focus, $c = 4$ is also the length of the diagonal of the rectangle that is formed by $b = 2$ and a. Now, applying the Pythagorean theorem to the shaded right triangle in Figure 6.41, we obtain

$$4^2 = a^2 + 2^2,$$

which implies $a = 2\sqrt{3}$. Since $(h, k) = (2, 1)$, $a^2 = (2\sqrt{3})^2 = 12$, and $b^2 = (2)^2 = 4$, the equation of the hyperbola in standard form is

$$\frac{(y - 1)^2}{4} - \frac{(x - 2)^2}{12} = 1.$$

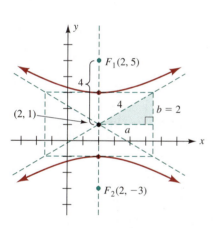

FIGURE 6.41
A hyperbola with foci $(2, 5)$ and $(2, -3)$ and eccentricity 2 has center $(2, 1)$ and vertical transverse axis $2b = 4$.

PROBLEM 4 Referring to Example 4, express the equation of the hyperbola in general form with integer coefficients.

◆ **Classification and Common Definition of the Conic Sections**

In the first three sections of this chapter, we have examined the graphs of general quadratic equations in two unknowns of the form

$$Ax^2 + Cy^2 + Dx + Ey + F = 0,$$

where *A, C, D, E,* and *F* are real numbers. We can classify the graph of this equation as a parabola, circle, ellipse, or hyperbola by looking at the values of *A* and *C*.

Classification of the Conic Sections

If the graph of a general quadratic equation in two unknowns

$$Ax^2 + Cy^2 + Dx + Ey + F = 0$$

exists and is not degenerate, then the graph is

1. a **parabola** if either $A = 0$ or $C = 0$, but not both.
2. a **circle** if $A = C \neq 0$.
3. an **ellipse** if $A \neq C$ and $AC > 0$.
4. a **hyperbola** if $AC < 0$.

In the first three sections of this chapter, we have also given three separate geometric definitions of the parabola, ellipse, and hyperbola. We now present the *common definition of the conic sections* in terms of the eccentricity. We will use this new definition to develop the polar equations of the conic sections in Section 6.5.

Common Definition of the Conic Sections

Let a focus *F* be a fixed point not on a fixed line (the directrix). Also, let $P(x, y)$ be a point in the plane and *Q* a point on the fixed line such that $\overline{PQ}$ is perpendicular to the fixed line. The set of all points *P* for which

$$\frac{PF}{PQ} = e \text{ (the eccentricity)}$$

is

1. a **parabola** if $e = 1$.
2. a **circle** if $e = 0$.
3. an **ellipse** if $0 < e < 1$.
4. a **hyperbola** if $e > 1$.

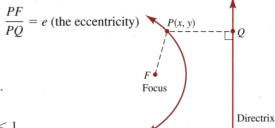

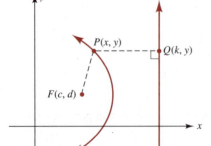

FIGURE 6.42

The set of all points $P(x, y)$ such that $PF/PQ = e$ defines a conic section.

We now show that this definition is consistent with our classification of the conic sections. Referring to the conic section in Figure 6.42 with focus $F(c, d)$ and corresponding directrix $x = k$, we have

$$\frac{PF}{PQ} = e$$

$$PF = e \cdot PQ \qquad \text{\textbf{Multiply both sides by } } PQ$$

SECTION 6.3 The Hyperbola and the Common Definition

$$\sqrt{(x-c)^2+(y-d)^2} = e|x-k| \quad \text{Apply the distance formula}$$

$$(x-c)^2+(y-d)^2 = e^2(x-k)^2 \quad \text{Square both sides}$$

$$x^2-2cx+c^2+y^2-2dy+d^2 = e^2x^2-2e^2kx+e^2k^2 \quad \text{Expand}$$

$$(1-e^2)x^2+y^2+(2ke^2-2c)x-2dy+(d^2-e^2k^2)=0 \quad \text{Collect like terms}$$

This last equation is a general quadratic equation of the form

$$Ax^2+Cy^2+Dx+Ey+F=0.$$

with $A = 1 - e^2$, $C = 1$, $D = 2ke^2 - 2c$, $E = -2d$ and $F = d^2 - e^2k^2$. We consider four cases:

Case 1: If $e = 1$, then $A = 1 - e^2 = 0$. When $A = 0$ and $C = 1$, we have a *parabola*.

Case 2: If $e = 0$, then $A = 1 - e^2 = 1$. When $A = C = 1$, we have *circle*.

Case 3: If $0 < e < 1$, then $A = 1 - e^2 > 0$. When $A > 0$ and $C = 1$, $AC > 0$. Hence, we have an *ellipse*.

Case 4: If $e > 1$, then $A = 1 - e^2 < 0$. When $A < 0$ and $C = 1$, $AC < 0$. Hence, we have a *hyperbola*.

EXAMPLE 5 Suppose a conic section has focus $F(4, 0)$, corresponding directrix $x = 9$, and eccentricity $e = \frac{2}{3}$. Find the equation of this conic section in general form and classify the equation as a parabola, circle, ellipse, or hyperbola.

SOLUTION Since $e = \frac{2}{3} < 1$, we expect to obtain the equation of an ellipse. Now referring to Figure 6.43 and applying our common definition, we have

$$e = \frac{2}{3} = \frac{PF}{PQ}$$

$$3PF = 2PQ \quad \text{Cross multiply}$$

$$3\sqrt{(x-4)^2+y^2} = 2|9-x| \quad \text{Apply the distance formula}$$

$$9[(x-4)^2+y^2] = 4(9-x)^2 \quad \text{Square both sides}$$

$$9(x^2-8x+16+y^2) = 4(81-18x+x^2) \quad \text{Expand the binomials}$$

$$5x^2 + 9y^2 - 180 = 0 \quad \text{Collect like terms}$$

[General form]

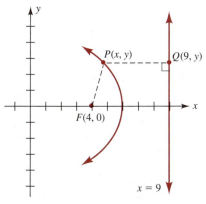

FIGURE 6.43
The conic section defined by $\dfrac{PF}{PQ} = \dfrac{2}{3} < 1$ should be an ellipse.

This last equation is the equation of an ellipse, since $A \neq C$ and $AC = (5)(9) = 45 > 0$. ◆

PROBLEM 5 Sketch the graph of the ellipse in Example 5 and label the endpoints of the major and minor axes. ◆

Exercises 6.3

Basic Skills

In Exercises 1–16, sketch the graph of each hyperbola. Label the center of the hyperbola if it is not at the origin, and label the vertices of the transverse axis.

1. $\dfrac{x^2}{4} - \dfrac{y^2}{16} = 1$
2. $\dfrac{y^2}{4} - \dfrac{x^2}{4} = 1$
3. $9y^2 - 4x^2 = 36$
4. $9x^2 - 16y^2 = 144$
5. $16x^2 - y^2 = 9$
6. $4y^2 - 16x^2 = 25$
7. $9x^2 - 25y^2 = -4$
8. $4y^2 - x^2 + 12 = 0$
9. $\dfrac{x^2}{9} - \dfrac{(y+2)^2}{4} = 1$
10. $\dfrac{(y-1)^2}{16} - (x+3)^2 = 1$
11. $(y-3)^2 - (x-1)^2 = 36$
12. $9(x+1)^2 - 25(y-4)^2 = 225$
13. $36x^2 - y^2 - 144x - 6y + 126 = 0$
14. $16y^2 - 9x^2 + 90x = 261$
15. $9y^2 - 3x^2 - 24x + 18y = 120$
16. $2x^2 - 4y^2 + 12x + 16y + 1 = 0$

Find the foci and eccentricity of the hyperbola whose equation is given in the indicated exercise.

17. Exercise 3
18. Exercise 4
19. Exercise 9
20. Exercise 12
21. Exercise 13
22. Exercise 16

In Exercises 23–32, determine the equation of the hyperbola with the given characteristics.

23. Vertices at $(0, \pm 4)$ and asymptotes $4y - x = 0$ and $4y + x = 0$.
24. Vertices at $(2, -2)$ and $(-2, -2)$ and endpoints of conjugate axis $(0, 0)$ and $(0, -4)$.
25. Asymptotes $y = \pm x$ and horizontal transverse axis of length 4.
26. Asymptotes $y = -x$ and $y = x + 2$ and vertical transverse axis of length 4.
27. Vertices at $(-2, -3)$ and $(-2, 3)$ and one focus at $(-2, 5)$.
28. Foci at $(-4, -1)$ and $(2, -1)$ and one of the vertices at $(1, -1)$.
29. Foci at $(\pm 2, 0)$ and transverse axis of length 2.
30. Foci at $(1, \pm 4)$ and conjugate axis of length 4.
31. Foci at $(0, 0)$ and $(3, 0)$ and eccentricity 3.
32. Foci at $(1, -3)$ and $(1, 5)$ and eccentricity $\tfrac{4}{3}$.

In Exercises 33–40, classify the graph of the equation as a parabola, circle, ellipse, or hyperbola. Assume that the graph of each equation exists and is not a degenerate conic section.

33. $x^2 + 4y^2 - 8y = 0$
34. $x^2 + y^2 + 16x = 0$
35. $y^2 - 2x - 4y + 10 = 0$
36. $y^2 - x^2 + 6x - 2y + 1 = 0$
37. $9x^2 - y^2 + 18x - 27 = 0$
38. $x^2 - 6x - 3y + 6 = 0$
39. $16x^2 + 9y^2 - 32x - 36y + 43 = 0$
40. $9x^2 + y^2 - 18x - 10y + 30 = 0$

In Exercises 41–46, a focus, its corresponding directrix, and the eccentricity of a conic section are given. Find the equation of the conic section in genral form, and classify the equation as a parabola, ellipse, or hyperbola.

41. Focus $(4, 0)$; directrix $x = 6$; eccentricity $e = 1$
42. Focus $(-2, 3)$; directrix $y = 5$; eccentricity $e = 1$
43. Focus $(3, 1)$; directrix $y = -3$; eccentricity $e = \tfrac{1}{2}$
44. Focus $(0, 8)$; directrix $x = -4$; eccentricity $e = \tfrac{4}{5}$
45. Focus $(2, -6)$; directrix x-axis; eccentricity $e = 2$
46. Focus $(-2, -3)$; directrix $x = 1$; eccentricity $e = \tfrac{3}{2}$

47. A Cassegrain telescope contains a primary parabolic mirror and a smaller hyperbolic mirror, as shown in the sketch. A ray of light from the moon strikes the parabolic mirror and is reflected toward the parabola's focus F_1, which is also a focus of the hyperbolic mirror. The ray of light is then reflected off the hyperbolic mirror toward the hyperbola's other focus F_2, which is the eyepiece of the telescope. If the equation of the hyperbolic mirror is $(x^2/441) - (y^2/400) = 1$, with x and y measured in centimeters, determine the distance from the hyperbolic mirror to the eyepiece.

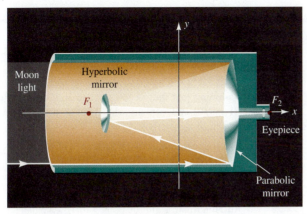

EXERCISE 47

comes within 6000 miles of the earth's surface before moving away. Assuming the radius of the earth is 4000 miles, determine the equation of the path of the UFO.

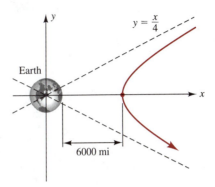

48. The figure shows the hyperbolic path of a UFO whose path moves closer toward earth along the line $y = x/4$ and

---

 Critical Thinking

49. Describe the general shape of a hyperbola whose eccentricity is (a) close to 1, or (b) very large.

50. What is the eccentricity of a hyperbola with asymptotes $y = \pm x$?

51. Suppose the transverse axis and conjugate axis of a hyperbola are both doubled. What effect does this have on the hyperbola's (a) eccentricity? (b) asymptotes?

52. The equation of a hyperbola with a horizontal transverse axis is $Ax^2 - Cy^2 = E$. Find the coordinates of its foci in terms of A, C, and E.

53. Describe the graph of each degenerate conic section.
    (a) $x^2 + y^2 - 6x - 8y + 25 = 0$
    (b) $9x^2 + 4y^2 + 54x - 16y + 97 = 0$
    (c) $4x^2 - y^2 - 16x - 4y + 12 = 0$
    (d) $y^2 - 9x^2 + 36x - 10y - 11 = 0$

54. When a supersonic jet breaks the sound barrier, it produces a conical shock wave, as shown in the sketch. Describe the shape of the intersection of the conical shock wave with the ground if the plane's path is parallel to the ground.

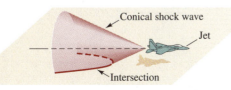

If a and b are positive real numbers, then solving

$$\frac{x^2}{a^2} - \frac{y^2}{b^2} = 1 \quad \text{and} \quad \frac{y^2}{b^2} - \frac{x^2}{a^2} = 1$$

for y yields

$$y = \pm \frac{b}{a}\sqrt{x^2 - a^2} \quad \text{and} \quad y = \pm \frac{b}{a}\sqrt{x^2 + a^2},$$

respectively. Taken separately, the equations

$$y = \frac{b}{a}\sqrt{x^2 - a^2} \qquad y = -\frac{b}{a}\sqrt{x^2 - a^2}$$

$$y = \frac{b}{a}\sqrt{x^2 + a^2} \qquad y = -\frac{b}{a}\sqrt{x^2 + a^2}$$

define y as a function of x, and the graph of each function is one half of a hyperbola. In Exercises 55–58, state the domain and range of each function, and sketch its graph.

55. $y = \frac{1}{2}\sqrt{x^2 - 4}$ 56. $y = -3\sqrt{x^2 - 25}$

57. $y = -\sqrt{x^2 + 9}$ 58. $y = \frac{3}{4}\sqrt{4x^2 + 1}$

Calculator Activities

In Exercises 59–62, find the coordinates of the vertices of each hyperbola. Round each coordinate to three significant digits.

59. $356x^2 - 129y^2 = 926$
60. $1.25x^2 - 2.70y^2 + 8.55y = 2.35$
61. $0.35x^2 - 0.12y^2 - 2.34y - 7.89 = 0$
62. $3.4x^2 - 6.4y^2 - 1.4x - 4.8y - 18.6 = 0$

If you have access to a graphing calculator, you can check the results of Exercises 59–62. To display the graph of the hyperbola $Ax^2 - Cy^2 + Dx + Ey + F = 0$ on a graphing calculator, first solve the equation of the hyperbola for y, and then enter the two functions that result. To do this, rewrite the equation as a quadratic in y, $-Cy^2 + Ey + (Ax^2 + Dx + F) = 0$, and then apply the quadratic formula to obtain

$$y = \frac{-E \pm \sqrt{E^2 + 4C(Ax^2 + Dx + F)}}{-2C}$$

In Exercise 63–66, use a graphing calculator to generate the graph of each hyperbola in the viewing rectangle. Then estimate the coordinates of the vertices by tracing to these points. Round each coordinate to three significant digits.

63. The hyperbola in Exercise 59.
64. The hyperbola in Exercise 60.
65. The hyperbola in Exercise 61.
66. The hyperbola in Exercise 62.

In Exercises 67–70, determine the equation in standard form of the hyperbola with the given characteristics.

67. Asymptotes $y = \pm 0.34x$ and horizontal transverse axis of length 6.24.
68. Vertices at $(3.32, -4.12)$ and $(3.32, 2.58)$ and conjugate axis of length 2.10.
69. Foci at $(\pm 21.9, 0)$ and transverse axis of length 34.6.
70. Foci at $(1.23, -3.67)$ and $(1.23, 5.33)$ and eccentricity 1.80.
71. Allied tanks are located at fixed points A and B, as shown in the sketch. An enemy tank located at point P fires its gun, and 1.3 seconds after the shot is heard at point B, the shot is heard at point A.

    (a) Assuming that the speed of sound is 1100 ft/s, find $PA - PB$ and explain why point P must lie on a branch of a hyperbola.

    (b) Determine the equation of the branch of the hyperbola described in part (a) if the coordinates (in feet) of points A and B are $(-1500, 0)$ and $(1500, 0)$, respectively.

    (c) An allied patrol unit at point $C(0, 0)$ calculates that the distance from them to the enemy tank is 950 feet. Determine the distances from the allied tanks to the enemy tank.

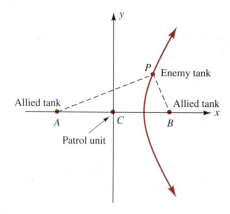

72. In an engineer's design for the horn on a megaphone a hyperbola is rotated about the x-axis (see figure). If the distance from A to B is 10.26 cm, determine the radius of each end of the megaphone.

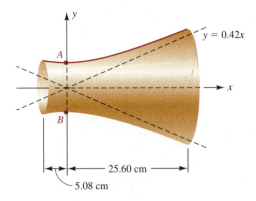

SECTION 6.4 Rotation of Axes 361

6.4 Rotation of Axes

Introductory Comments

In the previous three sections, we looked at the general quadratic equation in two unknowns,

$$Ax^2 + Bxy + Cy^2 + Dx + Ey + F = 0,$$

with $B = 0$. As we observed, if the graph of this type of equation exists and is not a degenerate case, then its graph is one of the conic sections—parabola, circle, ellipse, or hyperbola—with axes parallel to the xy-axes. In this section, we look at general quadratic equations in two unknowns in which $B \neq 0$. The presence of an xy-term rotates the conic section through some positive acute angle θ so that the axes of the conic section are no longer parallel to the xy-axes. However, note in Figure 6.44 that angle θ determines a new pair of axes, called the uv-axes, that share the same origin as the xy-axes and are *parallel* to the axes of the conic section. Hence, the equation of the conic section in the uv-coordinate system must be of the form

$$A'u^2 + C'v^2 + D'u + E'v + F' = 0.$$

We can sketch the graph of this equation by using the techniques developed in the preceding section, that is, by writing the equation in standard form and noting the important characteristics of the conic section.

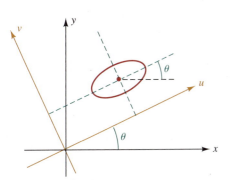

FIGURE 6.44
A conic section rotated through a positive acute angle θ with uv-axes that are parallel to the axes of the conic section.

Rotation Formulas

We begin by developing a set of formulas that express the old coordinates x and y in terms of the new coordinates u and v. Referring to Figure 6.45, note that the point P has coordinates (x, y) in the xy-coordinate system and (u, v) in the uv-coordinate system. Using right triangle OMP with (u, v) the coordinates of point P, we find

$$\cos \alpha = \frac{u}{r} \quad \text{and} \quad \sin \alpha = \frac{v}{r}$$

$$u = r \cos \alpha \qquad v = r \sin \alpha \qquad \textbf{Multiply both sides by } r$$

Now, referring to right triangle ONP with (x, y) the coordinates of point P, we have

$$\cos(\theta + \alpha) = \frac{x}{r}$$

$$x = r \cos(\theta + \alpha) \qquad \textbf{Multiply both sides by } r$$

$$x = r(\cos \theta \cos \alpha - \sin \theta \sin \alpha) \qquad \textbf{Apply sum formula for cosine}$$

$$x = (r \cos \alpha) \cos \theta - (r \sin \alpha) \sin \theta \qquad \textbf{Rewrite}$$

$$x = u \cos \theta - v \sin \theta \qquad \textbf{Substitute}$$

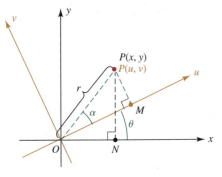

FIGURE 6.45
Point P has coordinates (x, y) in the xy-coordinate system and (u, v) in the uv-coordinate system.

and

$$\sin(\theta + \alpha) = \frac{y}{r}$$

$$y = r\sin(\theta + \alpha) \quad \text{Multiply both sides by } r$$
$$y = r(\sin\theta\cos\alpha + \cos\theta\sin\alpha) \quad \text{Apply sum formula for sine}$$
$$y = (r\cos\alpha)\sin\theta + (r\sin\alpha)\cos\theta \quad \text{Rewrite}$$
$$y = u\sin\theta + v\cos\theta \quad \text{Substitute}$$

We refer to these results as the **rotation formulas.**

 Rotation Formulas

If point P has coordinates (x, y) in the xy-coordinate system and (u, v) in the uv-coordinate system, then

$$x = u\cos\theta - v\sin\theta$$

and

$$y = u\sin\theta + v\cos\theta,$$

where θ is the angle of rotation from the positive x-axis to the positive u-axis.

EXAMPLE 1

Find the coordinates of point P in the xy-coordinate system if the coordinates of point P in the uv-coordinate system are $(5, \sqrt{3})$ and the angle of rotation from the positive x-axis to the positive u-axis is 30°.

SOLUTION Using the rotation formulas with $u = 5$, $v = \sqrt{3}$, and $\theta = 30°$, gives us the coordinates:

$$x = u\cos\theta - v\sin\theta \qquad \text{and} \qquad y = u\sin\theta + v\cos\theta$$
$$= 5\cos 30° - \sqrt{3}\sin 30° \qquad\qquad = 5\sin 30° + \sqrt{3}\cos 30°$$
$$= 5\left(\frac{\sqrt{3}}{2}\right) - \sqrt{3}\left(\frac{1}{2}\right) \qquad\qquad = 5\left(\frac{1}{2}\right) + \sqrt{3}\left(\frac{\sqrt{3}}{2}\right)$$
$$= 2\sqrt{3} \qquad\qquad\qquad\qquad\qquad = 4$$

Hence, the coordinates of point P in the xy-coordinate system are $(2\sqrt{3}, 4)$, as shown in Figure 6.46. ◆

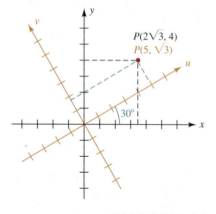

FIGURE 6.46
Point P has coordinates $(5, \sqrt{3})$ in the uv-coordinate system and $(2\sqrt{3}, 4)$ in the xy-coordinate system.

If a rotation of the xy-coordinate system through an angle θ is used to obtain the uv-coordinate system, then a rotation of the uv-coordinate system through an angle

SECTION 6.4 Rotation of Axes

$(-\theta)$ must yield the *xy*-coordinate system. If we replace x with u, y with v, and θ with $(-\theta)$ in the rotation formulas, we obtain

$$u = x\cos(-\theta) - y\sin(-\theta) \quad \text{and} \quad v = x\sin(-\theta) + y\cos(-\theta).$$

However, since $\cos(-\theta) = \cos\theta$ and $\sin(-\theta) = -\sin\theta$, we have

$$\boxed{u = x\cos\theta + y\sin\theta}$$

and

$$\boxed{v = -x\sin\theta + y\cos\theta}$$

These two formulas may be used to obtain the coordinates of a point P in the *uv*-coordinate system when given the coordinates of point P in the *xy*-coordinate system.

PROBLEM 1 Find the coordinates of point P in the *uv*-coordinate system if the coordinates of point P in the *xy*-coordinate system are $(2\sqrt{3}, 4)$ and the angle of rotation from the positive *x*-axis to the positive *u*-axis is $30°$. Refer to Example 1 to verify your answer. ◆

◆ **Rotation Angle that Eliminates the *xy*-Term**

To change the equation $Ax^2 + Bxy + Cy^2 + Dx + Ey + F = 0$ with $B \neq 0$ to an equivalent equation in the *uv*-coordinate system, we use the rotation formulas and replace x with $u\cos\theta - v\sin\theta$ and y with $u\sin\theta + v\cos\theta$ to obtain the following equation:

$$A(u^2\cos^2\theta - 2uv\sin\theta\cos\theta + v^2\sin^2\theta)$$
$$+ B(u^2\sin\theta\cos\theta + uv\cos^2\theta - uv\sin^2\theta - v^2\sin\theta\cos\theta)$$
$$+ C(u^2\sin^2\theta + 2uv\sin\theta\cos\theta + v^2\cos^2\theta)$$
$$+ D(u\cos\theta - v\sin\theta)$$
$$+ E(u\sin\theta + v\cos\theta)$$
$$+ F = 0.$$

Now combining like powers of u and v gives us

$$(A\cos^2\theta + B\sin\theta\cos\theta + C\sin^2\theta)u^2$$
$$+ (-2A\sin\theta\cos\theta + B\cos^2\theta - B\sin^2\theta + 2C\sin\theta\cos\theta)uv$$
$$+ (A\sin^2\theta - B\sin\theta\cos\theta + C\cos^2\theta)v^2$$
$$+ (D\cos\theta + E\sin\theta)u$$
$$+ (-D\sin\theta + E\cos\theta)v$$
$$+ F = 0,$$

which is an equation of the form

$$A'u^2 + B'uv + C'v^2 + D'u + E'v + F' = 0.$$

In order to sketch the graph of this equation by the methods we developed in the preceding section, the coefficient of the *uv*-term must be zero, that is,

$-2A \sin \theta \cos \theta + B \cos^2\theta - B \sin^2\theta + 2C \sin \theta \cos \theta = 0$

$(C - A) 2 \sin \theta \cos \theta + B(\cos^2 \theta - \sin^2 \theta) = 0$ **Rewrite**

$(C - A) \sin 2\theta + B \cos 2\theta = 0$ **Apply double-angle formulas**

$(C - A) + B \cot 2\theta = 0$ **Divide both sides by sin 2θ**

$\cot 2\theta = \dfrac{A - C}{B}$ **Solve for cot 2θ**

In summary, the general quadratic equation in two unknowns,

$$Ax^2 + Bxy + Cy^2 + Dx + Ey + F = 0$$

with $B \neq 0$, may be written in the form

$$A'u^2 + C'v^2 + D'u + E'v + F' = 0$$

by rotating the *xy*-coordinate system through an angle θ that satisfies the equation

$$\boxed{\cot 2\theta = \dfrac{A - C}{B}}$$

We refer to the positive acute angle θ that satisfies the equation $\cot 2\theta = (A - C)/B$ as the **rotation angle that eliminates the *xy* term.**

EXAMPLE 2   Find the rotation angle θ that eliminates the *xy* term in each equation.

(a) $xy - 2 = 0$

(b) $13x^2 + 6\sqrt{3}\,xy + 7y^2 - 32\sqrt{3}\,x - 32y + 48 = 0$

SOLUTION

(a) The equation $xy - 2 = 0$ is a general quadratic equation in two unknowns with $A = 0$, $B = 1$, $C = 0$, $D = 0$, $E = 0$, and $F = -2$. To eliminate the *xy* term, we must rotate the *xy*-axes through an angle θ that satisfies

$$\cot 2\theta = \dfrac{A - C}{B} = \dfrac{0 - 0}{1} = 0.$$

Now, $\cot 2\theta = 0$ implies $2\theta = 90°$, or $\theta = 45°$. Thus, the rotation angle θ that eliminates the *xy* term is $45°$.

(b) The equation $13x^2 + 6\sqrt{3}\,xy + 7y^2 - 32\sqrt{3}\,x - 32y + 48 = 0$ is a general quadratic equation in two unknowns with $A = 13$, $B = 6\sqrt{3}$, $C = 7$, $D = -32\sqrt{3}$, $E = -32$, and $F = 48$. To eliminate the xy term, we must rotate the xy-axes through an angle θ that satisfies

$$\cot 2\theta = \frac{A - C}{B} = \frac{13 - 7}{6\sqrt{3}} = \frac{1}{\sqrt{3}}.$$

Now, $\cot 2\theta = 1/\sqrt{3}$ implies $2\theta = 60°$, or $\theta = 30°$. Thus, the rotation angle θ that eliminates the xy term is $30°$. ◆

PROBLEM 2 Repeat Example 2 for the equation $3x^2 - \sqrt{3}\,xy + 2y^2 + 3x - 7 = 0$. ◆

The rotation angle θ is always a positive acute angle. Hence, we must have $0° < 2\theta < 180°$. Since the range of the inverse cosine function covers all angles from $0°$ to $180°$, we may use the inverse cosine key on a calculator set in degree mode to find the angle 2θ that satifies the equation $\cot 2\theta = (A - C)/B$.

EXAMPLE 3 Find the approximate value of the rotation angle θ that eliminates the xy term in the equation

$$4x^2 - 4xy + y^2 - 5\sqrt{5}\,x + 5 = 0.$$

SOLUTION This is a general quadratic equation in two unknowns with $A = 4$, $B = -4$, $C = 1$, $D = -5\sqrt{5}$, $E = 0$, and $F = 5$. To eliminate the xy term, we must rotate the xy-axes through an angle θ that satisfies

$$\cot 2\theta = \frac{A - C}{B} = \frac{4 - 1}{-4} = -\frac{3}{4}.$$

By the definition of cotangent, we conclude that 2θ is a second-quadrant angle whose terminal side passes through the point $(-3, 4)$. Now, referring to Figure 6.47, we have

$$r = \sqrt{(-3)^2 + (4)^2} = 5.$$

Hence, by the definition of cosine,

$$\cos 2\theta = -\frac{3}{5}$$

$$2\theta = \cos^{-1}\left(-\frac{3}{5}\right) \qquad \text{Solve for } 2\theta$$

$$\theta = \frac{\cos^{-1}\left(-\frac{3}{5}\right)}{2} \approx 63.4° \qquad \text{Evaluate by using the } \boxed{\cos^{-1}} \text{ key on a calculator set in degree mode}$$

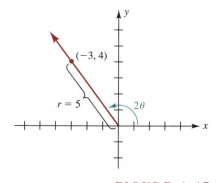

FIGURE 6.47
If $\cot 2\theta = -\frac{3}{4}$, then the point $(-3, 4)$ is on the terminal side of 2θ.

Thus, the rotation angle θ that eliminates the xy term is 63.4° (to the nearest tenth of a degree). ◆

PROBLEM 3 Repeat Example 3 for the equation $8y^2 + 6xy - 9 = 0$. ◆

◆ Classification of the Conic Sections

We can classify the graph of

$$Ax^2 + Bxy + Cy^2 + Dx + Ey + F = 0,$$

with $B \neq 0$ as a parabola, ellipse, or hyperbola, by evaluating its *discriminant*. We define the **discriminant** of a general quadratic equation in two unknowns as

$$B^2 - 4AC.$$

In Exercise 44, you are asked to show that the discriminant is an *invariant* (an expression that remains constant) under any rotation of axes. That is, whatever rotation angle we select in changing

$$Ax^2 + Bxy + Cy^2 + Dx + Ey + F = 0$$

to

$$A'u^2 + B'uv + C'v^2 + D'u + E'v + F' = 0,$$

we have

$$B^2 - 4AC = (B')^2 - 4A'C'.$$

Now, if the rotation angle is chosen so that $B' = 0$, then

$$B^2 - 4AC = -4A'C'.$$

Using this last equation, we consider three cases:

Case 1: If $B^2 - 4AC$ is *zero*, then $-4A'C'$ must be zero. This implies that either $A' = 0$ or $C' = 0$. Recall from Section 6.3 that the graph of $A'u^2 + C'v^2 + D'u + E'v + F' = 0$ is a *parabola* if either $A' = 0$ or $C' = 0$.

Case 2: If $B^2 - 4AC$ is *negative*, then $-4A'C'$ must be negative. This implies that $A'C' > 0$. Recall from Section 6.3 that the graph of $A'u^2 + C'v^2 + D'u + E'v + F' = 0$ is an *ellipse* (or a circle, which is a special case of an ellipse) if $A'C' > 0$.

Case 3: If $B^2 - 4AC$ is *positive*, then $-4A'C'$ must be positive. This implies that $A'C' < 0$. Recall from the preceding section that the graph of $A'u^2 + C'v^2 + D'u + E'v + F' = 0$ is a *hyperbola* if $A'C' < 0$.
We now summarize our findings.

Classification of the Conic Sections

If the graph of the general quadratic equation in two unknowns,

$$Ax^2 + Bxy + Cy^2 + Dx + Ey + F = 0$$

with $B \neq 0$, exists and is not degenerate, then its graph is

1. a *parabola* if $B^2 - 4AC = 0$.
2. an *ellipse* (or a circle, which is a special case of an ellipse) if $B^2 - 4AC < 0$.
3. a *hyperbola* if $B^2 - 4AC > 0$.

EXAMPLE 4 Classify each equation as a parabola, ellipse, or hyperbola. Assume that the graph of each equation exists and is not degenerate.

(a) $xy - 2 = 0$

(b) $13x^2 + 6\sqrt{3}\,xy + 7y^2 - 32\sqrt{3}\,x - 32y + 48 = 0$

(c) $4x^2 - 4xy + y^2 - 5\sqrt{5}\,x + 5 = 0$

SOLUTION

(a) In the equation $xy - 2 = 0$, we have $A = 0$, $B = 1$, and $C = 0$. Hence, the discriminant is

$$B^2 - 4AC = 1^2 - 4(0)(0) = 1 > 0.$$

Since the discriminant is greater than zero, we conclude that the graph of this equation is a hyperbola.

(b) In the equation $13x^2 + 6\sqrt{3}\,xy + 7y^2 - 32\sqrt{3}\,x - 32y + 48 = 0$, we have $A = 13$, $B = 6\sqrt{3}$, and $C = 7$. Hence, the discriminant is

$$B^2 - 4AC = (6\sqrt{3})^2 - 4(13)(7) = -256 < 0.$$

Since the discriminant is less than zero, we conclude that the graph of this equation is an ellipse.

(c) In the equation $4x^2 - 4xy + y^2 - 5\sqrt{5}\,x + 5 = 0$, we have $A = 4$, $B = -4$, and $C = 1$. Hence, the discriminant is

$$B^2 - 4AC = (-4)^2 - 4(4)(1) = 0.$$

Since the discriminant is zero, we conclude that the graph of this equation is a parabola. ◆

PROBLEM 4 Repeat Example 4 for the equation $8y^2 + 6xy - 9 = 0$. ◆

◆ **Graphing $Ax^2 + Bxy + Cy^2 + Dx + Ey + F = 0$ with $B \neq 0$**

If we know the rotation angle θ that eliminates the xy term, then we can use the rotation formulas to write the general quadratic equation in two unknowns, with $B \neq 0$, in the form

$$A'u^2 + C'v^2 + D'u + E'v + F' = 0.$$

We can sketch the graph of this equation by using the techniques developed in the preceding section, that is, by writing it in standard form and noting the important characteristics of the conic section.

EXAMPLE 5 Sketch the graph of each equation.

(a) $xy - 2 = 0$

(b) $13x^2 + 6\sqrt{3}\,xy + 7y^2 - 32\sqrt{3}\,x - 32y + 48 = 0$

SOLUTION

(a) From Example 4(a) and Example 2(a), the graph of this equation is a hyperbola, and the rotation angle θ that eliminates the xy term is $45°$. Thus, by the rotation formulas, we have

$$x = u\cos 45° - v\sin 45° = \frac{u-v}{\sqrt{2}}$$

and

$$y = u\sin 45° + v\cos 45° = \frac{u+v}{\sqrt{2}}.$$

Hence,

$$xy - 2 = 0$$

$\left(\dfrac{u-v}{\sqrt{2}}\right)\left(\dfrac{u+v}{\sqrt{2}}\right) = 2$ Substitute and add 2 to both sides

$\dfrac{u^2 - v^2}{2} = 2$ Simplify the left-hand side

$u^2 - v^2 = 4$ Multiply both sides by 2

$\dfrac{(u-0)^2}{2^2} - \dfrac{(v-0)^2}{2^2} = 1$ Divide both sides by 4 and write in standard form

We recognize this last equation as the equation of a hyperbola in standard-form with center $(0, 0)$, transverse axis of length $2a = 2(2) = 4$, and conjugate axis of length $2b = 2(2) = 4$. The graph of this hyperbola in the uv-coordinate system represents the graph of $xy - 2 = 0$ in the

SECTION 6.4 Rotation of Axes

xy-coordinate system, as shown in Figure 6.48. Note that the asymptotes for $xy - 2 = 0$ are the x- and y-axes.

(b) From Example 4(b) and Example 2(b), the graph of this equation is an ellipse, and the rotation angle θ that eliminates the xy term is 30°. Thus, by the rotation formulas, we have

$$x = u \cos 30° - v \sin 30° = \frac{\sqrt{3}\,u - v}{2}$$

and

$$y = u \sin 30° + v \cos 30° = \frac{u + \sqrt{3}\,v}{2}.$$

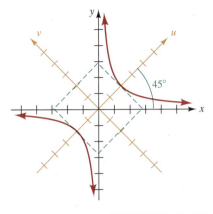

FIGURE 6.48
The graph of $xy - 2 = 0$ is a hyperbola with its axes rotated 45°.

Hence,

$$13x^2 + 6\sqrt{3}\,xy + 7y^2 - 32\sqrt{3}\,x - 32y + 48 = 0$$

$$13\left(\frac{\sqrt{3}\,u - v}{2}\right)^2 + 6\sqrt{3}\left(\frac{\sqrt{3}\,u - v}{2}\right)\left(\frac{u + \sqrt{3}\,v}{2}\right) + 7\left(\frac{u + \sqrt{3}\,v}{2}\right)^2 -$$

$$32\sqrt{3}\left(\frac{\sqrt{3}\,u - v}{2}\right) - 32\left(\frac{u + \sqrt{3}\,v}{2}\right) + 48 = 0 \qquad \text{Substitute}$$

$$16u^2 + 4v^2 - 64u + 48 = 0 \qquad \text{Simplify and combine like terms}$$

$$4u^2 + v^2 - 16u + 12 = 0 \qquad \text{Divide both sides by 4}$$

$$4(u^2 - 4u \quad) + v^2 = -12 \qquad \text{Regroup and factor}$$

$$4(u^2 - 4u + 4) + v^2 = 4 \qquad \text{Complete the square, add 16 to both sides}$$

$$\frac{(u - 2)^2}{1^2} + \frac{(v - 0)^2}{2^2} = 1 \qquad \text{Divide both sides by 4 and write in standard form}$$

We recognize this last equation as the equation of an ellipse in standard form with center $(2, 0)$, major axis of length $2b = 2(2) = 4$, and minor axis of length $2a = 2(1) = 2$. The graph of this ellipse in the uv-coordinate system represents the graph of
$13x^2 + 6\sqrt{3}\,xy + 7y^2 - 32\sqrt{3}\,x - 32y + 48 = 0$ in the xy-coordinate system, as shown in Figure 6.49. ◆

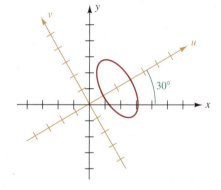

FIGURE 6.49
The graph of $13x^2 + 6\sqrt{3}\,xy + 7y^2 - 32\sqrt{3}\,x - 32y - 48 = 0$ is an ellipse with its axes rotated 30°.

PROBLEM 5 Referring to Example 5(b), the vertices of the ellipse in the uv-coordinate system are $(2, 2)$ and $(2, -2)$. What are the coordinates of the vertices of the ellipse in the xy-coordinate system? ◆

If the rotation angle θ that eliminates the xy term is not one of the special angles (30°, 45°, or 60°), then we may use the half-angle formulas (from Section 4.4) to determine the exact values of $\sin \theta$ and $\cos \theta$ for the rotation formulas. Since the rotation angle θ is always in quadrant I, we choose the positive square root for each half-angle formula and write

$$\sin\theta = \sqrt{\frac{1 - \cos 2\theta}{2}} \quad \text{and} \quad \cos\theta = \sqrt{\frac{1 + \cos 2\theta}{2}}$$

EXAMPLE 6 Sketch the graph of the equation $4x^2 - 4xy + y^2 - 5\sqrt{5}\,x + 5 = 0$.

SOLUTION From and Example 4(c) and Example 3, the graph of this equation is a parabola, and the rotation angle θ that eliminates the xy term is approximately $63.4°$. Using this value of θ in the rotation formulas would be very cumbersome. Instead, we use the half-angle formulas to determine the exact values of $\sin\theta$ and $\cos\theta$. Recall from Example 3 that

$$\cos 2\theta = -\frac{3}{5}.$$

Using this value for $\cos 2\theta$ in the half-angle formulas gives us

$$\sin\theta = \sqrt{\frac{1 - \cos 2\theta}{2}} = \sqrt{\frac{1 - (-\tfrac{3}{5})}{2}} = \frac{2}{\sqrt{5}}$$

and

$$\cos\theta = \sqrt{\frac{1 + \cos 2\theta}{2}} = \sqrt{\frac{1 + (-\tfrac{3}{5})}{2}} = \frac{1}{\sqrt{5}}.$$

Replacing $\sin\theta$ and $\cos\theta$ in the rotation formulas with these exact values gives us

$$x = \frac{u - 2v}{\sqrt{5}} \quad \text{and} \quad y = \frac{2u + v}{\sqrt{5}}.$$

Hence,

$$4x^2 - 4xy + y^2 - 5\sqrt{5}\,x + 5 = 0$$

$4\left(\dfrac{u - 2v}{\sqrt{5}}\right)^2 - 4\left(\dfrac{u - 2v}{\sqrt{5}}\right)\left(\dfrac{2u + v}{\sqrt{5}}\right) + \left(\dfrac{2u + v}{\sqrt{5}}\right)^2 - 5\sqrt{5}\left(\dfrac{u - 2v}{\sqrt{5}}\right) + 5 = 0$ Substitute

$$5v^2 - 5u + 10v + 5 = 0 \quad \text{Simplify and combine like terms}$$

$$v^2 - u + 2v + 1 = 0 \quad \text{Divide both-sides by 5}$$

$$v^2 + 2v + 1 = u \quad \text{Add } u \text{ to both sides}$$

$$u = (v + 1)^2 + 0 \quad \text{Factor and write in standard form}$$

We recognize this last equation as the equation of a parabola in standard form with vertex $(0, -1)$ and axis of symmetry $v = -1$. The graph of this parabola in the uv-coordinate system represents the graph of

FIGURE 6.50

The graph of $4x^2 - 4xy + y^2 - 5\sqrt{5}x + 5 = 0$ is a parabola with its axis rotated 63.4°.

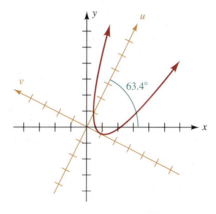

$4x^2 - 4xy + y^2 - 5\sqrt{5}\,x + 5 = 0$ in the xy-coordinate system, as shown in Figure 6.50. ◆

PROBLEM 6 Sketch the graph of the equation $8y^2 + 6xy - 9 = 0$. ◆

Exercises 6.4

Basic Skills

In Exercises 1–6, the coordinates of a point $P(u, v)$ and the rotation angle θ from the positive x-axis to the positive u-axis are given. Find the coordinates of point P in the xy-coordinate system.

1. $P(2\sqrt{3}, 2); \quad \theta = 30°$
2. $P(3, -\sqrt{3}); \quad \theta = 60°$
3. $P(-3\sqrt{2}, \sqrt{2}); \quad \theta = 45°$
4. $P(6, -2); \quad \theta = 45°$
5. $P(-5, -10); \quad \theta = \arctan \tfrac{4}{3}$
6. $P(13, 26); \quad \theta = \arcsin \tfrac{12}{13}$

In Exercises 7–12, the coordinates of a point $P(x, y)$ and the rotation angle θ from the positive x-axis to the positive u-axis are given. Find the coordinates of point P in the uv-coordinate system.

7. $P(-7, 3\sqrt{3}); \quad \theta = 60°$
8. $P(0, -8); \quad \theta = 30°$
9. $P(-8, -4); \quad \theta = 45°$
10. $P(-5\sqrt{2}, 3\sqrt{2}); \quad \theta = 45°$
11. $P(15, 5); \quad \theta = \cos^{-1}(\tfrac{3}{5})$
12. $P(4, -3); \quad \theta = \tan^{-1}(\tfrac{12}{5})$

Given that the graph of each equation in Exercises 13–36 exists and is not degenerate,

(a) classify each equation as a parabola, ellipse, or hyperbola,

(b) find the rotation angle θ that eliminates the xy term in the equation,

(c) use the rotation formulas with angle θ from part (b) to transform the equation to a uv-equation in standard form, and

(d) sketch the graph of the equation.

13. $xy + 8 = 0$
14. $2xy - 1 = 0$
15. $x^2 + xy + y^2 - 6 = 0$
16. $3x^2 - 10xy + 3y^2 + 8 = 0$
17. $7x^2 - 6\sqrt{3}\,xy + 13y^2 - 16 = 0$
18. $21x^2 + 10\sqrt{3}\,xy + 31y^2 - 144 = 0$
19. $x^2 + 2\sqrt{3}\,xy + 3y^2 - 2\sqrt{3}\,x + 2y = 0$
20. $\sqrt{2}x^2 + 2\sqrt{2}\,xy + \sqrt{2}\,y^2 - 4x + 4y = 0$
21. $2x^2 + 4\sqrt{3}\,xy - 2y^2 - 4 = 0$
22. $11x^2 - 10\sqrt{3}\,xy + y^2 - 16 = 0$
23. $5x^2 + 6xy + 5y^2 - 8\sqrt{2}\,x - 8\sqrt{2}\,y = 0$
24. $x^2 - 2xy + y^2 + 3\sqrt{2}\,x - 5\sqrt{2}\,y + 4 = 0$
25. $x^2 - 2xy + y^2 - 3\sqrt{2}\,x + \sqrt{2}\,y - 2 = 0$
26. $xy + \sqrt{2}\,y = 2$

27. $x^2 + 4xy + y^2 - 2x - 10y - 11 = 0$

28. $5x^2 - 8xy + 5y^2 + 2x + 2y - 7 = 0$

29. $2x^2 + 3xy - 2y^2 - 5 = 0$

30. $2x^2 + 4xy - y^2 - 6 = 0$

31. $6x^2 - 24xy - y^2 - 30 = 0$

32. $5x^2 - 12xy + 10y^2 - 14 = 0$

33. $9x^2 - 24xy + 16y^2 - 40x - 30y = 0$

34. $x^2 + 4xy - 2y^2 - 2x + 20y - 29 = 0$

35. $73x^2 + 72xy + 52y^2 + 30x - 40y - 75 = 0$

36. $9x^2 + 24xy + 16y^2 + 90x - 130y = 0$

37. Referring to the conic section in Exercise 17, find the coordinates in the xy-coordinate system for (a) the vertices and (b) the foci.

38. Referring to the conic section in Exercise 24, find the coordinates in the xy-coordinate system for (a) the vertices and (b) the foci.

Critical Thinking

39. Using the rotation formulas with $\theta = 90°$, transform the equation $(x^2/a^2) + (y^2/b^2) = 1$, with $a > 0$ and $b > 0$, to a uv-equation. Compare the graph of the uv equation to the xy equation.

40. Using the rotation formulas with a general angle θ, transform the equation $x^2 + y^2 = a^2$, with $a > 0$, to a uv-equation. Compare the graph of the uv equation to the xy equation.

41. By isolating radicals and squaring both sides, convert the equation $\sqrt{x} + \sqrt{y} = 1$ to a general quadratic equation in two unknowns. Identify the conic section it represents, and sketch its graph. What portion of this conic section represents the graph of $\sqrt{x} + \sqrt{y} = 1$?

42. Find the uv-equation of the hyperbola $3x^2 - 4y^2 = 12$ if it is rotated through a positive acute angle θ whose terminal side corresponds to an asymptote of $3x^2 - 4y^2 = 12$.

43. Describe the graph of each degenerate conic section.

    (a) $13x^2 - 6\sqrt{3}\,xy + 7y^2 = 0$

    (b) $x^2 - 10\sqrt{3}\,xy + 11y^2 = 0$

    (c) $x^2 + 2xy + y^2 - 8 = 0$

    (d) $9x^2 - 6xy + y^2 - 10 = 0$

44. On page 363 we used the rotation formulas to write the equation

$$Ax^2 + Bxy + Cy^2 + Dx + Ey + F = 0$$

in the form

$$A'u^2 + B'uv + C'v^2 + D'u + E'v + F' = 0,$$

where

$A' = A\cos^2\theta + B\sin\theta\cos\theta + C\sin^2\theta,$

$B' = -2A\sin\theta\cos\theta + B\cos^2\theta -$
$\qquad B\sin^2\theta + 2C\sin\theta\cos\theta,$

$C' = A\sin^2\theta - B\sin\theta\cos\theta + C\cos^2\theta,$

$D' = D\cos\theta + E\sin\theta,$

$E' = -D\sin\theta + E\cos\theta,$ and

$F' = F$

Show that $(B')^2 - 4A'C' = B^2 - 4AC$. This shows that the discriminant is invariant under any rotation of axes.

Calculator Activities

To display the graph of the conic section

$$Ax^2 + Bxy + Cy^2 + Dx + Ey + F = 0 \text{ with } C \neq 0$$

on a graphing calculator, first solve the equation for y and then enter the two functions that result. To do this, rewrite the equation as a quadratic in y,

$$Cy^2 + (Bx + E)y + (Ax^2 + Dx + F) = 0,$$

and then apply the quadratic formula to obtain

$$y = \frac{-(Bx + E) \pm \sqrt{(Bx + E)^2 - 4C(Ax^2 + Dx + F)}}{2C}$$

In Exercises 45–50, use a graphing calculator to generate the graph of each conic section in the viewing rectangle. Identify the graph and estimate the coordinates of the vertices by tracing to these points. Round each coordinate to three significant digits.

45. $32y^2 - 48xy + 23x - 120 = 0$

46. $27x^2 - 36xy + 12y^2 + 180x = 0$

47. $25x^2 - 32xy + 12y^2 - 42y + 120 = 0$

48. $1.5x^2 + 2.2xy - 3.5y^2 - 0.89x + 1.8y = 0$

49. $1.8x^2 - 4.8xy + 3.2y^2 - 12.0x + 3.5y + 20.0 = 0$

50. $3.6x^2 + 2.4xy + 2.9y^2 + 7.2x + 20.4y + 18.0 = 0$

6.5 Polar Equations

◆ Introductory Comments

Throughout this text, we have plotted points and graphed equations in the Cartesian coordinate system. In this section, we describe another coordinate system that is used to locate points in a plane and to graph equations—the *polar coordinate system*. Many equations, especially those representing conic sections, are easier to work with in the polar coordinate system than in the Cartesian coordinate system.

◆ Plotting Points in the Polar Coordinate System

The **polar coordinate system** consists of a fixed point O, called the **pole,** and the **polar axis,** a reference ray with endpoint at the pole. It is customary to let the pole coincide with the origin in the Cartesian coordinate system and the polar axis coincide with the positive x-axis. A point P in the polar coordinate system is assigned an ordered pair $P(r, \theta)$, where r and θ are called the **polar coordinates** of point P. If $r > 0$, then r is the distance from the pole to point P, and θ is the measure of the angle formed by ray OP and the polar axis (see Figure 6.51). If $r < 0$, then the absolute value of r is the distance from the pole to point P, and $\theta + 180°$ (or $\theta + \pi$ if θ is in radians) is the measure of the angle formed by ray OP and the polar axis (see Figure 6.52).

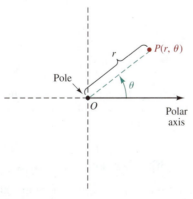

FIGURE 6.51

Plot of the point $P(r, \theta)$ when $r > 0$.

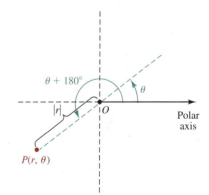

FIGURE 6.52

Plot of the point $P(r, \theta)$ when $r < 0$.

EXAMPLE 1 Plot the points with the given polar coordinates.

(a) $P(3, 60°)$ (b) $Q(-3, 60°)$ (c) $R\left(-4, -\dfrac{\pi}{4}\right)$ (d) $S(0, \pi)$

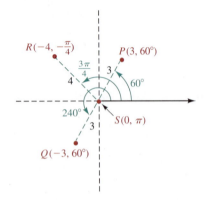

FIGURE 6.53
Plotting points in the polar coordinate system.

SOLUTION

(a) For $P(3, 60°)$, we have $r = 3$ and $\theta = 60°$. Hence, the distance from the pole O to point P is 3 units. Since $r > 0$, the measure of the angle formed by ray OP and the polar axis is $60°$, as shown in Figure 6.53.

(b) For $Q(-3, 60°)$, we have $r = -3$ and $\theta = 60°$. Hence, the distance from the pole O to point Q is $|-3| = 3$ units. Since $r < 0$, the measure of the angle formed by ray OQ and the polar axis is $60° + 180° = 240°$, as shown in Figure 6.53.

(c) For $R\left(-4, -\dfrac{\pi}{4}\right)$, we have $r = -4$ and $\theta = -\dfrac{\pi}{4}$, or $-45°$. Hence the distance from the pole O to point R is $|-4| = 4$ units. Since $r < 0$, the measure of the angle formed by ray OR and the polar axis is $-\dfrac{\pi}{4} + \pi = \dfrac{3\pi}{4}$ or $135°$, as shown in Figure 6.53.

(d) For $S(0, \pi)$, we have $r = 0$ and $\theta = \pi$, or $180°$. Since $r = 0$, the point S must be located at the pole O, as shown in Figure 6.53. In fact, all points of the form $(0, \theta)$ are located at the pole. ◆

For each point P in the Cartesian plane there corresponds a unique ordered pair of real numbers (x, y), and for each ordered pair of real numbers (x, y), there corresponds a unique point P in the Cartesian plane. As we see in Example 1(d), this *one-to-one correspondence* between ordered pairs and points in the polar coordinate system does not hold true at the pole. Even points not at the pole have infinitely many polar coordinate representations. For example, the point $P(r, \theta)$, with θ in radians, may also be represented by

$$P(r, \theta + 2\pi n) \quad \text{or} \quad P(-r, (\theta + \pi) + 2\pi n)$$

for any integer n.

PROBLEM 1 Find two other polar coordinate representations for $P(3, \pi)$, one with a positive value of r and the other with a negative value of r. ◆

◆ **Relationship Between (r, θ) and (x, y)**

From our discussion of trigonometry in Chapter 2, we note the following four relationships between a point P with polar coordinates (r, θ) and the same point P with Cartesian coordinates (x, y).

SECTION 6.5 Polar Equations

Relationship between (r, θ) and (x, y)

If point P has coordinates (x, y) in the Cartesian coordinate system and coordinates (r, θ) in the polar coordinate system, then

1. $x = r \cos \theta$
2. $y = r \sin \theta$
3. $r^2 = x^2 + y^2$
4. $\tan \theta = \dfrac{y}{x},\ x \neq 0$

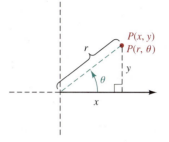

EXAMPLE 2 Convert the given coordinates to the indicated coordinates.

(a) $(-2, 135°)$ to (x, y)

(b) $(-2, 2\sqrt{3})$ to (r, θ) with $r > 0$ and $0 \leq \theta < 2\pi$

SOLUTION

(a) We can convert the polar coordinates $(-2, 135°)$ to Cartesian coordinates (x, y) by using $x = r \cos \theta$ and $y = r \sin \theta$:

$$x = r \cos \theta \qquad\qquad y = r \sin \theta$$
$$= -2 \cos 135° \qquad\qquad = -2 \sin 135°$$
$$= -2\left(-\dfrac{1}{\sqrt{2}}\right) \qquad\qquad = -2\left(\dfrac{1}{\sqrt{2}}\right)$$
$$= \sqrt{2} \approx 1.414 \qquad\qquad = -\sqrt{2} \approx -1.414$$

Hence, a point with polar coordinates $(-2, 135°)$ has Cartesian coordinates $(\sqrt{2}, -\sqrt{2})$, as illustrated in Figure 6.54.

(b) We can convert the Cartesian coordinates $(-2, 2\sqrt{3})$ to polar coordinates (r, θ) by using $r^2 = x^2 + y^2$ and $\tan \theta = \dfrac{y}{x}$:

$$r^2 = x^2 + y^2 \qquad\qquad \tan \theta = \dfrac{y}{x}$$
$$= (-2)^2 + (2\sqrt{3})^2 \qquad\qquad = \dfrac{2\sqrt{3}}{-2}$$
$$= 4 + 12$$
$$= 16 \qquad\qquad = -\sqrt{3}$$

which implies $r = 4$, since we want $r > 0$.

which implies the reference angle $\theta' = \pi/3$.

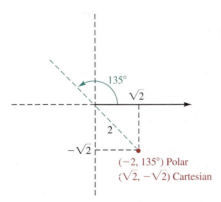

FIGURE 6.54
A point with polar coordinates $(-2, 135°)$ has Cartesian coordinates $(\sqrt{2}, -\sqrt{2})$.

Now, since (x, y) is a point in quadrant II and we want $r > 0$, we know that θ must be in quadrant II. Hence, $\theta = \pi - (\pi/3) = 2\pi/3$. Thus, a point with Cartesian coordinates $(-2, 2\sqrt{3})$ has polar coordinates $(4, 2\pi/3)$, as illustrated in Figure 6.55.

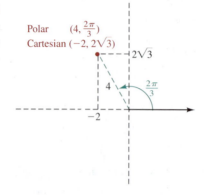

FIGURE 6.55
A point with Cartesian coordinates $\left(-2, 2\sqrt{3}\right)$ has polar coordinates $(4, 2\pi/3)$.

PROBLEM 2 Referring to Example 2(b), convert $(-2, 2\sqrt{3})$ to (r, θ) with $r < 0$ and $0 \leq \theta < 2\pi$.

◆ Sketching the Graph of a Polar Equation

We refer to any equation that is written in terms of r and θ as a **polar equation.** The **graph** of a polar equation is the set of all ordered pairs (r, θ) in the polar plane that satisfy the given equation. The point-plotting method (from Section 1.2) for sketching the graph of an elementary Cartesian equation in x and y may also be used to help sketch the graph of a simple polar equation.

EXAMPLE 3 Sketch the graph of each polar equation.

(a) $r = 3$ (b) $\theta = \dfrac{\pi}{4}$

SOLUTION

(a) We can think of the equation $r = 3$ as

$$r + 0 \cdot \theta = 3.$$

Regardless of what value we choose for θ, the quantity $0 \cdot \theta$ is always zero. This means that the ordered pairs satisfying the equation $r = 3$ can have any θ-value, provided the r-value is 3. The following table of values shows some points that satisfy the equation $r = 3$.

θ	0	$\pi/4$	$\pi/2$	$3\pi/4$	π	$5\pi/4$	$3\pi/2$	$7\pi/4$
r	3	3	3	3	3	3	3	3

SECTION 6.5 Polar Equations

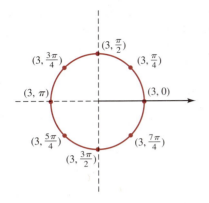

FIGURE 6.56
The graph of $r = 3$ is a circle with center at the pole and radius 3.

Plotting these points in the polar coordinate system, we conclude that the graph of $r = 3$ is a circle with radius 3, as shown in Figure 6.56.

(b) We can think of the equation $\theta = \pi/4$ as

$$0 \cdot r + \theta = \frac{\pi}{4}.$$

Regardless of what value we choose for r, the quantity $0 \cdot r$ is always zero. This means that the ordered pairs satisfying the equation $\theta = \pi/4$ can have any r-value, provided the θ-value is $\pi/4$. The following table of values shows some points that satisfy the equation $\theta = \pi/4$.

r	0	1	2	3	-1	-2	-3
θ	$\pi/4$	$\pi/4$	$\pi/4$	$\pi/4$	$\pi/4$	$\pi/4$	$\pi/4$

Plotting these points in the polar coordinate system, we conclude that the graph of $\theta = \pi/4$ is a line that passes through the pole and makes an angle of $\pi/4$ (45°) with the polar axis, as shown in Figure 6.57. ◆

We can verify the graphs in Example 3 by converting the polar equations to equivalent Cartesian equations in x and y. Referring to the polar equation $r = 3$ in Example 3(a), if we square both sides, we obtain $r^2 = 9$. Now replacing r^2 with $x^2 + y^2$ gives us

$$x^2 + y^2 = 9,$$

which we recognize as the equation of a circle with center at the origin and radius 3. In general, the graph of

$$\boxed{r = a}$$

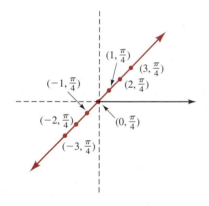

FIGURE 6.57
The graph of $\theta = \pi/4$ is a line that passes through the pole and makes an angle of $\pi/4$ with the polar axis.

with a a nonzero constant, is a circle with center at the pole and radius $|a|$.
For the polar equation $\theta = \pi/4$ in Example 3(b), we can write

$$\tan \theta = \tan \frac{\pi}{4} = 1.$$

Now replacing $\tan \theta$ with $\dfrac{y}{x}$ gives us

$$\frac{y}{x} = 1 \quad \text{or} \quad y = x,$$

which we recognize as the equation of a line that passes through the origin and splits quadrant I and quadrant III in half. In general, the graph of

$$\theta = k$$

with k a constant, is a line that passes through the pole and makes an angle of k radians with the polar axis.

PROBLEM 3 Describe the graph of each polar equation.

(a) $r = 5$ (b) $\theta = \dfrac{2\pi}{3}$ ◆

EXAMPLE 4 Sketch the graph of each polar equation and give its equivalent Cartesian equation

(a) $r = \dfrac{3}{\sin \theta}$ (b) $r = -6 \sin \theta$

SOLUTION

(a) The following table of values shows some points that satisfy the equation $r = 3/\sin \theta$.

θ	0	$\pi/6$	$\pi/2$	$5\pi/6$	π	$7\pi/6$	$3\pi/2$	$11\pi/6$
r	undefined	6	3	6	undefined	-6	-3	-6

By plotting these points in the polar coordinate system, we see that the graph of $r = 3/\sin \theta$ is a line, parallel to the polar axis and 3 units above the pole, as shown in Figure 6.58. We can confirm that the graph is a horizontal line by converting the equation to an equivalent Cartesian equation as follows:

$$r = \dfrac{3}{\sin \theta}$$

$r \sin \theta = 3$ *Multiply both sides by $\sin \theta$*

$y = 3$ *Replace $r \sin \theta$ with y*

which is a horizontal line 3 units above the x-axis.

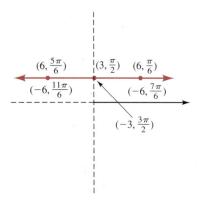

FIGURE 6.58
The graph of $r = 3/\sin \theta$ is a horizontal line 3 units above the polar axis.

(b) The following table of values shows some points that satisfy the equation $r = -6 \sin \theta$.

SECTION 6.5 Polar Equations

θ	0	$\pi/6$	$\pi/3$	$\pi/2$	$2\pi/3$	$5\pi/6$	π
r	0	-3	$-3\sqrt{3}$	-6	$-3\sqrt{3}$	-3	0

By plotting these points in the polar coordinate system, we see that the graph of $r = -6 \sin \theta$ is a circle with center $(-3, \pi/2)$, as shown in Figure 6.59.

Note: Our table of values contains only values of θ between 0 and π. Values of θ between π and 2π simply trace the circle again.

We can confirm that this graph is a circle by converting the equation to an equivalent Cartesian equation as follows:

$$r = -6 \sin \theta$$

$$r^2 = -6r \sin \theta \qquad \text{Multiply both sides by } r$$

$$x^2 + y^2 = -6y \qquad \text{Replace } r^2 \text{ with } x^2 + y^2 \text{ and } r \sin \theta \text{ with } y$$

$$x^2 + (y^2 + 6y +) = 0 \qquad \text{Add } 6y \text{ to both sides}$$

$$x^2 + (y^2 + 6y + 9) = 9 \qquad \text{Complete the square by adding 9 to both sides}$$

$$(x - 0)^2 + (y + 3)^2 = 3^2 \qquad \text{Write in standard form}$$

We recognize this last equation as the equation of a circle in standard form with center $(0, -3)$ and radius 3. ◆

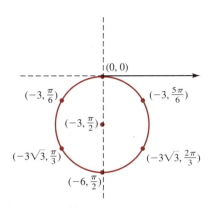

FIGURE 6.59
The graph of $r = -6 \sin \theta$ is a circle with center $(-3, \pi/2)$ and radius 3.

PROBLEM 4 Sketch the graph of each polar equation and state its equivalent Cartesian equation.

(a) $r = \dfrac{3}{\cos \theta}$ (b) $r = -6 \cos \theta$ ◆

Based on the results of Example 4 and Problem 4, we make the following conclusions. (See Exercises 65 and 66 for the proofs of these statements.)

1. The graph of

$$r = \dfrac{a}{\sin \theta}$$

with a a nonzero constant is a horizontal line, a units above the pole if $a > 0$ or $|a|$ units below the pole if $a < 0$.

2. The graph of

$$r = \frac{a}{\cos \theta}$$

with a a nonzero constant is a vertical line a units to the right of the pole is $a > 0$ or $|a|$ units to the left of the pole if $a < 0$.

3. The graph of

$$r = a \sin \theta$$

with a a nonzero constant, is a circle with radius $|a|/2$ and center $(a/2, 90°)$.

4. The graph of

$$r = a \cos \theta$$

with a a nonzero constant is a circle with radius $|a|/2$ and center $(a/2, 0°)$.

Many polar equations convert to a familar Cartesian equation, and we can use that Cartesian equation to help sketch the graph of the polar equation. However, consider the polar equation

$$r = 2 \sin 2\theta.$$

We can convert this equation to an equivalent Cartesian equation as follows:

$r = 2 \sin 2\theta$	
$r = 2(2 \sin \theta \cos \theta)$	**Apply the double angle formula for sine**
$r^3 = 2r^2(2 \sin \theta \cos \theta)$	**Multiply both sides by r^2**
$(r^2)^{3/2} = 4(r \cos \theta)(r \sin \theta)$	**Rewrite both sides**
$(x^2 + y^2)^{3/2} = 4xy$	**Replace r^2 with $x^2 + y^2$, $r \cos \theta$ with x, and $r \sin \theta$ with y**
$(x^2 + y^2)^3 = 16x^2y^2$	**Square both sides**

Note how much simpler the polar equation is than its equivalent Cartesian equation. In order to sketch the graph of either of these equations, which equation would be easier?

EXAMPLE 5 Sketch the graph of $r = 2 \sin 2\theta$.

SOLUTION The following table of values shows some points that satisfy the equation $r = 2 \sin 2\theta$.

θ	0	$\pi/6$	$\pi/4$	$\pi/3$	$\pi/2$	$2\pi/3$	$3\pi/4$	$5\pi/6$	π
r	0	$\sqrt{3}$	2	$\sqrt{3}$	0	$-\sqrt{3}$	-2	$-\sqrt{3}$	0

Since the period of $\sin 2\theta$ is π, values of θ between π and 2π, such as $7\pi/6$, $5\pi/4$, $4\pi/3$, $3\pi/2$, $5\pi/3$, $5\pi/4$, and $11\pi/6$, must yield the same r-values as shown in our table. As illustrated in Figure 6.60, the graph of $r = 2 \sin 2\theta$ [or $(x^2 + y^2)^3 = 16x^2y^2$] is a *four-leafed rose*. ◆

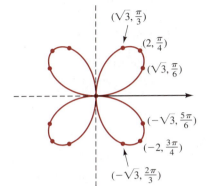

FIGURE 6.60
The graph of $r = 2 \sin 2\theta$ is a four-leafed rose.

Note: The vertical line test for functions (Section 1.3) is not valid for graphs of equations in the polar coordinate system. The equation $r = 2 \sin 2\theta$ defines r as a function of θ, even though its graph appears to fail the vertical line test (see Figure 6.60). Note that for each value of θ that we select, we obtain one and only one value of r. To emphasize the functional relationship, we write $r = f(\theta) = 2 \sin 2\theta$.

PROBLEM 5 Sketch the graph of $r = 2 \sin 3\theta$, a three-leafed rose. ◆

Polar Equations of the Conic Sections

We can develop the polar equations of the parabola, ellipse, and hyperbola from the common definition of the conic sections, which we gave at the end of Section 6.3. Suppose a focus F of a conic section is located at the pole, and its corresponding directrix is *perpendicular* to the polar axis and d units to the *right* of the pole, as shown in Figure 6.61. Then the set of all points $P(r, \theta)$ such that $PF/PQ = e$, where e is the eccentricity, defines this conic section. Referring to Figure 6.61, we have $PF = r$ and $PQ = d - a$, where $a = r \cos \theta$. Thus,

$$\frac{PF}{PQ} = \frac{r}{d - r \cos \theta} = e$$

$r = ed - er \cos \theta$	**Multiply both sides by $(d - r \cos \theta)$**
$r + er \cos \theta = ed$	**Add $er \cos \theta$ to both sides**
$r(1 + e \cos \theta) = ed$	**Factor**
$r = \dfrac{ed}{1 + e \cos \theta}$	**Divide both sides by $(1 + e \cos \theta)$**

FIGURE 6.61
A conic section with its focus at the pole and corresponding directrix perpendicular to the polar axis

Using similar arguments, we can derive polar equations of the conic sections when the directrix is *perpendicular* to the polar axis and d units to the *left* of the pole; when the directrix is *parallel* to the polar axis and d units *above* the pole; and when the directrix is *parallel* to the polar axis and d units *below* the pole. We summarize the results as follows.

Polar Equations of the Conic Sections

The **polar equations of the conic sections** with a focus at the pole have four forms:

Form 1 $r = \dfrac{ed}{1 + e \cos \theta}$

Directrix *perpendicular* to the polar axis and d units to the *right* of the pole

Form 2 $r = \dfrac{ed}{1 - e \cos \theta}$

Directrix *perpendicular* to the polar axis and d units to the *left* of the pole

Form 3 $r = \dfrac{ed}{1 + e \sin \theta}$

Directrix *parallel* to the polar axis and d units *above* the pole

Form 4 $r = \dfrac{ed}{1 - e \sin \theta}$

Directrix *parallel* to the polar axis and d units *below* the pole

The conic section is a *parabola* if $e = 1$, an *ellipse* if $0 < e < 1$, or a *hyperbola* if $e > 1$.

When working with the polar equation of a conic section, it is essential to recognize whether the equation is of form 1, 2, 3, or 4.

EXAMPLE 6 Sketch the graph of each polar equation.

(a) $r = \dfrac{2}{1 - \sin \theta}$ (b) $r = \dfrac{3}{1 - 2 \cos \theta}$

SOLUTION

(a) We recognize the polar equation $r = 2/(1 - \sin \theta)$ as a conic section of *form 4*, with $e = 1$ and $ed = 2$, which implies $d = 2$. Hence, the directrix is *parallel* to the polar axis and 2 units *below* the pole. Since $e = 1$, we know that this conic section is a parabola with focus at the pole. The vertex of a parabola is always located halfway between its focus and directrix. Hence, we conclude that the vertex is at the point $(1, 3\pi/2)$. The following table of values shows some points that satisfy the equation $r = 2/(1 - \sin \theta)$.

θ	0	$\pi/6$	$5\pi/6$	π
r	2	4	4	2

Plotting these points in the polar coordinate system, we obtain the parabola shown in Figure 6.62.

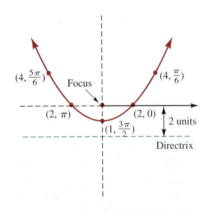

FIGURE 6.62
The graph of $r = 2/(1 - \sin \theta)$ is a parabola with vertex $(1, 3\pi/2)$.

(b) We recognize the polar equation $r = 3/(1 - 2 \cos \theta)$ as a conic section of *form 2*, with $e = 2$ and $ed = 3$, which implies $d = \frac{3}{2}$. Hence, the directrix is *perpendicular* to the polar axis and $\frac{3}{2}$ units to the *left* of the pole. Since

$e = 2 > 1$, we know that this conic section is a hyperbola with focus at the pole. Since the transverse axis is along the polar axis, the vertices of the hyperbola may be found by letting $\theta = 0$ and $\theta = \pi$. The following table of values shows some points that satisfy the equation $r = 3/(1 - 2\cos\theta)$.

θ	0	$\pi/2$	$2\pi/3$	π	$4\pi/3$	$3\pi/2$
r	-3	3	$\dfrac{3}{2}$	1	$\dfrac{3}{2}$	3

Plotting these points in the polar coordinate system and using symmetry to sketch the left branch, we obtain the hyperbola shown in Figure 6.63. ◆

We can verify the graphs in Example 6 by converting the polar equations to equivalent Cartesian equations in x and y. For Example 6(a), we have

$$r = \frac{2}{1 - \sin\theta}$$

$r - r\sin\theta = 2$	**Multiply both sides by $(1 - \sin\theta)$**
$r - y = 2$	**Replace $r\sin\theta$ with y**
$r = y + 2$	**Add y to both sides**
$r^2 = y^2 + 4y + 4$	**Square both sides**
$x^2 + y^2 = y^2 + 4y + 4$	**Replace r^2 with $x^2 + y^2$**
$x^2 - 4y - 4 = 0$	**Write in general form**
$y = \frac{1}{4}(x - 0)^2 - 1$	**Write in standard form**

We recognize this last equation as a parabola in standard form with vertex $(0, -1)$. Since $1/(4p) = 1/4$, we have $p = 1$. This implies that the focus is 1 unit above the vertex and the directrix is 1 unit below the vertex. Hence, the coordinates of the focus in the Cartesian plane are $(0, 0)$ and the directrix is the horizontal line $y = -2$ (see Figure 6.62).

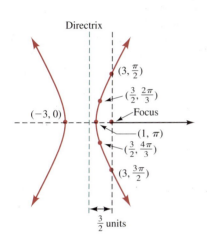

FIGURE 6.63
The graph of $r = 3/(1 - 2\cos\theta)$ is a hyperbola with vertices $(-3, 0)$ and $(1, \pi)$.

PROBLEM 6 Verify the graph in Example 6(b) by converting the polar equation to a Cartesian equation in x and y. ◆

Note that the first term in the denominator of the polar equations of the conic sections is always 1. If a number other than 1 appears in this position, we divide both numerator and denominator by this number before identifying the conic section.

EXAMPLE 7 Sketch the graph of the polar equation $r = \dfrac{3}{2 + \cos\theta}$.

SOLUTION The first term in the denominator of the polar equations of the conic sections is always 1. Thus for $r = 3/(2 + \cos \theta)$, we divide numerator and denominator by 2 to obtain

$$r = \frac{\frac{3}{2}}{1 + \frac{1}{2} \cos \theta}$$

We recognize this polar equation as a conic section of *form 1*, with $e = \frac{1}{2}$ and $ed = \frac{3}{2}$, which implies $d = 3$. Hence, the directrix is *perpendicular* to the polar axis and 3 units to the *right* of the pole. Since $e = \frac{1}{2}$, we know that this conic section is an ellipse with focus at the pole. Since the major axis is along the polar axis, the vertices of the ellipse may be found by letting $\theta = 0$ and $\theta = \pi$. The following table of values shows some points that satisfy the equation $r = 3/(2 + \cos \theta)$.

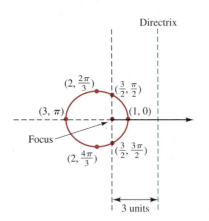

FIGURE 6.64
The graph of $r = 3/(2 + \cos \theta)$ is an ellipse with vertices $(3, \pi)$ and $(1, 0)$

θ	0	$\pi/2$	$2\pi/3$	π	$4\pi/3$	$3\pi/2$
r	1	$\frac{3}{2}$	2	3	2	$\frac{3}{2}$

Plotting these points in the polar coordinate system, we obtain the ellipse shown in Figure 6.64. ◆

PROBLEM 7 Verify the graph in Example 7 by converting the polar equation to a Cartesian equation in x and y. ◆

◆ Application: Elliptical Orbit of the Earth

In the early seventeenth century, the German astronomer Johannes Kepler observed that the planets in our solar system travel in elliptical orbits with the sun at one focus. This fact, which today is known as *Kepler's first law*, laid the foundation of modern astronomy. Later in the same century, Sir Isaac Newton gave a physical explanation for Kepler's observation and was able to predict the elliptical paths of the planets with great accuracy. The polar equations of the conic sections are used extensively in astronomy.

EXAMPLE 8 Earth is closest to the sun (at its perihelion) in January, when the distance between them is approximately 9.14×10^7 miles, and is furthest from the sun (aphelion) in July, when the distance between them is approximately 9.44×10^7 miles.

(a) Determine the polar equation of the elliptical orbit of the earth if the sun (focus) is at the pole, and its corresponding directrix is perpendicular to the polar axis and to the right of the pole.

(b) State the eccentricity of the orbit.

SOLUTION

(a) From the sketch of the elliptical orbit in Figure 6.65, we know that the polar equation is of *form 1:*

$$r = \frac{ed}{1 + e \cos \theta}$$

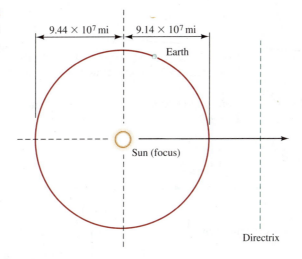

FIGURE 6.65
Sketch of the elliptical orbit of the earth.

Now, since $r = 9.14 \times 10^7$ when $\theta = 0$, we have

$$9.14 \times 10^7 = \frac{ed}{1 + e \cos 0} = \frac{ed}{1 + e},$$

which implies

$$ed = 9.14 \times 10^7 (1 + e).$$

Also, since $r = 9.44 \times 10^7$ when $\theta = \pi$, we have

$$9.44 \times 10^7 = \frac{ed}{1 + e \cos \pi} = \frac{ed}{1 - e},$$

which implies

$$ed = 9.44 \times 10^7 (1 - e).$$

Equating these two values of *ed* gives us

$$9.14 \times 10^7 (1 + e) = 9.44 \times 10^7 (1 - e)$$
$$(18.58 \times 10^7)e = 0.30 \times 10^7$$
$$e \approx 0.016$$

and, by substitution,

$$ed = 9.14 \times 10^7(1 + e) = 9.14 \times 10^7(1 + 0.016)$$
$$\approx 9.29 \times 10^7.$$

Thus, the polar equation of the elliptical orbit of the earth is given by

$$r = \frac{ed}{1 + e \cos \theta} = \frac{9.29 \times 10^7}{1 + 0.016 \cos \theta}.$$

(b) The eccentricity of the elliptical orbit of the earth is $e \approx 0.016$, which is close to 0. This tells us that the earth's orbit is nearly circular.

PROBLEM 8 Referring to Example 8, find the distance r from the earth to the sun when $\theta = \pi/2$.

Exercises 6.5

Basic Skills

In Exercises 1–8, plot the points with the given polar coordinates.

1. $A(1, 180°)$, $B(2, 60°)$, $C(3, 225°)$, $D(0, 210°)$
2. $A(3, \pi/2)$, $B(1, 5\pi/6)$, $C(2, 2\pi/3)$, $D(4, 7\pi/4)$
3. $A(2, -\pi/2)$, $B(1, -\pi/6)$, $C(4, -3\pi/4)$, $D(0, -3)$
4. $A(2, -90°)$, $B(1, -30°)$, $C(4, -135°)$, $D(0, -330°)$
5. $A(-3, 0°)$, $B(-2, 45°)$, $C(-5, 150°)$, $D(-4, 300°)$
6. $A(-1, \pi/2)$, $B(-2, \pi/6)$, $C(-4, 3\pi/4)$, $D(-6, 0)$
7. $A(-4, -3\pi/2)$, $B(-1, -\pi/4)$, $C(-5, -7\pi/6)$, $D(-2, -5\pi/3)$
8. $A(-1, -270°)$, $B(-4, -60°)$, $C(-3, -315°)$, $D(-5, -240°)$

9. For each of the points in Exercise 5, find two other polar coordinate representations, one with a positive value of r and the other with a negative value of r. Answers are not unique.

10. For each of the points in Exercise 8, find two other polar coordinate representations, one with a positive value of r and the other with a negative value of r. Answers are not unique.

In Exercises 11–20, convert the given coordinates to the indicated coordinates.

11. $(1, 180°)$ to (x, y)
12. $(-2, \pi/2)$ to (x, y)
13. $(-4\sqrt{2}, 7\pi/4)$ to (x, y)
14. $(6, -30°)$ to (x, y)
15. $(0, 2)$ to (r, θ) with $r > 0$ and $0 \le \theta < 360°$
16. $(-3, 0)$ to (r, θ) with $r > 0$ and $0 \le \theta < 2\pi$
17. $(-5, -5)$ to (r, θ) with $r > 0$ and $0 \le \theta < 2\pi$
18. $(-4, 4\sqrt{3})$ to (r, θ) with $r > 0$ and $0 \le \theta < 360°$
19. $(3\sqrt{3}, 3)$ to (r, θ) with $r < 0$ and $0 \le \theta < 360°$
20. $(2\sqrt{2}, -2\sqrt{2})$ to (r, θ) with $r < 0$ and $-2\pi < \theta \le 0$

In Exercises 21–40, sketch the graph of each polar equation, and give its equivalent Cartesian equation.

21. $r = 1$
22. $r = 4$
23. $\theta = \dfrac{\pi}{3}$
24. $\theta = \dfrac{5\pi}{6}$
25. $r = \dfrac{-2}{\sin \theta}$
26. $r = 4 \csc \theta$
27. $r = \dfrac{4}{\cos \theta}$
28. $r = -\sec \theta$
29. $r = 10 \cos \theta$
30. $r = -8 \cos \theta$
31. $r = -5 \sin \theta$
32. $r = 3 \sin \theta$
33. $r = 4 \sin \theta + 2 \cos \theta$
34. $r = 6 \sin \theta - 2 \cos \theta$
35. $r = 8 \cos \theta - \sin \theta$
36. $r = -2 \sin \theta - 3 \cos \theta$
37. $r = \dfrac{2}{\cos \theta + \sin \theta}$
38. $r = \dfrac{12}{4 \sin \theta - 3 \cos \theta}$

39. $r = \dfrac{-15}{3\cos\theta + 5\sin\theta}$ **40.** $r = \dfrac{6}{2\cos\theta - \sin\theta}$

In Exercises 41–48, sketch the graph of each polar equation.

41. $r = \dfrac{\theta}{\pi}, \theta \geq 0$ (a spiral)

42. $r\theta = \pi, \theta > 0$ (a spiral)

43. $r = 1 - \sin\theta$ (a *cardioid*, or heart)

44. $r = 2 + 2\cos\theta$ (a cardioid)

45. $r = 2\cos 2\theta$ (a four-leafed rose)

46. $r = 2\cos 3\theta$ (a three-leafed rose)

47. $r^2 = 4\cos 2\theta$ (a *lemniscate*, or figure eight)

48. $r^2 = 4\sin 2\theta$ (a *lemniscate*)

For each polar equation in Exercises 49–60,

(a) identify the conic section it represents,

(b) sketch its graph, and

(c) give its equivalent Cartesian equation and use it to verify the sketch in part (b).

49. $r = \dfrac{4}{1 - \cos\theta}$ **50.** $r = \dfrac{2}{1 + \sin\theta}$

51. $r = \dfrac{6}{1 + 2\cos\theta}$ **52.** $r = \dfrac{4}{1 - 3\sin\theta}$

53. $r = \dfrac{2}{1 - \frac{1}{3}\sin\theta}$ **54.** $r = \dfrac{\frac{1}{4}}{1 + \frac{1}{4}\cos\theta}$

55. $r = \dfrac{5}{6 - 6\cos\theta}$ **56.** $r = \dfrac{3}{2 + 2\sin\theta}$

57. $r = \dfrac{1}{2 + \sin\theta}$ **58.** $r = \dfrac{2}{5 - 2\cos\theta}$

59. $r = \dfrac{3}{3 - 4\cos\theta}$ **60.** $r = \dfrac{5}{2 + 3\sin\theta}$

Critical Thinking

61. (a) State a polar equation for the *x*-axis in the Cartesian coordinate system.

(b) State a polar equation for the *y*-axis in the Cartesian coordinate system.

62. Find the polar form of the general linear equation $Ax + By = C$. Express the answer in the form $r = f(\theta)$.

63. Find the equivalent polar equation of each Cartesian equation. Express each answer in the form $r = f(\theta)$.

(a) $(x - 1)^2 + (y - 1)^2 = 2$ (b) $x^2 - 4y - 4 = 0$

(c) $4x^2 + 3y^2 + 2y = 1$ (d) $x^2 - 3y^2 + 4y = 1$

64. Describe the symmetry of the graph of the polar equation $r = f(\theta)$ if

(a) $f(-\theta) = f(\theta)$ (b) $f(\theta + \pi) = f(\theta)$

65. Given that a is a nonzero real number, show that

(a) the graph of $r = \dfrac{a}{\sin\theta}$ is a horizontal line, a units above the pole if $a > 0$ or $|a|$ units below the pole if $a < 0$.

(b) the graph of $r = \dfrac{a}{\cos\theta}$ is a vertical line, a units to the right of the pole if $a > 0$ or $|a|$ units to the left of the pole if $a < 0$.

66. Given that a is a nonzero real number, show that

(a) the graph of $r = a\sin\theta$ is a circle with radius $|a|/2$ and center $(a/2, 90°)$.

(b) the graph of $r = a\cos\theta$ is a circle with radius $|a|/2$ and center $(a/2, 0°)$.

67. Unlike Cartesian equations, when two polar equations are graphed in the same plane, the intersection points of the two graphs *may not* have coordinates that satisfy both equations.

(a) Determine the point of intersection of $r = 4\sin\theta$ and $r = 4\cos\theta$, with $0 \leq \theta \leq \pi$, by solving the equations simultaneously.

(b) Sketch the equations $r = 4\sin\theta$ and $r = 4\cos\theta$, with $0 \leq \theta \leq \pi$, in the same polar plane and note that they have another point of intersection. Do the polar coordinates of this other intersection point satisfy both equations? Explain.

68. In how many distinct points do the graphs of the polar equations $r = 2\sin 2\theta$ and $r = 1$ intersect? Which of these point(s) satisfy both equations?

69. Suppose a circle has center with polar coordinates, (a, α) and radius a, as shown in the sketch. Use the law of cosines (Section 3.3) to find the polar equation of this circle. Express the equation in the form $r = f(\theta)$.

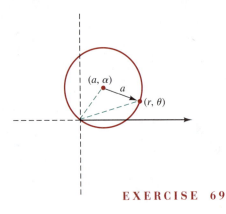

EXERCISE 69

70. Show that the distance d between two points $P(a, \alpha)$ and $Q(b, \beta)$ in the polar plane is given by

$$d = \sqrt{a^2 + b^2 - 2ab \cos(\alpha - \beta)}$$

Then use this *distance formula* to find the distance between the given points.

(a) $(3, 2\pi/3)$ and $(4, \pi/6)$
(b) $(-2, \pi/2)$ and $(4, -5\pi/6)$

Calculator Activities

In Exercises 71–74, use a calculator to convert the given polar coordinates to Cartesian coordinates. Round each of the coordinates x and y to three significant digits.

71. $(2.45, 32.8°)$
72. $(343, -128.2°)$
73. $(-0.943, 1.23)$
74. $(-5.56, -2.44)$

In Exercises 75–78, use a calculator to convert the given Cartesian coordinates to polar coordinates with $r > 0$ and $0 < \theta < 2\pi$. Round the coordinates r and θ to three significant digits.

75. $(12.4, 15.3)$
76. $(1.04, -4.78)$
77. $(-126, 236)$
78. $(-0.976, -0.561)$

The distances from a planet to the sun at the planet's perihelion (shortest distance from the sun) and aphelion (greatest distance from the sun) are given in Exercises 79–82.

(a) Determine the polar equation of the elliptical orbit of the planet if the sun (focus) is at the pole, and its corresponding directrix is perpendicular to the polar axis and to the right of the pole.

(b) State the eccentricity of the orbit.

79. Mercury: perihelion = 2.85×10^7 miles, aphelion = 4.32×10^7 miles

80. Pluto: perihelion = 2.75×10^9 miles; aphelion = 4.57×10^9 miles

81. Mars: perihelion = 1.28×10^8 miles, aphelion = 1.54×10^8 miles

82. Saturn: perihelion = 8.34×10^8 miles, aphelion = 9.33×10^8 miles

The polar equation of the elliptical orbit of Halley's comet with the sun (focus) at the pole and corresponding directrix perpendicular to the polar axis and to the right of the pole is given by

$$r = \frac{1.15}{1 + 0.968 \cos \theta}$$

with r measured in astronomical units (AU). Use this equation to answer the questions in Exercises 83 and 84.

83. What is the distance from the comet to the sun at its perihelion (shortest distance from the sun)? at its aphelion (greatest distance from the sun)?

84. What is the length (in miles) of the major axis of the elliptical orbit of the comet? the minor axis? (*Hint:* 1 AU ≈ 9.26×10^7 miles)

6.6 Parametric Equations

◆ Introductory Comments

Over a certain time interval, a spider crawls along the path that is shown in Figure 6.66. To describe this situation mathematically, we might find a Cartesian equation in x and y (or a polar equation in r and θ) for the path of the spider. However, the Cartesian (or polar) equation does not tell us the position of the spider at a particular time t, nor does it tell us the direction in which the spider travels along this path. Since the x- and y-coordinates of the spider vary with time t, both x and y are functions of t:

$$x = f(t) \quad \text{and} \quad y = g(t)$$

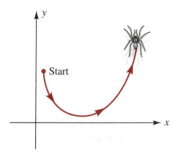

FIGURE 6.66
Path of a spider over a certain time interval.

where t ranges over some interval. We refer to the equations $x = f(t)$ and $y = g(t)$ as the **parametric equations** that describe the path of the spider and the auxilliary variable t, on which both x and y depend, as the **parameter.**

Parametric equations not only give a set of points (x, y) that describe the curve, but also describe the location on the curve at time t and the direction in which the curve is traced. In this section, we discuss the parametric equations of several curves. We begin by defining the parametric equations of a line.

◆ Parametric Equations of a Line

The set of equations

$$\boxed{x = x_1 + t \quad \text{and} \quad y = y_1 + mt}$$

are **parametric equations of a line** with t as the parameter. The line passes through the point (x_1, y_1) in the Cartesian plane and has slope m. To show that this is true, we eliminate the parameter t by solving for t in the first equation ($t = x - x_1$) and substituting this expression in the other equation:

$$y = y_1 + mt$$
$$y = y_1 + m(x - x_1) \qquad \text{Substitute } x - x_1 \text{ for } t$$
$$y - y_1 = m(x - x_1) \qquad \text{Subtract } y_1 \text{ from both sides}$$

We recognize the equation $y - y_1 = m(x - x_1)$ as the *point-slope form* of a line that passes through the point (x_1, y_1) and has slope m.

EXAMPLE 1 Identify and sketch the curve defined by the parametric equations

$$x = -1 + t \quad \text{and} \quad y = 3 + \tfrac{2}{3}t \quad \text{with } 0 \le t \le 6.$$

Indicate with arrows the direction of travel along the curve.

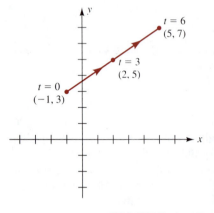

FIGURE 6.67
The parametric equations $x = -1 + t$ and $y = 3 + \frac{2}{3}t$, with $0 \le t \le 6$, define a line segment that passes through $(-1, 3)$ and has slope $\frac{2}{3}$.

SOLUTION We recognize this set of equations as parametric equations of a line. The line passes through the point $(-1, 3)$ and has slope $\frac{2}{3}$. To determine the portion of the line traversed in the given time interval and the direction of travel along that portion, we select values of t in the interval $[0, 6]$ and find the corresponding values of x and y.

t	0 (start)	3	6 (end)
$x = -1 + t$	-1	2	5
$y = 3 + \frac{2}{3}t$	3	5	7

Plotting these points (x, y) in the Cartesian plane, we observe that the graph starts at the point $(-1, 3)$ and travels a linear path until it reaches the point $(5, 7)$, as illustrated in Figure 6.67. ◆

PROBLEM 1 Show that the parametric equations

$$x = 5 - t \quad \text{and} \quad y = 7 - \tfrac{2}{3}t \quad \text{with } 0 \le t \le 6$$

define the same line segment as shown in Figure 6.67, but with opposite direction. ◆

◆ **Parametric Equations of a Parabola**

The set of equations

$$x = 2pt \quad \text{and} \quad y = pt^2$$

are **parametric equations of a parabola** with t as the parameter. The parabola has a *vertical axis* of symmetry, vertex $(0, 0)$, and focus $(0, p)$. To show that this is true, we eliminate the parameter t by solving for t in the first equation $[t = x/(2p)]$ and substituting this expression in the other equation:

$$\begin{aligned} y &= pt^2 \\ &= p\left(\frac{x}{2p}\right)^2 \quad &&\text{Substitute } x/(2p) \text{ for } t \\ &= \frac{1}{4p}x^2 \quad &&\text{Square the quotient and simplify} \end{aligned}$$

We recognize the equation $y = (1/4p)x^2$ as a parabola with vertical axis of symmetry, vertex at the origin, and focus $(0, p)$.

In a similar manner, we can show that the set of equations

$$x = pt^2 \quad \text{and} \quad y = 2pt$$

are **parametric equations of a parabola** with *horizontal axis* of symmetry, vertex $(0, 0)$, and focus $(p, 0)$.

EXAMPLE 2 Identify and sketch the curve defined by the parametric equations

$$x = -t \quad \text{and} \quad y = -\tfrac{1}{2}t^2 \quad \text{with } t \geq 0.$$

Indicate with arrows the direction of travel along the curve.

SOLUTION We recognize this set of equations as parametric equations of a parabola. The parabola has vertical axis of symmetry, vertex $(0, 0)$, and focus $(0, -\tfrac{1}{2})$. To determine the portion of the parabola traversed and the direction of travel along that portion, we select values of t in the interval $[0, \infty)$ and find the corresponding values of x and y.

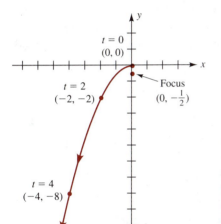

FIGURE 6.68
The parametric equations $x = -t$ and $y = -\tfrac{1}{2}t^2$ with $t \geq 0$ define the left-hand side of a parabola with vertical axis of symmetry, vertex $(0, 0)$, and focus $(0, -\tfrac{1}{2})$.

t	0 (start)	2	4	$+\infty$ (approach)
$x = -t$	0	-2	-4	$-\infty$
$y = -\tfrac{1}{2}t^2$	0	-2	-8	$-\infty$

Plotting these points (x, y) in the Cartesian plane, we observe that the graph starts at the point $(0, 0)$ and travels downward to the left along a parabolic path as $t \to \infty$ (see Figure 6.68). ◆

PROBLEM 2 Repeat Example 2 for the parametric equations

$$x = t^2 \quad \text{and} \quad y = 2t \quad \text{with } -2 \leq t \leq 2.$$

◆

◆ **Parametric Equations of an Ellipse and a Circle**

The set of equations

$$x = a \cos \theta \quad \text{and} \quad y = b \sin \theta$$

with $a > 0$, $b > 0$, and $0 \leq \theta < 2\pi$ are **parametric equations of an ellipse** (of a **circle** if $a = b$), with θ as the parameter. The center of the ellipse is the origin, the horizontal axis has length $2a$, and the vertical axis has length $2b$. To show that this is true, we eliminate the parameter θ by squaring both x/a and y/b, and then adding these expressions together as follows:

$$\left(\frac{x}{a}\right)^2 + \left(\frac{y}{b}\right)^2 = (\cos\theta)^2 + (\sin\theta)^2 \quad \text{Substitute}$$

$$\frac{x^2}{a^2} + \frac{y^2}{b^2} = \cos^2\theta + \sin^2\theta \quad \text{Square the expressions}$$

$$\frac{x^2}{a^2} + \frac{y^2}{b^2} = 1 \quad \text{Apply } \sin^2\theta + \cos^2\theta = 1$$

We recognize the equation $(x^2/a^2) + (y^2/b^2) = 1$ as the equation of an ellipse with center at the origin, horizontal axis of length $2a$, and vertical axis of length $2b$. Of course, if $a = b$ we have the equation of a circle with center at the origin and radius a.

EXAMPLE 3 Identify and sketch the curve defined by the parametric equations

$$x = 2\cos\pi t \quad \text{and} \quad y = 4\sin\pi t \quad \text{with } 0 \leq t \leq 2.$$

Indicate with arrows the direction of travel along the curve.

SOLUTION We recognize this set of equations as parametric equations of an ellipse with center at the origin, horizontal axis of length $2(2) = 4$, and vertical axis of length $2(4) = 8$. To determine the portion of the ellipse traversed and the direction of travel along that portion, we select values of t in the interval $[0, 2]$ and find the corresponding values of x and y.

t	0 (start)	$\frac{1}{2}$	1	$\frac{3}{2}$	2 (end)
$x = 2\cos\pi t$	2	0	-2	0	2
$y = 4\sin\pi t$	0	4	0	-4	0

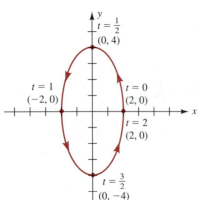

FIGURE 6.69
The parametric equations $x = 2\cos\pi t$ and $y = 4\sin\pi t$ with $0 \leq t \leq 2$ define an ellipse with center at the origin, horizontal axis of length 4, and vertical axis of length 8.

Plotting these points (x, y) in the Cartesian plane, we observe that the curve starts at the point $(2, 0)$ and travels counterclockwise in an elliptical path until it returns to the point where it started, at $(2, 0)$, as illustrated in Figure 6.69. ◆

The parametric equations

$$x = a\sin\theta \quad \text{and} \quad y = b\cos\theta$$

with $a > 0$, $b > 0$, and $0 \leq \theta < 2\pi$ also represent an ellipse with center at the origin, horizontal axis of length $2a$, and vertical axis of length $2b$. However, in this case, the direction of travel along the curve is clockwise, starting at $(0, b)$.

PROBLEM 3 Repeat Example 3 for the parametric equations

$$x = 2 \sin \pi t \quad \text{and} \quad y = 4 \cos \pi t \quad \text{with } 0 \leq t \leq 2.$$ ◆

◆ **Parametric Equations of a Hyperbola**

The set of equations

$$x = a \sec \theta \quad \text{and} \quad y = b \tan \theta$$

with $a > 0$, $b > 0$, and $0 \leq \theta < 2\pi$ $\left(\theta \neq \dfrac{\pi}{2}, \dfrac{3\pi}{2}\right)$ are **parametric equations of a hyperbola** with θ as the parameter. The center of the hyperbola is the origin, the transverse axis is horizontal with length $2a$, and the conjugate axis is vertical with length $2b$. To show that this is true, we eliminate the parameter θ by squaring both x/a and y/b, and then subtracting these expressions as follows:

$$\left(\frac{x}{a}\right)^2 - \left(\frac{y}{b}\right)^2 = (\sec \theta)^2 - (\tan \theta)^2 \quad \text{Substitute}$$

$$\frac{x^2}{a^2} - \frac{y^2}{b^2} = \sec^2 \theta - \tan^2 \theta \quad \text{Square the expressions}$$

$$\frac{x^2}{a^2} - \frac{y^2}{b^2} = 1 \quad \text{Apply } \sec^2 \theta = 1 + \tan^2 \theta$$

We recognize the equation $(x^2/a^2) - (y^2/b^2) = 1$ as the equation of a hyperbola with center at the origin, horizontal transverse axis of length $2a$, and vertical conjugate axis of length $2b$.

In a similar manner, we can show that the set of equations

$$x = a \tan \theta \quad \text{and} \quad y = b \sec \theta$$

are **parametric equations of a hyperbola** with center at the origin, vertical transverse axis of length $2b$, and horizontal conjugate axis of length $2a$.

EXAMPLE 4 Identify and sketch the curve defined by the parametric equations

$$x = 2 \sec \frac{\pi}{3} t \quad \text{and} \quad y = 4 \tan \frac{\pi}{3} t \quad \text{with } 0 \leq t < \frac{3}{2}.$$

Indicate with arrows the direction of travel along the curve.

SOLUTION We recognize this set of equations as parametric equations of a hyperbola with center at the origin, horizontal transverse axis of length $2(2) = 4$, and vertical conjugate axis of length $2(4) = 8$. To determine the portion of the hyperbola traversed and the direction of travel along that portion, we select values of t in the interval $[0, \frac{3}{2})$ and find the corresponding values of x and y.

t	0 (start)	$\frac{1}{2}$	1	$\frac{3}{2}$ (approach)
$x = 2\sec\frac{\pi}{3}t$	2	$\frac{4}{\sqrt{3}} \approx 2.3$	4	$+\infty$
$y = \tan\frac{\pi}{3}t$	0	$\frac{4}{\sqrt{3}} \approx 2.3$	$4\sqrt{3} \approx 6.9$	$+\infty$

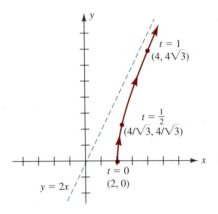

FIGURE 6.70

The parametric equations $x = 2\sec\frac{\pi}{3}t$ and $y = 4\tan\frac{\pi}{3}t$ with $0 \leq t < \frac{3}{2}$, define the upper right branch of a hyperbola with center at the origin, horizontal transverse axis of length 4, and vertical conjugate axis of length 8.

Plotting these points (x, y) in the Cartesian plane, we observe that the curve starts at the point $(2, 0)$ and travels upward to the right along a hyperbolic path as $t \to \frac{3}{2}$ (see in Figure 6.70). ◆

PROBLEM 4 Repeat Example 4 for the parametric equations

$$x = 2\tan\frac{\pi}{3}t \quad \text{and} \quad y = 4\sec\frac{\pi}{3}t \quad \text{with } 0 \leq t < \frac{3}{2}.$$
◆

◆ **Parametric Equations and the Graphing Calculator**

Most graphing calculators are capable of generating the curve defined by a pair of parametric equations. For example, to generate the curve defined by the parametric equations in Example 4,

$$x = 2\sec\frac{\pi}{3}t \quad \text{and} \quad y = 4\tan\frac{\pi}{3}t \quad \text{with } 0 \leq t < \frac{3}{2},$$

we begin by selecting parametric mode and radian mode on the calculator. Next, we press the RANGE key and enter maximum and minimum values of t, x, and y. From Figure 6.70, we choose a viewing rectangle

$$[-5, 5] \quad \text{by} \quad [-1, 8],$$

and since we want $0 \leq t < \frac{3}{2}$, we choose

$$t\text{-min} = 0 \qquad t\text{-max} = 1.5 \qquad t\text{-step} = 0.1$$

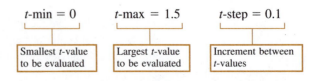

We can now enter the parametric equations $x = 2 \sec \frac{\pi}{3} t$ and $y = 4 \tan \frac{\pi}{3} t$ to generate the graph that is shown in Figure 6.70.

We can also generate the graph of a polar equation $r = f(\theta)$ by using the parametric feature on a graphing calculator. Recall from Section 6.5 that

$$x = r \cos \theta \quad \text{and} \quad y = r \sin \theta.$$

If we replace r with $f(\theta)$, we obtain the parametric equations

$$x = f(\theta) \cos \theta \quad \text{and} \quad y = f(\theta) \sin \theta$$

with parameter θ. Now replacing θ with t gives us the polar equation $r = f(\theta)$ in parametric form with parameter t.

Polar Equation in Parametric Form

The graph of the polar equation $r = f(\theta)$ is the curve defined by the parametric equations

$$x = f(t) \cos t \quad \text{and} \quad y = f(t) \sin t$$

with t as the parameter.

EXAMPLE 5   Generate the graph of the four-leafed rose $r = f(\theta) = 2 \sin 2\theta$ on a graphing calculator. (See Example 5 in Section 6.5.)

SOLUTION   The parametric equations with parameter t that describe this four-leafed rose are as follows:

$$x = f(t) \cos t \quad \text{and} \quad y = f(t) \sin t$$
$$x = 2 \sin 2t \cos t \qquad\qquad y = 2 \sin 2t \sin t$$

We begin by selecting parametric mode and radian mode on the calculator. Next, we press the [RANGE] key and enter maximum and minimum values of x, y, and t. Since the sine and cosine functions range from -1 to 1, we choose a viewing rectangle

$$[-2, 2] \quad \text{by} \quad [-2, 2],$$

and since we want the entire graph of the four-leafed rose, we choose

$$t\text{-min} = 0 \qquad t\text{-max} = 2\pi \approx 6.28 \qquad t\text{-step} = 0.1.$$

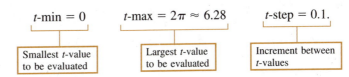

We can now enter the parametric equations $x = 2 \sin 2t \cos t$ and $y = 2 \sin 2t \sin t$ to generate the graph of the polar equation $r = 2 \sin 2\theta$ (see Figure 6.71).

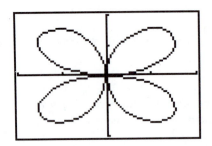

FIGURE 6.71
Graph of the polar equation $r = 2 \sin 2\theta$, as shown on a graphing calculator.

PROBLEM 5 Generate the graph of the three-leafed rose $r = f(\theta) = 2 \sin 3\theta$ on a graphing calculator. (See Problem 5 in Section 6.5.)

◆ **Application: Projectile Motion**

Nearly four hundred years ago, Galileo discovered the laws of motion for freely falling objects. He discovered that when a projectile is fired straight upward from the ground with an initial velocity v (in feet per second), its distance d (in feet) above the ground after time t (in seconds) is given by

$$d = vt - 16t^2.$$

We can use parametric equations to extend this idea to cases in which the initial velocity v is at an angle θ other than 90° from the horizontal. As shown in Figure 6.72, we may resolve the initial velocity v into its horizontal component v_x and its vertical component v_y, where

$$v_x = v \cos \theta \quad \text{and} \quad v_y = v \sin \theta.$$

The motion in the horizontal direction is one of constant velocity v_x. Hence, the position x of the projectile in the horizontal direction at any time t is

$$x = v_x t = (v \cos \theta)t.$$

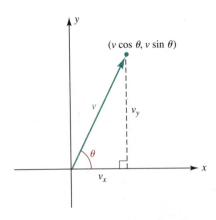

FIGURE 6.72
Initial velocity v resolved into its horizontal component v_x and vertical component v_y.

The motion in the vertical direction is the same as a projectile fired straight upward with initial velocity v_y. Thus, the position y of the projectile in the vertical direction at any time t is

$$y = v_y t - 16t^2 = (v \sin \theta)t - 16t^2.$$

Parametric Equations for Projectile Motion

If a projectile is fired from the ground with an initial velocity v at an angle θ with the horizontal, then the position (x, y) of the projectile t seconds after it is fired is given by the parametric equations

$$x = (v \cos \theta)t \quad \text{and} \quad y = (v \sin \theta)t - 16t^2,$$

where x and y are measured in feet.

EXAMPLE 6 A punter kicks a football with an initial velocity of 80 feet per second (ft/s) at an angle of 60° with the horizontal. Assuming that the ball is caught at the same height as it is kicked, determine

(a) the parametric equations that describe the path of the ball and

(b) the distance (in yards) downfield that the football is caught.

SOLUTION

(a) If $v = 80$ ft/s and $\theta = 60°$, then the parametric equations that describe the path of the ball are

$$x = (v \cos \theta)t \quad \text{and} \quad y = (v \sin \theta)t - 16t^2$$
$$= (80 \cos 60°)t \quad\quad\quad\quad = (80 \sin 60°)t - 16t^2$$
$$= 40t \quad\quad\quad\quad\quad\quad\quad = 40\sqrt{3}\, t - 16t^2$$

(b) The football is caught when $y = 0$, which implies

$$y = 40\sqrt{3}\, t - 16t^2 = 0$$
$$t(40\sqrt{3} - 16t) = 0$$
$$t = 0 \quad \text{or} \quad 40\sqrt{3} - 16t = 0$$
$$t = \frac{5\sqrt{3}}{2} \approx 4.33$$

In football jargon, $t = 4.33$ seconds is referred to as the "hang time" of the kick. Hence, the horizontal distance downfield at which the football is caught is given by

$$x = 40t = 40\left(\frac{5\sqrt{3}}{2}\right) \approx 173 \text{ ft} \quad \text{or} \quad 58 \text{ yd.}$$

PROBLEM 6 Referring to Example 6, identify the graph of the familiar Cartesian equation that develops when eliminating the parameter t, and then determine how high the ball rises in its path of motion. If you have access to a graphing calculator, check your answer by generating the graph of the parametric equations in the viewing rectangle and tracing to the maximum point.

Exercises 6.6

Basic Skills

In Exercises 1–20, identify and sketch the curve defined by each set of parametric equations, and indicate with arrows the direction of travel along the curve as t increases in value.

1. $x = t, y = 2t; \quad 0 \le t \le 4$
2. $x = t, y = -\frac{1}{2}t; \quad 0 \le t \le 8$
3. $x = -2 + t, y = 1 - 3t; \quad t \ge 0$
4. $x = 3 + t, y = -1 + 4t; \quad t \le 0$
5. $x = -2t, y = -t^2; \quad t \le 0$
6. $x = 2t^2, y = 4t; \quad t \ge 0$
7. $x = 3t^2, y = 6t; \quad -2 \le t \le 2$
8. $x = -\frac{2}{3}t, y = -\frac{1}{3}t^2; \quad 0 \le t \le 6$
9. $x = \cos \pi t, y = \sin \pi t; \quad 0 \le t \le 2$
10. $x = 2\cos \frac{2\pi}{3}t, y = 2\sin \frac{2\pi}{3}t; \quad 0 \le t \le 3$
11. $x = 3\sin \frac{\pi}{3}t, y = 3\cos \frac{\pi}{3}t; \quad 0 \le t \le 3$
12. $x = 5\sin \pi t, y = 5\cos \pi t; \quad 0 \le t \le \frac{1}{2}$
13. $x = \cos \frac{\pi}{2}t, y = 2\sin \frac{\pi}{2}t; \quad 0 \le t \le 4$
14. $x = 4\cos 3\pi t, y = 3\sin 3\pi t; \quad 0 \le t \le \frac{2}{3}$
15. $x = 3\sin 2\pi t, y = 2\cos 2\pi t; \quad 0 \le t \le \frac{1}{4}$
16. $x = \sin \frac{3\pi}{4}t, y = 3\cos \frac{3\pi}{4}t; \quad 0 \le t \le 2$
17. $x = \sec \pi t, y = \tan \pi t; \quad \frac{1}{2} < t < \frac{3}{2}$
18. $x = 2\tan \frac{\pi}{2}t, y = 3\sec \frac{\pi}{2}t; \quad 1 < t < 3$
19. $x = 3\tan \frac{3\pi}{4}t, y = 2\sec \frac{3\pi}{4}t; \quad 0 \le t < 2, t \ne \frac{2}{3}$
20. $x = 4\sec 4\pi t, y = 2\tan 4\pi t; \quad 0 \le t \le \frac{1}{4}, t \ne \frac{1}{8}$

In Exercises 21–30, find a set of parametric equations that describe a curve with the given characteristics. (Answers are not unique.)

21. Line: slope -4 and passing through $(-1, 4)$.
22. Line: passing through $(-3, -5)$ and $(0, 1)$.
23. Parabola: vertex $(0, 0)$ and focus $(0, 2)$.
24. Parabola: vertex $(0, 0)$ and directrix $x = 1$.
25. Circle: center $(0, 0)$ and radius 4.
26. Circle: center $(0, 0)$ and radius 3.
27. Ellipse: center $(0, 0)$, horizontal axis of length 6, and vertical axis of length 4.
28. Ellipse: center $(0, 0)$, horizontal axis of length 10, and eccentricity $\frac{3}{5}$.
29. Hyperbola: center $(0, 0)$, horizontal transverse axis of length 4, and vertical conjugate axis of length 4.
30. Hyperbola: center $(0, 0)$, vertical transverse axis of length 8, and eccentricity $\frac{5}{4}$.

In Exercises 31–40,

(a) *find the Cartesian equation in x and y for the given parametric equations and*

(b) *use the vertical and horizontal shift rules and the x-axis reflection rule (Section 1.4) to sketch the curve defined by the parametric equations. Label the x- and y-intercepts with their corresponding values of t.*

31. $x = t, y = |t| - 1$
32. $x = t, y = 2 - t^2$
33. $x = t - 2, y = \frac{1}{t}; t \ne 0$
34. $x = t + 4, y = |t|$
35. $x = t + 1, y = -t^3$
36. $x = t - 3, y = -\sqrt{t}; t \ge 0$
37. $x = t + 3, y = \sqrt{t} - 1; t \ge 0$
38. $x = t - 1, y = t^3 + 1$
39. $x = t - 5, y = 4 - t^2$
40. $x = t + 2, y = 1 - \sqrt[3]{t}$

41. A batter hits a baseball with an initial velocity of 90 feet per second (ft/s) at an angle of 45° with the horizontal, as shown in the sketch. Assuming that the ball is caught at the same height as it is hit, determine

(a) the parametric equations that describe the path of the ball,

(b) the distance d from the place where the ball is hit to the place where it is caught,

(c) the Cartesian equation in x and y that describes the path of the ball, and

(d) the maximum height h the ball rises in its path of motion.

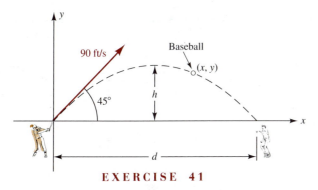

EXERCISE 41

42. A rocket is fired from the ground with an initial velocity of 1200 feet per second (ft/s) at an angle of 30° with the horizontal, as shown in the sketch. Assuming that the ground is level, determine

(a) the parametric equations that describe the path of the rocket,

(b) the distance d from where it is fired to where it hits the ground,

(c) the Cartesian equation in x and y that describes the path of the rocket, and

(d) the maximum height h the rocket rises in its path of motion.

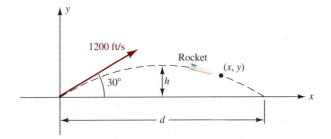

 Critical Thinking

43. Find the parametric equations of the circle $x^2 + y^2 - 4x = 0$ by using the parameter t, as defined.

(a) $t = \dfrac{y}{x}$, $x \neq 0$ (b) $t = \dfrac{x}{y}$, $y \neq 0$

44. Show that the set of equations

$$x = x_1 + (x_2 - x_1)t \quad \text{and} \quad y = y_1 + (y_2 - y_1)t$$

are parametric equations of a line through the points $P(x_1, y_1)$ and $Q(x_2, y_2)$. Using this set of equations, write parametric equations of a line through $(2, 3)$ and $(-4, 5)$ if the direction of travel along the line is as indicated.

(a) from $(2, 3)$ to $(-4, 5)$.   (b) from $(-4, 5)$ to $(2, 3)$.

45. Show that the set of equations

$$x = h + a \cos \theta \quad \text{and} \quad y = k + b \sin \theta$$

with $0 \leq \theta < 2\pi$ are parametric equations of an ellipse (or a circle if $a = b$) with center at (h, k), horizontal axis of length $2a$, and vertical axis of length $2b$.

46. Show that the set of equations

$$x = h + a \sec \theta \quad \text{and} \quad y = k + b \tan \theta$$

with $-\pi/2 < \theta < \pi/2$ and $\pi/2 < \theta < 3\pi/2$ are parametric equations of a hyperbola with center at (h, k), horizontal transverse axis of length $2a$, and vertical conjugate axis of length $2b$.

47. Suppose that $x = f(t)$ and $y = g(t)$ with $0 \leq t \leq 2$ are parametric equations of a curve C. Describe the graph of the following parametric equations in relation to curve C.

(a) $x = f(2t)$ and $y = g(2t)$; $0 \leq t \leq 1$

(b) $x = f(2 - t)$ and $y = g(2 - t)$; $0 \leq t \leq 2$

(c) $x = f(t) + 2$ and $y = g(t) - 1$; $0 \leq t \leq 2$

(d) $x = g(t)$ and $y = f(t)$; $0 \leq t \leq 2$

48. If a projectile is fired from the ground with an initial velocity v at an angle θ with the horizontal, then the position (x, y) of the projectile t seconds after it is fired is given by the parametric equations

$$x = (v \cos \theta)t \quad \text{and} \quad y = (v \sin \theta)t - 16t^2$$

with x and y measured in feet.

(a) Show that the path of the projectile is always a parabola.

(b) Show that the range R of the projectile (the horizontal distance from where it is fired to where it strikes the ground) is given by

$$R = \dfrac{v^2}{32} \sin 2\theta.$$

(c) Determine the angle θ that makes the range R as large as possible.

Calculator Activities

 In Exercises 49–52, use the results of Exercises 45 and 46 to find a set of parametric equations that describe the conic section with the given characteristics. Then use a graphing calculator to generate the graph of the conic section with the given characteristics.

49. Ellipse; center (2, 1), horizontal axis of length 4, and vertical axis of length 2.

50. Ellipse; center (−2, 3), horizontal axis of length 3, and vertical axis of length 5.

51. Hyperbola: center (−1, −2), horizontal transverse axis of length 2, and vertical conjugate axis of length 2.

52. Circle: center (2, −4) and radius 6.

 In Exercises 53–60, express each polar equation in parametric form then use a graphing calculator to generate the graph of the polar equation. Compare the graphs with those in Exercises 41–48 of Exercises 6.5.

53. $r = \dfrac{\theta}{\pi}$, $\theta \geq 0$ (a spiral)

54. $r\theta = \pi$, $\theta > 0$ (a spiral)

55. $r = 1 - \sin\theta$ (a *cardioid*, or heart)

56. $r = 2 + 2\cos\theta$ (a cardioid)

57. $r = 2\cos 2\theta$ (a four-leafed rose)

58. $r = 2\cos 3\theta$ (a three-leafed rose)

59. $r^2 = 4\cos 2\theta$ (a *lemniscate*, or figure eight)

60. $r^2 = 4\sin 2\theta$ (a lemniscate)

61. When a certain rocket is fired from the ground with an initial velocity v at an angle θ, as shown in the sketch, its position t seconds after it is fired is given by the parametric equations

$$x = 544t \quad \text{and} \quad y = 748t - 16t^2$$

with x and y measured in feet. Find the range R and height h to which the rocket rises.

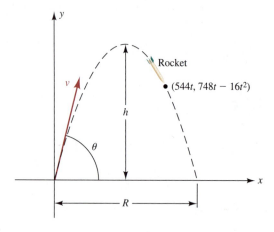

62. Referring to Exercise 61, find the initial velocity v and the angle θ at which the rocket is fired.

Chapter 6 Review

Questions for Group Discussion

1. Name the four main *conic sections*. Show how each is formed by slicing a double cone with a plane.

2. What is meant by *degenerate conic sections*? How are they formed?

3. What name is given to an equation of the form $Ax^2 + Bxy + Cy^2 + Dx + Ey + F = 0$? Is the graph of every equation of this form a conic section? Explain.

4. State the general form of the equation of a conic section if its axes of symmetry are parallel to the *xy*-axes.

5. State the geometric definition of a *parabola*. Illustrate with a sketch.

6. Under what conditions does the equation of a parabola define *y* as a function of *x*?
7. How can you tell from the equation of a parabola whether the curve opens upward, downward, to the left, or to the right?
8. State the geometric definition of an *ellipse*. Illustrate with a sketch.
9. What is meant by the *major axis* of an ellipse? How can you determine from the equation of an ellipse in standard form whether the major axis is horizontal or vertical?
10. Why is a *circle* a special case of an ellipse?
11. Explain the procedure for finding the coordinates of the *foci* of an ellipse from its equation. Illustrate with an example.
12. What is meant by the *eccentricity* of an ellipse? Describe the shape of an ellipse with eccentricity (a) close to 1 and (b) close to 0.
13. State the geometric definition of a *hyperbola*. Illustrate with a sketch.
14. What is meant by the *transverse axis* of a hyperbola? How can you determine from the equation of a hyperbola in standard form whether the transverse axis is horizontal or vertical?
15. How are *asymptotes* used as an aid in graphing a hyperbola?
16. What is meant by the *eccentricity* of a hyperbola? Describe the shape of a hyperbola with eccentricity (a) close to 1 and (b) very large.
17. Explain the procedure for finding the coordinates of the *foci* of a hyperbola from its equation. Illustrate with an example.
18. Discuss the *reflection property* of (a) a parabola, (b) an ellipse, and (c) a hyperbola.
19. State the *common definition of the conic sections*. Illustrate with a sketch.
20. What effect does the presence of an *xy* term have on the graph of a conic section? Explain the procedure for eliminating the *xy* term.
21. What is meant by the *discriminant* of a general quadratic equation in two unknowns? How is it used to classify the conic sections? Illustrate with examples.
22. Is there a one-to-one correspondence between ordered pairs and points in the *polar coordinate system*? Explain.
23. Explain the procedure for changing *polar coordinates* to Cartesian coordinates, and vice versa.
24. Given that *a* is a nonzero constant, describe the graph of each *polar equation:*

    (a) $r = a$ (b) $\theta = a$ (c) $r = \dfrac{a}{\sin \theta}$ (d) $r = a \cos \theta$

25. What is the *polar equation of a conic section* with focus at the pole and corresponding directrix parallel to the polar axis and *d* units above the pole? Under what conditions does the equation represents a parabola? an ellipse? a hyperbola?
26. What information does a set of parametric equations tell us about a curve? Are the parametric equations of a given curve unique? Explain.
27. Discuss the procedure for writing a polar equation in parametric form. Illustrate with an example.
28. What are the parametric equations for *projectile motion*? Discuss the path of a projectile.

Review Exercises

In Exercises 1–20, sketch the graph of each equation. If the graph is a parabola, label its vertex and any x- and y-intercept(s). If the graph is an ellipse or hyperbola, label its center and vertices.

1. $y^2 - 4x = 0$
2. $2x^2 + y = 0$
3. $x^2 + 9y^2 = 16$
4. $4x^2 + 3y^2 = 9$
5. $x^2 - y^2 = 25$
6. $4y^2 - x^2 = 49$
7. $y = (x - 2)^2 - 3$
8. $x = -2(y + 1)^2 + 2$
9. $9x^2 + (y - 3)^2 = 9$
10. $4(x + 2)^2 + 9(y - 1)^2 = 36$
11. $9x^2 - 16(y - 1)^2 = 144$
12. $(x + 2)^2 - (y - 1)^2 = -4$
13. $y^2 + x - 2y - 15 = 0$
14. $x^2 + 2x - 2y + 5 = 0$
15. $4x^2 + y^2 - 8x + 4y - 8 = 0$
16. $x^2 + 16y^2 - 32y - 9 = 0$
17. $4y^2 - 9x^2 - 8y + 5 = 0$
18. $16x^2 - 9y^2 - 32x - 36y - 24 = 0$
19. $2x^2 + 3y^2 - 4x + 12y + 14 = 0$
20. $x^2 - y^2 - 4x - 2y + 3 = 0$

21. Find the focus and directrix for the parabola in Exercise 1.
22. Find the foci and eccentricity for the ellipse in Exercise 4.
23. Find the foci and eccentricity for the hyperbola in Exercise 5.
24. Find the focus and directrix for the parabola in Exercise 14.
25. Find the foci and eccentricity for the ellipse in Exercise 15.
26. Find the foci and eccentricity for the hyperbola in Exercise 18.

In Exercises 27–40, determine the equation of the conic section from the given information.

27. Parabola with vertex $(0, -4)$ and x-intercepts ± 2.
28. Parabola with vertex $(2, 1)$ and y-intercepts -1 and 3.
29. Parabola passing through the origin with vertex $(-2, -3)$, and symmetric with respect to $y = -3$.
30. Parabola passing through $(-1, 0)$ with vertex $(2, 3)$, and symmetric with respect to $x = 2$.
31. Parabola with focus $(1, 0)$ and directrix $y = -1$.
32. Parabola with vertex $(3, -2)$ and focus $(4, -2)$.
33. Ellipse with vertices $(0, 0)$ and $(6, 0)$ and minor axis of length 4.
34. Ellipse with center $(2, 5)$, minor axis of length 2, and horizontal major axis of length 8.
35. Ellipse with foci $(0, 0)$ and $(4, 0)$ and eccentricity $\frac{1}{2}$.
36. Ellipse with vertices $(-1, -3)$ and $(-1, 3)$ and one focus at $(-1, 2)$.
37. Hyperbola with asymptotes $y = \pm\frac{2}{3}x$ and vertical transverse axis of length 6.
38. Hyperbola with center $(2, -1)$, horizontal transverse axis of length 10, and conjugate axis of length 4.
39. Hyperbola with vertices $(-3, -1)$ and $(3, -1)$ and one focus at $(4, -1)$.
40. Hyperbola with foci $(\pm 3, 0)$ and eccentricity 2.

Given that the graph of each equation in Exercises 41–48 exists and is not a degenerate conic section,

(a) *classify each equation as a parabola, ellipse, or hyperbola,*

(b) *find the rotation angle θ that eliminates the xy term in the equation,*

(c) *use the rotation formulas with angle θ from part (b) to transform the equation to a uv-equation in standard form, and*

(d) *sketch the graph of the equation.*

41. $xy - 4 = 0$
42. $4xy + 1 = 0$
43. $5x^2 - 6xy + 5y^2 - 8 = 0$
44. $x^2 + 2xy + y^2 + \sqrt{2}x - \sqrt{2}y + 2 = 0$
45. $3x^2 - 2\sqrt{3}xy + y^2 + 2x + 2\sqrt{3}y = 0$
46. $2x^2 + \sqrt{3}xy + y^2 - 50 = 0$
47. $17x^2 - 12xy + 8y^2 - 80 = 0$
48. $11x^2 - 24xy + 4y^2 + 30x + 40y - 45 = 0$

49. Referring to the conic section in Exercise 43, find the coordinates in the xy-coordinate system for (a) the vertices and (b) the foci.

50. Referring to the conic section in Exercise 48, find the coordinates in the *xy*-coordinate system for (a) the vertices and (b) the foci.

51. Convert the given polar coordinates to Cartesian coordinates.

(a) $(-2, 45°)$ \qquad (b) $\left(6, -\dfrac{\pi}{3}\right)$

52. Convert the given Cartesian coordinates to polar coordinates.

(a) $(-2, 2)$ with $r < 0$ and $0° < \theta < 360°$

(b) $(-\sqrt{3}, 1)$ with $r > 0$ and $-2\pi < \theta < 0$

In Exercises 53–64, sketch the graph of each polar equation, and give its equivalent Cartesian equation. If you have access to a graphing calculator, use the parametric mode to verify each sketch.

53. $r = 2$ \qquad **54.** $\theta = \pi$

55. $r = \dfrac{1}{\sin \theta}$ \qquad **56.** $r = -3 \sec \theta$

57. $r = 5 \cos \theta$ \qquad **58.** $r = -2 \sin \theta$

59. $r = 6 \sin \theta - 8 \cos \theta$ \qquad **60.** $r = \dfrac{12}{3 \cos \theta - 4 \sin \theta}$

61. $r = \dfrac{2}{1 + \cos \theta}$ \qquad **62.** $r = \dfrac{6}{1 - 3 \cos \theta}$

63. $r = \dfrac{1}{3 + \sin \theta}$ \qquad **64.** $r = \dfrac{3}{4 - 2 \sin \theta}$

In Exercises 65–74, identify and sketch the curve defined by each set of parametric equations. Indicate with arrows the direction of travel along the curve as t increases in value.

65. $x = 2 + t, \ y = 3 - 2t; \ 0 \le t \le 6$

66. $x = 3t, \ y = 1 + t; \ t \ge 0$

67. $x = -4t, \ y = -2t^2; \ t \ge 0$

68. $x = \tfrac{1}{2}t^2, \ y = t; \ -2 \le t \le 2$

69. $x = 2 \sin \pi t, \ y = 2 \cos \pi t; \ 0 \le t \le 1$

70. $x = \sin \dfrac{\pi}{3} t, \ y = \cos \dfrac{\pi}{3} t; \ 0 \le t \le 6$

71. $x = 3 \sin \dfrac{\pi}{2} t, \ y = 4 \cos \dfrac{\pi}{2} t; \ 0 \le t \le 4$

72. $x = \sin \pi t, \ y = 2 \cos \pi t; \ 0 \le t \le \tfrac{1}{2}$

73. $x = 2 \sec 2\pi t, \ y = 3 \tan 2\pi t; \ 0 \le t < \tfrac{1}{4}$

74. $x = \tan \dfrac{3\pi}{4} t, \ y = \sec \dfrac{3\pi}{4} t; \ 0 \le t \le \tfrac{4}{3}, \ t \ne \tfrac{2}{3}$

In Exercises 75–80, find a set of parametric equations that describe a curve with the given characteristics. (Answers are not unique.)

75. Line: passing through $(2, 3)$ and $(-1, 0)$.

76. Parabola: vertex $(0, 0)$ and directrix $y = -2$.

77. Hyperbola: center $(0, 0)$, horizontal transverse axis, and vertical conjugate axis of length 4.

78. Ellipse: center $(0, 0)$, horizontal major axis of length 20, and eccentricity $\tfrac{4}{5}$.

79. Circle: center $(-2, 4)$ and radius 4.

80. Hyperbola: center $(-1, 2)$, horizontal transverse axis of length 6, and eccentricuty $\tfrac{5}{3}$.

81. A satellite dish for receiving television signals is parabolic, with dimensions as shown in the sketch. At what distance d should the receiver be located in order to take advantage of the reflection property of a parabola?

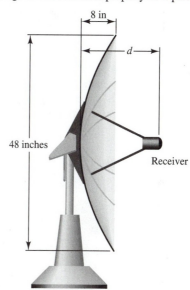

82. A simply supported beam with a concentrated load at midspan deflects in the shape of a parabola, as shown in the sketch. Determine the equation of the parabola (called the *elastic curve*). Write the equation in the form $y = ax^2 + bx$, with a and b rounded to three significant digits.

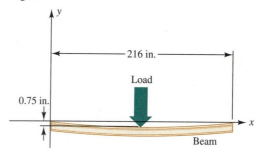

83. Ultrahigh-frequency (UHF) sound waves emanating from one focus of a lithotripter are reflected to the other focus of the device to break up kidney stones. Suppose a lithotripter is formed by rotating the portion of an ellipse below its minor axis about its major axis, as shown in the sketch. How far from the kidney stone should the opening of the lithotripter be placed in order to break up the stone?

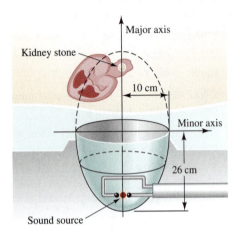

84. In an engineer's design for a cooling tower in a nuclear power plant, a hyperbola is rotated about the y-axis, as shown in the figure. If the distance from A to B is 12.3 meters, determine the radius of each end of the tower.

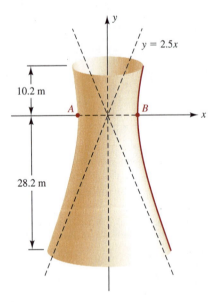

The distances from a planet to the sun at perihelion (shortest distance from the sun) and aphelion (greatest distance from the sun) are given in Exercises 85–86.

(a) Determine the polar equation of the elliptical orbit of the planet if the sun (focus) is at the pole, and its corresponding directrix is perpendicular to the polar axis and to the right of the pole.

(b) State the eccentricity of the orbit.

85. Jupitor: perihelion = 4.59×10^8 miles, aphelion = 5.05×10^8 miles

86. Uranus: perihelion = 18.3 AU, aphelion = 20.1 AU

87. The polar equation of the elliptical orbit of the comet Kohoutek with the sun (focus) at the pole and corresponding directrix perpendicular to the polar axis and to the right of the pole is given by

$$r = \frac{0.2689}{1 + 0.999925 \cos \theta}$$

with r measured in astronomical units (AU).

(a) What is the distance from the comet to the sun at perihelion (shortest distance from the sun)? at aphelion (greatest distance from the sun)?

(b) What is the length (in miles) of the major axis of the elliptical orbit of the comet? the minor axis? (*Hint:* 1 AU $\approx 9.26 \times 10^7$ miles.)

88. A golf ball is hit from a tee with an initial velocity of 150 ft/s at an angle of 35° with the horizontal. Assuming that the ball is hit straight down the fairway (no hook, no slice) and that the ground is level, determine

(a) the parametric equations that describe the path of the ball,

(b) the distance (in yards) from the tee to the place where the ball strikes the ground,

(c) the Cartesian equation in x and y that describes the path of the ball, and

(d) the maximum height the ball rises in its path of motion.

CHAPTER 7

Exponential and Logarithmic Functions

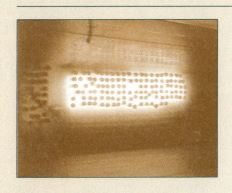

Strontium-90, a waste product from nuclear reactors, has a half-life of 28 years. It is estimated that a certain quantity of this material will be safe to handle when its mass is $\frac{1}{1000}$ of its original amount. Determine the time required to store strontium-90 until it reaches this level of safety.

(For the solution, see Example 9 in Section 7.2).

7.1   Properties and Graphs of Exponential Functions
7.2   Logarithmic Functions
7.3   Properties of Logarithms
7.4   Graphs of Logarithmic Functions
7.5   Logarithmic and Exponential Equations

7.1 Properties and Graphs of Exponential Functions

In this section we study a class of transcendental functions in which the variable appears as an exponent. We then define the inverse for this class of transcendental functions and discuss several applied problems in which these functions and their inverses appear. We begin by reviewing the idea of rational exponents and then extending this idea to include irrational exponents.

◆ Real Exponents

In basic algebra we evaluated expressions of the form b^x, where b is a *positive real number* and x is any *rational number*. Whether x is a positive integer, a negative integer, a common fraction, or a decimal fraction, we can evaluate b^x as a real number, as shown in the following examples:

$$4^2 = 4 \cdot 4 = 16$$

$$4^{-2} = \frac{1}{4^2} = \frac{1}{16}$$

$$4^{3/2} = (4^{1/2})^3 = (\sqrt{4})^3 = (2)^3 = 8$$

$$4^{1.4} = 4^{14/10} = 4^{7/5} \approx 6.9644$$

> By using the y^x key on a calculator, we can write the fifth root of 4 to the seventh power as an approximate decimal number.

In general, if b is a positive real number and x is any rational number, then b^x is a well-defined positive real number.

Can we assign any meaning to b^x where b is a positive real number and x is an *irrational* number? For example, can we evaluate $4^{\sqrt{2}}$? We can use the $\boxed{y^x}$ key on a calculator to find values of 4^x for rational numbers x that approach $\sqrt{2}$:

Values of 4^x for rational values of x that approach $\sqrt{2}$.

x	1	1.4	1.41	1.414	1.4142	1.41421
4^x	4	6.964405...	7.061624...	7.100891...	7.102860...	7.102958...

From these values, we see that as the exponent approaches $\sqrt{2}$, the expression 4^x appears to approach a unique positive real number whose decimal expansion begins with 7.10. Hence, we define the value of $4^{\sqrt{2}}$ as the real number approached by 4^x as x takes on rational values that get closer and closer to $\sqrt{2}$. In general, if b is a positive real number and x is irrational, then we define b^x as the number approached by b^r as r takes on rational values that get closer and closer to x. We conclude that *for each real number x, b^x is a unique positive real number.*

The laws of exponents given in basic algebra apply to all real exponents. We restate them here and refer to them as the **properties of real exponents.**

Properties of Real Exponents

If the bases a and b are positive real numbers and the exponents x and y represent any real numbers, then

1. $b^0 = 1$
2. $b^{-x} = \dfrac{1}{b^x}$
3. $b^x b^y = b^{x+y}$
4. $(b^x)^y = b^{xy}$
5. $\dfrac{b^x}{b^y} = b^{x-y}$
6. $(ab)^x = a^x b^x$
7. $\left(\dfrac{a}{b}\right)^x = \dfrac{a^x}{b^x}$
8. $\dfrac{a^{-x}}{b^{-y}} = \dfrac{b^y}{a^x}$

Since $1^x = 1$ for all real x, we may also express property 2 as

$$b^{-x} = \left(\dfrac{1}{b}\right)^x$$

EXAMPLE 1 Use the properties of real exponents to write each expression as a constant or as an expression in the form b^x with $b > 0$.

(a) $\dfrac{5^\pi}{5^{x+\pi}}$ \quad (b) $4^{-3x} \cdot 8^{2x+1}$

SOLUTION

(a) $\dfrac{5^\pi}{5^{x+\pi}} = 5^{\pi-(x+\pi)} = 5^{-x} = \left(\dfrac{1}{5}\right)^x$

(b) $4^{-3x} \cdot 8^{2x+1} = (2^2)^{-3x} \cdot (2^3)^{2x+1}$ \quad Write each factor with the same base

$\qquad = 2^{-6x} \cdot 2^{6x+3}$ \quad Multiply the exponents

$\qquad = 2^3 \text{ or } 8$ \quad Add the exponents and simplify ◆

PROBLEM 1 Repeat Example 1 for $\dfrac{125^{x-2}}{25^{2x-3}}$. ◆

Exponential Functions and Their Graphs

We now use the properties of real exponents to define the *exponential function* with base b.

Exponential Function

If b is a real number such that $b > 0$ and $b \neq 1$, then the function f defined by

$$f(x) = b^x$$

is called an **exponential function** with base b. The domain of f is $(-\infty, \infty)$ and the range is $(0, \infty)$.

Note: We exclude 1 as a base for exponential functions, since $f(x) = 1^x = 1$ is a constant function. We exclude zero as a base, since 0^x is undefined when $x \leq 0$. We exclude negative numbers as bases, since b^x for $b < 0$ is an imaginary number for infinitely many values of x, such as $x = \frac{1}{2}, \frac{1}{4}, \frac{1}{6}$, and so on.

EXAMPLE 2 Sketch the graph of each exponential function.

(a) $f(x) = 4^x$ (b) $g(x) = \left(\frac{1}{4}\right)^x$

SOLUTION

(a) Selecting convenient inputs for x, we find the corresponding outputs $f(x)$ as shown in the following table:

x	-2	$-\frac{3}{2}$	-1	$-\frac{1}{2}$	0	$\frac{1}{2}$	1	$\frac{3}{2}$	2
$f(x)$	$\frac{1}{16}$	$\frac{1}{8}$	$\frac{1}{4}$	$\frac{1}{2}$	1	2	4	8	16

Plotting the points $(x, f(x))$ and connecting them in a smooth curve, we graph the exponential function $f(x) = 4^x$ as shown in Figure 7.1. Notice that the graph of this function is always increasing and its y-intercept is 1. Also, observe that $f(x)$ approaches zero [$f(x) \to 0$] as x decreases without bound ($x \to -\infty$). Hence, the x-axis is a horizontal asymptote.

(b) By the properties of real exponents,

$$g(x) = \left(\frac{1}{4}\right)^x = 4^{-x}$$

Now, by the y-axis reflection rule (Section 1.4), the graph of g must be the same as the graph of $f(x) = 4^x$ reflected about the y-axis. The graph of g is shown in Figure 7.2. Notice that the graph of this function is always

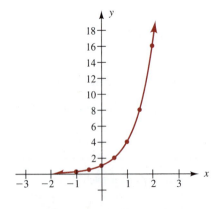

FIGURE 7.1
The graph of $f(x) = 4^x$ is an increasing function. Its y-intercept is 1 and the x-axis is a horizontal asymptote.

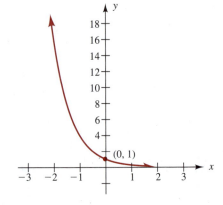

FIGURE 7.2
The graph of $g(x) = \left(\frac{1}{4}\right)^x = 4^{-x}$ is a decreasing function. Its y-intercept is 1 and the x-axis is a horizontal asymptote.

SECTION 7.1 Properties and Graphs of Exponential Functions

decreasing and its y-intercept is 1. Also, observe that $g(x)$ approaches zero $[g(x) \to 0]$ as x increases without bound $(x \to \infty)$. Hence, the x-axis is a horizontal asymptote. ◆

PROBLEM 2 On the same coordinate plane, sketch the graphs of the exponential functions

$$f(x) = 5^x \quad \text{and} \quad g(x) = (\tfrac{1}{5})^x = 5^{-x}.$$ ◆

Some important observations about the graph of the exponential function $f(x) = b^x$ will be useful in our following discussion:

Characteristics of the Graph of $f(x) = b^x$

1. The y-intercept is 1, and the graph has no x-intercept.
2. The x-axis is a horizontal asymptote.
3. If $b > 1$, the graph of $f(x) = b^x$ is always increasing.
4. If $0 < b < 1$, the graph of $f(x) = b^x$ is always decreasing.

Knowing the basic shape of the exponential function $f(x) = b^x$ enables us to graph several other related functions by applying the shift rules and axis reflections rules, which we discussed in Section 1.4.

EXAMPLE 3 Sketch the graph of each function. Label the horizontal asymptote and the y-intercept.

(a) $F(x) = 4^{x+2}$ (b) $G(x) = 3 - 4^{-x}$

SOLUTION

(a) By the horizontal shift rule (Section 1.4), the graph of $F(x) = 4^{x+2}$ is the same as the graph of $f(x) = 4^x$ shifted horizontally to the left 2 units. The y-intercept is

$$F(0) = 4^{0+2} = 4^2 = 16,$$

and the horizontal asymptote remains the x-axis. The graph of $F(x) = 4^{x+2}$ is shown in Figure 7.3.

(b) By the x-axis reflection rule and the vertical shift rule (Section 1.4), the graph of $G(x) = 3 - 4^{-x}$ is the same as the graph of $g(x) = 4^{-x} = (\tfrac{1}{4})^x$ reflected about the x-axis and then shifted vertically upward 3 units. The y-intercept is

$$G(0) = 3 - 4^0 = 3 - 1 = 2.$$

Note that $G(x) \to 3$ as $x \to \infty$. Thus, $y = 3$ is the horizontal asymptote. The graph of $G(x) = 3 - 4^{-x}$ is shown in Figure 7.4.

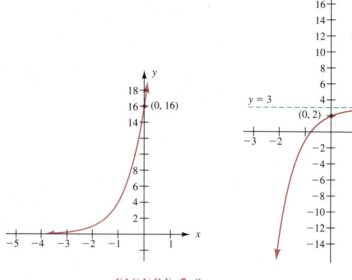

FIGURE 7.3
Graph of $F(x) = 4^{x+2}$

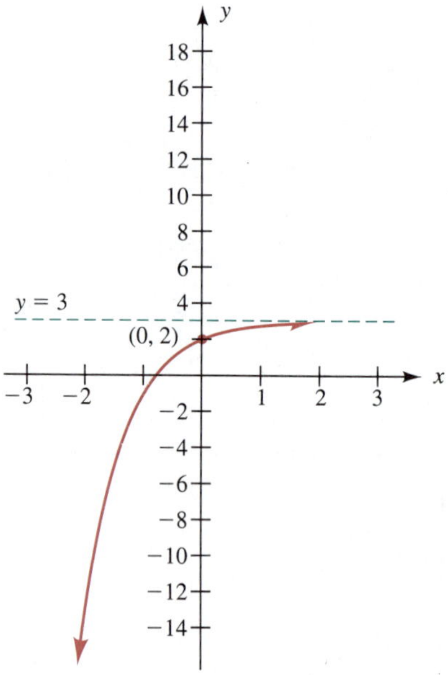

FIGURE 7.4
Graph of $G(x) = 3 - 4^{-x}$

Note: The graph of G shown in Figure 7.4 has an x-intercept between 0 and -1. To find the x-intercept, it is necessary to solve the equation

$$3 - 4^{-x} = 0 \quad \text{or} \quad 4^{-x} = 3$$

for x. However, none of our previous methods can be used to solve an equation in which the unknown appears as an exponent. In Section 7.2, we introduce the *logarithmic function,* which we can use to help solve this type of equation.

PROBLEM 3 State the domain and range of the function G defined in Example 3(b).

◆ **Base e**

In applications of exponential functions, one particular irrational number occurs frequently as the base. This irrational number is denoted by the letter e and its value is

$$e = 2.71828\ldots$$

The function f defined by

$$f(x) = e^x$$

SECTION 7.1 Properties and Graphs of Exponential Functions

is called the **exponential function with base *e*.** With a calculator, we can evaluate $f(x) = e^x$ for $x = 0, \pm 1, \pm 2,$ and ± 3 by using the $\boxed{e^x}$ key, the $\boxed{y^x}$ key with $y = 2.718$, or the exponential table in Appendix C.

Table of values for $f(x) = e^x$

x	-3	-2	-1	0	1	2	3
$f(x)$	$e^{-3} \approx 0.05$	$e^{-2} \approx 0.14$	$e^{-1} \approx 0.37$	1	$e \approx 2.72$	$e^2 \approx 7.39$	$e^3 \approx 20.1$

Plotting these points, we sketch the graph of $f(x) = e^x$, as shown in Figure 7.5. The graph of $g(x) = e^{-x}$, which is the same as the graph of $f(x) = e^x$ reflected about the *y*-axis, is shown in Figure 7.6.

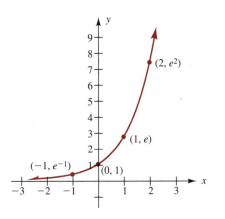

FIGURE 7.5
Graph of $f(x) = e^x$

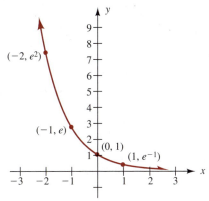

FIGURE 7.6
Graph of $g(x) = e^{-x}$

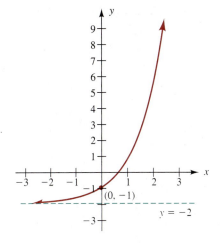

FIGURE 7.7
Graph of $F(x) = e^x - 2$

EXAMPLE 4 Sketch the graph of the function F defined by $F(x) = e^x - 2$. Label the horizontal asymptote and the *y*-intercept.

SOLUTION By the vertical shift rule (Section 1.4), the graph of $F(x) = e^x - 2$ is the same as the graph of $f(x) = e^x$ shifted vertically downward 2 units, as shown in Figure 7.7. The *y*-intercept is

$$F(0) = e^0 - 2 = 1 - 2 = -1.$$

Note that $F(x) \to -2$ as $x \to -\infty$. Thus, $y = -2$ is the horizontal asymptote. ◆

PROBLEM 4 Try to find the *x*-intercept for the graph of the function $F(x) = e^x - 2$ shown in Figure 7.7. What situation occurs? ◆

◆ **Application: Compound Interest**

Exponential functions occur in problems involving compound interest. The amount of simple interest i earned on a principal P invested at a certain rate of interest r per year over a time of t years is given by the formula

$$i = Prt.$$

Thus, the amount A in the account after t years is

$$A = P + Prt.$$

Compound interest is interest paid both on the principal and on any interest earned previously. If a certain principal P is deposited in a savings account at an interest rate r per year and interest is *compounded n times per year,* then interest is paid on both principal and any interest earned previously every $(1/n)$th of a year. The interest i accrued after the first compounding period is

$$i = Prt = Pr\frac{1}{n}.$$

Hence the amount A_1 in the account after the first compounding period is

$$A_1 = P + Prt = P + Pr\frac{1}{n} = P\left(1 + \frac{r}{n}\right).$$

The interest i accrued after the second compounding period is

$$i = A_1 r\frac{1}{n}.$$

and the amount A_2 in the account after the second compounding period is

$$A_2 = A_1 + A_1 r\frac{1}{n}.$$

Now, replacing A_1 with $P\left(1 + \frac{r}{n}\right)$, we have

$$A_2 = P\left(1 + \frac{r}{n}\right) + P\left(1 + \frac{r}{n}\right)r\frac{1}{n}$$

$$= P\left(1 + \frac{r}{n}\right)\left[1 + \frac{r}{n}\right] \qquad \text{\textbf{Factor out }} P\left(1 + \frac{r}{n}\right) \text{\textbf{ from each term}}$$

$$= P\left(1 + \frac{r}{n}\right)^2 \qquad \text{\textbf{Simplify}}$$

Continuing in this manner, we can show that after 1 year (after n compounding periods) the amount A in the account is

$$A = P\left(1 + \frac{r}{n}\right)^n$$

and after t years the amount in the account is

$$A = P\left(1 + \frac{r}{n}\right)^{nt}.$$

Compound Interest Formula for n Compoundings Per Year

If a certain principal P is deposited in a savings account at an interest rate r per year and interest is compounded n times per year, then the amount A in the account after t years is given by the formula

$$A = P\left(1 + \frac{r}{n}\right)^{nt}.$$

If the number of compoundings per year increases *without bound*, then we have what is called **continuous compounding.** In the formula

$$A = P\left(1 + \frac{r}{n}\right)^{nt}$$

suppose we let $n/r = m$. Then $n = mr$, and by direct substitution and the properties of real exponents, we have

$$A = P\left(1 + \frac{r}{n}\right)^{nt} = P\left(1 + \frac{1}{m}\right)^{mrt} = P\left[\left(1 + \frac{1}{m}\right)^{m}\right]^{rt}.$$

Using a calculator, we can show that as m gets larger and larger, the expression $\left(1 + \frac{1}{m}\right)^m$ approaches a number whose decimal expansion begins with 2.718.

Values of $\left(1 + \frac{1}{m}\right)^m$ as m gets larger and larger

m	1	10	100	1000	10,000	100,000
$\left(1 + \frac{1}{m}\right)^m$	2	2.59374...	2.70481...	2.71692...	2.71814...	2.71826...

We say that as m increases without bound ($m \to \infty$) the *limit* of $(1 + 1/m)^m$ is the number e and we write

$$\lim_{m \to \infty} \left(1 + \frac{1}{m}\right)^m = e.$$

Hence, for continuous compounding, we have

$$A = P\left[\left(1 + \frac{1}{m}\right)^m\right]^{rt} = Pe^{rt}$$

Compound Interest Formula for Continuous Compounding

If a certain principal P is deposited in a savings account at an interest rate r per year and interest is compounded continuously, then the amount A in the account after t years is given by the formula

$$A = Pe^{rt}.$$

EXAMPLE 5 Suppose that $1000 is deposited in a bank account at an interest rate of 8% per year. Assuming the depositor makes no subsequent deposit or withdrawal, find the amount after 5 years if the interest is compounded **(a)** quarterly or **(b)** continuously.

SOLUTION

(a) If the interest is compounded quarterly, then it is compounded 4 times per year. Using the formula $A = P\left(1 + \dfrac{r}{n}\right)^{nt}$ with $P = 1000$, $r = 0.08$, $n = 4$, and $t = 5$, we have

$$A = 1000\left(1 + \frac{0.08}{4}\right)^{4 \cdot 5} = 1000(1.02)^{20} \approx 1485.95$$

Thus, if the interest is compounded quarterly, the amount in the account after 5 years is $1485.95.

(b) To find the amount in the account when the interest is compounded continuously, we use the formula $A = Pe^{rt}$ with $P = 1000$, $r = 0.08$, and $t = 5$ as follows:

$$A = 1000e^{(0.08)5} = 1000e^{0.4} \approx 1491.82$$

Thus, if the interest is compounded continuously, the amount in the account after 5 years is $1491.82. ◆

PROBLEM 5 Repeat Example 5 if the interest is compounded monthly.

Exercises 7.1

Basic Skills

In Exercises 1–14, use the properties of real exponents to write each expression as a constant or as an expression in the form b^x, where $b > 0$.

1. 1^{5x}
2. $1^{x - \pi}$
3. $(6^x)^{-2}$
4. $\left(\dfrac{1}{3}\right)^{-2x}$

5. $(4^{x-1})^0$
6. $(2^{2x})^{-2}$
7. $e(e^{x-1})$
8. $e^x \cdot e^{-x}$
9. $(3)(3^{2x-1})$
10. $\dfrac{36}{6^{2-x}}$
11. $(9^{3x-2})(27^{1-2x})$
12. $\dfrac{8^{2-x}}{4^{x+3}}$
13. $\left(\dfrac{5^{2x}}{25^{x-1}}\right)^{1/2}$
14. $(2^{-x} \cdot 4^x)^2$

In Exercises 15 and 16, sketch the graphs of the given equations on the same coordinate plane. What is the point of intersection of the graphs?

15. $y = 2^x$, $y = 3^x$, and $y = e^x$
16. $y = 2^{-x}$, $y = 3^{-x}$, and $y = e^{-x}$

Use the graphs of the equations in Exercises 15 and 16 in conjunction with the shift rules and axis reflection rules (Section 1.4) to sketch the graphs of the functions in Exercises 17-30. In each case, label the y-intercept and horizontal asymptote.

17. $f(x) = -2^x$
18. $f(x) = -3^{-x}$
19. $h(x) = 2^{x-2}$
20. $h(x) = e^{x+1}$
21. $F(x) = -(3^x + 2)$
22. $F(x) = 2^x + 1$
23. $G(x) = 2(\tfrac{1}{2})^x$
24. $G(x) = 9 \cdot 3^{-x}$
25. $H(x) = 1 - e^{-x}$
26. $H(x) = 2 - (\tfrac{1}{3})^{-x}$
27. $A(x) = 2 + 3^{2-x}$
28. $A(x) = 2^{x+1} - 3$
29. $h(x) = \dfrac{e}{e^x}$
30. $h(x) = 2 - e^{1-x}$

In Exercises 31–38, find the amount in the given account.

31. $5000 invested at 9% per year compounded annually for 4 years
32. $3000 invested at 7% per year compounded semiannually for 8 years
33. $10,000 invested at $8\tfrac{1}{2}$% per year compounded quarterly for 12 years
34. $800 invested at $9\tfrac{3}{4}$% per year compounded daily for 10 years
35. $2000 invested at 6% compounded continuously for 9 years
36. $15,000 invested at 10% compounded continuously for 6 years
37. $120,000 invested at 8% compounded continuously for 6 months
38. $250,000 invested at $12\tfrac{3}{4}$% compounded continuously for 8 months

Critical Thinking

39. Perform the indicated operations and simplify.
    (a) $(2^x + 2^{-x})(2^x - 2^{-x})$
    (b) $(3^x + 3^{-x})^2 - (3^x - 3^{-x})^2$

40. Given the function f defined by $f(x) = e^x$, determine if f is a one-to-one function and if there exists an inverse function f^{-1}. Sketch the graph of f^{-1} if it exists.

In Exercises 41–46, use factoring to help find the zeros of each function.

41. $F(x) = 2e^{-2x} - xe^{-2x}$
42. $G(x) = 2x^2 e^{5x} - xe^{5x} - e^{5x}$
43. $g(x) = \dfrac{(x^2 - 4x + 4)e^x}{x + 1}$
44. $h(x) = \dfrac{x^2 e^x + 2x\, e^x}{x - 3}$
45. $H(x) = x^2 3^{2x} - 9^x$
46. $f(x) = x^2(\tfrac{1}{2})^{3x} - x8^{-x}$

47. The graph of the exponential function $y = a^x$ passes through the point $(-2, 16)$. Determine the value of the base a.

48. Given that f is an exponential function defined by $f(x) = e^x$, show that
    (a) $f(a)f(b) = f(a + b)$
    (b) $\dfrac{f(a)}{f(b)} = f(a - b)$
    (c) $[f(a)]^n = f(na)$

49. Complete the table, given that the function f is defined by $f(x) = (-2)^x$.

x	0	1	2	3	4	−1	−2	−3	−4
$f(x)$									

EXERCISE 49 (continued)

Why is it not possible to sketch the graph of f by plotting the points in this table and connecting them with a smooth curve?

50. The *hyperbolic sine function* f and the *hyperbolic cosine function* g are defined by

$$f(x) = \frac{e^x - e^{-x}}{2} \quad \text{and} \quad g(x) = \frac{e^x + e^{-x}}{2}$$

(a) Determine if f and g are even or odd functions.
(b) Sketch the graphs of f and g.

51. Suppose P is the air pressure (in pounds per square inch, or psi) at a certain height h in feet above sea level, and $P = ke^{-0.00004h}$, where k is a constant.

(a) Determine k if the air pressure at sea level is 15 psi.
(b) What is the pressure outside the cabin of an airplane whose altitude is 20,000 feet?

52. Suppose A is the number of moose living in a protected forest in Maine after a certain time t (in years since 1968), and $A = ke^{0.15t}$, where k is a constant.

(a) Determine k if 201 moose were present in 1988.
(b) What is the expected moose population in 1998?

53. The number A of bacteria present in a certain culture is a function of time t (in minutes) and is described by $A(t) = 18e^{0.2t}$.

(a) Find the number of bacteria initially present in the culture.
(b) Find the number of bacteria present in the culture after 10 minutes.
(c) Sketch the graph of the function A for $t \geq 0$.

54. The voltage V (in volts) in an electrical circuit is a function of time t (in milliseconds) and is given by $V(t) = 400e^{-t/10}$.

(a) Find the initial voltage in the circuit.
(b) Find the voltage in the circuit after 30 milliseconds.
(c) Sketch the graph of the function V for $t \geq 0$.
(d) Describe the behavior of $V(t)$ as t increases without bound ($t \to \infty$).

55. The value V in dollars of a certain piece of machinery is a function of its age t in years and is given by $V(t) = 2000 + 10{,}000e^{-0.35t}$.

(a) What is the value of the machine when it is purchased?
(b) What is the value of the machine 4 years after it is purchased?
(c) Sketch the graph of the function V for $t \geq 0$.
(d) What is the significance of the horizontal asymptote?

56. For a certain secretarial student, the number N of words typed (correctly) per minute on a new word processor is a function of the time t (in hours) that the student has practiced on the machine and is given by $N(t) = 60(1 - e^{-0.05t})$.

(a) What is the student's rate of typing N after 10 hours of practice?
(b) Sketch and graph of the function N for $t \geq 0$.
(c) What is the best rate of typing this student can hope to achieve on this word processor?

Calculator Activities

57. Use a calculator to evaluate each expression. Round each answer to three significant digits.

(a) $3^{-\sqrt{2}}$ (b) $\left(\frac{4}{17}\right)^{\sqrt{5}}$ (c) e^π (d) π^e

58. In calculus, it is shown that if $|x| \leq 1$, then

$$e^x \approx 1 + x + \frac{x^2}{2} + \frac{x^3}{6} + \frac{x^4}{24} + \frac{x^5}{120}.$$

Use this formula and a calculator to evaluate each expression.

(a) $e^{0.5}$ (b) $e^{-0.4}$ (c) e (d) e^{-1}

59. Given that $f(x) = e^x$,

(a) show that the *difference quotient*

$$\frac{f(x + \Delta x) - f(x)}{\Delta x} = e^x \left(\frac{e^{\Delta x} - 1}{\Delta x} \right).$$

(b) use a calculator to evaluate $\dfrac{e^{\Delta x} - 1}{\Delta x}$ for the values of Δx given in the table:

Δx	1	0.1	0.01	0.001	0.0001
$\dfrac{e^{\Delta x} - 1}{\Delta x}$					

(c) use your results from the table in part (b) to determine the number that $\dfrac{e^{\Delta x} - 1}{\Delta x}$ seems to approach as $\Delta x \to 0^+$.

60. In statistics, the *normal probability distribution function* is given by

$$f(x) = \dfrac{1}{\sigma \sqrt{\pi}} e^{-(x-\mu)^2/2\sigma^2}$$

where μ is the mean and σ the standard deviation of the distribution. The graph of this function is usually referred to as a "bell-shaped" curve. By plotting points, sketch the graph of this bell-shaped curve for $\mu = 0$ and $\sigma = 1$. If you have access to a graphing calculator, check your results.

7.2 Logarithmic Functions

◆ Introductory Comments

As we discussed in Section 7.1, the exponential function $f(x) = b^x$ is always increasing when $b > 1$ and is always decreasing when $0 < b < 1$. Recall from Section 1.5 that every function f that is either an increasing function or a decreasing function is also a one-to-one function and has an inverse function f^{-1}. Using the method suggested in Section 1.5, we can attempt to find the inverse function of $f(x) = b^x$ by replacing x with $f^{-1}(x)$ and solving for $f^{-1}(x)$. If we proceed in this manner, we obtain

$$f(x) = b^x$$
$$f(f^{-1}(x)) = b^{f^{-1}(x)}$$
$$x = b^{f^{-1}(x)}$$

However, we have no algebraic procedure that we can use to solve this last equation for $f^{-1}(x)$. By convention, we solve this equation for $f^{-1}(x)$ by writing

$$f^{-1}(x) = \log_b x \quad \longleftarrow \boxed{\text{Read "log base } b \text{ of } x\text{."}}$$

The function f^{-1} is called the *logarithmic function* and $\log_b x$ represents *the power to which b must be raised in order to obtain x*.

Recall from Section 1.5 that the graph of a function and its inverse are symmetric with respect to the line $y = x$. Hence, the graph of $f^{-1}(x) = \log_b x$ may be obtained from the graph of $f(x) = b^x$ by reflecting the graph of f in the line $y = x$, as shown in Figure 7.8.

Since the exponential function and logarithmic function are inverses of each other, the domain of the logarithmic function must be the range of the exponential function, namely, $(0, \infty)$. In other words, $\log_b x$ is defined only when x is positive.

◆ Logarithmic Function

For $b > 0$ and $b \neq 1$, the **logarithmic function** with base b is defined as

$$y = \log_b x \quad \text{if and only if} \quad b^y = x, \quad x > 0.$$

418 CHAPTER 7 Exponential and Logarithmic Functions

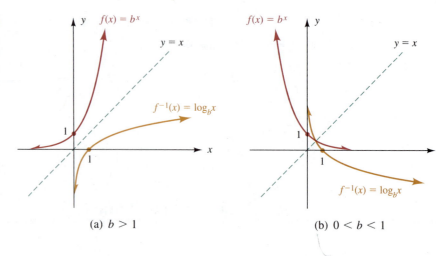

FIGURE 7.8
Like the graphs of any pair of inverse functions, the graphs of $f^{-1}(x) = \log_b x$ and $f(x) = b^x$ are symmetric with respect to the line $y = x$.

(a) $b > 1$ (b) $0 < b < 1$

◆ Evaluating Logarithms

Two logarithmic bases occur in most applied problems—base 10 and base e. We refer to a base 10 logarithm ($\log_{10}$) as a **common logarithm** and a base e logarithm ($\log_e$) as a **natural logarithm** or a **Napierian logarithm** (named after the Scottish mathematician John Napier, 1550–1617). The common logarithm of a positive number x is usually written **log x** (read "log of x"), and the natural logarithm of a positive number x is usually written **ln x** (read "el en of x"):

$$\log x \quad \text{means} \quad \log_{10} x$$

and

$$\ln x \quad \text{means} \quad \log_e x$$

Remember that *logarithms are exponents*.

EXAMPLE 1 Find the value of each logarithm, if it is defined.

(a) $\log_5 25$ (b) $\log \frac{1}{100}$ (c) $\log_9 27$ (d) $\ln(-1)$

SOLUTION

(a) Letting $y = \log_5 25$ and applying the definition of a logarithmic function, we have

$$5^y = 25.$$

Since $5^2 = 25$, we conclude that $y = 2$. Hence, $\log_5 25 = 2$.

(b) Letting $y = \log \frac{1}{100} = \log_{10} \frac{1}{100}$ and applying the definition of a logarithmic function, we have

$$10^y = \frac{1}{100}.$$

Since $10^{-2} = \frac{1}{100}$, we conclude that $y = -2$. Hence, $\log \frac{1}{100} = -2$.

(c) Letting $y = \log_9 27$ and applying the definition of a logarithmic function, we have

$$9^y = 27.$$

Since $9^{3/2} = 27$, we conclude that $y = \frac{3}{2}$. Hence, $\log_9 27 = \frac{3}{2}$.

(d) The domain of the logarithmic function $y = \log_b x$ is $(0, \infty)$. Thus, for $x \leq 0$ the logarithmic function is undefined. Hence, we conclude

$$\ln(-1) = \log_e(-1) \text{ is undefined.} \quad \blacklozenge$$

PROBLEM 1 Find the value of $\log_{1/2} 8$. $\quad \blacklozenge$

Most calculators have $\boxed{\text{LOG}}$ and $\boxed{\text{LN}}$ keys. We can use these keys to evaluate common logarithms and natural logarithms, respectively. If you do not have access to a calculator, use the logarithmic tables in Appendix C. In Section 7.3, we illustrate a procedure that uses a calculator to evaluate logarithms to bases other than 10 or e.

EXAMPLE 2 Use a calculator to find the approximate value of each logarithm.

(a) $\log 25$ **(b)** $\ln \frac{1}{2}$

SOLUTION

(a) The expression $\log 25$ represents the power x such that $10^x = 25$. We know that $10^1 = 10$ and $10^2 = 100$. Since 25 is between 10 and 100, we know that $\log 25$ must be some real number between 1 and 2. Using the $\boxed{\text{LOG}}$ key on a calculator, we approximate the value to four significant digits:

$$\log 25 \approx 1.398$$

(b) The expression $\ln \frac{1}{2}$ represents the power x such that $e^x = \frac{1}{2}$, or 0.5. Since $e^0 = 1$ and $e^{-1} = 1/e \approx 0.37$, we know that $\ln \frac{1}{2}$ must be some real number between 0 and -1. Using the $\boxed{\text{LN}}$ key on a calculator, we calculate that, to four significant digits, the approximate value of $\ln \frac{1}{2}$ is

$$\ln \tfrac{1}{2} = \ln 0.5 \approx -0.6931 \quad \blacklozenge$$

PROBLEM 2 Use a calculator to verify the results we obtained in Example 1(b) and 1(d). $\quad \blacklozenge$

◆ **Logarithmic Identities**

Next, we present four logarithmic identities. Each identity is a direct consequence of the definition of the logarithmic function, and in each case we assume $b > 0$ and $b \neq 1$.

1. If $y = \log_b 1$, then $b^y = 1$. Since $b^0 = 1$, we conclude that
   $\log_b 1 = 0$.
2. If $y = \log_b b$, then $b^y = b$. Since $b^1 = b$, we conclude that
   $\log_b b = 1$.
3. If $y = \log_b b^x$, then $b^y = b^x$. Since the exponential function is one-to-one, it follows that $y = x$. Hence, we conclude that **$\log_b b^x = x$.**
4. If $y = \log_b x$, then $b^y = x$. Replacing y with $\log_b x$ in the equation $b^y = x$, we obtain **$b^{\log_b x} = x$.**

In summary, we restate these identities.

Logarithmic Identities

For all bases b with $b > 0$ and $b \neq 1$,

1. $\log_b 1 = 0$
2. $\log_b b = 1$
3. $\log_b b^x = x$ for all real numbers x
4. $b^{\log_b x} = x$ provided $x > 0$

EXAMPLE 3 Find the value of each expression.

(a) $\log_{16} 1$ (b) $\ln e$ (c) $\log \sqrt{10}$ (d) $e^{\ln 6}$

SOLUTION

(a) Since $\log_b 1 = 0$ for any permissible base b, we have
$$\log_{16} 1 = 0.$$

(b) Since $\log_b b = 1$ for any permissible base b, we have
$$\ln e = \log_e e = 1.$$

(c) Since $\log_b b^x = x$ for any permissible base b and any real number x, we have
$$\log \sqrt{10} = \log_{10} 10^{1/2} = \tfrac{1}{2}.$$

(d) Since $b^{\log_b x} = x$ for any permissible base b and all positive numbers x, we have
$$e^{\ln 6} = e^{\log_e 6} = 6.$$ ◆

PROBLEM 3 Find the value of $\log_3 3^{12}$. ◆

SECTION 7.2 Logarithmic Functions

In the next example, we apply the logarithmic identities to simplify expressions containing logarithms. This procedure is often employed in calculus.

EXAMPLE 4 Simplify each expression.

(a) $e^{\ln(t+1)}$ (b) $\log_2 16^{t-2}$

SOLUTION

(a) Since $b^{\log_b x} = x$, provided $x > 0$, we have

$$e^{\ln(t+1)} = e^{\log_e(t+1)} = t + 1 \text{ provided } t > -1.$$

(b) We begin by applying the properties of real exponents (Section 7.1) and matching the base of the exponential expression to the base of the logarithm:

$$\log_2 16^{t-2} = \log_2 (2^4)^{t-2} = \log_2 2^{4t-8}.$$

Now, since $\log_b b^x = x$ for any permissible base b and any real number x, we have

$$\log_2 16^{t-2} = \log_2 2^{4t-8} = 4t - 8. \quad \blacklozenge$$

PROBLEM 4 Simplify $(\log_2 2)^{4t-8}$. $\quad \blacklozenge$

♦ **Equating Logarithmic and Exponential Forms**

From our definition of the logarithmic function, we can state that the equation

$$\log_b u = v \quad \text{is equivalent to} \quad b^v = u.$$

The equation $\log_b u = v$ is said to be in **logarithmic form** and the equation $b^v = u$ is said to be in **exponential form.** If we change an equation from logarithmic form to exponential form, or vice versa, the base b remains the same in each case. That is, in logarithmic form the base is the subscript b, and in exponential form the base is the number b being raised to a power.

EXAMPLE 5 Change each equation from logarithmic form to exponential form, then find the value of the unknown.

(a) $\log_b 9 = \frac{2}{3}$ (b) $3 \log_2 (x^2 - 2x + 1) = 12$

SOLUTION

(a) The equation

$$\underbrace{\log_b 9 = \tfrac{2}{3}}_{\text{Logarithmic form}} \quad \text{is equivalent to} \quad \underbrace{b^{2/3} = 9}_{\text{Exponential form}}.$$

For all logarithmic bases b, we must have $b > 0$. Thus, we can solve for b as follows:

$$b^{2/3} = 9$$

$$(b^{2/3})^{3/2} = 9^{3/2} \quad \text{Raise both sides to the 3/2 power}$$

$$b = 27 \quad \text{Simplify}$$

(b) We begin by dividing both sides of the equation by 3 to obtain

$$\log_2(x^2 - 2x + 1) = 4.$$

The equation

$$\underbrace{\log_2(x^2 - 2x + 1) = 4}_{\text{Logarithmic form}} \quad \text{is equivalent to} \quad \underbrace{2^4 = x^2 - 2x + 1.}_{\text{Exponential form}}$$

Solving for x, we find

$$x^2 - 2x + 1 = 2^4$$

$$x^2 - 2x - 15 = 0 \quad \text{Subtract 16 from both sides}$$

$$(x - 5)(x + 3) = 0 \quad \text{Factor the quadratic and solve}$$

$$x = 5 \quad \text{or} \quad x = -3 \quad \blacklozenge$$

As with any equation we solve, it is good practice to check the solution of a logarithmic equation. To check Example 5(a), we proceed as follows:

$$\text{Check:} \quad \log_{27} 9 = \frac{2}{3} \quad ? \quad \text{Replace } b \text{ with 27}$$

$$27^{2/3} = 9 \quad ? \quad \text{Change to exponential form}$$

$$9 = 9 \quad \checkmark \quad \text{Simplify by using the definition of a rational exponent}$$

PROBLEM 5 Referring to Example 5(b), show that both $x = 5$ and $x = -3$ satisfy the equation

$$3 \log_2 (x^2 - 2x + 1) = 12. \quad \blacklozenge$$

EXAMPLE 6 Change each equation from exponential form to logarithmic form, then find the value of the unknown.

(a) $10^{4x} = 30$ **(b)** $7e^{1-2t} = 28$

SOLUTION

(a) The equation

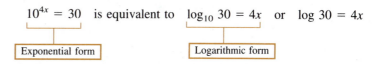

$10^{4x} = 30$ is equivalent to $\log_{10} 30 = 4x$ or $\log 30 = 4x$

Exponential form — Logarithmic form

Solving for x, we find

$$4x = \log 30$$

$$x = \frac{\log 30}{4} \approx 0.3693$$

Exact answer — Approximate answer using a calculator

(b) We begin by dividing both sides of the equation by 7 to obtain

$$e^{1-2t} = 4.$$

The equation

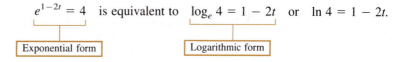

$e^{1-2t} = 4$ is equivalent to $\log_e 4 = 1 - 2t$ or $\ln 4 = 1 - 2t$.

Exponential form — Logarithmic form

Solving for t, we find

$$\ln 4 = 1 - 2t$$

$$2t = 1 - \ln 4$$

$$t = \frac{1 - \ln 4}{2} \approx -0.1931$$

Exact answer — Approximate answer using a calculator

To check Example 6(a), we proceed as follows:

Check: $10^{4[(\log 30)/4]} = 30$? **Replace x with $(\log 30)/4$**

$10^{\log 30} = 30$? **Simplify**

$30 = 30$ ✔ **Apply logarithmic identity 4**

PROBLEM 6 Check Example 6(b).

◆ Applications: Exponential Growth and Decay

Functions that change by a fixed multiple over the same increment of time have the form

$$A(t) = A_0 e^{kt}$$

where $A(t)$ is the amount present at time t, A_0 is the original amount present at time 0, and k is a constant related to the rate of growth or decay. If $k > 0$, then $A(t) = A_0 e^{kt}$ is an **exponential growth function,** and if $k < 0$, then $A(t) = A_0 e^{kt}$ is called an **exponential decay function.**

The compound interest formula

$$A = Pe^{rt}$$

is an example of an exponential growth function. In Section 7.1, we used this formula to find the amount A in a savings account when a certain principal P is invested at an interest rate r per year and interest is compounded continuously. Now that we have defined logarithms, we can also use this formula to find the time t or the interest rate r for a given deposit to reach a particular amount.

EXAMPLE 7 A certain amount of money is deposited in a savings account paying 8% interest per year compounded continuously. Assuming the depositor makes no subsequent deposit or withdrawal, how long will it take for the money to double?

SOLUTION If a certain principal P is deposited in a savings account at an interest rate of 8% per year and interest is compounded continuously, then the amount A in the account after t years is given by the formula

$$A = Pe^{0.08t}$$

We want to find the time t that it will take to *double P*. Thus, we replace A with $2P$ and solve for t as follows:

$$2P = Pe^{0.08t}$$
$$2 = e^{0.08t} \quad \text{Divide both sides by } P, P \neq 0$$
$$0.08t = \ln 2 \quad \text{Change to logarithmic form}$$
$$t = \frac{\ln 2}{0.08} \approx 8.66 \text{ years} \quad \text{Divide both sides by 0.08}$$

Thus, it takes approximately 8 years 8 months to double the investment. ◆

PROBLEM 7 At what interest rate, compounded continuously, must money be invested if the amount is to double in 5 years 6 months? ◆

SECTION 7.2 Logarithmic Functions

Under controlled laboratory conditions a population P of living organisms increases exponentially as a function of time t, and the growth pattern is described by the **Malthusian model**

$$P(t) = P_0 e^{kt}$$

where P_0 is the initial population and k is a positive constant. The Malthusian model, named after Englishman Thomas Malthus (1766–1834), is another example of an exponential growth function.

EXAMPLE 8   Suppose 18 bacteria are present initially in a culture, and twenty minutes later 270 bacteria are present.

(a) Determine the Malthusian model that describes this growth pattern.

(b) Determine the time t (to the nearest minute) when 360 bacteria are present in the culture.

SOLUTION

(a) Since 18 bacteria are present initially, we know that $P_0 = 18$. Thus, starting with the Malthusian model $P(t) = P_0 e^{kt}$, we can write

$$P(t) = 18 e^{kt}.$$

We must now determine the constant k. Using the fact that when $t = 20$, then $P(t) = 270$, we have

$$270 = 18 e^{20k}$$
$$15 = e^{20k} \qquad \text{Divide both sides by 18}$$
$$20k = \ln 15 \qquad \text{Change to logarithmic form}$$
$$k = \frac{\ln 15}{20} \approx 0.1354 \qquad \text{Solve for } k$$

Replacing k with 0.1354 gives us the Malthusian model that describes this growth pattern. Thus, we have

$$P(t) = 18 e^{0.1354 t}$$

(b) Using the exponential function developed in part (a), we replace $P(t)$ with 360 and solve for t as follows:

$$360 = 18 e^{0.1354 t}$$
$$20 = e^{0.1354 t} \qquad \text{Divide both sides by 18}$$
$$0.1354 t = \ln 20 \qquad \text{Change to logarithmic form}$$
$$t = \frac{\ln 20}{0.1354} \approx 22 \text{ minutes} \qquad \text{Solve for } t$$

PROBLEM 8 Referring to Example 8, determine the time t (to the nearest minute) when 540 bacteria are present in the culture. ◆

Radioactive materials decrease exponentially over time and their decay patterns can be described by exponential decay functions. When working with radioactive materials, the term *half-life* is often used. **Half-life** is defined as the time required for a given mass of a radioactive material to disintegrate to half its original mass.

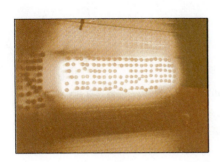

EXAMPLE 9 Strontium-90, a waste product from nuclear reactors, has a half-life of 28 years. It is estimated that a certain quantity of this material will be safe to handle when its mass is $\frac{1}{1000}$ of its original amount. Determine the time required to store strontium-90 until it reaches this level of safety.

SOLUTION If the half-life of strontium-90 is 28 years, then half of the original amount is present after 28 years. Thus, starting with the exponential decay function $A(t) = A_0 e^{kt}$, we replace $A(t)$ with $\frac{1}{2} A_0$ and t with 28, and then solve for k as follows:

$$\tfrac{1}{2} A_0 = A_0 e^{k(28)}$$

$$\tfrac{1}{2} = e^{k(28)} \qquad \text{\color{red}Divide both sides by } A_0$$

$$28k = \ln \tfrac{1}{2} \qquad \text{\color{red}Change to logarithmic form}$$

$$k = \frac{\ln \tfrac{1}{2}}{28} \approx -0.02476 \qquad \text{\color{red}Solve for } k$$

Thus, we have

$$A(t) = A_0 e^{-0.02476t}$$

To determine the time t required to store strontium-90 until its mass reaches $\frac{1}{1000}$ of its original amount, we replace $A(t)$ with $0.001 A_0$, and solve for t as follows:

$$0.001 A_0 = A_0 e^{-0.02476t}$$

$$0.001 = e^{-0.02476t} \qquad \text{\color{red}Divide both sides by } A_0, A_0 \neq 0$$

$$-0.02476t = \ln 0.001 \qquad \text{\color{red}Change to logarithmic form}$$

$$t = \frac{\ln 0.001}{-0.02476} \approx 279 \text{ years} \qquad \text{\color{red}Solve for } t$$

◆

PROBLEM 9 Referring to Example 9, how long does it take 30 grams of strontium-90 to decay to 27 grams? ◆

Exercises 7.2

Basic Skills

In Exercises 1–30, find the value of each expression, if it is defined.

1. $\log_6 36$
2. $\log_2 8$
3. $\log_5 125$
4. $\log_3 81$
5. $\log_3 3$
6. $\log_7 (-7)$
7. $\log_2 \frac{1}{16}$
8. $\log_3 \frac{1}{27}$
9. $\log_{1/3} 9$
10. $\log_{1/2} 16$
11. $\log 100$
12. $\log \frac{1}{1000}$
13. $\ln 1$
14. $\log 1$
15. $\ln e^2$
16. $\ln \sqrt{e}$
17. $\log_{64}(-4)$
18. $\log_9 3$
19. $\log_4 8$
20. $\log_8 32$
21. $\log_{1/8} 16$
22. $\log_{27} \frac{1}{9}$
23. $4^{\log_4 3}$
24. $3^{\log_3 10}$
25. $e^{\ln 5}$
26. $10^{\log 2}$
27. $10^{\ln e}$
28. $e^{\log 10}$
29. $\log (\ln e)$
30. $\ln (\log 10)$

In Exercises 31–46, simplify each expression. Assume that each variable is restricted to those real numbers that allow the logarithmic expression to be defined.

31. $\log_4 4^{x-3}$
32. $\log_3 3^{x+1}$
33. $6^{\log_6 (x+2)}$
34. $5^{\log_5 (x-4)}$
35. $e^{\ln (x^2+4)}$
36. $10^{\log (1/x)}$
37. $\ln e^{x^2+2x}$
38. $(\ln e)^{x^2+2x}$
39. $(\log 10)^{2-x^2}$
40. $\log 10^{2-x^2}$
41. $\log_2 8^{x+2}$
42. $\log_9 3^{2x-2}$
43. $e^{x+\ln x}$ [Hint: $a^{x+y} = a^x a^y$]
44. $e^{x-\ln x}$ [Hint: $a^{x-y} = a^x/a^y$]
45. $e^{2 \ln x}$ [Hint: $a^{xy} = (a^x)^y$]
46. $e^{-\ln x}$ [Hint: $a^{xy} = (a^x)^y$]

In Exercises 47–60, change each equation to exponential form, then find the value of the unknown.

47. $\log_b 9 = 2$
48. $\log_b 125 = -3$
49. $\log_b 8 = -\frac{3}{4}$
50. $\log_b 64 = \frac{2}{3}$
51. $\log x = -2$
52. $\ln x = 3$
53. $2 \log_4 x = 3$
54. $3 \log_{1/8} x = -4$
55. $\ln (x - e) = 1$
56. $-3 \log_8 (x + 9) = 1$
57. $\log_2 (2x^2 + 5x + 5) = 3$
58. $\ln (x^2 - 3x - 3) = 0$
59. $2 \log |x - 1| - 1 = 0$
60. $\log_3 \sqrt{2x^2 - 3x} = 1$

In Exercises 61–70, change each equation to logarithmic form, then find the value of the unknown.

61. $10^x = 5$
62. $e^x = 3$
63. $e^{3x} = \frac{2}{3}$
64. $10^{-2x} = 60$
65. $10^{3-2x} = 28$
66. $e^{2x-1} = 2$
67. $4e^{-7x} = 15$
68. $12(10^{1-x}) = 60$
69. $3 - 2(10^{-x}) = -117$
70. $1 + 2e^x = 9$

71. One thousand dollars is deposited in a savings account paying 7% interest per year compounded continuously. Assuming the depositor makes no subsequent deposit or withdrawal, find the time required (in years) for the investment to accumulate to the given amount:

    (a) $1500   (b) $6000   (c) $100,000

72. A certain amount of money is deposited into a savings account paying 10% interest per year compounded continuously. Assuming no subsequent deposit or withdrawal is made, find the time required for the amount to double.

73. When a child is born, his grandfather invests $10,000 for the child's college education. When the child is 18 years old, the balance in the account is $73,000. If interest is compounded continuously, find the interest rate (to the nearest percent) at which the investment is made. Assume no subsequent deposit or withdrawal is made and the interest rate does not change.

74. At what interest rate, compounded continuously, must money be invested if the amount is to quadruple in 18 years?

75. In 1982 foresters found 24 eagles living in a certain national park. In 1992 they counted 144 eagles living in this area.

    (a) Determine the Malthusian model that describes this growth pattern.

    (b) In what year is the eagle population expected to reach 300?

76. Approximately 10^4 bacteria are present in a culture. Five hours later, approximately 10^7 bacteria are present.

    (a) Determine the Malthusian model that describes this growth pattern.

EXERCISE 76 *(continued)*

(b) Approximately how many minutes does it take for this population to double in size?

77. When a plant or animal dies, the amount of carbon-14 in the plant or animal decreases exponentially according to the exponential decay function. Suppose in 1986 an archaeologist discovers a human skull in an ancient burial site and determines that 60% of the original carbon-14 is still present. If the half-life of carbon-14 is 5600 years, in approximately what year did this person die?

78. The radioactive substance iodine-131 decays from 30 grams to 25 grams in 50.5 hours.

(a) Determine the exponential decay function that describes this decay pattern.

(b) How many hours will it take for this radioactive substance to decay to 20 grams?

(c) Find the half-life (in days) of this radioactive substance.

Critical Thinking

79. Find the base b of the logarithmic function $f(x) = \log_b x$ if the graph of this function passes through the given point.

(a) $(81, 4)$ (b) $(\frac{1}{8}, -\frac{3}{2})$

80. Between what two consecutive integers does the value of each logarithm lie?

(a) $\log_3 99$ (b) $\log_5 150$

(c) $\log_8 7$ (d) $\log_{1/2} 40$

81. Use interval notation to describe the domain of each function.

(a) $f(x) = \dfrac{1}{1 - \ln x}$

(b) $F(x) = \sqrt{10 - \log x}$

82. Given that $f(x) = e^x$ and $g(x) = \ln x$, find $f(g(x))$ and $g(f(x))$. What does this tell us about the functions f and g?

83. In a particular electrical circuit, the current i (in amperes) is a function of the time t (in seconds) after the switch is closed and is given by $i(t) = 20(1 - e^{-4.5t})$.

(a) Discuss the behavior of $i(t)$ as t increases without bound ($t \to \infty$).

(b) Find the time in milliseconds (ms) when $i(t) = 6$ amperes.

84. When limited resources such as food supply and space are taken into account, the population P of a living organism as a function of time t is more accurately described by the *logistic law* than by the Malthusian model. For the logistic law,

$$P(t) = \frac{cP_0}{P_0 + (c - P_0)e^{-kt}},$$

where P_0 is the initial population and c and k are positive constants. Suppose the population of wolves in a certain forest follows the logistic law with $c = 1500$, $k = 0.2$, and time t in years.

(a) Discuss the behavior of $P(t)$ as t increases without bound ($t \to \infty$).

(b) If $P_0 = 200$, find the time (in years) when $P(t) = 600$ wolves.

If a heated object of temperature T_0 is placed in a cooler medium that has a constant temperature of T_1, then the temperature T of the object at any time t is given by the formula

$$T = T_1 + (T_0 - T_1)e^{-kt}$$

where k is a positive constant. The formula is **Newton's law of cooling**, named after Sir Isaac Newton (1642–1727). Use this formula to answer Exercises 85 and 86.

85. A pizza baked at 400 °F is removed from the oven and placed in a room with a constant temperature of 70 °F. The pizza cools to 300 °F in 2 minutes.

(a) State the formula that describes this cooling process.

(b) Find the temperature of the pizza after it cools for 3 minutes.

(c) How soon after it is removed from the oven does the pizza cool to 150 °F?

86. A bottle of white zinfandel wine is stored in a room with a temperature of 70 °F. The manufacturer recommends that the wine is best served slightly chilled to 40 °F. At 4:00 P.M. the wine is placed in a refrigerator that has a constant temperature of 35 °F, and after 30 minutes the temperature of the wine has cooled to 50 °F. At what time should the wine be removed from the refrigerator if it is to be served at 40 °F?

Calculator Activities

Given the functions f and g defined by

$$f(x) = \ln(x^2 - 2x + 3) \quad \text{and} \quad g(x) = \log \frac{1}{x^2 + 2}$$

use a calculator to compute the functional values given in Exercises 87–92. Round each value to three significant digits.

87. $f(8.2)$
88. $g(11.5)$
89. $(f + g)(-5.22)$
90. $(f \cdot g)(-0.34)$
91. $(f \circ g)(4.4)$
92. $(g \circ f)(21.8)$

93. Use the $\boxed{\text{LN}}$ key on your calculator to help complete the following table.

x	2	18	24	π
y	3	4	15	10
$\ln(xy)$				
$\ln x + \ln y$				
$\ln(x/y)$				
$\ln x - \ln y$				
$\ln(x^y)$				
$y \ln x$				

(a) Compare the values in the table for $\ln(xy)$ and $\ln x + \ln y$. What do you conclude?

(b) Compare the values in the table for $\ln(x/y)$ and $\ln x - \ln y$. What do you conclude?

(c) Compare the values in the table for $\ln(x^y)$ and $y \ln x$. What do you conclude?

94. In calculus, it is shown that if $|x| \leq 1$, then

$$\ln(x+1) \approx x - \frac{x^2}{2} + \frac{x^3}{3} - \frac{x^4}{4} + \frac{x^5}{5} - \frac{x^6}{6}.$$

Use this formula and a calculator to evaluate each expression.

(a) $\ln 1.8$ (b) $\ln 0.2$ (c) $\ln 1.25$ (d) $\ln 0.75$

95. Chemists describe the acidity or alkalinity of a liquid by denoting its *pH*. An acid has pH < 7 and an alkaline has pH > 7. By definition,

$$\text{pH} = -\log[\text{H}^+],$$

where $[\text{H}^+]$ is the liquid's concentration of hydrogen ions, measured in moles per liter, mol/L. Determine the pH of the given liquid.

(a) milk: $[\text{H}^+] \approx 4.1 \times 10^{-7}$ mol/L
(b) apple juice: $[\text{H}^+] \approx 6.2 \times 10^{-4}$ mol/L

96. Refer to Exercise 95. Determine the hydrogen ion concentration in each liquid.

(a) vinegar: pH ≈ 2.5
(b) human blood: pH ≈ 7.4

7.3 Properties of Logarithms

In this section, we develop some important properties of logarithms. These properties enable us to rewrite certain functions that contain logarithms so that we may sketch the graph of these functions (Section 7.4). Also, these properties enable us to rewrite certain logarithmic equations so that we may solve these equations (Section 7.5).

◆ **Rewriting Logarithmic Expressions**

Recall from Section 7.1 three properties of real exponents for real numbers m and n:

1. $b^m b^n = b^{m+n}$
2. $\dfrac{b^m}{b^n} = b^{m-n}$
3. $(b^m)^n = b^{mn}$

By using logarithmic identity 4 from Section 7.2 ($b^{\log_b u} = u$, $u > 0$) and these three properties of real exponents, we can obtain three corresponding **properties of logarithms:**

1. $b^{\log_b xy} = xy$ **Apply log identity 4 with $x > 0, y > 0$**

    $= b^{\log_b x} \cdot b^{\log_b y}$ **Rewrite using log identity 4**

    $= b^{\log_b x + \log_b y}$ **Add exponents**

Since the exponential function is one-to-one, we conclude that the exponents $\log_b xy$ and $\log_b x + \log_b y$ must be equal; that is

$$\log_b xy = \log_b x + \log_b y$$

The logarithm of a product is the sum of the logarithms of the factors.

2. $b^{\log_b (x/y)} = \dfrac{x}{y}$ **Apply log identity 4 with $x > 0, y > 0$**

    $= \dfrac{b^{\log_b x}}{b^{\log_b y}}$ **Rewrite using log identify 4**

    $= b^{\log_b x - \log_b y}$ **Subtract exponents**

Since the exponential function is one-to-one, we conclude that the exponents $\log_b (x/y)$ and $\log_b x - \log_b y$ must be equal; that is,

$$\log_b \frac{x}{y} = \log_b x - \log_b y$$

The logarithm of a quotient is the logarithm of the numerator minus the logarithm of the denominator.

3. $b^{\log_b x^n} = x^n$ **Apply log identity 4 with $x > 0$**

    $= (b^{\log_b x})^n$ **Rewrite using log identity 4**

    $= b^{n \log_b x}$ **Multiply exponents**

Since the exponential function is one-to-one, we conclude that the exponents $\log_b x^n$ and $n \log_b x$ must be equal; that is

$$\log_b x^n = n \log_b x$$

The logarithm of a quantity raised to a power is the product of the power and the logarithm of that quantity.

We now summarize these properties.

Properties of Logarithms

If $b > 0$, $b \neq 1$, then for positive real numbers x and y.

1. $\log_b xy = \log_b x + \log_b y$
2. $\log_b \dfrac{x}{y} = \log_b x - \log_b y$
3. $\log_b x^n = n \log_b x$

In the following example, we use the properties of logarithms to write a single logarithmic expression as the sum and difference of simpler logarithmic expressions. This procedure will be used in Section 7.4 to sketch the graphs of functions that contain logarithmic expressions.

EXAMPLE 1 Write each logarithmic expression as sums and differences of simpler logarithmic expressions without logarithms of products, quotients, and powers. Then simplify, if possible.

(a) $\log \sqrt[3]{10x^2}$ (b) $\ln \dfrac{\sqrt{x^2+4}}{xe^{3x}}$

SOLUTION

(a) Assuming that $x > 0$, we have

$\log \sqrt[3]{10x^2} = \log (10x^2)^{1/3}$ **Change to a rational exponent**

$= \tfrac{1}{3} \log (10x^2)$ **Apply log property 3**

$= \tfrac{1}{3}(\log 10 + \log x^2)$ **Apply log property 1**

$= \tfrac{1}{3}(\log 10 + 2 \log x)$ **Apply log property 3**

$= \tfrac{1}{3}(1 + 2 \log x)$ **Evaluate log 10**

(b) Assuming that $x > 0$, we have

$\ln \dfrac{\sqrt{x^2+4}}{xe^{3x}} = \ln \dfrac{(x^2+4)^{1/2}}{xe^{3x}}$ **Rewrite**

$= \ln (x^2+4)^{1/2} - \ln xe^{3x}$ **Apply log property 2**

$= \ln (x^2+4)^{1/2} - (\ln x + \ln e^{3x})$ **Apply log property 1**

        Be sure to remember the parentheses.

$= \tfrac{1}{2} \ln (x^2+4) - (\ln x + 3x \ln e)$ **Apply log propety 3**

$= \tfrac{1}{2} \ln (x^2+4) - \ln x - 3x$ **Evaluate ln e**

Note: The logarithmic expression in Example 1(a) is defined for all real numbers except 0. We make the assumption that $x > 0$ so that the properties of logarithms may be applied. If we want to allow for the possibility that $x < 0$, then we write

$$\log \sqrt[3]{10x^2} = \tfrac{1}{3}(1 + 2 \log |x|).$$

Avoid using the three properties of logarithms in situations where they do not apply. Referring to Example 1(b), to apply log property 3 and write

$$\ln xe^{3x} \quad \text{as} \quad 3x \ln xe \quad \text{is WRONG}$$

$$\ln (xe)^{3x} = 3x \ln xe \quad \text{by log property 3.}$$

Several other common errors that occur when working with the properties of logarithms are mentioned in Exercises 1–4.

PROBLEM 1 Show by using a specific example that $(\log_b x)^n$ and $n \log_b x$ are *not* equivalent logarithmic expressions. ◆

The properties of logarithms can also be used in the reverse sense to write sums and differences of logarithmic expressions as a single logarithmic expression. For problems of this type, we begin by applying logarithmic property 3 and change any numerical coefficient to an exponent. Once we have removed numerical coefficients, we can apply the other two properties. The technique of writing sums and differences of logarithmic expressions as a single logarithm will be used in Section 7.5 to solve logarithmic equations.

EXAMPLE 2 Express each as a single logarithmic expression with a coefficient of 1. Then simplify, if possible.

(a) $\log 50 + 2 \log 4 - 3 \log 2$ **(b)** $\ln (x^2 - 1) - 2 \ln (x + 1)$

SOLUTION

(a) Beginning with logarithmic property 3, we have

$$\log 50 + 2 \log 4 - 3 \log 2$$
$$= \log 50 + \log 4^2 - \log 2^3 \quad \textbf{Apply log property 3}$$
$$= \log 50 + \log 16 - \log 8 \quad \textbf{Simplify}$$
$$= \log (50 \cdot 16) - \log 8 \quad \textbf{Apply log property 1}$$
$$= \log \frac{50 \cdot 16}{8} \quad \textbf{Apply log property 2}$$
$$= \log 100 = 2 \quad \textbf{Reduce and evaluate}$$

(b) Assuming that $x > 1$, we have

$$\ln (x^2 - 1) - 2 \ln (x + 1)$$
$$= \ln (x^2 - 1) - \ln (x + 1)^2 \quad \textbf{Apply log property 3}$$

$$= \ln \frac{x^2 - 1}{(x + 1)^2} \qquad \text{Apply log property 2}$$

$$= \ln \frac{(x + 1)(x - 1)}{(x + 1)^2} \qquad \text{Factor the numerator}$$

$$= \ln \frac{x - 1}{x + 1} \qquad \text{Reduce}$$

PROBLEM 2 Use the [LOG] key on a calculator to evaluate $\log 50 + 2 \log 4 - 3 \log 2$. Your answer should agree with the result we obtained in Example 2(a).

◆ Change of Base Formula

In some instances we may find it useful to change the base b logarithm, $\log_b x$, to an expression involving logarithms of another base a. If we let

$$\log_b x = y,$$

then we can change to exponential form and write

$$b^y = x.$$

Taking the base a logarithm of both sides of this equation, we obtain

$$\log_a b^y = \log_a x \qquad \text{Take the base } a \text{ logarithm of both sides}$$

$$y \log_a b = \log_a x \qquad \text{Apply log property 3}$$

$$y = \frac{\log_a x}{\log_a b} \qquad \text{Divide both sides by } \log_a b$$

Now, replacing y with $\log_b x$ gives us the **change of base formula.**

◆ Change of Base Formula

If $\log_b x$ is defined, then

$$\log_b x = \frac{\log_a x}{\log_a b}, \qquad a > 0, \quad a \neq 1.$$

We can use a calculator to evaluate logarithms to bases other than 10 or e by applying the change of base formula. This procedure is illustrated in the next example.

EXAMPLE 3 Use the change of base formula to find the approximate value of $\log_4 24$.

SOLUTION The expression $\log_4 24$ represents the power to which 4 must be raised to get 24. We know that $4^2 = 16$ and $4^3 = 64$. Since 24 is between 16 and 64, $\log_4 24$ must be a real number between 2 and 3. Using the change of base formula with common logarithms, we write

$$\log_4 24 = \frac{\log 24}{\log 4}.$$

Now, using the $\boxed{\text{LOG}}$ key on a calculator, we obtain

$$\log_4 24 \approx 2.292$$

PROBLEM 3 Repeat Example 3 using the change of base formula with natural logarithms. You should obtain the same result.

If we replace x with a in the change of base formula, we obtain

$$\log_b a = \frac{1}{\log_a b}$$

or, equivalently,

$$(\log_b a)(\log_a b) = 1$$

EXAMPLE 4 Simplify each logarithmic expression.

(a) $\dfrac{1}{\log_2 12} + \dfrac{1}{\log_6 12}$ (b) $(\log_3 16)(\log_2 27)$

SOLUTION

(a) $\dfrac{1}{\log_2 12} + \dfrac{1}{\log_6 12} = \log_{12} 2 + \log_{12} 6$ Apply $\dfrac{1}{\log_a b} = \log_b a$

$\qquad\qquad\qquad\qquad\quad = \log_{12}(2 \cdot 6)$ Apply log property 1

$\qquad\qquad\qquad\qquad\quad = \log_{12} 12$ Simplify

$\qquad\qquad\qquad\qquad\quad = 1$ Apply $\log_b b = 1$

(b) $(\log_3 16)(\log_2 27) = (\log_3 2^4)(\log_2 3^3)$ Rewrite

$\qquad\qquad\qquad\quad = (4 \log_3 2)(3 \log_2 3)$ Apply log property 3

$\qquad\qquad\qquad\quad = 4 \cdot 3[(\log_3 2)(\log_2 3)]$ Rearrange the factors

$\qquad\qquad\qquad\quad = 12[1]$ Apply $(\log_b a)(\log_a b) = 1$

$\qquad\qquad\qquad\quad = 12$ Simplify

PROBLEM 4 Repeat Example 4 for $(\log_2 5)(\log_5 8)$.

◆ Application: Logarithmic Scales

When physical quantities vary over a large range of values, it is convenient to work with **logarithmic scales** in order to obtain a more manageable set of numbers. Some examples of logarithmic scales are the **decibel scale** for measuring the magnitude of sound, the **brightness scale** for measuring the magnitude of a star, and the **Richter scale** (named after seismologist Charles F. Richter) for measuring the magnitude of an earthquake.

On the Richter scale, the magnitude R of an earthquake is given by

$$R = \log \frac{I}{I_0}$$

where I is the intensity of the earthquake and I_0 is the intensity of a zero-level earthquake having magnitude $R = 0$.

EXAMPLE 5 Seismologists estimate that the San Francisco earthquake of 1906 measured 8.3 on the Richter scale. How many times more intense was this earthquake than the Loma Prieta quake, which occurred during the 1989 World Series and measured 7.1 on the Richter scale?

SOLUTION Let

$$I_a = \text{intensity of the 1906 earthquake}$$

and

$$I_b = \text{intensity of the 1989 earthquake.}$$

Now, using the formula $R = \log \frac{I}{I_0}$ and logarithmic property 2, we have

$$8.3 = \log \frac{I_a}{I_0} \qquad \text{and} \qquad 7.1 = \log \frac{I_b}{I_0}$$
$$8.3 = \log I_a - \log I_0 \qquad\qquad 7.1 = \log I_b - \log I_0.$$

Solving these equations simultaneously, we find

$$8.3 = \log I_a - \log I_0$$
$$\underline{7.1 = \log I_b - \log I_0}$$
$$1.2 = \log I_a - \log I_b \qquad \textbf{Subtract}$$
$$1.2 = \log \frac{I_a}{I_b} \qquad \textbf{Apply log property 2}$$
$$\frac{I_a}{I_b} = 10^{1.2} \approx 16 \qquad \textbf{Change to exponential form}$$

Hence, the earthquake in 1906 was about 16 times as intense as the one in 1989.

PROBLEM 5 One of the strongest earthquakes ever recorded measured 8.9 on the Richter scale. It occurred in Japan in 1933. How many times more intense was this earthquake than the one that occurred in California during the 1989 World Series?

Exercises 7.3

Basic Skills

In Exercises 1–4, show by using specific examples that the given logarithmic expressions are not equivalent.

1. $\log_b (x + y)$ and $\log_b x + \log_b y$
2. $\log_b (x - y)$ and $\log_b x - \log_b y$
3. $\dfrac{\log_b x}{\log_b y}$ and $\log_b \dfrac{x}{y}$
4. $(\log_b x)(\log_b y)$ and $\log_b xy$

In Exercises 5–18, write each expression as sums and differences of simpler logarithmic expressions without logarithms of products, quotients, and powers. Then simplify if possible. Assume that each variable is restricted to those real numbers that allow the properties of logarithms to be applied.

5. $\log_3 (27^2 \cdot 81)$
6. $\log 100x^2$
7. $\log_2 [8x(x + 2)]$
8. $\log_5 \sqrt[4]{25x}$
9. $\ln \dfrac{xe^2}{10}$
10. $\log \dfrac{(x + 4)^2}{2x}$
11. $\log_3 \sqrt[3]{\dfrac{x^2}{27}}$
12. $\log_b \left(\dfrac{x^2 + 2}{x + 3}\right)^3$
13. $\log_b \dfrac{x^2}{\sqrt[3]{(x + 1)^2}}$
14. $\ln \dfrac{1}{2\sqrt{3x^2 + 2}}$
15. $\log_3 \dfrac{1}{9\sqrt{x^3 y^{2/3}}}$
16. $\log \dfrac{\sqrt[3]{1 + y^2}}{(x - 2)(x + 2)^{1/2}}$
17. $\ln \left(\dfrac{e^{x^2}}{e^x + 1}\right)^2$
18. $\ln \dfrac{xe^{-2x}}{\sqrt{2e^x - 1}}$

In Exercises 19–30, write each expression as a single logarithm with a coefficient of 1. Then simplify if possible. Assume that each variable is restricted to those real numbers that allow the properties of logarithms to be applied.

19. $\ln 2 + \ln 3 + \ln 4$
20. $\ln 6 - \ln 3 + \ln 5$
21. $\log 40 - (3 \log 2 - \log 20)$
22. $2 \log 6 - (2 \log 3 + \log 4)$
23. $-2 \log_3 6 + 4 \log_3 2 - \frac{1}{2} \log_3 16$
24. $5 \log_6 2 + \frac{2}{3} \log_6 27 + \frac{3}{2} \log_6 4$
25. $\log_5 (x - 1) - 2 \log_5 x + \log_5 (x + 3)$
26. $3 \ln x + \frac{2}{3} \ln (x - 1) + \ln 2$
27. $2 \ln (x - 2) - \ln (x^2 - 4)$
28. $2 \ln (x^3 + 1) - \ln (x^2 - x + 1)$
29. $\frac{1}{2}[\ln (x - 3) + \ln (x + 3)] - 2(\ln x - \ln 3)$
30. $3 [\log (x + \sqrt{x^2 - 1}) + \log (x - \sqrt{x^2 - 1})] - \log 2$

In Exercises 31–38, use the change of base formula to find the approximate value of each logarithm. Round the answer to four significant digits.

31. $\log_2 12$
32. $\log_4 9$
33. $\log_{12} 945$
34. $\log_{20} 1250$
35. $\log_3 \frac{2}{3}$
36. $\log_5 \dfrac{17}{5}$
37. $\log_{2/3} 34$
38. $\log_{1/2} \frac{3}{4}$

In Exercises 39–46, simplify each logarithmic expression.

39. $(\log_4 5)(\log_5 4)$
40. $(\log 7)(\log_7 10)$
41. $(\log_3 4)(\log_4 81)$
42. $(\log_5 16)(\log_2 125)$
43. $\log 20 + \dfrac{1}{\log_5 10}$
44. $\dfrac{1}{\log_4 10} + \dfrac{1}{\log_{25} 10}$

45. $\dfrac{1}{\log_a ab} + \dfrac{1}{\log_b ab}$

46. $\dfrac{\log_b x}{\log_{ab} x} - \dfrac{\log_b x}{\log_a x}$

47. The strongest earthquake ever recorded in the United States measured 8.6 on the Richter scale. It occurred in Alaska on Good Friday, March 27, 1964. How many times more intense was this earthquake than the one that occurred in California during the 1989 World Series (magnitude 7.1)?

48. One of the aftershocks from the earthquake during the 1989 World Series was 60 times less intense than the major earthquake (magnitude 7.1). What was the magnitude of the aftershock?

Critical Thinking

49. Suppose the functions f and g are defined by

$$f(x) = \ln x^2 \quad \text{and} \quad g(x) = 2 \ln x.$$

Are f and g the same function? Explain.

50. Show that if $f(x) = \ln x$, then $f(1/x) = -f(x)$.

51. Given that $\log_b A = 2$ and $\log_b B = 3$, find the value of each logarithm.

  (a) $\log_A b$ (b) $\log_B b$
  (c) $\log_A b^2$ (d) $\log_B \sqrt{b}$
  (e) $\log_{AB} b$ (f) $\log_{AB} (1/b)$
  (g) $\log_{A/B} b$ (h) $\log_{B/A} b$

52. Find the fallacy in the following argument:

$$\begin{aligned} 3 &= \log_2 8 \\ &= \log_2 (4 + 4) \\ &= \log_2 4 + \log_2 4 \\ &= 2 + 2 \\ &= 4 \end{aligned}$$

The magnitude m of a star is given by

$$m = -2.5 \log \dfrac{B}{B_0}$$

where B is the brightness of the star and B_0 is the brightness of a zero-level star having magnitude $m = 0$. Use this formula to answer Exercises 53 and 54.

53. The magnitude of our sun is -26.8 and the magnitude of Sirius, the brightest star in the heavens, is -1.5. How many times brighter is the sun than Sirius?

54. What is the difference in magnitude between two stars if the brightness of one is 50 times the brightness of the other?

The loudness L in decibels of a sound is given by

$$L = 10 \log \dfrac{I}{I_0}$$

where I is the intensity of the sound and I_0 is the intensity of the faintest sound that can be heard. Use this formula to answer Exercises 55 and 56.

55. The loudness of a whisper is 30 decibels and the loudness of a rock concert is 120 decibels. How many times more intense is the rock concert than the whisper?

56. If two sounds differ by 20 decibels, how many times more intense is the loudness of the more audible sound than the other sound?

Calculator Activities

57. Use the [LOG] key on your calculator to find the approximate values of

  log 6.2, log 62, log 620, and log 6200.

Compare the values you obtain. Do you observe a pattern? Show why this pattern develops by expressing 62, 620, and 6200 in scientific notation and then finding the common logarithms of these expressions by using the properties of logarithms.

58. Use the [LOG] key on your calculator to find the approximate values of

$$\log 3.5, \quad \log 0.35, \quad \log 0.035, \quad \text{and} \quad \log 0.0035.$$

Compare the values you obtain. Do you observe a pattern? Show why this pattern develops by expressing 0.35, 0.035, and 0.0035 in scientific notation and then finding the common logarithms of these expressions by using the properties of logarithms.

59. Evaluate the following expression by each of the given procedures:

$$\ln \pi + \ln\left(\frac{\sqrt{2}}{\pi}\right) + \frac{1}{2}\ln\left(\frac{3}{2}\right) - \ln\left(\frac{\sqrt{3}}{e}\right)$$

(a) Use the [LN] key on your calculator.
(b) Apply the properties of logarithms.

60. Use the change of base formula and a calculator to help complete the following table.

a	b	$\log_b (1/a)$	$\log_{1/b} a$
2	3		
18	4		
24	15		
π	10		

Compare the values of $\log_b (1/a)$ and $\log_{1/b} a$. Prove the relationship that you observe.

7.4 Graphs of Logarithmic Functions

◆ Introductory Comments

In Figure 7.8 of Section 7.2, we reflected the graph of the exponential function in the line $y = x$ to obtain the graph of its inverse, the logarithmic function. The graphs in Figure 7.9 show the basic shape of the logarithmic function $f(x) = \log_b x$ for $b > 1$ and for $0 < b < 1$.

FIGURE 7.9
Graph of the logarithmic function
$f(x) = \log_b x$

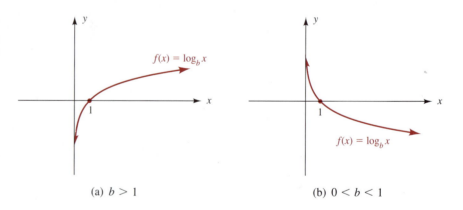

(a) $b > 1$ (b) $0 < b < 1$

In this section, we study the characteristics of the graph of $f(x) = \log_b x$ and use these characteristics along with the properties of logarithms (Section 7.3) to help sketch the graphs of several related functions.

◆ Characteristics of the Graph of $f(x) = \log_b x$

In Figure 7.9(a) the graph of $f(x) = \log_b x$ for $b > 1$ is always increasing and its x-intercept is 1. Also, for $b > 1$, observe that $f(x)$ decreases without bound $[f(x) \to -\infty]$ as x approaches 0 from the right ($x \to 0^+$). Hence, the y-axis is a vertical asymptote. In Figure 7.9(b) the graph of $f(x) = \log_b x$ for $0 < b < 1$ is always decreasing although its x-intercept remains 1. Also, for $0 < b < 1$, observe that $f(x)$ increases without bound $[f(x) \to \infty]$ as x approaches 0 from the right $(x \to 0^+)$. Again, the y-axis is a vertical asymptote. We can summarize these features of the graph of the logarithmic function as follows.

◆ Characteristics of the Graph of $f(x) = \log_b x$

1. The x-intercept is 1, and the graph has no y-intercept.
2. The y-axis is a vertical asymptote.
3. If $b > 1$, the graph of $f(x) = \log_b x$ is always increasing.
4. If $0 < b < 1$, the graph of $f(x) = \log_b x$ is always decreasing.

To sketch the graph of a particular logarithmic function, we may simply use the basic characteristics of the general graph and plot a couple points.

EXAMPLE 1 Sketch the graph of each logarithmic function.

(a) $f(x) = \log_4 x$ (b) $g(x) = \log_{1/4} x$

SOLUTION

(a) Since the base is 4, and $b = 4 > 1$, the graph of $f(x) = \log_4 x$ is always increasing. The x-intercept is 1, and since $f(x) \to -\infty$ as $x \to 0^+$, the y-axis is a vertical asymptote. Selecting a few convenient inputs for x, we find their corresponding outputs $f(x)$ as shown in the table:

x	$\frac{1}{4}$	1	4
$f(x)$	-1	0	1

Plotting the points associated with these ordered pairs and connecting them to form a smooth curve gives us the graph of $f(x) = \log_4 x$, as shown in Figure 7.10.

FIGURE 7.10
The function $f(x) = \log_4 x$ is an increasing function.

(b) Since the base b is $\frac{1}{4}$ and $0 < \frac{1}{4} < 1$, the graph of $g(x) = \log_{1/4} x$ is always decreasing. The x-intercept is 1, and, since $g(x) \to \infty$ as $x \to 0^+$, the y-axis is a vertical asymptote. Selecting a few convenient inputs for x, we find their corresponding outputs $g(x)$ as follows:

x	$\frac{1}{4}$	1	4
$g(x)$	1	0	-1

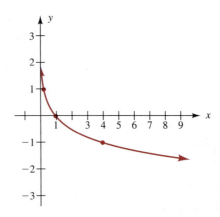

FIGURE 7.11
The function $g(x) = \log_{1/4} x$ is a *decreasing function*.

Plotting the points associated with these ordered pairs and connecting them to form a smooth curve gives us the graph of $g(x) = \log_{1/4} x$, as shown in Figure 7.11. ◆

Note: Observe that the graph of $g(x) = \log_{1/4} x$ is the same as the graph $f(x) = \log_4 x$ reflected about the *x*-axis. In general, if *b* is a real number such that $b > 0$ and $b \neq 1$, then the graph of $g(x) = \log_{1/b} x$ is the same as the graph of $f(x) = \log_b x$ reflected about the *x*-axis.

PROBLEM 1 Sketch the graph of the functions $f(x) = \log_5 x$ and $g(x) = \log_{1/5} x$ on the same coordinate plane. ◆

◆ **Graphing Related Functions**

Knowing the basic shape of the logarithmic function $f(x) = \log_b x$ enables us to graph several other related functions by applying the shift rules and axis reflection rules that we discussed in Section 1.4.

EXAMPLE 2 Sketch the graph of each function. Label the vertical asymptote and any *x*- and *y*-intercepts.

(a) $F(x) = \log_4 (x + 2)$ (b) $G(x) = 1 + \log_4 (-x)$

SOLUTION

(a) By the horizontal shift rule (Section 1.4), the graph of $F(x) = \log_4 (x + 2)$ is the same as the graph of $f(x) = \log_4 x$ shifted to the left 2 units, as shown in Figure 7.12. We find the *y*-intercept by evaluating $F(0)$, as follows:

$$F(0) = \log_4 (0 + 2) = \log_4 2 = \tfrac{1}{2}.$$

We find the *x*-intercept by solving the equation $F(x) = 0$, as follows:

$$\log_4 (x + 2) = 0$$

$$4^0 = x + 2 \qquad \text{Change to exponential form}$$

$$x = -1 \qquad \text{Solve for } x$$

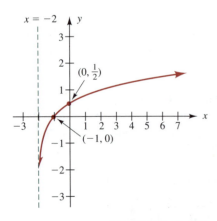

FIGURE 7.12
Graph of $F(x) = \log_4 (x + 2)$

Note that $F(x) \to -\infty$ as x approaches -2 from the right ($x \to -2^+$). Thus, the vertical asymptote is the line $x = -2$.

(b) By the y-axis reflection rule and the vertical shift rule (Section 1.4), the graph of $G(x) = 1 + \log_4(-x)$ is the same as the graph of $f(x) = \log_4 x$ reflected about the y-axis and then shifted vertically upward 1 unit, as shown in Figure 7.13. We find the x-intercept by solving the equation $G(x) = 0$:

$$1 + \log_4(-x) = 0$$
$$\log_4(-x) = -1 \qquad \text{Subtract 1 from both sides}$$
$$4^{-1} = -x \qquad \text{Change to exponential form}$$
$$x = -\tfrac{1}{4} \qquad \text{Solve for } x$$

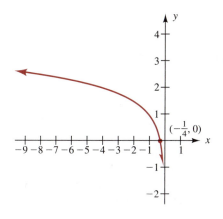

FIGURE 7.13
Graph of $G(x) = 1 + \log_4(-x)$

Note that $G(x) \to -\infty$ as x approaches 0 from the left ($x \to 0^-$). Hence, the y-axis is a vertical asymptote. ◆

PROBLEM 2 Repeat Example 2 for $H(x) = 1 - \log_4 x$. ◆

To sketch the graphs of other logarithmic functions, we apply the properties of logarithms (Section 7.3) in order to obtain an equivalent function whose graph is easily plotted.

EXAMPLE 3 Sketch the graph of $H(x) = \log_4 \dfrac{1}{x-1}$. Label the vertical asymptote and any x- and y-intercepts.

SOLUTION We begin by rewriting this function as follows:

$$H(x) = \log_4 \frac{1}{x-1} = \log_4 1 - \log_4(x-1) \qquad \text{Apply log property 2}$$
$$= 0 - \log_4(x-1) \qquad \text{Evaluate } \log_4 1$$
$$= -\log_4(x-1) \qquad \text{Simplify}$$

Now by the horizontal shift rule and x-axis reflection rule (Section 1.4), the graph of $H(x) = -\log_4(x-1)$ is the same as the graph of $f(x) = \log_4 x$, shifted horizontally to the right 1 unit and then reflected about the x-axis, as shown in Figure 7.14. We find the x-intercept by solving the equation $H(x) = 0$:

$$\log_4 \frac{1}{x-1} = 0$$
$$4^0 = \frac{1}{x-1} \qquad \text{Change to exponential form}$$
$$x - 1 = 1 \qquad \text{Solve for } x$$
$$x = 2$$

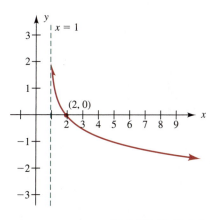

FIGURE 7.14
Graph of $H(x) = \log_4 \dfrac{1}{x-1}$

Note that $H(x) \to \infty$ as $x \to 1^+$. Hence $x = 1$ is a vertical asymptote. ◆

PROBLEM 3 Repeat Example 3 for $F(x) = \log_4 16x$. ◆

When applying the properties of logarithms to a logarithmic function, we must always preserve the domain of the original function. For example, consider the functions F and G defined by

$$F(x) = \log_4 x^2 \quad \text{and} \quad G(x) = 2 \log_4 x.$$

The domain of the function F is the set of all real numbers x with $x \neq 0$. However, the domain of the function G is the set of all positive real numbers x. Since the functions F and G have different domains, they are not the same function. To apply logarithmic property 3 to function F and preserve the same function, we must use the absolute value of x, and we write

$$F(x) = \log_4 x^2 = \log_4 |x|^2 = 2 \log_4 |x|.$$

EXAMPLE 4 Sketch the graph of $F(x) = \log_4 x^2$.

SOLUTION As we discussed in the preceding paragraph,

$$F(x) = \log_4 x^2 = 2 \log_4 |x|.$$

If $x > 0$, then $|x| = x$. Hence, by the vertical stretch rule (Section 1.4), the graph of F for $x > 0$ is the same as the graph of $f(x) = \log_4 x$ stretched vertically by a factor of 2. If $x < 0$, then $|x| = -x$. Hence, by the vertical stretch rule and y-axis reflection rule (Section 1.4), the graph of F for $x < 0$ is the same as the graph of $f(x) = \log_4 x$ stretched vertically by a factor of 2 and then reflected about the y-axis. The graph of F is shown in Figure 7.15. ◆

FIGURE 7.15
Graph of $F(x) = \log_4 x^2$

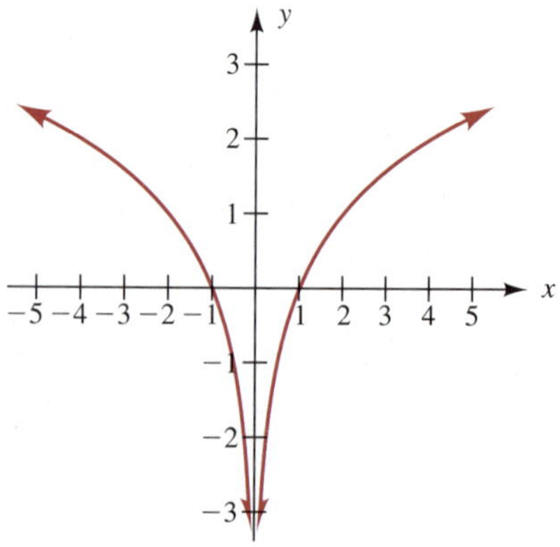

PROBLEM 4 Sketch the graph of $F(x) = \log_4 x^3$. ◆

◆ **Application: The Retention Function**

In psychological experiments it has been shown that most of what we learn is rapidly forgotten, and the remainder of the learned material slowly recedes from memory. The function R that describes the percentage of the learned material we retain over a period of time t is called the **retention function.** The retention function is often defined in terms of logarithms, and its graph is called the **retention curve.**

EXAMPLE 5 In a psychological experiment, a student is asked to memorize a list of nonsense syllables by studying the list until one perfect repetition is performed from memory. At various times over the next month, the student is asked to recall the list of syllables. It is found that the retention function R associated with this experiment is given by

$$R(t) = 100 - 29 \log (27t + 1), \quad 0 \le t \le 30,$$

where t is time in days.

(a) What percentage of the learned material is retained initially (at $t = 0$)?

(b) What percentage of the learned material is retained after 8 hours?

(c) How much time has elapsed when 42% of the learned material is still retained?

(d) Sketch the graph of the retention curve associated with this function.

SOLUTION

(a) The percentage of the learned material that is retained initially is

$$R(0) = 100 - 29 \log [27(0) + 1]$$
$$= 100 - 29 \log 1 = 100\%$$

(b) The percentage of the learned material retained after 8 hours ($t = \frac{1}{3}$ day) is

$$R(\tfrac{1}{3}) = 100 - 29 \log [27(\tfrac{1}{3}) + 1]$$
$$= 100 - 29 \log 10 = 71\%$$

(c) To determine the time when 42% of the learned material is retained, we solve the equation $R(t) = 42$:

$$42 = 100 - 29 \log (27t + 1)$$
$$-58 = -29 \log (27t + 1) \quad \text{Subtract 100 from both sides}$$
$$2 = \log (27t + 1) \quad \text{Divide both sides by } -29$$
$$10^2 = 27t + 1 \quad \text{Change to exponential form}$$
$$t = 3\tfrac{2}{3} \text{ days} \quad \text{Solve for } t$$

(d) The retention curve associated with this function is shown in Figure 7.16. Observe from the graph that forgetting is at first rapid and then slow. ◆

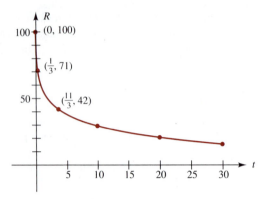

FIGURE 7.16
Retention curve of
$R(t) = 100 - 29 \log (27t + 1)$
for $0 \leq t \leq 30$

PROBLEM 5 Referring to Example 5, determine the percentage of the learned material that is retained after 10 days, 20 days, and 30 days. Check these values with the retention curve in Figure 7.16. ◆

Exercises 7.4

 Basic Skills

In Exercises 1 and 2, sketch the graphs of the equations on the same coordinate plane.

1. $y = \log_2 x$, $y = \log_3 x$, $y = \log x$, and $y = \ln x$
2. $y = \log_{1/2} x$, $y = \log_{1/3} x$, $y = \log_{1/10} x$, and $y = \log_{1/e} x$

Use the graphs of the equations in Exercises 1 and 2 in conjunction with the shift rules and axis reflection rules (Section 1.4) to sketch the graphs of the functions in Exercises 3–16. Label any x- and y-intercepts and the vertical asymptote.

3. $F(x) = -\log_2 x$
4. $g(x) = -\log_{1/3} x$
5. $h(x) = \log_{1/10} (-x)$
6. $f(x) = -\ln (-x)$
7. $g(x) = \log_3 x + 2$
8. $G(x) = \log_2 x - 1$
9. $G(x) = \log_3 (x + 2)$
10. $H(x) = \log_2 (x - 1)$
11. $f(x) = -3 + \ln (x - 1)$
12. $f(x) = 2 + \log (x + 2)$
13. $H(x) = 2 - \log_2 (x + 3)$
14. $F(x) = 1 - \log (x - 4)$
15. $f(x) = \ln (3 - x) + 2$
16. $f(x) = 2 - \log_3 (2 - x)$

Determine which of the functions in Exercises 17–22 are the same function.

17. $f(x) = \ln x^2$, $g(x) = 2 \ln x$, and $h(x) = 2 \ln |x|$
18. $f(x) = \ln x^3$, $g(x) = 3 \ln x$, and $h(x) = 3 \ln |x|$
19. $f(x) = \log \sqrt{x + 1}$, $g(x) = \frac{1}{2} \log (x + 1)$, and $h(x) = \frac{1}{2} \log |x + 1|$
20. $f(x) = \log (x - 1)^4$, $g(x) = 4 \log (x - 1)$, and $h(x) = 4 \log |x - 1|$
21. $f(x) = \ln \frac{1}{x + 3}$, $g(x) = -\ln (x + 3)$, and $h(x) = -\ln |x + 3|$
22. $f(x) = \log \frac{10}{(x - 1)^2}$, $g(x) = 1 - 2 \log (x - 1)$, and $h(x) = 1 - 2 \log |x - 1|$

In Exercises 23–30, use the properties of logarithms to help sketch the graph of each function. Label any x- and y-intercepts and the vertical asymptote.

23. $f(x) = \frac{1}{2} \log x^4$
24. $f(x) = \ln x^3$
25. $f(x) = \ln \frac{1}{x + 3}$
26. $f(x) = \log_3 \frac{9}{x}$
27. $f(x) = \log_{1/2} [8(x - 1)]$
28. $f(x) = \log_2 (16x + 32)$
29. $f(x) = \log_3 \frac{1}{18 - 9x}$
30. $f(x) = \log \frac{10}{(x - 1)^2}$

31. In a psychological experiment, a student is asked to memorize a list of telephone numbers by studying the list

until one perfect repetition is performed from memory. At various times over the next week, the student is asked to recall the list of numbers. It is found that the retention function R associated with this experiment is given by

$$R(t) = 100 - 31 \log (36t + 1), \quad 0 \leq t \leq 7,$$

where t is time in days.

(a) What percentage of the learned material is retained initially (at $t = 0$)?

(b) What percentage of the learned material is retained after 6 hours?

(c) How much time has elapsed when 33% of the learned material is still retained?

(d) Sketch the graph of the retention curve associated with this function.

32. Students taking their last calculus course are given a final exam and the average score of the group is recorded. Each month thereafter, the students are given an equivalent exam and each time their average score is recorded. It is found that the average score S is a function of time t (in months) and is defined by

$$S(t) = 72 - 22 \log (t + 1), \quad 0 \leq t \leq 12.$$

(a) What is the average score on the original exam?

(b) What is the average score six months later?

(c) How much time elapses before the average score decreases to 50?

(d) Sketch the graph of the function S.

Critical Thinking

In Exercises 33–44,

(a) find the inverse of each function.

(b) sketch the graphs of the function and its inverse on the same coordinate plane.

33. $f(x) = 6^x$
34. $g(x) = (\frac{1}{6})^x$
35. $h(x) = e^{x/2}$
36. $F(x) = 10^{2x/3}$
37. $G(x) = 1 - 10^{2x}$
38. $H(x) = 3e^{2x+1}$
39. $f(x) = \log_8 x$
40. $g(x) = \log_{1/8} x$
41. $h(x) = -\ln \dfrac{x}{2}$
42. $F(x) = 2 \log \dfrac{3x}{2}$
43. $G(x) = 1 + 2 \log 3x$
44. $H(x) = 3 \ln (2x - 1)$

45. The number N of computers sold by a certain company is a function of the amount x (in thousand of dollars) that is spent on advertising and is given by

$$N(x) = 1000[1 + \ln (x + 1)].$$

(a) If no money is spent on advertising, how many computers are sold?

(b) If $100,000 is spent on advertising, approximately how many computers are sold?

(c) How much money in advertising must be spent to sell 4000 computers?

(d) Sketch the graph of the function N.

46. The time t in minutes it takes for a cup of hot tea to cool to a temperature of T degrees Fahrenheit when it is placed in a room whose temperature is maintained at 70 °F is given by

$$t(T) = -8 \ln \dfrac{T - 70}{100}, \quad 70 < T \leq 170.$$

(a) How long does it take for the tea to cool to 140 °F?

(b) How long does it take for the tea to cool to 100 °F?

(c) What is the temperature of the tea (to the nearest degree) when $t = 4$ minutes?

(d) Sketch the graph of the function t.

Calculator Activities

 47. Given the function $f(x) = \dfrac{\ln x}{x}$,

(a) determine the domain of the function.

(b) determine any zeros of the function.

(c) explain the behavior of $f(x)$ as $x \to \infty$.

(d) explain the behavior of $f(x)$ as $x \to 0^+$.

(e) sketch the graph of the function.

EXERCISE 47 (continued)

(f) use a graphing calculator to generate the graph of this function, and trace to the relative maximum point. Zoom in on this maximum point to estimate the value of x where the maximum value of the function seems to occur.

48. Given the function $f(x) = \dfrac{1}{x \ln x}$,

   (a) determine the domain of the function.
   (b) determine any zeros of the function.

(c) explain the behavior of $f(x)$ as $x \to \infty$ and as $x \to 1^+$.
(d) explain the behavior of $f(x)$ as $x \to 0^+$ and as $x \to 1^-$.
(e) sketch the graph of the function.
(f) use a graphing calculator to generate the graph of this function, and trace to the relative maximum point in the interval $(0, 1)$. Zoom in on this point to estimate the maximum value of this function in the interval $(0, 1)$.

7.5 Logarithmic and Exponential Equations

◆ Introductory Comments

Equations that contain logarithmic expressions are referred to as **logarithmic equations**. In Section 7.2, we solved logarithmic equations containing a single logarithmic expression by changing the equation to exponential form. For example,

| Logarithmic form | | Exponential form |

$$\log_3 (x - 12) = 2 \quad \text{is equivalent to} \quad 3^2 = x - 12$$
$$x = 3^2 + 12$$
$$x = 21$$

An equation in which the variable appears in an exponent is referred to as an **exponential equation**. In Section 7.2, we solved some exponential equations with base 10 and base e by changing the equation to logarithmic form. For example,

| Exponential form | | Logarithmic form |

$$e^{x/2} = 9 \quad \text{is equivalent} \quad \ln 9 = \dfrac{x}{2}$$
$$x = 2 \ln 9$$
$$x = \ln 81 \approx 4.394$$

In this section, we look at

1. A logarithmic equation that contains more than one logarithmic expression.
2. An exponential equation in which the base is different from 10 or e.

◆ Logarithmic Equations

To solve many logarithmic equations that contain more than one logarithmic expression, we use the following procedure.

SECTION 7.5 Logarithmic and Exponential Equations   447

Procedure for Solving Logarithmic Equations

1. Isolate the logarithmic expressions on one side of the equation.
2. Apply the properties of logarithms, and write the equation in logarithmic form.
3. Change to exponential form, and solve for the unknown.
4. Check the solutions. *This procedure may produce extraneous roots.*

EXAMPLE 1 Solve each logarithmic equation.

(a) $2 - \log x = \log 3$ (b) $\log_4 (x - 2) + 2 \log_4 x = 1 + \log_4 2x$

SOLUTION

(a) Using the given procedure, we have

$$2 - \log x = \log 3$$
$$2 = \log 3 + \log x \qquad \text{Isolate the logarithms on one side}$$
$$2 = \log 3x \qquad \text{Apply log property 1}$$
$$10^2 = 3x \qquad \text{Change to exponential form}$$
$$x = \frac{100}{3} \qquad \text{Solve for } x$$

Check: Replacing x with $\frac{100}{3}$ in the original equation, we have

$$2 - \log \tfrac{100}{3} = \log 3 \quad ?$$
$$2 - (\log 100 - \log 3) = \log 3 \quad ?$$
$$2 - (2 - \log 3) = \log 3 \quad ?$$
$$\log 3 = \log 3 \quad ✔$$

Thus, $x = \frac{100}{3}$ is a solution.

(b) Using the given procedure, we have

$$\log_4 (x - 2) + 2 \log_4 x = 1 + \log_4 2x$$
$$\log_4 (x - 2) + 2 \log_4 x - \log_4 2x = 1 \qquad \text{Isolate the logarithmic expressions}$$
$$\log_4 \frac{x^2(x - 2)}{2x} = 1 \qquad \text{Apply the log properties and write in logarithmic form}$$
$$\log_4 \frac{x(x - 2)}{2} = 1 \qquad \text{Reduce, } x \neq 0$$
$$4^1 = \frac{x(x - 2)}{2} \qquad \text{Change to exponential form}$$

$$x^2 - 2x - 8 = 0 \qquad \text{\textbf{Write in quadratic form and solve for } } x$$

$$(x - 4)(x + 2) = 0$$

$$x = 4 \quad \text{or} \quad x = -2$$

Check I: $x = 4$

$\log_4 2 + 2 \log_4 4 = 1 + \log_4 8$?

$\frac{1}{2} + 2 = 1 + \frac{3}{2}$?

$\frac{5}{2} = \frac{5}{2}$ ✓

Check II: $x = -2$

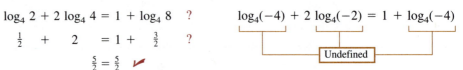

Thus, $x = 4$ is the only solution. ◆

PROBLEM 1 Solve the logarithmic equation $2 \ln x - \ln 9 = 4$. ◆

Many literal equations and formulas that contain logarithmic expressions may be solved by using this procedure.

EXAMPLE 2 Solve the literal equation $\ln (y + a) - 2 \ln (x + b) = c$ for y. Assume a, b, and c are constants.

SOLUTION We solve for y as follows:

$$\ln (y + a) - 2 \ln (x + b) = c$$

$$\ln \frac{y + a}{(x + b)^2} = c \qquad \text{\textbf{Apply the log properties and write in logarithmic form}}$$

$$e^c = \frac{y + a}{(x + b)^2} \qquad \text{\textbf{Change to exponential form}}$$

$$y = e^c(x + b)^2 - a \qquad \text{\textbf{Solve for } } y$$

Since c is a constant, e^c is also a constant. Relabeling e^c as the constant k gives us

$$y = k(x + b)^2 - a$$

where $k = e^c$. ◆

PROBLEM 2 Solve the literal equation $x + \ln y = \ln c$, where c is a constant, for y. ◆

◆ Exponential Equations

We now look at exponential equations in which the bases are different from 10 or e. If both sides of an exponential equation can be written as powers of the same base, then the equation can be solved by equating the powers; that is,

$$\text{if} \quad b^x = b^y, \quad \text{then} \quad x = y.$$

EXAMPLE 3
Solve the exponential equation $27^{x-2} = 9$.

SOLUTION Both 27 and 9 can be written in terms of a base of 3. Hence,

$$27^{x-2} = 9$$
$$(3^3)^{x-2} = 3^2 \qquad \text{Write both sides with the same base}$$
$$3^{3(x-2)} = 3^2 \qquad \text{Multiply exponents}$$
$$3(x - 2) = 2 \qquad \text{Equate exponents and solve for } x$$
$$3x = 8$$
$$x = \tfrac{8}{3}$$

Check: $27^{[(8/3)-2]} = 27^{2/3} = 9$ ✔

PROBLEM 3
Solve the exponential equation $4^{3x+2} = 8^{4x}$.

An alternate method for solving the exponential equation in Example 3 is to begin by taking the common (or natural) logarithm of both sides of the equation:

$$27^{x-2} = 9$$
$$\log 27^{x-2} = \log 9 \qquad \text{Take the common logarithm of both sides}$$
$$(x - 2) \log 27 = \log 9 \qquad \text{Apply log property 3}$$
$$x - 2 = \frac{\log 9}{\log 27} \qquad \text{Divide both sides by log 27}$$
$$x = 2 + \frac{\log 9}{\log 27} \qquad \text{Add 2 to both sides}$$

Now, since $9 = 3^2$ and $27 = 3^3$, we apply log property 3 and write

$$x = 2 + \frac{\log 9}{\log 27} = 2 + \frac{2 \log 3}{3 \log 3} = 2 + \frac{2}{3} = \frac{8}{3},$$

which agrees with the answer we obtained in Example 3.

This procedure is particularly useful when both sides of an exponential equation *cannot* be written as powers of the same base.

Procedure for Solving Exponential Equations

1. Take the common (or natural) logarithm of both sides of the equation.
2. Apply the properties of logarithms, and write the powers as coefficients of logarithms.
3. Solve for the unknown, and check the solution.

EXAMPLE 4 Solve each exponential equation.

(a) $6^x = 50$ (b) $3^{x-2} = 2^{-x}$

SOLUTION

(a) Since both sides of this equation cannot be written as powers of the same base, we apply the given procedure:

$$6^x = 50$$

$$\log 6^x = \log 50 \qquad \text{Take the common logarithm of both sides}$$

$$x \log 6 = \log 50 \qquad \text{Apply log property 3}$$

$$x = \frac{\log 50}{\log 6} \approx 2.183 \qquad \text{Divide both sides by log 6}$$

Use the $\boxed{\text{LOG}}$ key on a calculator to find the approximate value.

You can check this solution with a calculator by using the $\boxed{Y^x}$ key.

(b) Using the given procedure, we have

$$3^{x-2} = 2^{-x}$$

$$\log 3^{x-2} = \log 2^{-x} \qquad \text{Take the common logarithm of both sides}$$

$$(x - 2) \log 3 = -x \log 2 \qquad \text{Apply log property 3}$$

Be sure to remember the parentheses.

$$x \log 3 - 2 \log 3 = -x \log 2 \qquad \text{Multiply}$$

$$x \log 3 + x \log 2 = 2 \log 3 \qquad \text{Group the } x \text{ terms on one side}$$

$$x(\log 3 + \log 2) = 2 \log 3 \qquad \text{Factor out } x$$

$$x = \frac{2 \log 3}{\log 3 + \log 2} \qquad \text{Divide both sides by } (\log 3 + \log 2)$$

$$x = \frac{\log 9}{\log 6} \approx 1.226$$

Use the $\boxed{\text{LOG}}$ key on a calculator to find the approximate value.

Note: To solve Example 4(a), we may also proceed as in Section 7.2 and change the exponential form $6^x = 50$ to the logarithmic form

$$x = \log_6 50.$$

SECTION 7.5 Logarithmic and Exponential Equations

Now to evaluate $\log_6 50$, we use the change of base formula (Section 7.3) with common logarithms to obtain

$$x = \log_6 50 = \frac{\log 50}{\log 6} \approx 2.183$$

PROBLEM 4 Solve the equation $5^{x-1} = 325$. ◆

◆ Equations of Quadratic Type

We can solve other logarithmic and exponential equations by recognizing them as equations of quadratic type.

EXAMPLE 5 Solve each equation.

(a) $(\log x)^2 = \log x^3$ (b) $e^x + 3e^{-x} = 4$

SOLUTION

(a) We begin by rewriting the equation as follows:

$$(\log x)^2 = \log x^3$$
$$(\log x)^2 = 3 \log x \qquad \text{Apply log property 3}$$
$$(\log x)^2 - 3 \log x = 0 \qquad \text{Subtract } 3 \log x \text{ from both sides}$$

Now observe that this last equation is of quadratic type. Letting $u = \log x$, we obtain

$$u^2 - 3u = 0$$
$$u(u - 3) = 0 \qquad \text{Factor}$$

$u = 0$	or $u = 3$	Apply the zero product property
$\log x = 0$	$\log x = 3$	Replace u with $\log x$
$x = 10^0$	$x = 10^3$	Change to exponential form
$x = 1$	$x = 1000.$	Simplify

You can check both solutions.

(b) We begin by rewriting the equation as follows:

$$e^x + 3e^{-x} = 4$$
$$e^{2x} + 3 = 4e^x \qquad \text{Multiply both sides by } e^x$$
$$e^{2x} - 4e^x + 3 = 0 \qquad \text{Subtract } 4e^x \text{ from both sides}$$

Now observe that this last equation is of quadratic type. Letting $u = e^x$, we obtain

$$u^2 - 4u + 3 = 0$$

$$(u - 1)(u - 3) = 0 \quad \text{Factor}$$

$$u = 1 \quad \text{or} \quad u = 3 \quad \text{Apply the zero product property}$$

$$e^x = 1 \quad\quad\quad e^x = 3 \quad \text{Replace } u \text{ with } e^x$$

$$x = \ln 1 \quad\quad x = \ln 3 \quad \text{Change to logarithmic form}$$

$$x = 0 \quad\quad\quad x \approx 1.099 \quad \text{Evaluate}$$

You can check both solutions. ◆

PROBLEM 5 Solve the equation $\log \sqrt{x} = \sqrt{\log x}$ by squaring both sides. ◆

◆ **Graphical Methods of Solutions**

If the techniques that we have discussed in this section are insufficient to solve a particular logarithmic or exponential equation, we can use a calculator or computer with graphing capabilities to find the approximate solutions.

EXAMPLE 6 Use a graphing calculator to find the approximate solutions of the equation $2^x + 3^x = 9$.

SOLUTION The equation $2^x + 3^x = 9$ is equivalent to

$$2^x = 9 - 3^x.$$

To solve this equation graphically, we graph the equations

$$y = 2^x \quad \text{and} \quad y = 9 - 3^x$$

on the same set of coordinate axes and determine the x-coordinate(s) of their intersection point(s). Using the techniques of graphing exponential functions (Section 7.1), we obtain the graphs shown in Figure 7.17. Note that the graphs intersect once, and the x-coordinate of the intersection point appears to be between $x = 1$ and $x = 2$.

To display these graphs on a calculator, we begin by choosing a viewing rectangle by pressing the [RANGE] key. Although we can select any Range values, Figure 7.17 suggests a reasonable viewing rectangle:

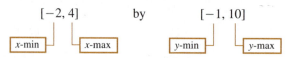

By entering the equations $y = 2^x$ and $y = 9 - 3^x$, we obtain the two graphs in the viewing rectangle. Next, we press the [TRACE] key and move the blinking cursor to the approximate point of intersection of the two curves. Reading the

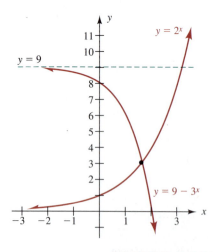

FIGURE 7.17
The solution of the equation $2^x + 3^x = 9$ is between $x = 1$ and $x = 2$.

SECTION 7.5 Logarithmic and Exponential Equations

x-coordinate of this point, we find $x \approx 1.6$ to the nearest tenth. To obtain more precise values of *x*, we may zoom in on this intersection point by activating the Zoom feature and tracing to this point once again. By repeating this process of tracing and zooming, we find that the *x*-coordinate of the point of intersection is $x \approx 1.620$ to the nearest thousandth. In summary, we conclude that the solution to the equation $2^x + 3^x = 9$ is approximately 1.620. ◆

PROBLEM 6 Use the $\boxed{y^x}$ key on your calculator to check the approximate solution of the equation in Example 6. ◆

Exercises 7.5

Basic Skills

In Exercises 1–12, solve each logarithmic equation.

1. $\log x + \log 5 = 1$
2. $\ln x = 1 + \ln 5$
3. $3 \ln (x + 1) - \ln 27 = 3$
4. $2 \log_6 (2 - x) = 2 - \log_6 3$
5. $\log_4 (x + 12) - \log_4 (x - 3) = 2$
6. $\log_3 x + \log_3 (x - 6) = 3$
7. $\log_5 x + \log_5 (x + 2) = \log_5 (x + 6)$
8. $\log 2 + \log (2x - 3) = \log 3x - \log 3$
9. $\log (x + 6) + 1 = 2 \log (3x - 2)$
10. $\log_2 (x - 1) + 2 \log_2 x = 2 + \log_2 3x$
11. $\frac{1}{2} \ln (3 - 2x) - \ln x = 0$
12. $\frac{1}{2} \log (x + 3) + \log 2 = 1$

In Exercises 13—20, solve each literal equation for y. Assume a, b, and c are constants.

13. $\ln y + 2x = \ln c$
14. $2(x + \ln x) = \ln y - \ln c$
15. $2 \log (x + a) = 1 - \log y$
16. $\log y = \log (x + y) + 2$
17. $\ln (x + y) - \ln (x - y) = c$
18. $\ln (x + 2y) + 2 \ln x = \ln y + c$
19. $\frac{1}{a} \ln y - \frac{1}{a} \ln (a - by) = c$
20. $-\frac{1}{4} \ln (y + 2) + \frac{1}{4} \ln (y - 2) = x + c$

In Exercises 21–36, solve each exponential equation.

21. $9^x = 27$
22. $16^x = \frac{1}{8}$
23. $2^{x-1} = 16^x$
24. $5^{x+3} = 25^{x-2}$
25. $7^x = 35$
26. $3^x = 36$
27. $3^{2x-1} = 40$
28. $6^{x+5} = 75$
29. $10^{2x-1} = 4^{-x}$
30. $5^{x-2} = 4^{2x+1}$
31. $e^{x/2} = 2^{x-1}$
32. $4^{-2x} = e^{x+1}$
33. $3 \cdot 2^{x-2} = 6^x$
34. $5 \cdot 3^{-x} = 4 \cdot 2^{x+2}$
35. $2^x - 6 \cdot 8^{3-x} = 0$
36. $\frac{10^{x-1}}{e^{2x}} - 4 = 0$

Rewrite each equation in Exercises 37–46 as an equation of quadratic type, then solve.

37. $(\ln x)^2 = \ln x^2$
38. $2[\log (x + 1)]^2 - \log (x + 1)^3 = 5$
39. $e^{4x} - 3e^{2x} = 4$
40. $2^{3x} + 4 \cdot 2^{-3x} = 5$
41. $3 \cdot 10^x - 10^{-x} = 2$
42. $\frac{1}{2}(e^x - 9e^{-x}) = 4$
43. $\frac{e^x + e^{-x}}{2} = 10$
44. $10^x + 10^{-x} = 6$
45. $2 \log_x 3 + \log_3 x = 3$
46. $2 \log_4 x - 3 \log_x 4 = 5$

Critical Thinking

47. In Section 7.1 we developed the compound interest formula

$$A = P\left(1 + \frac{r}{n}\right)^{nt}$$

Solve for t and express the answer in terms of natural logarithms.

48. Solve for x in terms of b.
(a) $\log_b (x - 3) = -1 + \log_b 5$
(b) $2 + \log_b (4x + 1) = \log_b (2 - x)$

49. If y varies directly as the mth power of x, then

$$y = kx^m,$$

where k is the variation constant.

(a) Take the natural logarithm of both sides of this equation and show that

$$\ln y = m \ln x + \ln k.$$

(b) Construct the graph of $\ln y = m \ln x + \ln k$ in an XY-plane, where $X = \ln x$ and $Y = \ln y$. What type of graph do we obtain?

50. Find the fallacy in the following argument.

$1 < 2$	
$\frac{1}{4} < \frac{1}{2}$	**Divide both sides by 4**
$\ln \frac{1}{4} < \ln \frac{1}{2}$	**Take the natural logarithm of both sides**
$\ln \left(\frac{1}{2}\right)^2 < \ln \frac{1}{2}$	**Rewrite $\frac{1}{4}$**
$2 \ln \frac{1}{2} < \ln \frac{1}{2}$	**Apply log property 3**
$2 < 1$	**Divide both sides by $\ln \frac{1}{2}$**

51. Because of acid rain, the population P of fish in a pond in New Hampshire is decreasing according to the equation

$$\log_2 P = -\tfrac{1}{3}t + \log_2 P_0,$$

where t is time in years after 1985 and P_0 is the original population in 1985.

(a) Solve the equation for P.
(b) In the year 1991, what percent of the original fish population remained?

52. In a series circuit containing a capacitor C, a resistance R, and a battery source E, the instantaneous current i at any time t is given by

$$\ln i = -\frac{t}{RC} + \ln E - \ln R.$$

(a) Solve the equation for i.
(b) What happens to the current i as t increases without bound ($t \to \infty$)?

Calculator Activities

In Exercises 53–58, use a graphing calculator to find the approximate solutions of the equation. Round each answer to four significant digits.

53. $2x + \ln x = 5$
54. $e^{-x} - x^3 = 0$
55. $xe^x = 4$
56. $2^{x+1} + 5^x = 1$
57. $2^x = 3 - x^2$
58. $\ln (x + 1) = x^2 - 4x$

59. When a cable or rope is suspended between two points at the same height and allowed to hang under its own weight, it forms a curve called a **catenary** (after the Latin

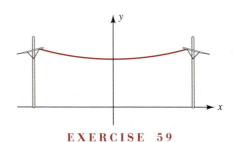

EXERCISE 59

word for chain). The power line shown in the figure is an example of a catenary, and its equation is given by

$$y = 25(e^{x/50} + e^{-x/50}),$$

where x and y are measured in feet.
(a) Find the height of the power line at $x = 0$.
(b) Find x (to the nearest tenth of a foot) when $y = 75$ ft.

60. A student infected with the flu returns to a college campus of 1000 students. If no student leaves campus, then the time t (in days) it takes for N students to become infected with the flu is given by

$$t = \ln 999N - \ln (1000 - N).$$

(a) How many days does it take for 20 students to become infected?
(b) Determine the number of infected students after 6 days.

Chapter 7 Review

Questions for Group Discussion

1. If $b > 0$, how is b^r defined when r is an irrational number?
2. Under what conditions is the *exponential function* $f(x) = b^x$ an increasing function? a decreasing function?
3. How is the graph of $f(x) = b^x$ related to the graph of $g(x) = b^{-x}$? Discuss the behavior of $f(x)$ and $g(x)$ as $x \to \infty$.
4. Are the functions $f(x) = e^{x+1}$ and $g(x) = e \cdot e^x$ identical? Explain.
5. Give an example of an *exponential growth function* and an *exponential decay function*.
6. What is the meaning of the *base b logarithm*, $\log_b x$? When is $\log_b x$ undefined?
7. State the four *logarithmic identities*.
8. What is the base for *common logarithms*? for *natural logarithms*?
9. How are the expressions $\log_b a$ and $\log_a b$ related?
10. State the procedure for changing an equation from *logarithmic form* to *exponential form*, and vice versa.
11. Explain the general procedure for solving an *exponential equation*. Illustrate with an example.
12. List some characteristics of the graph of the *logarithmic function* $f(x) = \log_b x$.
13. Are the functions $f(x) = \ln (x + 1)^2$ and $g(x) = 2 \ln (x + 1)$ identical? Explain.
14. How is the graph of $f(x) = \log_b x$ related to the graph of $g(x) = \log_b (1/x)$?
15. What is the domain and range of the *natural logarithmic function* $f(x) = \ln x$?
16. What is the *inverse function* of the natural logarithmic function?
17. How can we rewrite the logarithm of a product? of a quotient? of a quantity raised to a power?
18. What is the value of $(1 + 1/n)^n$ as n increases without bound?

Review Exercises

In Exercises 1–6, use the properties of real exponents to write each expression as a constant or as an expression in the form b^x, where $b > 0$.

1. 1^{3x-2}
2. $(3^{x+\pi})^0$
3. $e^{2x} \cdot e^{1-2x}$
4. $(2^{-3x})^2$
5. $\dfrac{4^{x+3}}{8^{x+2}}$
6. $(3^{-x} \cdot 9^x)^2$

In Exercises 7–20, find the value of each expression.

7. $\log_7 49$
8. $\log_2 32$
9. $\log \sqrt[3]{100}$
10. $\ln e^3$
11. $\log_{1/2} 8$
12. $\log_{1/3} 9$
13. $\log_8 2$
14. $\log_{25} \frac{1}{5}$
15. $\log_8 \frac{1}{4}$
16. $\log_{16} 64$
17. $e^{\ln 4}$
18. $10^{\log 8}$
19. $3^{\log_7 7}$
20. $8^{\log_2 1}$

In Exercises 21–30, simplify each expression. Assume the variables are restricted to those real numbers that allow the logarithmic expressions to be defined.

21. $\log^7 7^{2x}$
22. $5^{\log_5 (2-x)}$
23. $e^{\ln (x^2+1)}$
24. $\ln e^{2-3x}$
25. $\log_3 27^{1-x}$
26. $\log_4 2^{4x-4}$
27. $e^{2x+\ln x}$
28. $e^{-\ln (x-1)}$
29. $(\log_3 x)(\log_x 3)$
30. $\dfrac{1}{\log_x \frac{x}{2}} - \dfrac{1}{\log_2 \frac{x}{2}}$

In Exercises 31–36, write each expression as sums and differences of simpler logarithmic expressions without logarithms of products, quotients, and powers. Simplify whenever possible. Assume that any variable is restricted to those real numbers that allow for the properties of logarithms to be applied.

31. $\log_3 [9x^2(x-1)]$
32. $\log_4 \left[4x\sqrt{x+2}\right]$
33. $\log \dfrac{1}{x\sqrt{2x-3}}$
34. $\log \dfrac{1000}{xy^{1/3}}$
35. $\ln \dfrac{xe^{-x^2}}{e^x - 1}$
36. $\ln \dfrac{4e^{2x}}{\sqrt{e^x + 2}}$

In Exercises 37–42, write each expression as a single logarithm with a coefficient of 1. Simplify whenever possible. Assume that any variable is restricted to those real numbers that allow for the properties of logarithms to be applied.

37. $\ln 12 - \ln 3 + \ln 2$
38. $-2 \log_2 6 + 3 \log_2 3 - \frac{1}{2} \log_2 9$
39. $3 \log_3 (x+1) - \log_3 (x^2-1) + \log_3 (x-1)$
40. $2 \ln x + \frac{1}{2} \ln (x+1) + \ln 2$
41. $\log \dfrac{1}{\sqrt{x} - \sqrt{x-1}} - \log (\sqrt{x} + \sqrt{x-1})$
42. $\log (\sqrt{x+1} + 1) + \log (\sqrt{x+1} - 1)$

In Exercises 43–64, solve each equation.

43. $\log_5 x = 1$
44. $\ln x = 10$
45. $\ln |x+2| - 3 = 0$
46. $\log_3 (x^2 - 3x + 5) = 2$
47. $\log_3 (x-7) - \log_3 (x+1) = 2$
48. $\log 3 - \log (2-x) = \log 2 - \log 2x$
49. $\frac{1}{2} \log (12-x) - \log x = 0$
50. $\log_2 (3x-2) + 2 = 2 \log_2 (x+2)$
51. $\ln y = c + 2 \ln |x|$, where c is a constant, solve for y
52. $3(x - \ln x) = \ln (2y - 3) + \ln c$, where c is a constant, solve for y
53. $10^{2x} = 80$
54. $e^{2x-1} = 9$
55. $16^x = \frac{1}{8}$
56. $3^{x+3} = 27^{1-x}$
57. $3^{2x} = 4^{1-x}$
58. $e^{2x} = 2^{x+1}$
59. $2e^{-x} = 3^{2x-1}$
60. $5 \cdot 2^{-x} = 3 \cdot 5^{x+2}$
61. $x^2 e^{-x} - e^{-x} = 0$
62. $x \ln x - 3 \ln x = 0$
63. $\log_x 3 + 2 \log_3 x = 3$
64. $e^x + 1 = 2e^{-x}$

In Exercises 65–80, sketch the graph of each function. Label any x- or y-intercept(s) and horizontal or vertical asymptote.

65. $f(x) = 5^x$
66. $g(x) = 6^{-x}$
67. $h(x) = e^{x-2}$
68. $F(x) = 2 + e^x$
69. $G(x) = -(2 + e^{x+1})$
70. $H(x) = 3^x + 3^{-x}$
71. $f(x) = 2^{|x+1|}$
72. $g(x) = x - e^x$
73. $h(x) = \log_{1/5} x$
74. $F(x) = \log_6 (-x)$
75. $G(x) = -(2 + \log_4 x)$
76. $H(x) = \log_4 (x+2)$
77. $f(x) = \ln \dfrac{1}{x+1}$
78. $g(x) = \ln \sqrt{1-x}$

79. $h(x) = \dfrac{\ln |x|}{x}$ **80.** $F(x) = x \ln x - x$

In Exercises 81–84, use the change of base formula and a calculator to evaluate each logarithm to the nearest thousandth.

81. $\log_3 45$ **82.** $\log_6 30$

83. $\log_{1/2} 100$ **84.** $\log_{2/3} \tfrac{1}{2}$

In Exercises 85–92, find the inverse of each function. Then sketch the graphs of the function and its inverse on the same coordinate plane.

85. $f(x) = 8^x$ **86.** $g(x) = e^{-2x}$

87. $h(x) = 2e^{3x-2}$ **88.** $F(x) = 1 - 8^{1-x}$

89. $G(x) = \log_{1/2}(2x)$ **90.** $H(x) = -\ln(x-1)$

91. $f(x) = 1 - \ln x^2,\ x > 0$

92. $g(x) = \ln(x-1) - \ln(2x)$

93. Suppose that $5000 is invested at an interest rate of 9% per year. Assuming the investor makes no subsequent deposit or withdrawal, find the balance after 10 years if the interest is compounded (a) semiannually, (b) monthly, or (c) continuously.

94. Suppose that $100,000 is invested at an interest rate of $7\tfrac{1}{2}\%$ per year. Assuming the investor makes no subsequent deposit or withdrawal, find the balance after 9 months if the interest is compounded (a) quarterly, (b) daily, or (c) continuously.

95. Five hundred dollars is deposited into a savings account paying $6\tfrac{3}{4}\%$ interest per year compounded continuously. Assuming the depositor makes no subsequent deposit or withdrawal, find the time it takes for the deposit to accumulate to $800.

96. At what interest rate compounded continuously must money be invested if the amount is to triple in 12 years?

97. Initially, approximately 10^3 bacteria are present in a culture. Three hours later approximately 10^5 bacteria are present.

(a) Determine the Mathusian model that describes this growth pattern.

(b) Approximately how many minutes does it take for this population to double in size?

98. A radioactive substance decays from 60 grams to 50 grams in 4 years.

(a) Determine the exponential decay function that describes this decay pattern.

(b) How many years will it take for this radioactive substance to decay to 10 grams?

(c) Find the half-life of this radioactive substance.

99. One of the strongest earthquakes ever recorded measured 8.9 on the Richter scale. It occurred in Japan in 1933. How many times more intense was this earthquake than one that measures 6.0 on the Richter scale?

100. Students taking their last French course are given a final exam and the average score of the group is recorded. Each month thereafter, the students are given an equivalent exam, and each time their average score is recorded. It is found that the average score S is a function of time t (in months) and is given by $S(t) = 77 - 17 \ln(t+1)$.

(a) What is the average score on the original final exam?

(b) What is the average score three months later?

(c) How much time elapses before the average score decreases to 40?

(d) Sketch the graph of the function S.

101. When a hot metal object with a temperature of 400 °F is placed in a room whose temperature is 70 °F, the object cools at a rate such that its temperature T is a function of the time t (in minutes) that the object has been in the room and this rate of cooling is given by Newton's cooling law

$$T(t) = 70 + (400 - 70)e^{-kt},$$

where k is a constant.

(a) Determine the value of the constant k if the temperature of the object after 10 minutes is 290 °F.

(b) Find $T(60)$, the temperature of the object after one hour.

(c) How much time has elapsed when $T(t) = 150$ °F?

(d) Sketch the graph of the function T.

102. In a certain forest, the population P of foxes after t years is described by the logistic law

$$P(t) = \dfrac{25000}{100 + 150e^{-0.1t}}.$$

(a) Find $P(10)$, the number of foxes in the forest after ten years.

(b) Discuss the behavior of $P(t)$ as t increases without bound ($t \to \infty$).

(c) Find the time in years when $P(t) = 210$ foxes.

(d) Sketch the graph of the function P.

Cumulative Review Exercises

Chapter 4, 5, 6, and 7

1. Find the zeros of the function f defined by
   $f(x) = a + \dfrac{b}{\ln x}$, where a and b are constants.

2. Find all the roots of the equation $x^5 = 243$ and plot the roots in the complex plane.

3. Identify the graph of each equation.
   (a) $x + y = 9$ (b) $x + y^2 = 9$
   (c) $x^2 + y^2 = 9$ (d) $x^2 - y^2 = 9$
   (e) $2x^2 + y^2 = 9$ (f) $xy = 9$

4. Find the equations of the asymptotes for each of the following hyperbolas.
   (a) $4x^2 - y^2 = 16$ (b) $y^2 - x^2 + 2x = 5$

5. Solve for x:
   (a) $\log_x 3 - \log_4 16 - \log_3 \tfrac{1}{9} - \ln e = 2 \log 1$
   (b) $x^{\ln x} - e^2 x = 0$

6. Use a power reduction formula to help sketch the graph of $y = \sin^2 x$.

7. Simplify each expression to a constant.
   (a) $\sin^2 \theta + \sin \theta \cot \theta \cos \theta$
   (b) $\sin\left(x - \dfrac{\pi}{6}\right) - \cos\left(\dfrac{2\pi}{3} - x\right)$
   (c) $\cot^2 \theta - \dfrac{2 \cot \theta}{\tan 2\theta}$
   (d) $\dfrac{\cos^2(x/2)}{\cos x + \sin^2(x/2)}$

8. Use the properties of logarithms (Section 7.3) to rewrite each expression as a single logarithmic expression. Then apply the fundamental trigonometric identities to simplify your answer. Assume that θ is a first quadratnt angle.
   (a) $\log (\sin \theta) + \log (\csc \theta)$
   (b) $\ln (\cos \theta) - \ln (\sin \theta)$
   (c) $2 \ln (\sin \theta) - \ln (\tan \theta) - \ln (\cos \theta)$
   (d) $\tfrac{1}{2} \log (1 - \sin^2 \theta) + \log (\sec \theta)$

9. Sketch the graphs of the functions $f(x) = 3x$ and $g(x) = xe^x$ on the same set of coordinate axes. Then determine the intersection points of the two graphs.

10. Sketch the graph of the following parabolas. Label the vertex and x- and y-intercepts.
    (a) $x^2 - y - 4x + 3 = 0$
    (b) $y^2 + 4y + x - 1 = 0$
    (c) $x^2 - 2xy + y^2 - 2x - 2y + 1 = 0$

11. Sketch the graphs of the ellipses $x^2 + 2y^2 = 18$ and $9x^2 + 4y^2 - 24y = 0$ on the same coordinate plane. Express the coordinates of their intersection points to three significant digits.

12. When the base is 4, what are the logarithms of the following numbers.
    (a) 64 (b) $\tfrac{1}{16}$ (c) 32 (d) $\tfrac{1}{8}$

13. Show that each trigonometric equation is an identity.
    (a) $(\tan x + \sec x)^2 = \dfrac{1 + \sin x}{1 - \sin x}$
    (b) $\sec^2 x \sin^2 x + (\sin x + \cos x)^2 - \sec^2 x = \sin 2x$
    (c) $1 + \tan x \tan \dfrac{x}{2} = \sec x$
    (d) $\dfrac{1 - \cos 6x}{1 + \cos 6x} = \tan^2 3x$
    (e) $\sqrt{\dfrac{\csc x - 1}{\csc x + 1}} = \dfrac{1 - \sin x}{|\cos x|}$
    (f) $\dfrac{\sin 4x - \sin 2x}{\cos 4x + \cos 2x} = \tan x$

14. For $a > 0$, show that $a^x = e^{x \ln a}$. Use this fact to express each of the following expressions as an exponential with base e.
    (a) 10^x (b) 3^{-x} (c) 5^{3x} (d) 2^{x+1}

15. Given the functions $f(x) = \ln(x + 1)$ and $g(x) = e^{-x} - 1$, find
    (a) $f \circ g$ (b) $g \circ f$
    Then determine the domain of these functions.

16. Given $f(x) = \ln x$, show that the difference quotient
    $\dfrac{f(x + \Delta x) - f(x)}{\Delta x} = \ln\left(1 + \dfrac{\Delta x}{x}\right)^{1/\Delta x}$

17. For each equation, find the solution in the interval $[0, 2\pi)$.
    (a) $\tan 3x = \sqrt{3}$
    (b) $2 \sin x - 3 \csc x + 5 = 0$
    (c) $3 \cos x - \cos 2x = 1$
    (d) $\sin 3x + \sin x + \cos x = 0$
    (e) $\cot 4x + \cot 2x = 0$
    (f) $(25^{\sin x})^{\cot x} = \dfrac{1}{5}$

18. Given the function $f(x) = 2 - e^{-x}$.
    (a) show that f is one-to-one.

459

(b) find the inverse function f^{-1} and state its domain.

19. Expand $(\cos\theta + i\sin\theta)^3$ by using DeMoivre's theorem and also by cubing the expression as you would a binomial. Equate the two results and obtain formulas for $\sin 3\theta$ in terms of $\sin\theta$ and $\cos 3\theta$ in terms of $\cos\theta$.

20. A hot cup of coffee, with an initial temperature of 200 °F, cools in a room with a temperature of 70 °F. After t minutes, the temperature T of the coffee is $T = 70 + 130e^{-0.38t}$. How many *seconds* does it take for the temperature of the coffee to reach 120 °F?

21. Solve each equation for y.

(a) $2\ln x - \dfrac{1}{2}\ln y = 1$ (b) $x + 2 = e^{x - \ln y}$

(c) $2\log_3(y + 1) = 2 - \log_3(y - 1)$

22. Simplify each expression.

(a) $e^{-\ln(\sin x)}$, $0 < x \le \pi/2$

(b) $\ln|\cot x| + \ln|\sin x| - \ln|\cos x|$

23. Another form of a complex number, which is used frequently in calculus, is called the **exponential form.** It is written $re^{i\theta}$, where $e \approx 2.718$, and r and θ have the same meaning as in the *trigonometric form* with θ expressed in radians. The *exponential form* of a complex number is defined as follows:

$$re^{i\theta} = r(\cos\theta + i\sin\theta).$$

For $r = 1$, we have

$$e^{i\theta} = (\cos\theta + i\sin\theta).$$

Show that the usual *laws of exponents* continue to apply to imaginary exponents. That is, show that

(a) $e^{i\theta_1} \cdot e^{i\theta_2} = e^{i(\theta_1 + \theta_2)}$

(b) $\dfrac{e^{i\theta_1}}{e^{i\theta_2}} = e^{i(\theta_1 - \theta_2)}$

(c) $(e^{i\theta})^n = e^{i(n\theta)}$

24. Using the definition of the *exponential form* of a complex number in Exercise 23, show that

(a) $\dfrac{e^{i\theta} + e^{-i\theta}}{2} = \cos\theta$ (b) $\dfrac{e^{i\theta} - e^{-i\theta}}{2} = i\sin\theta$

These identities are usually referred to as **Euler's formula** (named after mathematician Leonhard Euler, 1707–1783).

25. Perform the indicated operations and record the results in standard form.

(a) $(1 - i)^3 (1 + i)^5$ (b) $\dfrac{(-1 + i)^{12}}{(1 + \sqrt{3}\,i)^3 (-\sqrt{3} - i)^6}$

26. Show that each equation is an identity.

(a) $\log(1 - \cos x) + \log(1 + \cos x) = 2\log|\sin x|$

(b) $2\log|\sec x| - \log|\tan x| = \log|\sec x \csc x|$

(c) $-\ln|\csc x - \cot x| = \ln|\csc x + \cot x|$

(d) $-\ln|\sec x - \tan x| = \ln|\sec x + \tan x|$

27. Find the general form of the solution of each equation.

(a) $\ln(2 - \cos^2 x) = 0$

(b) $\log_2(1 - \sin x) = 1 + 2\log_2(\sin x)$

28. Find a polar equation that has the same graph as the given Cartesian equation. Express the answer in the form $r = f(\theta)$.

(a) $2x - 3y = 5$ (b) $y^2 - 6x - 9 = 0$

29. Find the Cartesian equation in x and y for the given parametric equations. Express the answer in the form $y = f(x)$.

(a) $x = e^{t-1}$, $y = 3t$ (b) $x = \ln 2t$, $y = \sin 2t$

30. The number N of golf balls (in dozens) sold by a certain company is a function of the amount x in thousand of dollars that is spent on advertising and is given by

$$N(x) = 2500[1 + \ln(x + 1)].$$

(a) If no money is spent on advertising, how many dozens of balls are sold?

(b) If $500,000 is spent on advertising, approximately how many dozens of balls are sold?

(c) How much money in advertising must be spent to sell 40,000 dozen balls?

(d) Sketch the graph of the function N.

31. Two circular stove pipes that are each 8 inches in diameter are cut at an angle of 45° to form an elbow as shown in the sketch. Find the length of the major and minor axis of the elliptical intersection

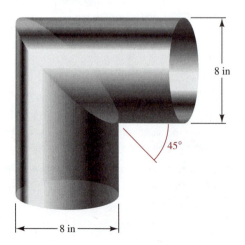

32. The planet Mercury travels in an elliptical orbit with the sun at one focus, as shown in the sketch. The eccentricity of the orbit is 0.205, and the distance from the sun to Mercury at perihelion is 2.85×10^7 miles.

(a) Determine the distance from Mercury to the sun at aphelion.

(b) Find the polar equation of the orbit if the sun is at the pole and its corresponding directrix is perpendicular to the polar axis and to the right of the pole.

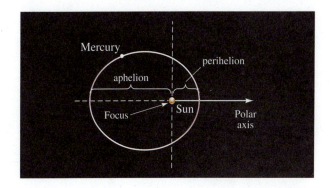

APPENDIX A

Significant Digits

Consider measuring the width of the wooden block in Figure A.1 with a ruler marked in intervals of 0.1-inch. The width of the block appears to be 1.8 inches. However, does the end of the block fall exactly in the middle of the marking for 1.8 inches, or slightly to the left or right of this marking? If we use a powerful magnifying glass, we might attempt to answer this question, but we could never determine the *exact* width of the block. For this reason, we say that every number found by a measuring process is an *approximate number*.

FIGURE A.1
Measuring the width of a block with a ruler marked in 0.1-inch intervals

The number 8 in the tenths position of the number 1.8 inches does have some *significance*, since the width seems to be closer to 1.8 inches than to either 1.7 inches or 1.9 inches. If a digit contributes to our knowledge of how good an approximation is, it is called a **significant digit**. Throughout this text, you are asked to round answers to a certain number of significant digits. The following may be used as a guide for this purpose.

Significant Digits

1. All nonzero digits are significant.

   Examples: 475 has 3 significant digits, 12.827 has 5 significant digits.

2. Zeros between nonzeros digits are significant.

   Examples: 6506 has 4 significant digits, 42.0072 has 6 significant digits.

3. Zeros appearing at the end of a decimal fraction are significant.

   Examples: 7.00 has 3 significant digits, 76.40 has 4 significant digits.

4. Zeros at the beginning of a decimal fraction are *not* significant and serve only to locate the decimal point correctly.

   Examples: 0.002 has 1 significant digit, 0.023 has 2 significant digits.

5. Zeros at the end of an integer are *not* significant unless a tilde (~) is placed above one of the zeros. The tilde is placed over the last zero that is significant.

   Examples: 22,000 has 2 significant digits, 22,0$\tilde{0}$0 has 4 significant digits.

If a number is written in scientific notation as

$$k \times 10^n, \quad \text{where} \quad 1 \leq |k| < 10 \text{ and } n \text{ is an integer,}$$

then the number of significant digits in that number is the same as the number of significant digits in k.

Examples: 6.2×10^3 = 6200 has 2 significant digits.

6.20×10^3 = 62$\tilde{0}$0 has 3 significant digits.

6.2×10^{-3} = 0.0062 has 2 significant digits.

6.20×10^{-3} = 0.00620 has 3 significant digits.

To avoid *rounding errors* when working with approximate numbers, we carry along a few extra significant digits through the calculating process and then round the final answer to the desired accuracy. For example, suppose we wish to find the area of the gable end of the house shown in Figure A.2 and round this answer to three significant digits. First, we find the area of the rectangular part of the gable, without rounding the answer, as follows:

$$A = lw = (24.26 \text{ ft})(9.78 \text{ ft}) \approx 237.2628 \text{ sq ft.}$$

Next, we find the area of the triangular part of the gable, without rounding the answer, as follows:

$$A = \tfrac{1}{2}bh = \tfrac{1}{2}(24.26 \text{ ft})(8.52 \text{ ft}) \approx 103.3476 \text{ sq ft.}$$

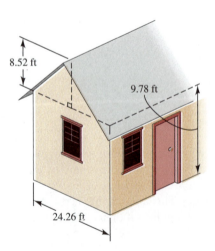

FIGURE A.2
Dimension of the gable end of a house

Finally, we add these areas and then round to the desired accuracy of three significant digits. Hence, the area of the gable end of the house is

$$237.2628 \text{ sq ft} + 103.3476 \text{ sq ft} = 340.6104 \text{ sq ft} \approx 341 \text{ sq ft.}$$

Round to 3 significant digits

If we round the areas of the rectangular and triangular parts to three significant digits and then add these areas, we accumulate a *rounding error*. Note that

$$237 \text{ sq ft} + 103 \text{ sq ft} = 34\tilde{0} \text{ sq ft,}$$

which is not the desired answer. When working with approximate numbers, we usually use a calculator to perform the basic operations. It is best to let the calculator store all the digits and then round the final display on the calculator to the desired accuracy. By using this procedure, you will be sure to avoid rounding errors.

Some exercises in this text that deal with approximate numbers may not specify a required number of significant digits in the final answer. For exercises of this nature, look at the data given in the problem and determine which approximate number has the *fewest* number of significant digits. It is common practice to round the final answer to the same number of significant digits as the approximate number with the fewest number of significant digits. For example, note in Figure A.2 that both 8.52 ft and 9.78 ft have the fewest number of significant digits, namely, three. Hence, it is acceptable to round the area of the gable to three significant digits as well.

APPENDIX B

Asymptotes

◆ Vertical Asymptotes

The reciprocal function $f(x) = 1/x$ (see Section 1.4) is one of the simplest rational functions. Its domain is the set of all real numbers except 0. Using interval notation, we write the domain as $(-\infty, 0) \cup (0, \infty)$. Figure A.3 shows the graph of the reciprocal function. We say that the y-axis (the line $x = 0$) is a *vertical asymptote* of the graph of f, since the graph approaches the y-axis as x gets closer and closer to zero. As x approaches zero from the left ($x \to 0^-$), the value of $f(x)$ decreases without bound $[f(x) \to -\infty]$, and as x approaches zero from the right ($x \to 0^+$), the value of $f(x)$ increases without bound $[f(x) \to \infty]$.

FIGURE A.3
Graph of $f(x) = \frac{1}{x}$

We define a vertical asymptote as follows:

Vertical Asymptotes

The line $x = a$ is a **vertical asymptote** of the graph of a function f if at least one of the following statements is true.

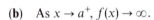

(a) As $x \to a^-$, $f(x) \to \infty$.   (b) As $x \to a^+$, $f(x) \to \infty$.

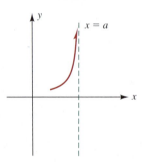 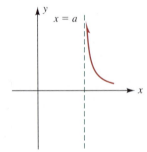

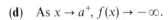

(c) As $x \to a^-$, $f(x) \to -\infty$.   (d) As $x \to a^+$, $f(x) \to -\infty$.

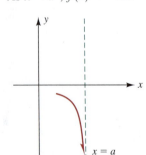 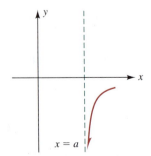

◆ Horizontal Asymptotes

Returning to the reciprocal function, $f(x) = 1/x$, sketched in Figure A.3, we say that the x-axis (the line $y = 0$) is a *horizontal asymptote* of the graph of f, since the graph approaches the x-axis as $|x|$ gets larger and larger. As x increases without bound ($x \to \infty$), $f(x)$ approaches zero through values *greater than* zero $[f(x) \to 0^+]$ and as x decreases without bound ($x \to -\infty$), $f(x)$ approaches zero through values less than zero $[f(x) \to 0^-]$.

We define a horizontal asymptote as follows:

◆ Horizontal Asymptotes

The line $y = b$ is a **horizontal asymptote** of the graph of a function f if at least one of the following statements is true.

(a) As $x \to \infty$, $f(x) \to b^+$.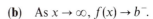

(b) As $x \to \infty$, $f(x) \to b^-$.

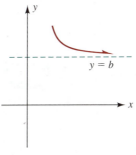
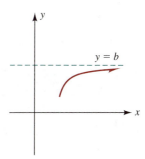

(c) As $x \to -\infty$, $f(x) \to b^+$.

(d) As $x \to -\infty$, $f(x) \to b^-$.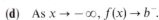

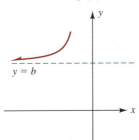
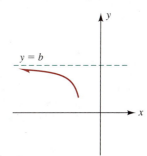

APPENDIX C

Tables

Table 1 Values of e^x and e^{-x}

x	e^x	e^{-x}	x	e^x	e^{-x}	x	e^x	e^{-x}	x	e^x	e^{-x}
0.00	1.0000	1.0000	0.25	1.2840	0.7788	2.0	7.3891	0.1353	4.5	90.017	0.0111
0.01	1.0101	0.9901	0.30	1.3499	0.7408	2.1	8.1662	0.1225	4.6	99.484	0.0101
0.02	1.0202	0.9802	0.35	1.4191	0.7047	2.2	9.0250	0.1108	4.7	109.95	0.0091
0.03	1.0305	0.9704	0.40	1.4918	0.6703	2.3	9.9742	0.1003	4.8	121.51	0.0082
0.04	1.0408	0.9608	0.45	1.5683	0.6376	2.4	11.023	0.0907	4.9	134.29	0.0074
0.05	1.0513	0.9512	0.50	1.6487	0.6065	2.5	12.182	0.0821	5.0	148.41	0.0067
0.06	1.0618	0.9418	0.55	1.7333	0.5769	2.6	13.464	0.0743	5.5	244.69	0.0041
0.07	1.0725	0.9324	0.60	1.8221	0.5488	2.7	14.880	0.0672	6.0	403.43	0.0025
0.08	1.0833	0.9231	0.65	1.9155	0.5220	2.8	16.445	0.0608	6.5	665.14	0.0015
0.09	1.0942	0.9139	0.70	2.0138	0.4966	2.9	18.174	0.0550	7.0	1096.6	0.0009
0.10	1.1052	0.9048	0.75	2.1170	0.4724	3.0	20.086	0.0498	7.5	1808.0	0.0006
0.11	1.1163	0.8958	0.80	2.2255	0.4493	3.1	22.198	0.0450	8.0	2981.0	0.0003
0.12	1.1275	0.8869	0.85	2.3396	0.4274	3.2	24.533	0.0408	8.5	4914.8	0.0002
0.13	1.1388	0.8781	0.90	2.4596	0.4066	3.3	27.113	0.0369	9.0	8103.1	0.0001
0.14	1.1503	0.8694	0.95	2.5857	0.3867	3.4	29.964	0.0334	10.0	22026	0.00005
0.15	1.1618	0.8607	1.0	2.7183	0.3679	3.5	33.115	0.0302			
0.16	1.1735	0.8521	1.1	3.0042	0.3329	3.6	36.598	0.0273			
0.17	1.1853	0.8437	1.2	3.3201	0.3012	3.7	40.447	0.0247			
0.18	1.1972	0.8353	1.3	3.6693	0.2725	3.8	44.701	0.0224			
0.19	1.2092	0.8270	1.4	4.0552	0.2466	3.9	49.402	0.0202			
0.20	1.2214	0.8187	1.5	4.4817	0.2231	4.0	54.598	0.0183			
0.21	1.2337	0.8106	1.6	4.9530	0.2019	4.1	60.340	0.0166			
0.22	1.2461	0.8025	1.7	5.4739	0.1827	4.2	66.686	0.0150			
0.23	1.2586	0.7945	1.8	6.0496	0.1653	4.3	73.700	0.0136			
0.24	1.2712	0.7866	1.9	6.6859	0.1496	4.4	81.451	0.0123			

Table 2 Values of ln x

x	ln x	x	ln x	x	ln x
		4.5	1.5041	9.0	2.1972
0.1	−2.3026	4.6	1.5261	9.1	2.2083
0.2	−1.6094	4.7	1.5476	9.2	2.2192
0.3	−1.2040	4.8	1.5686	9.3	2.2300
0.4	−0.9163	4.9	1.5892	9.4	2.2407
0.5	−0.6931	5.0	1.6094	9.5	2.2513
0.6	−0.5108	5.1	1.6292	9.6	2.2618
0.7	−0.3567	5.2	1.6487	9.7	2.2721
0.8	−0.2231	5.3	1.6677	9.8	2.2824
0.9	−0.1054	5.4	1.6864	9.9	2.2925
1.0	0.0000	5.5	1.7047	10	2.3026
1.1	0.0953	5.6	1.7228	11	2.3979
1.2	0.1823	5.7	1.7405	12	2.4849
1.3	0.2624	5.8	1.7579	13	2.5649
1.4	0.3365	5.9	1.7750	14	2.6391
1.5	0.4055	6.0	1.7918	15	2.7081
1.6	0.4700	6.1	1.8083	16	2.7726
1.7	0.5306	6.2	1.8245	17	2.8332
1.8	0.5878	6.3	1.8405	18	2.8904
1.9	0.6419	6.4	1.8563	19	2.9444
2.0	0.6931	6.5	1.8718	20	2.9957
2.1	0.7419	6.6	1.8871	25	3.2189
2.2	0.7885	6.7	1.9021	30	3.4012
2.3	0.8329	6.8	1.9169	35	3.5553
2.4	0.8755	6.9	1.9315	40	3.6889
2.5	0.9163	7.0	1.9459	45	3.8067
2.6	0.9555	7.1	1.9601	50	3.9120
2.7	0.9933	7.2	1.9741	55	4.0073
2.8	1.0296	7.3	1.9879	60	4.0943
2.9	1.0647	7.4	2.0015	65	4.1744
3.0	1.0986	7.5	2.0149	70	4.2485
3.1	1.1314	7.6	2.0281	75	4.3175
3.2	1.1632	7.7	2.0412	80	4.3820
3.3	1.1939	7.8	2.0541	85	4.4427
3.4	1.2238	7.9	2.0669	90	4.4998
3.5	1.2528	8.0	2.0794	95	4.5539
3.6	1.2809	8.1	2.0919	100	4.6052
3.7	1.3083	8.2	2.1041		
3.8	1.3350	8.3	2.1163		
3.9	1.3610	8.4	2.1282		
4.0	1.3863	8.5	2.1401		
4.1	1.4110	8.6	2.1518		
4.2	1.4351	8.7	2.1633		
4.3	1.4586	8.8	2.1748		
4.4	1.4816	8.9	2.1861		

Table 3 Values of log x

x	0	1	2	3	4	5	6	7	8	9
1.0	.0000	.0043	.0086	.0128	.0170	.0212	.0253	.0294	.0334	.0374
1.1	.0414	.0453	.0492	.0531	.0569	.0607	.0645	.0682	.0719	.0755
1.2	.0792	.0828	.0864	.0899	.0934	.0969	.1004	.1038	.1072	.1106
1.3	.1139	.1173	.1206	.1239	.1271	.1303	.1335	.1367	.1399	.1430
1.4	.1461	.1492	.1523	.1553	.1584	.1614	.1644	.1673	.1703	.1732
1.5	.1761	.1790	.1818	.1847	.1875	.1093	.1931	.1959	.1987	.2014
1.6	.2041	.2068	.2095	.2122	.2148	.2175	.2201	.2227	.2253	.2279
1.7	.2304	.2330	.2355	.2380	.2405	.2430	.2455	.2480	.2504	.2529
1.8	.2553	.2577	.2601	.2625	.2648	.2672	.2695	.2718	.2742	.2765
1.9	.2788	.2810	.2833	.2856	.2878	.2900	.2923	.2945	.2967	.2989
2.0	.3010	.3032	.3054	.3075	.3096	.3118	.3139	.3160	.3181	.3201
2.1	.3222	.3243	.3263	.3284	.3304	.3324	.3345	.3365	.3385	.3404
2.2	.3424	.3444	.3464	.3483	.3502	.3522	.3541	.3560	.3579	.3598
2.3	.3617	.3636	.3655	.3674	.3692	.3711	.3729	.3747	.3766	.3784
2.4	.3802	.3820	.3838	.3856	.3874	.3892	.3909	.3927	.3945	.3962
2.5	.3979	.3997	.4014	.4031	.4048	.4065	.4082	.4099	.4116	.4133
2.6	.4150	.4166	.4183	.4200	.4216	.4232	.4249	.4265	.4281	.4298
2.7	.4314	.4330	.4346	.4362	.4378	.4393	.4409	.4425	.4440	.4456
2.8	.4472	.4487	.4502	.4518	.4533	.4548	.4564	.4579	.4594	.4609
2.9	.4624	.4639	.4654	.4669	.4683	.4698	.4713	.4728	.4742	.4757
3.0	.4771	.4786	.4800	.4814	.4829	.4843	.4857	.4871	.4886	.4900
3.1	.4914	.4928	.4942	.4955	.4969	.4983	.4997	.5011	.5024	.5038
3.2	.5051	.5065	.5079	.5092	.5105	.5119	.5132	.5145	.5159	.5172
3.3	.5185	.5198	.5211	.5224	.5237	.5250	.5263	.5276	.5289	.5302
3.4	.5315	.5328	.5340	.5353	.5366	.5378	.5391	.5403	.5416	.5428
3.5	.5441	.5453	.5465	.5478	.5490	.5502	.5514	.5527	.5539	.5551
3.6	.5563	.5575	.5587	.5599	.5611	.5623	.5635	.5647	.5658	.5670
3.7	.5682	.5694	.5705	.5717	.5729	.5740	.5752	.5763	.5775	.5786
3.8	.5798	.5809	.5821	.5832	.5843	.5855	.5866	.5877	.5888	.5899
3.9	.5911	.5922	.5933	.5944	.5955	.5966	.5977	.5988	.5999	.6010
4.0	.6021	.6031	.6042	.6053	.6064	.6075	.6085	.6096	.6107	.6117
4.1	.6128	.6138	.6149	.6160	.6170	.6180	.6191	.6201	.6212	.6222
4.2	.6232	.6243	.6253	.6263	.6274	.6284	.6294	.6304	.6314	.6325
4.3	.6335	.6345	.6355	.6365	.6375	.6385	.6395	.6405	.6415	.6425
4.4	.6435	.6444	.6454	.6464	.6474	.6484	.6493	.6503	.6513	.6522
4.5	.6532	.6542	.6551	.6561	.6571	.6580	.6590	.6599	.6609	.6618
4.6	.6628	.6637	.6646	.6656	.6665	.6675	.6684	.6693	.6702	.6712
4.7	.6721	.6730	.6739	.6749	.6758	.6767	.6776	.6785	.6794	.6803
4.8	.6812	.6821	.6830	.6839	.6848	.6857	.6866	.6875	.6884	.6893
4.9	.6902	.6911	.6920	.6928	.6937	.6946	.6955	.6964	.6972	.6981
5.0	.6990	.6998	.7007	.7016	.7024	.7033	.7042	.7050	.7059	.7067
5.1	.7076	.7084	.7093	.7101	.7110	.7118	.7126	.7135	.7143	.7152
5.2	.7160	.7168	.7177	.7185	.7193	.7202	.7210	.7218	.7226	.7235
5.3	.7243	.7251	.7259	.7267	.7275	.7284	.7292	.7300	.7308	.7316
5.4	.7324	.7332	.7340	.7348	.7356	.7364	.7372	.7380	.7388	.7396

Table 3 Values of log x (continued)

x	0	1	2	3	4	5	6	7	8	9
5.5	.7404	.7412	.7419	.7427	.7435	.7443	.7451	.7459	.7466	.7474
5.6	.7482	.7490	.7497	.7505	.7513	.7520	.7528	.7536	.7543	.7551
5.7	.7559	.7566	.7574	.7582	.7589	.7597	.7604	.7612	.7619	.7627
5.8	.7634	.7642	.7649	.7657	.7664	.7672	.7679	.7686	.7694	.7701
5.9	.7709	.7716	.7723	.7731	.7738	.7745	.7752	.7760	.7767	.7774
6.0	.7782	.7789	.7796	.7803	.7810	.7818	.7825	.7832	.7839	.7846
6.1	.7853	.7860	.7868	.7875	.7882	.7889	.7896	.7903	.7910	.7917
6.2	.7924	.7931	.7938	.7945	.7952	.7959	.7966	.7973	.7980	.7987
6.3	.7993	.8000	.8007	.8014	.8021	.8028	.8035	.8041	.8048	.8055
6.4	.8062	.8069	.8075	.8082	.8089	.8096	.8102	.8109	.8116	.8122
6.5	.8129	.8136	.8142	.8149	.8156	.8162	.8169	.8176	.8182	.8189
6.6	.8195	.8202	.8209	.8215	.8222	.8228	.8235	.8241	.8248	.8254
6.7	.8261	.8267	.8274	.8280	.8287	.8293	.8299	.8306	.8312	.8319
6.8	.8325	.8331	.8338	.8344	.8351	.8357	.8363	.8370	.8376	.8382
6.9	.8388	.8395	.8401	.8407	.8414	.8420	.8426	.8432	.8439	.8445
7.0	.8451	.8457	.8463	.8470	.8476	.8482	.8488	.8494	.8500	.8506
7.1	.8513	.8519	.8525	.8531	.8537	.8543	.8549	.8555	.8561	.8567
7.2	.8573	.8579	.8585	.8591	.8597	.8603	.8609	.8615	.8621	.8627
7.3	.8633	.8639	.8645	.8651	.8657	.8663	.8669	.8675	.8681	.8686
7.4	.8692	.8698	.8704	.8710	.8716	.8722	.8727	.8733	.8739	.8745
7.5	.8751	.8756	.8762	.8768	.8774	.8779	.8785	.8791	.8797	.8802
7.6	.8808	.8814	.8820	.8825	.8831	.8837	.8842	.8848	.8854	.8859
7.7	.8865	.8871	.8876	.8882	.8887	.8893	.8899	.8904	.8910	.8915
7.8	.8921	.8927	.8932	.8938	.8943	.8949	.8954	.8960	.8965	.8971
7.9	.8976	.8982	.8987	.8993	.8998	.9004	.9009	.9015	.9020	.9025
8.0	.9031	.9036	.9042	.9047	.9053	.9058	.9063	.9069	.9074	.9079
8.1	.9085	.9090	.9096	.9101	.9106	.9112	.9117	.9122	.9128	.9133
8.2	.9138	.9143	.9149	.9154	.9159	.9165	.9170	.9175	.9180	.9186
8.3	.9191	.9196	.9201	.9206	.9212	.9217	.9222	.9227	.9232	.9238
8.4	.9243	.9248	.9253	.9258	.9263	.9269	.9274	.9279	.9284	.9289
8.5	.9294	.9299	.9304	.9309	.9315	.9320	.9325	.9330	.9335	.9340
8.6	.9345	.9350	.9355	.9360	.9365	.9370	.9375	.9380	.9385	.9390
8.7	.9395	.9400	.9405	.9410	.9415	.9420	.9425	.9430	.9435	.9440
8.8	.9445	.9450	.9455	.9460	.9465	.9469	.9474	.9479	.9484	.9489
8.9	.9494	.9499	.9504	.9509	.9513	.9518	.9523	.9528	.9533	.9538
9.0	.9542	.9547	.9552	.9557	.9562	.9566	.9571	.9576	.9581	.9586
9.1	.9590	.9595	.9600	.9605	.9609	.9614	.9619	.9624	.9628	.9633
9.2	.9638	.9643	.9647	.9652	.9657	.9661	.9666	.9671	.9675	.9680
9.3	.9685	.9689	.9694	.9699	.9703	.9708	.9713	.9717	.9722	.9727
9.4	.9731	.9736	.9741	.9745	.9750	.9754	.9759	.9763	.9768	.9773
9.5	.9777	.9782	.9786	.9791	.9795	.9800	.9805	.9809	.9814	.9818
9.6	.9823	.9827	.9832	.9836	.9841	.9845	.9850	.9854	.9859	.9863
9.7	.9868	.9872	.9877	.9881	.9886	.9890	.9894	.9899	.9903	.9908
9.8	.9912	.9917	.9921	.9926	.9930	.9934	.9939	.9943	.9948	.9952
9.9	.9956	.9961	.9965	.9969	.9974	.9978	.9983	.9987	.9991	.9996

| x | 0 | 1 | 2 | 3 | 4 | 5 | 6 | 7 | 8 | 9 |

Note: To evaluate log x for $0 < x < 1$ or $x > 10$, rewrite x using scientific notation and then apply the properties of logarithms. For instance,
1. log 0.25 = log (2.5 × 10^{-1}) = log 2.5 + log 10^{-1} = 0.3979 − 1 = −0.6021
2. log 25 = log (2.5 × 10^1) = log 2.5 + log 10^1 = 0.3979 + 1 = 1.3979

Table 4 Values of the Trigonometric Functions

Angle θ		sin θ	csc θ	tan θ	cot θ	sec θ	cos θ		
Degrees	Radians								
0° 00′	.0000	.0000	No value	.0000	No value	1.000	1.0000	1.5708	90° 00′
10	029	029	343.8	029	343.8	000	000	679	50
20	058	058	171.9	058	171.9	000	000	650	40
30	087	087	114.6	087	114.6	000	1.0000	621	30
40	116	116	85.95	116	85.94	000	.9999	592	20
50	145	145	68.76	145	68.75	000	999	563	10
1° 00′	.0175	.0175	57.30	.0175	57.29	1.000	.9998	1.5533	89° 00′
10	204	204	49.11	204	49.10	000	998	504	50
20	233	233	42.98	233	42.96	000	997	475	40
30	262	262	38.20	262	38.19	000	997	446	30
40	291	291	34.38	291	34.37	000	996	417	20
50	320	320	31.26	320	31.24	001	995	388	10
2° 00′	.0349	.0349	28.65	.0349	28.64	1.001	.9994	1.5359	88° 00′
10	378	378	26.45	378	26.43	001	993	330	50
20	407	407	24.56	407	24.54	001	992	301	40
30	436	436	22.93	437	22.90	001	990	272	30
40	465	465	21.49	466	21.47	001	989	243	20
50	495	494	20.23	495	20.21	001	988	213	10
3° 00′	.0524	.0523	19.11	.0524	19.08	1.001	.9986	1.5184	87° 00′
10	553	552	18.10	553	18.07	002	985	155	50
20	582	581	17.20	582	17.17	002	983	126	40
30	611	610	16.38	612	16.35	002	981	097	30
40	640	640	15.64	641	15.60	002	980	068	20
50	669	669	14.96	670	14.92	002	978	039	10
4° 00′	.0698	.0698	14.34	.0699	14.30	1.002	.9976	1.5010	86° 00′
10	727	727	13.76	729	13.73	003	974	981	50
20	756	756	13.23	758	13.20	003	971	952	40
30	785	785	12.75	787	12.71	003	969	923	30
40	814	814	12.29	816	12.25	003	967	893	20
50	844	843	11.87	846	11.83	004	964	864	10
5° 00′	.0873	.0872	11.47	.0875	11.43	1.004	.9962	1.4835	85° 00′
10	902	901	11.10	904	11.06	004	959	806	50
20	931	929	10.76	934	10.71	004	957	777	40
30	960	958	10.43	963	10.39	005	954	748	30
40	.0989	.0987	10.13	.0992	10.08	005	951	719	20
50	.1018	.1016	9.839	.1022	9.788	005	948	690	10
6° 00′	.1047	.1045	9.567	.1051	9.514	1.006	.9945	1.4661	84° 00′
10	076	074	9.309	080	9.255	006	942	632	50
20	105	103	9.065	110	9.010	006	939	603	40
30	134	132	8.834	139	8.777	006	936	573	30
40	164	161	8.614	169	8.556	007	932	544	20
50	193	190	8.405	198	8.345	007	929	515	10
7° 00′	.1222	.1219	8.206	.1228	8.144	1.008	.9925	1.4486	83° 00′
		cos θ	sec θ	cot θ	tan θ	csc θ	sin θ	Radians	Degrees
								Angle θ	

APPENDIX C Tables

Table 4 Values of the Trigonometric Functions *(continued)*

Angle θ		sin θ	csc θ	tan θ	cot θ	sec θ	cos θ		
Degrees	Radians								
7° 00'	.1222	.1219	8.206	.1228	8.144	1.008	.9925	1.4486	83° 00'
10	251	248	8.016	257	7.953	008	922	457	50
20	280	276	7.834	287	7.770	008	918	428	40
30	309	305	7.661	317	7.596	009	914	399	30
40	338	334	7.496	346	7.429	009	911	370	20
50	367	363	7.337	376	7.269	009	907	341	10
8° 00'	.1396	.1392	7.185	.1405	7.115	1.010	.9903	1.4312	82° 00'
10	425	421	7.040	435	6.968	010	899	283	50
20	454	449	6.900	465	827	011	894	254	40
30	484	478	765	495	691	011	890	224	30
40	513	507	636	524	561	012	886	195	20
50	542	536	512	554	435	012	881	166	10
9° 00'	.1571	.1564	6.392	.1584	6.314	1.012	.9877	1.4137	81° 00'
10	600	593	277	614	197	013	872	108	50
20	629	622	166	644	6.084	013	868	079	40
30	658	650	6.059	673	5.976	014	863	050	30
40	687	679	5.955	703	871	014	858	1.4021	20
50	716	708	855	733	769	015	853	1.3992	10
10° 00'	.1745	.1736	5.759	.1763	5.671	1.015	.9848	1.3963	80° 00'
10	774	765	665	793	576	016	843	934	50
20	804	794	575	823	485	016	838	904	40
30	833	822	487	853	396	017	833	875	30
40	862	851	403	883	309	018	827	846	20
50	891	880	320	914	226	018	822	817	10
11° 00'	.1920	.1908	5.241	.1944	5.145	1.019	.9816	1.3788	79° 00'
10	949	937	164	.1974	5.066	019	811	759	50
20	.1978	965	089	.2004	4.989	020	805	730	40
30	.2007	.1994	5.016	035	915	020	799	701	30
40	036	.2022	4.945	065	843	021	793	672	20
50	065	051	876	095	773	022	787	643	10
12° 00'	.2094	.2079	4.810	.2126	4.705	1.022	.9781	1.3614	78° 00'
10	123	108	745	156	638	023	775	584	50
20	153	136	682	186	574	024	769	555	40
30	182	164	620	217	511	024	763	526	30
40	211	193	560	247	449	025	757	497	20
50	240	221	502	278	390	026	750	468	10
13° 00'	.2269	.2250	4.445	.2309	4.331	1.026	.9744	1.3439	77° 00'
10	298	278	390	339	275	027	737	410	50
20	327	306	336	370	219	028	730	381	40
30	356	334	284	401	165	028	724	352	30
40	385	363	232	432	113	029	717	323	20
50	414	391	182	462	061	030	710	294	10
14° 00'	.2443	.2419	4.134	.2493	4.011	1.031	.9703	1.3265	76° 00'
		cos θ	sec θ	cot θ	tan θ	csc θ	sin θ	Radians	Degrees
								Angle θ	

Table 4 Values of the Trigonometric Functions *(continued)*

Angle θ		sin θ	csc θ	tan θ	cot θ	sec θ	cos θ		Angle θ
Degrees	Radians								
14° 00'	.2443	.2419	4.134	.2493	4.011	1.031	.9703	1.3265	76° 00'
10	473	447	086	524	3.962	031	696	235	50
20	502	476	4.039	555	914	032	689	206	40
30	531	504	3.994	586	867	033	681	177	30
40	560	532	950	617	821	034	674	148	20
50	589	560	906	648	776	034	667	119	10
15° 00'	.2618	.2588	3.864	.2679	3.732	1.035	.9659	1.3090	75° 00'
10	647	616	822	711	689	036	652	061	50
20	676	644	782	742	647	037	644	032	40
30	705	672	742	773	606	038	636	1.3003	30
40	734	700	703	805	566	039	628	1.2974	20
50	763	728	665	836	526	039	621	945	10
16° 00'	.2793	.2756	3.628	.2867	3.487	1.040	.9613	1.2915	74° 00'
10	822	784	592	899	450	041	605	886	50
20	851	812	556	931	412	042	596	857	40
30	880	840	521	962	376	043	588	828	30
40	909	868	487	.2944	340	044	580	799	20
50	938	896	453	.3026	305	045	572	770	10
17° 00'	.2967	.2924	3.420	.3057	3.271	1.046	.9563	1.2741	73° 00'
10	.2996	952	388	089	237	047	555	712	50
20	.3025	.2979	357	121	204	048	546	683	40
30	054	.3007	326	153	172	048	537	654	30
40	083	035	295	185	140	049	528	625	20
50	113	062	265	217	108	050	520	595	10
18° 00'	.3142	.3090	3.236	.3249	3.078	1.051	.9511	1.2566	72° 00'
10	171	118	207	281	047	052	502	537	50
20	200	145	179	314	3.018	053	492	508	40
30	229	173	152	346	2.989	054	483	479	30
40	258	201	124	378	960	056	474	450	20
50	287	228	098	411	932	057	465	421	10
19° 00'	.3316	.3256	3.072	.3443	2.904	1.058	.9455	1.2392	71° 00'
10	345	283	046	476	877	059	446	363	50
20	374	311	3.021	508	850	060	436	334	40
30	403	338	2.996	541	824	061	426	305	30
40	432	365	971	574	798	062	417	275	20
50	462	393	947	607	773	063	407	246	10
20° 00'	.3491	.3420	2.924	.3640	2.747	1.064	.9397	1.2217	70° 00'
10	520	448	901	673	723	065	387	188	50
20	549	475	878	706	699	066	377	159	40
30	578	502	855	739	675	068	367	130	30
40	607	529	833	772	651	069	356	101	20
50	636	557	812	805	628	070	346	072	10
21° 00'	.3665	.3584	2.790	.3839	2.605	1.071	.9336	1.2043	69° 00'
		cos θ	sec θ	cot θ	tan θ	csc θ	sin θ	Radians	Degrees
								Angle θ	

Table 4   Values of the Trigonometric Functions *(continued)*

Angle θ									
Degrees	Radians	sin θ	csc θ	tan θ	cot θ	sec θ	cos θ		
21° 00'	.3665	.3584	2.790	.3839	2.605	1.071	.9336	1.2043	69° 00'
10	694	611	769	872	583	072	325	1.2014	50
20	723	638	749	906	560	074	315	1.1985	40
30	752	665	729	939	539	075	304	956	30
40	782	692	709	.3973	517	076	293	926	20
50	811	719	689	.4006	496	077	283	897	10
22° 00'	.3840	.3746	2.669	.4040	2.475	1.079	.9272	1.1868	68° 00'
10	869	773	650	074	455	080	261	839	50
20	898	800	632	108	434	081	250	810	40
30	927	827	613	142	414	082	239	781	30
40	956	854	595	176	394	084	228	752	20
50	985	881	577	210	375	085	216	723	10
23° 00'	.4014	.3907	2.559	.4245	2.356	1.086	.9205	1.1694	67° 00'
10	043	934	542	279	337	088	194	665	50
20	072	961	525	314	318	089	182	636	40
30	102	.3987	508	348	300	090	171	606	30
40	131	.4014	491	383	282	092	159	577	20
50	160	041	475	417	264	093	147	548	10
24° 00'	.4189	.4067	2.459	.4452	2.246	1.095	.9135	1.1519	66° 00'
10	218	094	443	487	229	096	124	490	50
20	247	120	427	522	211	097	112	461	40
30	276	147	411	557	194	099	100	432	30
40	305	173	396	592	177	100	088	403	20
50	334	200	381	628	161	102	075	374	10
25° 00'	.4363	.4226	2.366	.4663	2.145	1.103	.9063	1.1345	65° 00'
10	392	253	352	699	128	105	051	316	50
20	422	279	337	734	112	106	038	286	40
30	451	305	323	770	097	108	026	257	30
40	480	331	309	806	081	109	013	228	20
50	509	358	295	841	066	111	.9001	199	10
26° 00'	.4538	.4384	2.281	.4877	2.050	1.113	.8988	1.1170	64° 00'
10	567	410	268	913	035	114	975	141	50
20	596	436	254	950	020	116	962	112	40
30	625	462	241	.4986	2.006	117	949	083	30
40	654	488	228	.5022	1.991	119	936	054	20
50	683	514	215	059	977	121	923	1.1025	10
27° 00'	.4712	.4540	2.203	.5095	1.963	1.122	.8910	1.0996	63° 00'
	cos θ	sec θ	cot θ	tan θ	csc θ	sin θ	Radians	Degrees	
							Angle θ		

Table 4 Values of the Trigonometric Functions (continued)

Angle θ		sin θ	csc θ	tan θ	cot θ	sec θ	cos θ		
Degrees	Radians								
27° 00'	.4712	.4540	2.203	.5095	1.963	1.122	.8910	1.0996	63° 00'
10	741	566	190	132	949	124	897	966	50
20	771	592	178	169	935	126	884	937	40
30	800	617	166	206	921	127	870	908	30
40	829	643	154	243	907	129	857	879	20
50	858	669	142	280	894	131	843	850	10
28° 00'	.4887	.4695	2.130	.5317	1.881	1.133	.8829	1.0821	62° 00'
10	916	720	118	354	868	134	816	792	50
20	945	746	107	392	855	136	802	763	40
30	.4974	772	096	430	842	138	788	734	30
40	.5003	797	085	467	829	140	774	705	20
50	032	823	074	505	816	142	760	676	10
29° 00'	.5061	.4848	2.063	.5543	1.804	1.143	.8746	1.0647	61° 00'
10	091	874	052	581	792	145	732	617	50
20	120	899	041	619	780	147	718	588	40
30	149	924	031	658	767	149	704	559	30
40	178	950	020	696	756	151	689	530	20
50	207	.4975	010	735	744	153	675	501	10
30° 00'	.5236	.5000	2.000	.5774	1.732	1.155	.8660	1.0472	60° 00'
10	265	025	1.990	812	720	157	646	443	50
20	294	050	980	851	709	159	631	414	40
30	323	075	970	890	698	161	616	385	30
40	352	100	961	930	686	163	601	356	20
50	381	125	951	.5969	675	165	587	327	10
31° 00'	.5411	.5150	1.942	.6009	1.664	1.167	.8572	1.0297	59° 00'
10	440	175	932	048	653	169	557	268	50
20	469	200	923	088	643	171	542	239	40
30	498	225	914	128	632	173	526	210	30
40	527	250	905	168	621	175	511	181	20
50	556	275	896	208	611	177	496	152	10
32° 00'	.5585	.5299	1.887	.6249	1.600	1.179	.8480	1.0123	58° 00'
10	614	324	878	289	590	181	465	094	50
20	643	348	870	330	580	184	450	065	40
30	672	373	861	371	570	186	434	036	30
40	701	398	853	412	560	188	418	1.0007	20
50	730	422	844	453	550	190	403	.9977	10
33° 00'	.5760	.5446	1.836	.6494	1.540	1.192	.8387	.9948	57° 00'
		cos θ	sec θ	cot θ	tan θ	csc θ	sin θ	Radians	Degrees
								Angle θ	

Table 4   Values of the Trigonometric Functions *(continued)*

Angle θ Degrees	Angle θ Radians	sin θ	csc θ	tan θ	cot θ	sec θ	cos θ		
33° 00′	.5760	.5446	1.836	.6494	1.540	1.192	.8387	.9948	57° 00′
10	789	471	828	536	530	195	371	919	50
20	818	495	820	577	520	197	355	890	40
30	847	519	812	619	511	199	339	861	30
40	876	544	804	661	501	202	323	832	20
50	905	568	796	703	492	204	307	803	10
34° 00′	.5934	.5592	1.788	.6745	1.483	1.206	.8290	.9774	56° 00′
10	963	616	781	787	473	209	274	745	50
20	.5992	640	773	830	464	211	258	716	40
30	.6021	644	766	873	455	213	241	687	30
40	050	688	758	916	446	216	225	657	20
50	080	712	751	.6959	437	218	208	628	10
35° 00′	.6109	.5736	1.743	.7002	1.428	1.221	.8192	.9599	55° 00′
10	138	760	736	046	419	223	175	570	50
20	167	783	729	089	411	226	158	541	40
30	196	807	722	133	402	228	141	512	30
40	225	831	715	177	393	231	124	483	20
50	254	854	708	221	385	233	107	454	10
36° 00′	.6283	.5878	1.701	.7265	1.376	1.236	.8090	.9425	54° 00′
10	312	901	695	310	368	239	073	396	50
20	341	925	688	355	360	241	056	367	40
30	370	948	681	400	351	244	039	338	30
40	400	972	675	445	343	247	021	308	20
50	429	.5995	668	490	335	249	.8004	279	10
37° 00′	.6458	.6018	1.662	.7536	1.327	1.252	.7986	.9250	53° 00′
10	487	041	655	581	319	255	969	221	50
20	516	065	649	627	311	258	951	192	40
30	545	088	643	673	303	260	934	163	30
40	574	111	636	720	295	263	916	134	20
50	603	134	630	766	288	266	898	105	10
38° 00′	.6632	.6157	1.624	.7813	1.280	1.269	.7880	.9076	52° 00′
10	661	180	618	860	272	272	862	047	50
20	690	202	612	907	265	275	844	.9018	40
30	720	225	606	.7954	257	278	826	.8988	30
40	749	248	601	.8002	250	281	808	959	20
50	778	271	595	050	242	284	790	930	10
39° 00′	.6807	.6293	1.589	.8098	1.235	1.287	.7771	.8901	51° 00′
		cos θ	sec θ	cot θ	tan θ	csc θ	sin θ	Radians	Degrees
								Angle θ	

Table 4 Values of the Trigonometric Functions *(continued)*

Angle θ								
Degrees	Radians	sin θ	csc θ	tan θ	cot θ	sec θ	cos θ	
39° 00′	.6807	.6293	1.589	.8098	1.235	1.287	.7771	.8901	51° 00′
10	836	316	583	146	228	290	753	872	50
20	865	338	578	195	220	293	735	843	40
30	894	361	572	243	213	296	716	814	30
40	923	383	567	292	206	299	698	785	20
50	952	406	561	342	199	302	679	756	10
40° 00′	.6981	.6428	1.556	.8391	1.192	1.305	.7660	.8727	50° 00′
10	.7010	450	550	441	185	309	642	698	50
20	039	472	545	491	178	312	623	668	40
30	069	494	540	541	171	315	604	639	30
40	098	517	535	591	164	318	585	610	20
50	127	539	529	642	157	322	566	581	10
41° 00′	.7156	.6561	1.524	.8693	1.150	1.325	.7547	.8552	49° 00′
10	185	583	519	744	144	328	528	523	50
20	214	604	514	796	137	332	509	494	40
30	243	626	509	847	130	335	490	465	30
40	272	648	504	899	124	339	470	436	20
50	301	670	499	.8952	117	342	451	407	10
42° 00′	.7330	.6691	1.494	.9004	1.111	1.346	.7431	.8378	48° 00′
10	359	713	490	057	104	349	412	348	50
20	389	734	485	110	098	353	392	319	40
30	418	756	480	163	091	356	373	290	30
40	447	777	476	217	085	360	353	261	20
50	476	799	471	271	079	364	333	232	10
43° 00′	.7505	.6820	1.466	.9325	1.072	1.367	.7314	.8203	47° 00′
10	534	841	462	380	066	371	294	174	50
20	563	862	457	435	060	375	274	145	40
30	592	884	453	490	054	379	254	116	30
40	621	905	448	545	048	382	234	087	20
50	650	926	444	601	042	386	214	058	10
44° 00′	.7679	.6947	1.440	.9657	1.036	1.390	.7193	.8029	46° 00′
10	709	967	435	713	030	394	173	.7999	50
20	738	.6988	431	770	024	398	153	970	40
30	767	.7009	427	827	018	402	133	941	30
40	796	030	423	884	012	406	112	912	20
50	825	050	418	.9942	006	410	092	883	10
45° 00′	.7854	.7071	1.414	1.000	1.000	1.414	.7071	.7854	45° 00′
		cos θ	sec θ	cot θ	tan θ	csc θ	sin θ	Radians	Degrees
								Angle θ	

Solutions to Problems and Answers to Odd-Numbered Exercises

CHAPTER 1

Section 1.1

Problems

1. (a) $a \leq 8$ (b) $-4 \leq b < 0$

2. (a) $BA = \left|\dfrac{5}{3} - \dfrac{-13}{4}\right| = \left|\dfrac{59}{12}\right| = \dfrac{59}{12}$

   (b) $BA = |\sqrt{3} - 2| = -(\sqrt{3} - 2) = 2 - \sqrt{3}$, since $\sqrt{3} - 2 < 0$

3. (a) The interval $(3, \infty)$ represents all real numbers greater than 3. We show the graph of this set of real numbers as follows:

   (b) The interval $[4, 9)$ represents all real numbers greater than or equal to 4 but less than 9. We show the graph of this set of real numbers as follows:

4. (a) The graph of the set of real numbers described by $[-2, -1) \cup (-1, 2]$ is as follows:

   (b) The graph of the set of real numbers described by $(-\infty, -3] \cup [3, 5) \cup (5, \infty)$ is as follows:

Exercises

1. $x < 0$ 3. $a \leq 7$ 5. $2 < p \leq 10$
7. $0 < c < 8$ 9. $-2 \leq t \leq 0$ 11. 15   13. 1.9
15. $\dfrac{5}{2}$ 17. $9\dfrac{1}{6}$ 19. $\pi - \sqrt{2}$ 21. $[-4, -2]$
23. $(2, 6)$ 25. $[-1, 4)$ 27. $(-\infty, -1]$
29. $(10, \infty)$ 31. $(-\infty, 0) \cup (0, \infty)$
33. $(-\infty, -7] \cup [-4, \infty)$ 35. $[0, 2) \cup (2, \infty)$
37. $(-\infty, 1) \cup (1, 6) \cup (6, \infty)$
39. $(-\infty, 0) \cup [2, 3) \cup (3, \infty)$

A17

Answers to Odd Numbered Exercises

41.
43. (graph)
45. (graph)
47. (graph)
49. (graph)

51. $x - \pi$ 53. 1 55. $-5 < x < 5$
57. $|a - 7| \geq 3$ 59. $|1 - d| < |d|$ 61. The interval notation $[a, a]$ represents the single point a on the real number line. The interval notation (a, a) is meaningless.

63. $(-\infty, -1) \cup (0, 1)$ 65. $\dfrac{157}{50}, \pi, \dfrac{22}{7}, 3.145, \sqrt{10}, 3.2$

67. 21,311,963 69. 3.4299×10^{13}

Section 1.2

Problems

1.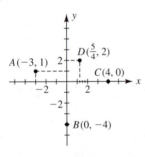

2. (a) $\left(\dfrac{-3+1}{2}, \dfrac{1+(-2)}{2}\right) = \left(-1, -\dfrac{1}{2}\right)$

   (b) $AB = \sqrt{[1-(-3)]^2 + (-2-1)^2} = \sqrt{25} = 5$

3. $AB = \sqrt{[2-(-2)]^2 + (5-3)^2} = \sqrt{20} = 2\sqrt{5}$

   $AC = \sqrt{(4-2)^2 + (1-5)^2} = \sqrt{20} = 2\sqrt{5}$

   $BC = \sqrt{[4-(-2)]^2 + (1-3)^2} = \sqrt{40} = 2\sqrt{10}$

   Since $AB = AC$, we conclude that triangle ABC is isosceles. Also, since $(AB)^2 + (AC)^2 = (BC)^2$, we conclude that triangle ABC is a right triangle with hypotenuse $\overline{BC}$.

   Area $= \dfrac{1}{2}(2\sqrt{5})(2\sqrt{5}) = 10$ square units.

4.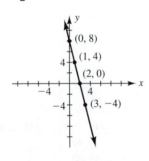

5.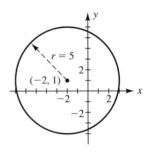

6.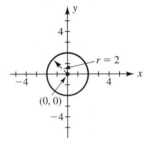

7. Completing the square, we have
$$(x+2)^2 + (y+3)^2 = 0$$
Since $r^2 = 0$, the graph of this equation is the single point $(-2, -3)$.

Exercises

1. III 3. positive, negative 5. x-axis
7. $A(2, 1), B(-4, 4), C(-3, -1), D(4, -2), E(3, 0), F(0, -3)$
9. (a) 5 (b) $\left(\dfrac{5}{2}, 4\right)$ 11. (a) $\sqrt{34}$
   (b) $\left(-\dfrac{1}{2}, \dfrac{11}{2}\right)$ 13. (a) $\sqrt{113}$ (b) $\left(-1, -\dfrac{3}{2}\right)$
15. (a) $\dfrac{5}{12}$ (b) $\left(-\dfrac{5}{8}, \dfrac{5}{6}\right)$ 17. (a) $3\sqrt{2}$
   (b)   19. $d = \sqrt{x^2 + y^2}$
21. $3\sqrt{13} + 3\sqrt{5}$ 23. $(AB)^2 = (AC)^2 + (BC)^2$, where $AB = 5\sqrt{2}, AC = 2\sqrt{10}, BC = \sqrt{10}$; area $= 10$ square units
25. $AB = AC = BC = 2\sqrt{2}$; area $= 2\sqrt{3}$ square units
27. $\left(-1, \dfrac{3}{2}\right), (2, -3), \left(5, -\dfrac{15}{2}\right)$ 29. $2\sqrt{17}, 5\sqrt{2}, \sqrt{74}$
31.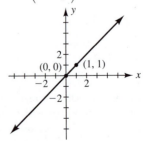

33.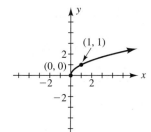

35.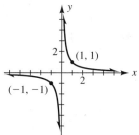

37.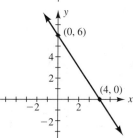

39.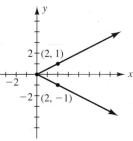

41.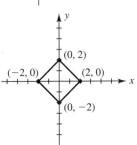

43.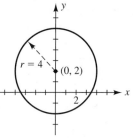

45.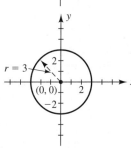

47.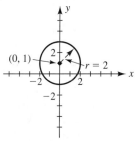

49.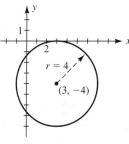

51.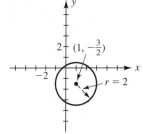

53. The equation does not define a circle and has no graph.
55. $x^2 + y^2 = 4$ 57. $x^2 + y^2 - 4x + 6y + 11 = 0$
59. $x^2 + y^2 = 25$ 61. $x^2 + y^2 - 8y + 11 = 0$
63. $\left(\dfrac{a}{2}, \dfrac{b}{2}\right)$ 65. $\dfrac{\sqrt{a^2 + b^2}}{2}$ 67. $(a + b, c)$
69. $\left(\dfrac{a+b}{2}, \dfrac{c}{2}\right)$ 71. The graph of $y = |x| + c$, where $c > 0$, is the same as the graph of $y = |x|$, but is shifted vertically upward c units.   73. The graph of $y = \sqrt{x + c}$, where $c < 0$, is the same as the graph of $y = \sqrt{x}$, but is shifted horizontally to the right $|c|$ units.
75. (a) $x^2 + y^2 - 10x + 6y + 25 = 0$ (b) 27π
77. (a) Approximately 4.10 (b) (1.28, 1.60)
79. (a) Approximately 90.0 (b) $(-23.45, -38.4)$
81. Radius ≈ 4.33 units, area ≈ 59.0 square units
83. Center is (3.61, 2.42); radius is 2.56; x-intercepts are 2.78 and 4.44; no y-intercept   85. (a) $(x - 8.2)^2 + y^2 = 14.3^2$ (b) 12.4 ft

Section 1.3

Problems

1. Solving the equation for y, we obtain $y = 2 - \frac{2}{3}x$. Since there is one and only one output y for each input value x that we choose, we conclude that $2x + 3y = 6$ defines y as a function of x. Also, note that the graph of this equation passes the vertical line test.

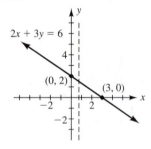

2. $G(x - h) = (x - h)^2 - 4(x - h) + 3$
$= x^2 - 2xh + h^2 - 4x + 4h + 3$

3. The radicand $2 - x$ must be positive. Thus,
$$2 - x > 0$$
$$-x > -2$$
$$x < 2.$$
Hence, the domain is $(-\infty, 2)$.

4. $h(-x) = (-x) - (-x)^2 = -x - x^2$.
Since $h(-x) \neq h(x)$ and $h(-x) \neq -h(x)$, we conclude that the function h is neither even nor odd.

5. Domain $(-\infty, \infty)$; range $(-\infty, \infty)$

Exercises

1. Defines y as a function of x 3. Does not define y as a function of x 5. Defines y as a function of x 7. Does not define y as a function of x 9. Defines y as a function of x 11. Defines y as a function of x 13. Defines y as a function of x 15. 13   17. 5   19. $9 - 2\sqrt{2}$
21. $|2ab + 3|$ 23. $t - 4$ 25. $n^2 - n + 1$
27. $\sqrt{x}$ 29. $\sqrt{1 + x^2} - 4$ 31. $4x - 2\sqrt{x - 2} - 7$
33. $4x - 34\sqrt{x} + 73$ 35. $(-\infty, \infty)$ 37. $(-\infty, \infty)$
39. $(-\infty, 4]$ 41. $[-4, 4]$ 43. $(-\infty, -2) \cup (-2, \infty)$
45. $(-\infty, -2) \cup (-2, 2) \cup (2, \infty)$
47. $(-\infty, -5) \cup (-5, 2) \cup (2, \infty)$
49. $\left[\dfrac{1}{2}, 4\right) \cup (4, \infty)$ 51. Odd   53. Even
55. Odd   57. Even   59. Neither   61. $\dfrac{5}{3}$
63. $8, -5$ 65. $4 \pm 2\sqrt{3}$ 67. $\dfrac{3}{2}$
69. ± 2 71. (a) $(-\infty, \infty)$ (b) $[-2, \infty)$ (c) ± 2
(d) Even   73. (a) $(-\infty, \infty)$ (b) $[-1, \infty)$
(c) $0, -2$ (d) Neither   75. (a) $[-\sqrt{5}, \sqrt{5}]$
(b) $[0, \sqrt{5}]$ (c) $\pm\sqrt{5}$ (d) Even   77. (a) $(-\infty, \infty)$
(b) $[-2, 0)$ (c) None (d) Even   79. (a) $[-2, 2]$
(b) $[-22, 22]$ (c) $0, \pm\sqrt[4]{5}$ (d) Odd   81. 2
83. $2x + \Delta x$ 85. $2x + 2 + \Delta x$ 87. $\dfrac{-1}{x(x + \Delta x)}$
89. (a) $r(25) = 7$ m, which represents the radius of the oil spill after 25 min (b) 13 min   91. The population is one-fourth as large   93. 90.6   95. -0.915
97. 2.40   99. ± 1.83 101. (a) $C(12.4) = \$19.48$, which represents the cost of a taxi cab fare when the cab is driven 12.4 miles (b) 22.4 mi

Section 1.4

Problems

1. The graph of $h(x) = x^3 - 1$ is obtained from the graph of $f(x) = x^3$ shifted vertically *downward* 1 unit:

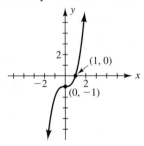

2. The graph of $h(x) = (x - 1)^3$ is obtained from the graph of $f(x) = x^3$ shifted horizontally to the *right* 1 unit:

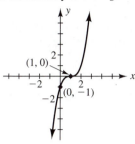

3. The graph of $F(x) = (x + 1)^2 - 4$ is obtained from the graph of $f(x) = x^2$ shifted vertically downward 4 units and horizontally to the left 1 unit:

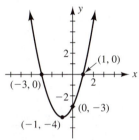

4. The graph of $F(x) = -|x + 2|$ is obtained by reflecting the graph of $f(x) = |x|$ about the x-axis and then shifting this graph to the left 2 units:

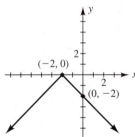

5. The graph of $F(x) = 4x - 1$ is obtained by stretching the graph of $f(x) = x$ by a factor of 4 and then shifting this graph downward 1 unit.

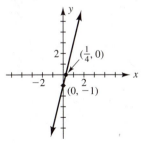

6. Referring to Figure 1.39, note that as x increases, $F(x)$ decreases for all x in the domain of F. Hence, we conclude that the function F is a decreasing function.

Exercises

1.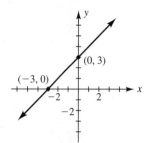

3.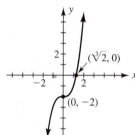

5.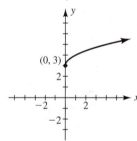

7.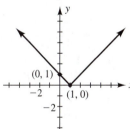

9.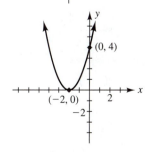

11.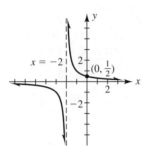

13.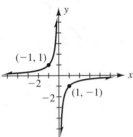

15.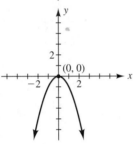

17.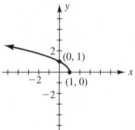

19.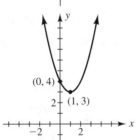

21.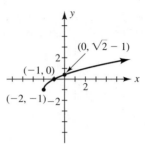

23.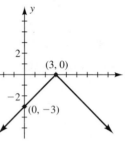

25.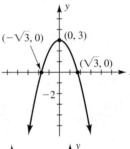

27.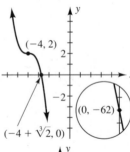

29.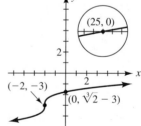

31.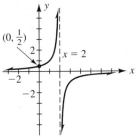

33.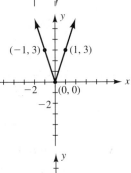

35.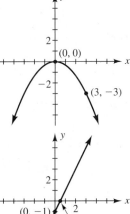

37.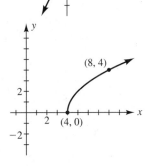

39.

41.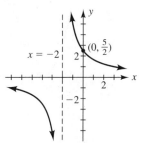

43. Increasing **45.** Decreasing **47.** Neither
49. Increasing
51.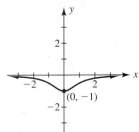

53.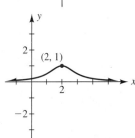

55. $y = 1 + \sqrt{4 - x^2}$ **57.** $y = -2 + \sqrt{4 - (x+1)^2}$
59. (a)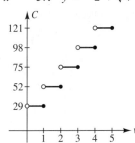

(b) 75¢
61.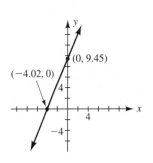

63.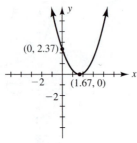

65.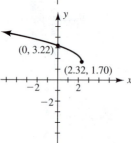

67. Horizontal shift rule

Section 1.5

Problems

1. $(f \circ g)(0) = f(g(0)) = f(0) = -2$

2. (a) $f(9) = 3$. Hence,
$(g \circ f)(9) = g(f(9)) = g(3) = 3^2 - 4 = 5$
(b) $(g \circ f)(9) = 9 - 4 = 5$

3. $f(g(x)) = f(\sqrt[5]{x} + 2) = [(\sqrt[5]{x} + 2) - 2]^5 = (\sqrt[5]{x})^5 = x$
$g(f(x)) = g[(x - 2)^5] = \sqrt[5]{(x - 2)^5} + 2 = (x - 2) + 2 = x$

4. $H(a) = H(b)$ implies $a^2 - 1 = b^2 - 1$
$a^2 = b^2$
$a = b$, if $a \geq 0, b \geq 0$

Also, note that the graph of $H(x) = x^2 - 1$ with $x \geq 0$ passes the horizontal line test.

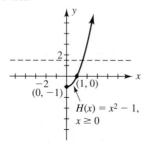

5. $f(f^{-1}(x)) = f(\sqrt[3]{x+1})$
$= (\sqrt[3]{x+1})^3 - 1 = (x + 1) - 1 = x$
$f^{-1}(f(x)) = f^{-1}(x^3 - 1) = \sqrt[3]{(x^3 - 1) + 1} = \sqrt[3]{x^3} = x$

6. Interchanging x and y and solving for y, we obtain
$$x = \frac{1}{y + 1}$$
$$y + 1 = \frac{1}{x}$$
$$y = \frac{1}{x} - 1 \text{ is the inverse function.}$$

7. $f[f^{-1}(x)] = 4 - [f^{-1}(x)]^2$
$x = 4 - [f^{-1}(x)]^2$
$[f^{-1}(x)]^2 = 4 - x$
$f^{-1}(x) = \sqrt{4 - x}$

Exercises

1. 0   **3.** Undefined   **5.** 2   **7.** 1

9. (a) $(f \circ g)(x) = 2x^2 + 3$ (b) $(-\infty, \infty)$

11. (a) $(G \circ f)(x) = \dfrac{1}{2x + 1}$ (b) $\left(-\infty, -\dfrac{1}{2}\right) \cup \left(-\dfrac{1}{2}, \infty\right)$

13. (a) $(F \circ h)(x) = \sqrt{x - 4}$ (b) $[4, \infty)$

15. (a) $(G \circ H)(x) = \dfrac{x - 1}{2}$ (b) $(-\infty, 1) \cup (1, \infty)$

17. (a) $(H \circ f)(x) = \dfrac{1}{x}$ (b) $(-\infty, 0) \cup (0, \infty)$

19. (a) $(f \circ f)(x) = 4x + 3$ (b) $(-\infty, \infty)$ **21.** $\dfrac{1}{2}$

33. (a) One-to-one   **35.** (a) Not one-to-one
(b) $[0, \infty)$ is one restriction.   **37.** (a) One-to-one

39. (a) Not one-to-one   (b) $[-2, \infty)$ is one restriction.

41. (a) Not one-to-one   (b) $[2, \infty)$ is one restriction.

43. (a) $f^{-1}(x) = \dfrac{x}{2}$

(b)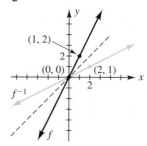

45. (a) $f^{-1}(x) = 3 - x$

(b)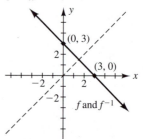

47. (a) $f^{-1}(x) = \sqrt[3]{2-x}$
(b)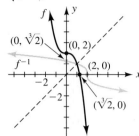

49. (a) $f^{-1}(x) = x^3 + 4$
(b)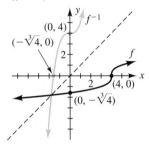

51. (a) $f^{-1}(x) = \dfrac{1}{x} - 1$
(b)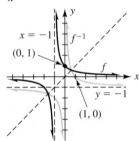

53. (a) $f^{-1}(x) = (x-1)^2 + 4$, $x \geq 1$
(b)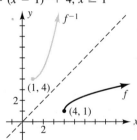

55. (a) $f^{-1}(x) = \sqrt{x} + 1$
(b)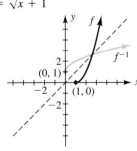

57. (a) $f^{-1}(x) = \sqrt{9-x^2}$, $0 \leq x \leq 3$
(b)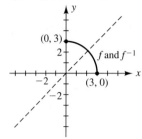

59. (a) $f^{-1}(x) = -\sqrt{x-3}$
(b)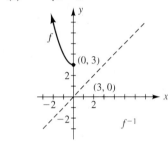

61. 4 **63.** 3 **65.** 2

67.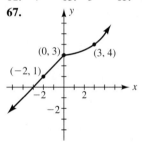

x	-2	0	3
$f^{-1}(x)$	1	3	4

69.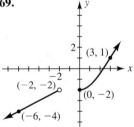

x	-2	0	3
$f^{-1}(x)$	undefined	-2	1

71. $f^{-1}(4) = 14$ **73.** $(f^{-1} \circ g)(-2) = -13$

75. $(f \circ g)^{-1}(x) = (g^{-1} \circ f^{-1})(x) = \dfrac{3x+1}{3}$

77. $(-\infty, 0) \cup (0, \infty)$ **79.** $\left(-\infty, \dfrac{2}{3}\right) \cup \left(\dfrac{2}{3}, \infty\right)$

81. $f(x) = \dfrac{1}{3}x - 2$ **83.** $g^{-1}(r) = \dfrac{\sqrt{\pi r}}{2\pi}$, $r > 0$

85. $(f \circ g^{-1})(S) = \dfrac{S\sqrt{\pi S}}{6\pi}$, the volume of a sphere as a function of its surface area. **87.** -0.684 **89.** 1.92 **91.** 0.173 **93.** (a) f and g are inverses, since $(f \circ g)(x) = x$. (b) f and g are inverses, since $(f \circ g)(x) = x$.

Section 1.6

 Problems

1. The maximum area of 2500 square feet occurs when $l = 50$ ft and $w = 100 - l = 50$ ft.

2. The volume V of water (in cubic inches) is given by $V = 8t$, where t is time (in minutes).
The time t it takes to fill the funnel is $\dfrac{64\pi \text{ cu in}}{8 \text{ cu in/min}} = 8\pi$ min.
Hence, the domain of $V = 8t$ is $[0, 8\pi]$.

3. $T = \dfrac{168{,}000}{40} = \$4200$

4. $t = \dfrac{240}{50} = 4.8$ hr

5. Since the electrical resistance R of a wire varies directly as its length l and inversely as the square of its radius r, we write $R = \dfrac{kl}{r^2}$. Doubling both l and r, we obtain

$$R = \dfrac{k(2l)}{(2r)^2} = \dfrac{1}{2} \cdot \dfrac{kl}{r^2}.$$

Thus, the resistance becomes one half as large.

 Exercises

1. $d = \dfrac{c}{\pi}$ **3.** (a) $C = 40 + \dfrac{n}{5}$ **5.** (a) $S = \dfrac{d}{\sqrt{2}}$
(b) $A = \dfrac{d^2}{2}$ (c) $P = 2\sqrt{2}\,d$ **7.** $W = 125{,}000 + 4000t$
9. (a) $V = \dfrac{4}{3}\pi h^3$ (b) $V = \dfrac{1}{6}\pi r^3$
11. (a) $d = \sqrt{2 - 2x}$ (b) $[-1, 1]$
13. (a) $A = \dfrac{x\sqrt{x}}{2}$ (b) $P = x + \sqrt{x} + \sqrt{x^2 + x}$
15. $S = 2x^2 + \dfrac{256}{x}$ **17.** (a) $V = 9h^2$
(b) Domain $[0, 4]$

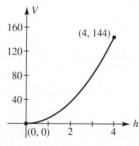

19. (a) $d = 30\sqrt{9 + t^2}$ (b) Domain $[0, 3]$; range $[90, 90\sqrt{2}\,]$ **21.** (a) $d = \dfrac{F}{50}$
(b) Domain $[0, 2400]$

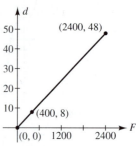

23. (a) $W = 125l$
(b) Domain $[0, \infty)$

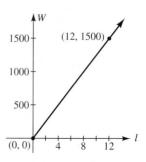

25. (a) $N = \dfrac{10{,}000}{d}$
(b) Domain $(0, \infty)$

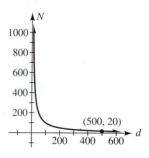

27. (a) $w = \dfrac{300}{f}$
(b) Domain $(0, \infty)$

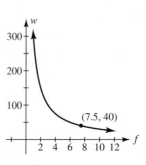

29. (a) $F = \dfrac{160{,}000}{d^2}$

(b) Domain $(0, \infty)$

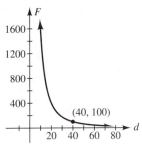

23.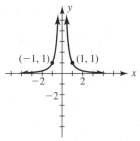

31. The force is tripled.   **33.** $A = 30r - 2r^2 - \dfrac{\pi r^2}{2}$

35. $d = \begin{cases} 30t & \text{, if } t \le 3 \\ 30\sqrt{9 + (t-3)^2} & \text{, if } 3 < t \le 6 \end{cases}$ **37.** 25 sec

39. (a) $A = \dfrac{x\sqrt{262.44 - x^2}}{2}$ (b) Approximately 64.6 sq units   **41.** (a) $A = 1000x - 2x^2$ (b) $(0, 500)$
(c) 250 ft by 500 ft   **43.** (a) $T \approx 2.007\sqrt{l}$
(b) Approximately 3.11 seconds

25.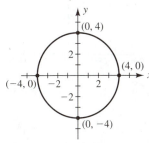

27.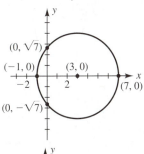

Chapter 1 Review Exercises

1. $[3, 7]$ **3.** $(-1, \infty)$ **5.** $(-\infty, -6] \cup [-1, \infty)$
7. $[0, 4) \cup (4, \infty)$ **9.** (a) 5   (b) $\left(2, \dfrac{9}{2}\right)$
(c) $(x - 2)^2 + \left(y - \dfrac{9}{2}\right)^2 = \dfrac{25}{4}$ **11.** (a) $2\sqrt{5}$
(b) $(0, 2)$ (c) $x^2 + (y - 2)^2 = 5$ **13.** (a) $\sqrt{97}$
(b) $\left(\dfrac{3}{2}, -3\right)$ (c) $\left(x - \dfrac{3}{2}\right)^2 + (y + 3)^2 = \dfrac{97}{4}$
15. $AB = BC = CD = DA = 2\sqrt{2}$; perimeter $= 8\sqrt{2}$; area $= 8$ **17.** $\dfrac{a}{2}$, half as long as OA
19. $\dfrac{\sqrt{(a - b)^2 + c^2}}{2}$, half as long as AB
21.

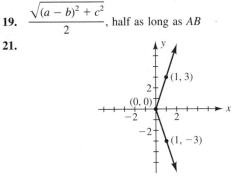

29.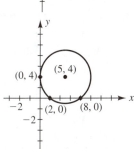

31. 1   **33.** 5   **35.** $2x^2 - 13x + 21$ **37.** -5
39. 5   **41.** $18x^4 + 45x^2 + 28$ **43.** $\dfrac{3x + 1}{x - 1}$
45. $\dfrac{x - 2}{3}$ **47.** 2   **49.** $4x - 1 + 2\Delta x$
51. (a) $(-\infty, \infty)$ (b) -3 (c) Neither
(d)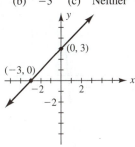

(e) Increasing on $(-\infty, \infty)$ (f) One-to-one
(g) $(-\infty, \infty)$
53. (a) $(-\infty, \infty)$ (b) 0   (c) Odd
(d)
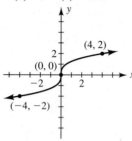

(e) Increasing on $(-\infty, \infty)$ (f) One-to-one
(g) $(-\infty, \infty)$
55. (a) $(-\infty, \infty)$ (b) ± 3 (c) Even
(d)
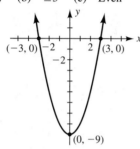

(e) Decreasing: $(-\infty, 0)$; increasing: $(0, \infty)$ (f) Not one-to-one   (g) $[-9, \infty)$
57. (a) $(-\infty, \infty)$ (b) -3 (c) Neither
(d)
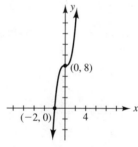

(e) Decreasing: $(-\infty, -3)$; increasing: $(-3, \infty)$ (f) Not one-to-one   (g) $[0, \infty)$
59. (a) $(-\infty, \infty)$ (b) -2 (c) Neither
(d)

(e) Increasing on $(-\infty, \infty)$ (f) One-to-one
(g) $(-\infty, \infty)$
61. (a) $(-\infty, \infty)$ (b) 2 and 4   (c) Neither
(d)
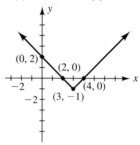

(e) Decreasing: $(-\infty, 3)$; increasing: $(3, \infty)$ (f) Not one-to-one   (g) $[-1, \infty)$
63. (a) $(-\infty, \infty)$ (b) 0 and 6   (c) Neither
(e) Increasing: $(-\infty, 3)$; decreasing: $(3, \infty)$ (f) Not one-to-one   (g) $(-\infty, 9]$
65. (a) $[-10, 10]$ (b) ± 10 (c) Even
(e) Increasing: $[-10, 0)$; decreasing: $(0, 10]$ (f) Not one-to-one   (g) $[0, 10]$
67. (a) $(-\infty, \infty)$ (b) 0 and $\pm\sqrt{2}$ (c) Even
(e) Increasing: $(-\infty, -1) \cup (0, 1)$; decreasing: $(-1, 0) \cup (1, \infty)$ (f) Not one-to-one   (g) $(-\infty, 1]$
69. (a) $(-\infty, 3]$ (b) None   (c) Neither
(d)
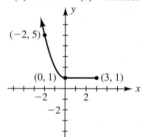

(e) Decreasing: $(-\infty, 0]$; constant: $(0, 3]$ (f) Not one-to-one   (g) $[1, \infty)$
71.
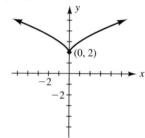

73.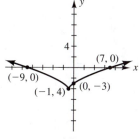

75.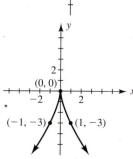

77. (a) $f^{-1}(x) = \dfrac{x}{2}$

(b)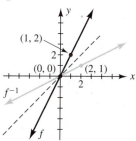

79. (a) $f^{-1}(x) = \sqrt[3]{3-x}$

(b)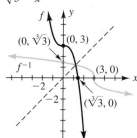

81. (a) $f^{-1}(x) = x^2 + 2,\ x \geq 0$

(b)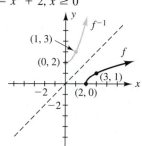

83. (a) $f^{-1}(x) = -\sqrt{4-x}$

(b)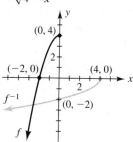

85. $(-\infty, 3]$ 87. 3 89. 4 91. Yes; (a) -2

(b) 0 (c) -6 93. $f^{-1}(x) = \dfrac{a - xb}{x - 1}$

95. (a) $63.43 (b) 425.6 miles

97. $C = \begin{cases} 0.015n \text{ if } n \leq 1000 \\ 15 + 0.02(n - 1000) \text{ if } n > 1000 \end{cases}$

99. (a) $A = 3x - x^3$ (b) Domain $[0, \sqrt{3}]$

101. (a) $T = \dfrac{V}{25}$ (b) Domain $[0, \infty)$

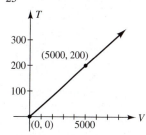

103. (a) $G = \dfrac{n(2000 - n)}{19{,}200}$ (b) Approximately 39 field mice per acre

CHAPTER 2

Section 2.1

Problems

1. Angles coterminal with 600° are all of the form $(600 + 360n)°$, where n is an integer. Some *positive* angles coterminal with 600° are 240°, 960°, 1320°, and so on. Some *negative* angles coterminal with 600° are $-120°$, $-480°$, $-840°$, and so on.

2. (a) Note that $0.4075° = 0.4075°\left(\dfrac{60'}{1°}\right) = 24.45'$ and $0.45' = 0.45'\left(\dfrac{60''}{1'}\right) = 27''$. Hence, $16.4075° = 16°24'27''$.

(b) $123°25'23'' = 123° + 25'\left(\dfrac{1°}{60'}\right) + 23''\left(\dfrac{1°}{3600''}\right)$
$\approx 123.423°$.

3. Angles coterminal with $\frac{9\pi}{4}$ are of the form $\frac{9\pi}{4} + 2\pi n$, where n is an integer. By letting $n = -1$, we find the smallest positive coterminal angle to be $\pi/4$.

4. Converting to radians, we have
$$125° = 125° \left(\frac{\pi}{180°}\right) = \frac{25\pi}{36}$$

Thus,
$$s = \theta r = \left(\frac{25\pi}{36}\right)(48.0) \approx 104.7 \text{ ft.}$$

5. We first convert to feet per second:
$$45 \text{ mph} = 45\, \frac{\text{mi}}{\text{h}} \left(\frac{5280 \text{ ft}}{1 \text{ mi}}\right)\left(\frac{1 \text{ h}}{3600 \text{ sec}}\right) = 66 \text{ ft/s.}$$

Thus, the angular speed ω is
$$\omega = \frac{66 \text{ ft/s}}{1.25 \text{ ft}} = 52.8 \text{ rad/s.}$$

Hence,
$$\theta = \left(52.8\,\frac{\text{rad}}{\text{sec}}\right)\left(\frac{60 \text{ sec}}{1 \text{ min}}\right)\left(\frac{1 \text{ rev}}{2\pi \text{ rad}}\right) \approx 504 \text{ rpm.}$$

◆ Exercises

1.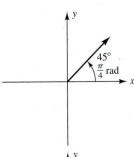

3.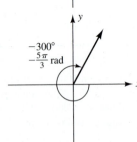

5.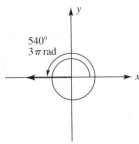

7.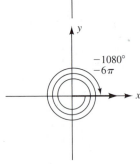

9. 220°   11. 336°   13. 280°   15. 54°38′
17. π 19. $\frac{\pi}{4}$ 21. $\frac{\pi}{3}$ 23. $8 - 2\pi$ 25. $\frac{\pi}{3}$
27. $\frac{5\pi}{4}$ 29. $\frac{12\pi}{5}$ 31. $-\frac{25\pi}{12}$ 33. 225°
35. $-10°$ 37. $292\frac{1}{2}°$ 39. 2160°
41. (a) 6π cm   (b) 14π cm   (c) 4π cm   (d) 9π cm
43. 600π ft   45. 1 rad   47. (a) $\frac{\pi}{30}$ rad/s
(b) $\frac{\pi}{1800}$ rad/s   49. (a) $\frac{200\pi}{3}$ rad/min   (b) $\frac{5\pi}{9}$ ft/s
51. Since 1 rad $\approx$ 57.3°, we conclude that 1 radian is larger than 1°.   53. 10π cm
55. (c) sector (i) $\frac{27\pi}{2}$ sq m   sector (ii) 27 sq ft
57. (a) 22.6°   (b) 0.39   59. (a) 215.48°   (b) 3.76
61. (a) 32°18′36″   (b) 0.56   63. (a) 306°07′21″
(b) 5.34   65. (a) 85.9437°   (b) 85°56′37″
67. (a) 572.9578°   (b) 572°57′28″
69. about 84.9 cm   71. 86.4 cm

Section 2.2

◆ Problems

1. If $x = -2$ and $y = 4$, then
$r = \sqrt{(-2)^2 + (4)^2} = \sqrt{20} = 2\sqrt{5}$. Hence,
$$\sin\theta = \frac{y}{r} = \frac{4}{2\sqrt{5}} = \frac{2}{\sqrt{5}}$$
$$\cos\theta = \frac{x}{r} = \frac{-2}{2\sqrt{5}} = -\frac{1}{\sqrt{5}}$$

$$\tan \theta = \frac{y}{x} = \frac{4}{-2} = -2$$
$$\csc \theta = \frac{1}{\sin \theta} = \frac{\sqrt{5}}{2}$$
$$\sec \theta = \frac{1}{\cos \theta} = -\sqrt{5}$$
$$\cot \theta = \frac{1}{\tan \theta} = -\frac{1}{2}$$

2. If $\cot \theta = \frac{3}{4} > 0$ and $\cos \theta < 0$, then θ is in quadrant III, where $x < 0$ and $y < 0$. Thus, for $\cot \theta = \frac{3}{4} = x/y$, we choose $x = -3$, $y = -4$, and thus $r = \sqrt{(-3)^2 + (-4)^2} = \sqrt{25} = 5$. Hence, we have

$$\sin \theta = \frac{y}{r} = -\frac{4}{5} \qquad \cos \theta = \frac{x}{r} = -\frac{3}{5}$$
$$\tan \theta = \frac{y}{x} = \frac{4}{3} \qquad \csc \theta = \frac{1}{\sin \theta} = -\frac{5}{4}$$
$$\sec \theta = \frac{1}{\cos \theta} = -\frac{5}{3}$$

3. From the identity $1 + \tan^2 \theta = \sec^2 \theta$, we have $\sec \theta = \pm\sqrt{1 + \tan^2 \theta}$. Since $\tan \theta = -1 < 0$ and $\cos \theta < 0$, we know that θ is in quadrant II, and hence $\sec \theta < 0$. Thus,

$$\sec \theta = -\sqrt{1 + (-1)^2} = -\sqrt{2} \quad \text{and} \quad \cos \theta = \frac{1}{\sec \theta} = -\frac{1}{\sqrt{2}}.$$

4. (a) $\sin B = \dfrac{\text{side opposite } B}{\text{hypotenuse}} = \dfrac{8}{12} = \dfrac{2}{3}$

(b) $\cos B = \dfrac{\text{side adjacent to } B}{\text{hypotenuse}} = \dfrac{4\sqrt{5}}{12} = \dfrac{\sqrt{5}}{3}$

(c) $\tan B = \dfrac{\text{side opposite } B}{\text{side adjacent to } B} = \dfrac{8}{4\sqrt{5}} = \dfrac{2}{\sqrt{5}}$

5. $\sec 12°32' = \csc (90° - 12°32')$
$= \csc (89°60' - 12°32') = \csc 77°28'$

Exercises

1. $\sin \theta = \frac{4}{5}$, $\cos \theta = \frac{3}{5}$, $\tan \theta = \frac{4}{3}$, $\csc \theta = \frac{5}{4}$, $\sec \theta = \frac{5}{3}$, $\cot \theta = \frac{3}{4}$ **3.** $\sin \theta = -\frac{\sqrt{3}}{2}$, $\cos \theta = -\frac{1}{2}$, $\tan \theta = \sqrt{3}$, $\csc \theta = -\frac{2}{\sqrt{3}}$, $\sec \theta = -2$, $\cot \theta = \frac{1}{\sqrt{3}}$ **5.** $\sin \theta = 1$, $\cos \theta = 0$, $\tan \theta$ is undefined, $\csc \theta = 1$, $\sec \theta$ is undefined, $\cot \theta = 0$ **7.** quadrant III **9.** quadrant II **11.** quadrant I **13.** quadrant IV **15.** $\csc \theta = \frac{5}{3}$, $\cos \theta = \frac{4}{5}$, $\sec \theta = \frac{5}{4}$, $\tan \theta = \frac{3}{4}$, $\cot \theta = \frac{4}{3}$ **17.** $\cot \theta = -\frac{3}{2}$, $\sin \theta = \frac{2}{\sqrt{13}}$, $\csc \theta = \frac{\sqrt{13}}{2}$, $\cos \theta = -\frac{3}{\sqrt{13}}$, $\sec \theta = -\frac{\sqrt{13}}{3}$

19. $\sec \theta = \frac{2}{\sqrt{3}}$, $\sin \theta = -\frac{1}{2}$, $\csc \theta = -2$, $\tan \theta = -\frac{1}{\sqrt{3}}$, $\cot \theta = -\sqrt{3}$ **21.** $\tan \theta = -1$, $\sin \theta = -\frac{1}{\sqrt{2}}$, $\csc \theta = -\sqrt{2}$, $\cos \theta = \frac{1}{\sqrt{2}}$, $\sec \theta = \sqrt{2}$ **23.** $\cos \theta = -\frac{5}{6}$, $\sin \theta = -\frac{\sqrt{11}}{6}$, $\csc \theta = -\frac{6}{\sqrt{11}}$, $\tan \theta = \frac{\sqrt{11}}{5}$, $\cot \theta = \frac{5}{\sqrt{11}}$ **25.** $\sin \theta = \frac{1}{3}$ **27.** $\cot \theta = -5$ **29.** $\tan \theta = 3$ **31.** $\sin \theta = \frac{5}{13}$ **33.** $\sin \theta = \frac{4}{5}$ **35.** $\sec \theta = -\sqrt{3}$ **37.** $\tan \theta = \frac{1}{\sqrt{2}}$ **39.** (a) $\sin \alpha = \frac{15}{17}$ (b) $\cos \alpha = \frac{8}{17}$ (c) $\cot \beta = \frac{15}{8}$ **41.** (a) $\tan \beta = \frac{4\sqrt{2}}{7}$ (b) $\csc \alpha = \frac{9}{7}$ (c) $\sec \alpha = \frac{9}{4\sqrt{2}}$ **43.** (a) $\sin \beta = \frac{3}{\sqrt{10}}$ (b) $\cos \alpha = \frac{3}{\sqrt{10}}$ (c) $\tan \beta = 3$ **45.** $\cos 60°$ **47.** $\cot \dfrac{\pi}{4}$ **49.** $\tan 55°17'$

51.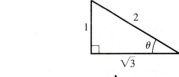

53.

55. No. Reciprocals can't have different algebraic signs.
57. $\cos \theta = \pm\sqrt{1 - \sin^2 \theta}$, $\tan \theta = \dfrac{\pm \sin \theta}{\sqrt{1 - \sin^2 \theta}}$, $\csc \theta = \dfrac{1}{\sin \theta}$, $\sec \theta = \dfrac{\pm 1}{\sqrt{1 - \sin^2 \theta}}$, $\cot \theta = \dfrac{\pm\sqrt{1 - \sin^2 \theta}}{\sin \theta}$
59. By similar triangles, the sine of the angle that passes through the point $(2a, 2b)$ must be the same as θ.
61. $\sin \theta = -0.522$, $\tan \theta = -0.612$, $\sec \theta = 1.17$, $\csc \theta = -1.92$, $\cot \theta = -1.63$ **63.** $\sin \theta = 0.444$, $\cos \theta = -0.896$, $\tan \theta = -0.496$, $\sec \theta = -1.12$, $\cot \theta = -2.02$ **65.** $\sin \theta = -0.612$, $\cos \theta = -0.791$, $\csc \theta = -1.64$, $\sec \theta = -1.26$, $\cot \theta = 1.29$
67. $\sin \theta = 0.461$ **69.** $\sin \theta = -0.394$
71. $\cot \theta = -0.949$ **73.** (a) $\sin \alpha = 0.868$
(b) $\cos \alpha = 0.497$ (c) $\cot \beta = 1.75$

Section 2.3

Problems

1. (a) Since 450° is coterminal with 90° and sin 90° = 1, we conclude that sin 450° = 1.
   (b) Since 7π is coterminal with π and cos $\pi = -1$, we conclude that cos $7\pi = -1$.
   (c) Since $-\dfrac{\pi}{2}$ is coterminal with $\dfrac{3\pi}{2}$ and tan $\dfrac{3\pi}{2}$ is undefined, we conclude that $\tan\left(-\dfrac{\pi}{2}\right)$ is undefined.
   (d) Since $-1080°$ is coterminal with 0° and sec 0° = 1, we conclude that sec $(-1080°) = 1$.

2. (a) Since 750° is coterminal with 30° and sin 30° = $\tfrac{1}{2}$, we conclude that sin 750° = $\tfrac{1}{2}$.
   (b) Since $\dfrac{25\pi}{4}$ is coterminal with $\dfrac{\pi}{4}$ and tan $\dfrac{\pi}{4} = 1$, we conclude that tan $\dfrac{25\pi}{4} = 1$.
   (c) Since $-660°$ is coterminal with 60° and cos 60° = $\tfrac{1}{2}$, we conclude that cos $(-660°) = \tfrac{1}{2}$.
   (d) Since $-\dfrac{5\pi}{3}$ is coterminal with $\dfrac{\pi}{3}$ and csc $\dfrac{\pi}{3} = \dfrac{2}{\sqrt{3}}$, we conclude that $\csc\left(-\dfrac{5\pi}{3}\right) = \dfrac{2}{\sqrt{3}}$.

3. (a) Since $\dfrac{5\pi}{6}$ is between $\dfrac{\pi}{2}$ and π, we have $\theta' = \pi - \dfrac{5\pi}{6} = \dfrac{\pi}{6}$.
   (b) Since $-495°$ is coterminal with 225° and 225° is between 180° and 270°, we have $\theta' = 225° - 180° = 45°$.

4. (a) sin 150° = sin 30° = $\tfrac{1}{2}$
   (b) tan $\dfrac{3\pi}{4} = -\tan\dfrac{\pi}{4} = -1$
   (c) sec $(-420°) = $ sec 60° = 2

5. $g(5) = 2\tan\left[\dfrac{\pi}{4}(5-1)\right] = 2\tan\pi = 2(0) = 0$

6. (a) With a calculator set in degree mode, a typical keying sequence is [237] [SIN] [1/x], and we obtain csc 237° ≈ −1.192.
   (b) With a calculator set in degree mode, a typical keying sequence is [121.5] [TAN], and we obtain tan 121°30′ ≈ −1.632.
   (c) With a calculator set in radian mode, a typical keying sequence is [43] [×] [π] [÷] [24] [=] [cos] [1/x], and we obtain sec $\dfrac{43\pi}{24}$ ≈ 1.260.

Exercises

1.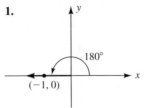

sin 180° = 0
cos 180° = −1
tan 180° = 0
csc 180° = undefined
sec 180° = −1
cot 180° = undefined

3.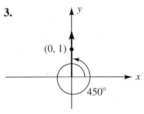

sin 450° = 1
cos 450° = 0
tan 450° = undefined
csc 450° = 1
sec 450° = undefined
cot 450° = 0

5.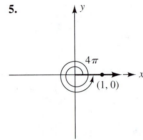

sin 4π = 0
cos 4π = 1
tan 4π = 0
csc 4π = undefined
sec 4π = 1
cot 4π = undefined

7.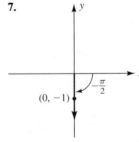

sin $(-\pi/2) = -1$
cos $(-\pi/2) = 0$
tan $(-\pi/2)$ = undefined
csc $(-\pi/2) = -1$
sec $(-\pi/2)$ = undefined
cot $(-\pi/2) = 0$

9.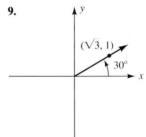

sin 30° = 1/2
cos 30° = $\sqrt{3}/2$
tan 30° = $1/\sqrt{3}$
csc 30° = 2
sec 30° = $2/\sqrt{3}$
cot 30° = $\sqrt{3}$

11.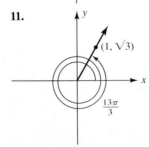

sin $(13\pi/3) = \sqrt{3}/2$
cos $(13\pi/3) = 1/2$
tan $(13\pi/3) = \sqrt{3}$
csc $(13\pi/3) = 2/\sqrt{3}$
sec $(13\pi/3) = 2$
cot $(13\pi/3) = 1/\sqrt{3}$

13.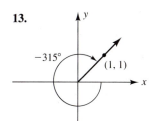

sin (−315°) = $1/\sqrt{2}$
cos (−315°) = $1/\sqrt{2}$
tan (−315°) = 1
csc (−315°) = $\sqrt{2}$
sec (−315°) = $\sqrt{2}$
cot (−315°) = 1

15.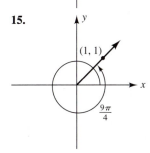

sin ($9\pi/4$) = $1/\sqrt{2}$
cos ($9\pi/4$) = $1/\sqrt{2}$
tan ($9\pi/4$) = 1
csc ($9\pi/4$) = $\sqrt{2}$
sec ($9\pi/4$) = $\sqrt{2}$
cot ($9\pi/4$) = 1

17. 60° **19.** 54° **21.** $\dfrac{\pi}{6}$ **23.** $\dfrac{\pi}{5}$ **25.** $\pi - 2$
27. 53°38′ **29.** $-\tfrac{1}{2}$ **31.** $-\tfrac{1}{2}$ **33.** $\sqrt{3}$
35. −1 **37.** $-\sqrt{2}$ **39.** $\dfrac{\sqrt{3}}{2}$ **41.** $\dfrac{1}{\sqrt{2}}$
43. $-\dfrac{2}{\sqrt{3}}$ **45.** $-\dfrac{1}{\sqrt{3}}$ **47.** $-\dfrac{\sqrt{3}}{2}$ **49.** 2
51. −2 **53.** 3 **55.** 4 **57.** 0 **59.** 2
61. $\tfrac{3}{2}$ **63.** $\dfrac{3\sqrt{3}}{2}$ **65.** 0 **67.** $-\tfrac{3}{2}$
69. (a) 12 cm (b) −12 cm (c) 0 cm (d) $-6\sqrt{2}$ cm
71. It appears that sin $(-\theta) = -\sin \theta$. **73.** Since csc π and cot π are undefined, (sin π)(csc π) ≠ 1, (tan π)(cot π) ≠ 1, cot $\pi \ne \dfrac{\cos \pi}{\sin \pi}$, and $1 + \cot^2 \pi \ne \csc^2 \pi$.
75. sin x = csc x at $x = \dfrac{\pi}{2}, \dfrac{3\pi}{2}$; cos x = sec x at $x = 0, \pi$;
tan x = cot x at $x = \dfrac{\pi}{4}, \dfrac{3\pi}{4}, \dfrac{5\pi}{4}, \dfrac{7\pi}{4}$ **77.** (a) cos θ
(b) sin θ (c) $\sin^2 \theta + \cos^2 \theta = 1$ **79.** 0.4226
81. −0.2588 **83.** −0.2250 **85.** −1.236
87. 1.236 **89.** 0.2679 **91.** 1.557 **93.** −1.101
95. 0.7274 **97.** 0.4901 **99.** $\dfrac{\sin t}{t}$ seems to approach 1 as $t \to 0^+$ **101.** (a) about −15.1 volts (b) about −10.7 volts (c) about −15.0 volts (d) about 1.34 volts

Section 2.4

 Problems

1. The graph of $y = -\tfrac{1}{2} \cos x$ is the same as the graph of $y = \tfrac{1}{2} \cos x$ reflected about the x-axis:

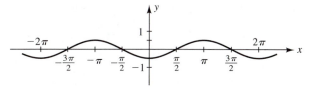

2. Since the cosine function is *even*, we have $y = \tfrac{1}{2} \cos\left(-\dfrac{\pi}{4}x\right) = \tfrac{1}{2} \cos \dfrac{\pi}{4} x$. Hence, the graph of $y = \tfrac{1}{2} \cos\left(-\dfrac{\pi}{4}x\right)$ is exactly the graph of $y = \tfrac{1}{2} \cos \dfrac{\pi}{4} x$ (see Figure 2.50).

3. The graph of $y = 3 \sin 2x + \pi$ is the same as the graph of $y = 3 \sin 2x$ shifted *vertically upward* π units as shown in the sketch:

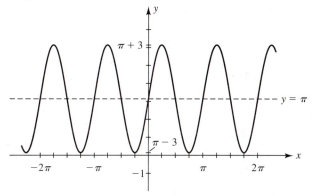

4. We can think of the graph in Figure 2.53 as a cosine curve with amplitude 2, period $4\pi/3$, and no phase shift. Hence, using $a = 2$, $b = \tfrac{3}{2}$ and $c = 0$, we have the equation $y = 2 \cos \tfrac{3}{2} x$.

5. The population P at the beginning of the ninth month (September) is

$$P = 70 + 50 \sin\left(\dfrac{9\pi}{6} - \dfrac{\pi}{3}\right)$$
$$= 70 + 50 \sin \dfrac{7\pi}{6}$$
$$= 70 + 50 \left(-\tfrac{1}{2}\right) = 45 \text{ hawks}$$

 Exercises

1.

x	0	$\dfrac{\pi}{4}$	$\dfrac{\pi}{2}$	$\dfrac{3\pi}{4}$	π	$\dfrac{5\pi}{4}$	$\dfrac{3\pi}{2}$	$\dfrac{7\pi}{4}$	2π
$f(x) = \sin x$	0	$\dfrac{1}{\sqrt{2}}$	1	$\dfrac{1}{\sqrt{2}}$	0	$-\dfrac{1}{\sqrt{2}}$	−1	$-\dfrac{1}{\sqrt{2}}$	0
$g(x) = \cos x$	1	$\dfrac{1}{\sqrt{2}}$	0	$-\dfrac{1}{\sqrt{2}}$	−1	$-\dfrac{1}{\sqrt{2}}$	0	$\dfrac{1}{\sqrt{2}}$	1

See Figure 2.42 for the graphs of f and g.

A34 Answers to Odd Numbered Exercises

3. $0, \pm\pi, \pm 2\pi, \pm 3\pi, \pm 4\pi$ 5. $\pm\pi, \pm 3\pi$

7. $\pm\dfrac{\pi}{4}, \pm\dfrac{7\pi}{4}, \pm\dfrac{9\pi}{4}, \pm\dfrac{15\pi}{4}$

9. $\dfrac{7\pi}{6}, \dfrac{11\pi}{6}, \dfrac{19\pi}{6}, \dfrac{23\pi}{6}, -\dfrac{\pi}{6}, -\dfrac{5\pi}{6}, -\dfrac{13\pi}{6}, -\dfrac{17\pi}{6}$

11. amplitude: 5
    period: 2π

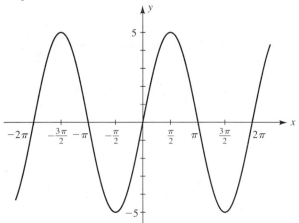

13. amplitude: 4
    period: $2\pi/3$

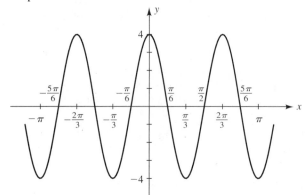

15. amplitude: $\tfrac{1}{2}$
    period: 12π

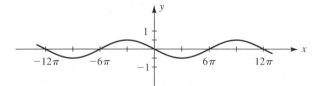

17. amplitude: $\tfrac{4}{3}$
    period: 2

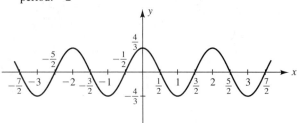

19. amplitude: 1
    period: $4\pi/3$

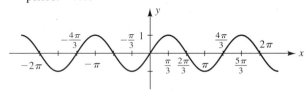

21. amplitude: 1
    period: $\tfrac{16}{3}$

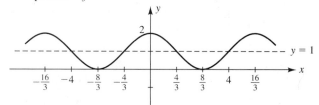

23. amplitude: 1
    period: 2π
    phase shift: left $\pi/6$

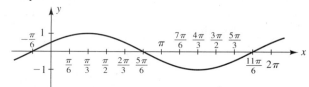

25. amplitude: 4
    period: 4π
    phase shift: right $3\pi/4$

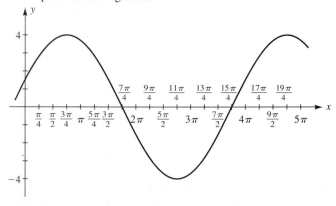

27. amplitude: $\frac{1}{2}$
    period: 2
    phase shift: left $\frac{3}{4}$

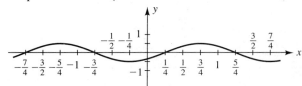

29. amplitude: 1
    period: π
    phase shift: right $\pi/6$

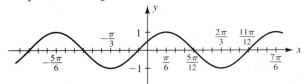

31. amplitude: 1.5
    period: $8\pi/3$
    phase shift: left $4\pi/3$

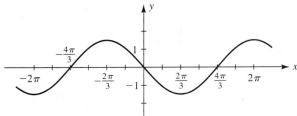

33. amplitude: 2
    period: $2\pi/3$
    phase shift: right $\pi/12$

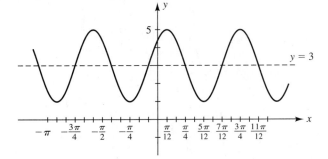

35. amplitude: 1
    period: $\pi/3$
    phase shift: left $\pi/6$

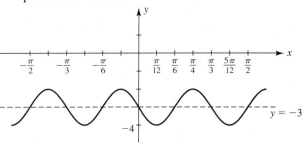

37. (a) $y = 3\sin\frac{\pi}{2}x$ (b) $y = 3\cos\left(\frac{\pi}{2}x + \frac{3\pi}{2}\right)$

39. (a) $y = 2\sin\left(x + \frac{\pi}{2}\right)$ (b) $y = 2\cos x$

41. (a) $y = \frac{3}{2}\sin\left(x + \frac{3\pi}{4}\right)$ (b) $y = \frac{3}{2}\cos\left(x + \frac{\pi}{4}\right)$

43. (a)

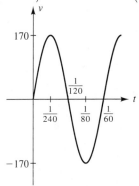

amplitude: 170
period: $\frac{1}{60}$ (b) 60 Hz

45. (a) February, $7700 (b) August, $1300

47. (a) $d = -120 + 70\sin\frac{\pi x}{240}$ (b) $d = -155$ ft

49. $\sin\left(x - \frac{\pi}{2}\right) = -\cos x$ 51. $\cos\left(x + \frac{3\pi}{2}\right) = \sin x$

53. $\sin(x - \pi) = -\sin x$ 55. $f(x) = -1 + 4\sin\frac{\pi x}{4}$

57. If c is the period of a periodic function f, then $f(t + c) = f(t)$ for all t in the domain of f. Since the two inputs $t + c$ and t give the same output value, we conclude that f is not one-to-one.   59. amplitude: 125; period: 0.0184; no phase shift   61. amplitude: 26.6; period: 28.2; phase shift: 5.56 units right   63. amplitude: 1.00; period: 0.503; phase shift: 0.824 units left   65. greatest during the year 1993; least during the year 2000   67. $x \approx 0.739$
69. (a) $f(x) = \sin x$ (b) $P(1) \approx 0.841, f(1) \approx 0.841$

Section 2.5

Problems

1. Period: $\dfrac{\pi}{|-2|} = \dfrac{\pi}{2}$ The graph completes one cycle from

$$2x = -\dfrac{\pi}{2} \quad \text{to} \quad 2x = \dfrac{\pi}{2}$$
$$x = -\dfrac{\pi}{4} \quad \text{to} \quad x = \dfrac{\pi}{4}$$

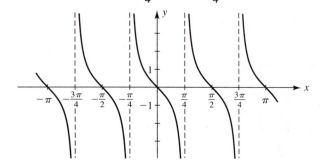

2. Period: $\dfrac{\pi}{|-\pi|} = 1$ The graph completes one cycle from

$$\pi x = 0 \quad \text{to} \quad \pi x = \pi$$
$$x = 0 \quad \text{to} \quad x = 1$$

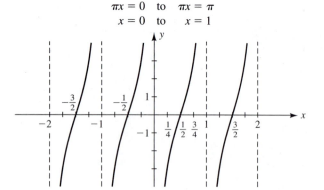

3. Period: $\dfrac{2\pi}{|-1/3|} = 6\pi$ The graph completes one cycle from

$$\dfrac{x}{3} = -\dfrac{\pi}{2} \quad \text{to} \quad \dfrac{x}{3} = \dfrac{3\pi}{2}$$
$$x = -\dfrac{3\pi}{2} \quad \text{to} \quad x = \dfrac{9\pi}{2}$$

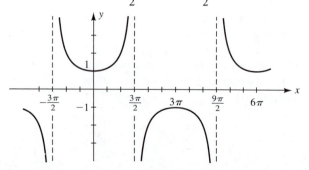

Exercises

1.

x	$-\dfrac{\pi}{2}$	$-\dfrac{\pi}{3}$	$-\dfrac{\pi}{4}$	$-\dfrac{\pi}{6}$	0	$\dfrac{\pi}{6}$	$\dfrac{\pi}{4}$	$\dfrac{\pi}{3}$	$\dfrac{\pi}{2}$
$f(x) = \tan x$	undef.	$-\sqrt{3}$	-1	$\dfrac{-1}{\sqrt{3}}$	0	$\dfrac{1}{\sqrt{3}}$	1	$\sqrt{3}$	undef.

See Figure 2.60 for the graph of $f(x) = \tan x$.

3.

x	0	$\dfrac{\pi}{4}$	$\dfrac{\pi}{2}$	$\dfrac{3\pi}{4}$	π	$\dfrac{5\pi}{4}$	$\dfrac{3\pi}{2}$	$\dfrac{7\pi}{4}$	2π
$f(x) = \csc x$	undef.	$\sqrt{2}$	1	$\sqrt{2}$	undef.	$-\sqrt{2}$	-1	$-\sqrt{2}$	undef.

See Figure 2.68 for the graph of $f(x) = \csc x$.

5. $0, \pm\pi, \pm 2\pi, \pm 3\pi, \pm 4\pi$

7. $\dfrac{3\pi}{4}, \dfrac{7\pi}{4}, \dfrac{11\pi}{4}, \dfrac{15\pi}{4}, -\dfrac{\pi}{4}, -\dfrac{5\pi}{4}, -\dfrac{9\pi}{4}, -\dfrac{13\pi}{4}$

9. $\dfrac{\pi}{6}, \dfrac{7\pi}{6}, \dfrac{13\pi}{6}, \dfrac{19\pi}{6}, -\dfrac{5\pi}{6}, -\dfrac{11\pi}{6}, -\dfrac{17\pi}{6}, -\dfrac{23\pi}{6}$

11. $\dfrac{2\pi}{3}, \dfrac{5\pi}{3}, \dfrac{8\pi}{3}, \dfrac{11\pi}{3}, -\dfrac{\pi}{3}, -\dfrac{4\pi}{3}, -\dfrac{7\pi}{3}, -\dfrac{10\pi}{3}$

13. $\dfrac{\pi}{2}, \dfrac{5\pi}{2}, -\dfrac{3\pi}{2}, -\dfrac{7\pi}{2}$ 15. $\pm\dfrac{2\pi}{3}, \pm\dfrac{4\pi}{3}, \pm\dfrac{8\pi}{3}, \pm\dfrac{10\pi}{3}$

17. $\pm\dfrac{\pi}{6}, \pm\dfrac{11\pi}{6}, \pm\dfrac{13\pi}{6}, \pm\dfrac{23\pi}{6}$ 19. No solution

21. period: $\pi/3$

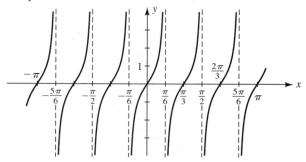

23. period: $4\pi/3$

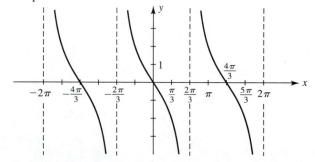

25. period: $\pi/4$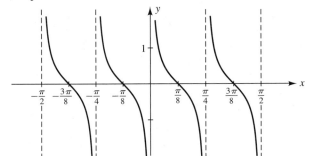

27. period: $3\pi/2$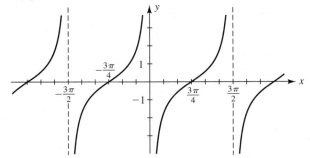

29. period: 2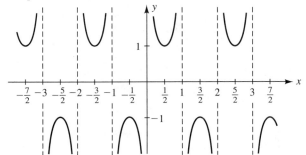

31. period: $2\pi/3$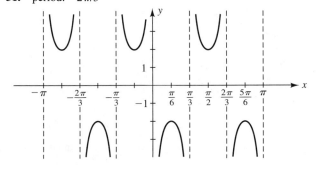

33. period: $5\pi/2$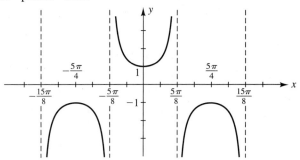

35. period: $16\pi/3$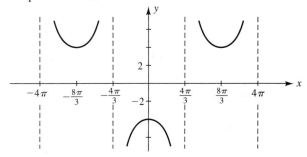

37. period: π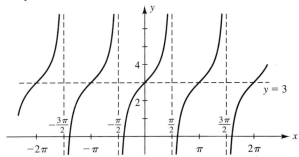

39. period: π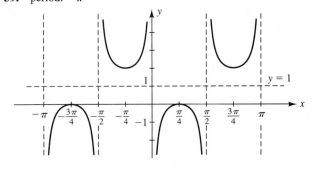

41. period: $\pi/3$

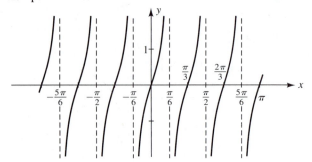

43. period: 1

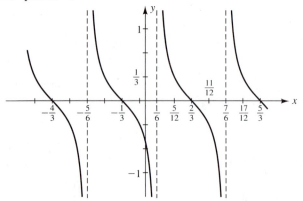

45. period: 2π

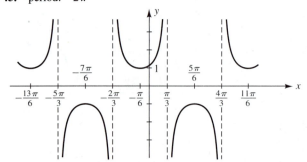

47. period: 2

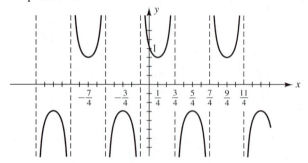

49. $y = 3 \tan \frac{1}{6}x$ 51. (a) $y = \frac{2}{3} \csc 8x$

53. $\tan\left(x + \frac{\pi}{2}\right) = -\cot x$ 55. $\cot(\pi - x) = -\cot x$

57. $\csc\left(x + \frac{\pi}{2}\right) = \sec x$ 59. $\sec\left(x - \frac{3\pi}{2}\right) = -\csc x$

61. $f(x) = 2 + 3 \csc \frac{\pi x}{4}$ 63. 0.726   65. 7.22

67. 2.38   69. $x \approx \pm 1.11, \pm 2.77$

Section 2.6

Problems

1. (a) For a calculator set in radian mode, a typical keying sequence is $\boxed{0.53}\ \boxed{\text{INV}}\ \boxed{\text{SIN}}$. We obtain

$$\arcsin(0.53) \approx 0.5586.$$

(b) For a calculator set in radian mode, a typical keying sequence is $\boxed{0.766}\ \boxed{\pm}\ \boxed{\text{INV}}\ \boxed{\text{SIN}}$. We obtain

$$\sin^{-1}(-0.766) \approx -0.8726.$$

2. (a) For a calculator set in radian mode, a typical keying sequence is $\boxed{0.809}\ \boxed{\text{INV}}\ \boxed{\text{COS}}$. We obtain

$$\arccos(0.809) \approx 0.6283.$$

(b) For a calculator set in radian mode, a typical keying sequence is $\boxed{0.42}\ \boxed{\pm}\ \boxed{\text{INV}}\ \boxed{\text{COS}}$. We obtain

$$\cos^{-1}(-0.42) \approx 2.004.$$

3. (a) For a calculator set in the radian mode, a typical keying sequence is $\boxed{0.977}\ \boxed{\text{INV}}\ \boxed{\text{TAN}}$. We obtain

$$\arctan(0.977) \approx 0.7738.$$

(b) For a calculator set in radian mode, a typical keying sequence is $\boxed{1.6}\ \boxed{\pm}\ \boxed{\text{INV}}\ \boxed{\text{TAN}}$. We obtain

$$\tan^{-1}(-1.6) \approx -1.012.$$

4. (a) Since 2 is in the domain of the inverse tangent function, we have $\tan(\arctan 2) = 2$.

(b) Since $3\pi/2$ is coterminal with $-\pi/2$ and $-\pi/2$ is in the domain of the restricted sine function, we have

$$\sin^{-1}\left(\sin \frac{3\pi}{2}\right) = \sin^{-1}\left[\sin\left(-\frac{\pi}{2}\right)\right] = -\frac{\pi}{2}.$$

5. We begin by rewriting the expression:

$$\cos\left[\arctan\left(-\tfrac{1}{2}\right)\right] = \cos\left[-\arctan \tfrac{1}{2}\right] = \cos\left[\arctan \tfrac{1}{2}\right].$$

Now let $\theta = \arctan \tfrac{1}{2}$. Then $\tan \theta = \tfrac{1}{2} = \dfrac{\text{opp}}{\text{adj}}$, and by the

Pythagorean theorem, the hypotenuse is $\sqrt{5}$. From the right triangle shown, we have

$$\cos\left[\arctan \tfrac{1}{2}\right] = \cos\theta = \frac{\text{adj}}{\text{hyp}} = \frac{2}{\sqrt{5}}.$$

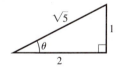

6. For $\cos(\sin^{-1} 2x)$, we let $\theta = \sin^{-1} 2x$. Then $\sin\theta = \dfrac{\text{opp}}{\text{hyp}} = \dfrac{2x}{1}$, and by the Pythagorean theorem, the adjacent side is $\sqrt{1 - 4x^2}$. From the right triangle shown, we have

$$\cos(\sin^{-1} 2x) = \cos\theta = \sqrt{1 - 4x^2},\ -\tfrac{1}{2} \le x \le \tfrac{1}{2}.$$

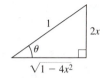

7. Domain: $[90, 90\sqrt{2}]$; range: $[0, \pi/4]$

Exercises

1. $\dfrac{\pi}{6}$ **3.** $\dfrac{\pi}{3}$ **5.** $\dfrac{\pi}{4}$ **7.** $-\dfrac{\pi}{3}$ **9.** $-\dfrac{\pi}{4}$

11. $\dfrac{3\pi}{4}$ **13.** $\dfrac{\pi}{2}$ **15.** undefined **17.** $\tfrac{2}{3}$

19. undefined **21.** $\dfrac{\pi}{8}$ **23.** -1 **25.** π

27. $\dfrac{\pi}{4}$ **29.** $\dfrac{\pi}{10}$ **31.** $-\dfrac{3\pi}{8}$ **33.** $\dfrac{3}{\sqrt{10}}$

35. $\dfrac{\sqrt{5}}{3}$ **37.** $-\dfrac{\sqrt{13}}{2}$ **39.** $-\dfrac{1}{\sqrt{3}}$ **41.** $\sqrt{1-x^2}$, $-1 \le x \le 1$ **43.** $\dfrac{|x|}{x\sqrt{x^2-1}}$, $x < -1$ or $x > 1$

45. $\dfrac{1}{x-1}$, $0 \le x < 1$ or $1 < x \le 2$ **47.** $\dfrac{|x-1|}{\sqrt{x^2-2x+2}}$, $x \ne 1$

49.

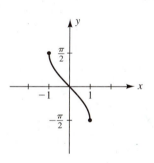

51.

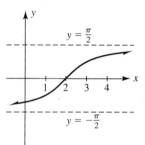

53.

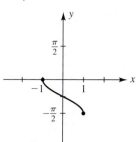

55. (a) $\theta = \arctan\dfrac{h}{550}$ (b) Domain: $[0, \infty)$; range: $\left[0, \dfrac{\pi}{2}\right)$ (c) $\dfrac{\pi}{4}$

57. (a)

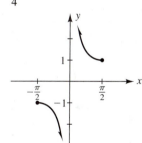

(b)

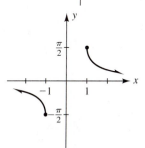

Domain: $(-\infty, -1] \cup [1, \infty)$; range: $[-\pi/2, 0) \cup (0, \pi/2]$

59. (a)

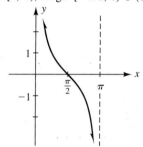

59. (b)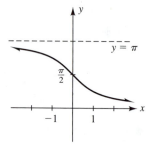

Domain: $(-\infty, \infty)$; range: $(0, \pi)$
61. (a) $\cos(-t + \pi) = -\cos t$ **63.** One example is $\tan^{-1}(\tan \pi) = 0 \neq \pi$. This does not contradict the inverse function concept, since π is not in the domain of the *restricted tangent function*. **65.** The notation $\sin^{-1} x$ denotes the inverse sine and $(\sin x)^{-1}$ denotes the reciprocal of the sine, which is the cosecant. **67.** $\frac{5}{6}$ **69.** $1, \frac{1}{2}$ **71.** 0.2838 **73.** 1.446 **75.** 1.706 **77.** 1.254 **79.** -0.3948 **81.** undefined **83.** 0.887 **85.** -0.7168
87. (a) $\theta = \arctan \frac{8}{t}$ (b) Domain: $(0, \infty)$; range $\left(0, \frac{\pi}{2}\right)$
(c) 33.7° **89.** $x \approx \pm 1.16$

Chapter 2 Review Exercises

1. (a)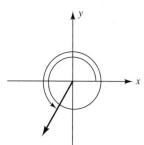

(b) $600°, \frac{10\pi}{3}$ (c) $240°, \frac{4\pi}{3}$ (d) $60°, \frac{\pi}{3}$

3. (a)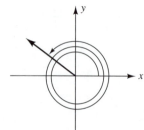

(b) $864°, \frac{24\pi}{5}$ (c) $144°, \frac{4\pi}{5}$ (d) $36°, \frac{\pi}{5}$

5. (a)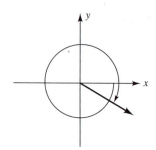

(b) $-390°, -\frac{13\pi}{6}$ (c) $330°, \frac{11\pi}{6}$ (d) $30°, \frac{\pi}{6}$

7. $\frac{10\pi}{9}$ **9.** $\frac{13\pi}{3}$ **11.** 330° **13.** $-1440°$
15. (a) 123.705 (b) 2.159 **17.** (a) 151.834°
(b) 151°50′02″ **19.** $\sin \theta = \frac{1}{\sqrt{2}}, \cos \theta = \frac{1}{\sqrt{2}},$
$\tan \theta = 1, \csc \theta = \sqrt{2}, \sec \theta = \sqrt{2}, \cot \theta = 1$
21. $\sin \theta = -\frac{3}{\sqrt{10}}, \cos \theta = -\frac{1}{\sqrt{10}}, \tan \theta = 3,$
$\csc \theta = -\frac{\sqrt{10}}{3}, \sec \theta = -\sqrt{10}, \cot \theta = \frac{1}{3}$ **23.** 2
25. $-\frac{3}{5}$ **27.** $-\sqrt{3}$ **29.** $-\frac{12}{5}$ **31.** $\frac{8}{17}$ **33.** $\frac{15}{8}$
35. $-\frac{1}{2}$ **37.** $\sqrt{3}$ **39.** $\sqrt{2}$ **41.** -1 **43.** $\frac{\pi}{3}$
45. Undefined **47.** $\frac{\pi}{3}$ **49.** $\frac{2\sqrt{2}}{3}$ **51.** 0.3090
53. -0.9147 **55.** 1.057 **57.** -0.3827
59. 1.107 **61.** 2.044 **63.** $\sin 2 \approx 0.9093$
65. $\tan 3 \approx -0.1425$ **67.** $(f \circ g)(x) = \sqrt{1 - 16x^2},$
$-\frac{1}{4} \leq x \leq \frac{1}{4}$ **69.** $(f \circ g)(x) = \frac{1}{\sqrt{x-1}}, x > 1$
71.

73.

75.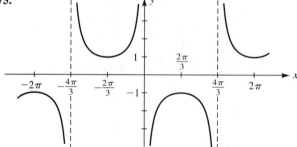

77.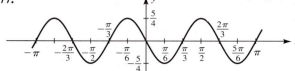

79.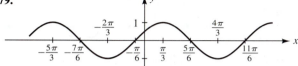

81.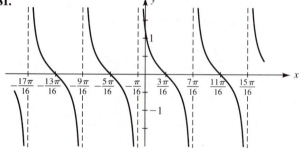

83.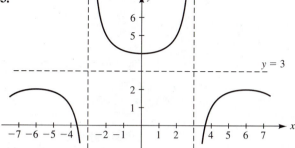

85.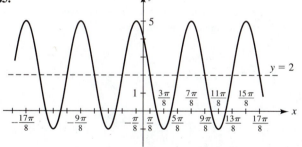

87.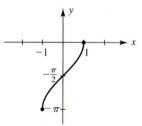

89.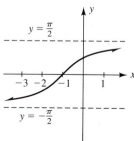

91. about 5400 mi **93.** $L = 10 \sin \dfrac{\theta}{2}$;

θ	1	$\pi/12$	$\pi/3$	1.85
L	4.79	1.31	5	8

95. (a) $\alpha = \arctan \dfrac{x}{3}$ (b) $\beta = \pi - \arctan \dfrac{x}{3}$

(c) $\theta = \arctan \dfrac{x}{3} + \arctan \dfrac{7}{x} - \dfrac{\pi}{2}$

CHAPTER 3

Section 3.1

Problems

1. $\alpha = 90° - 47° = 43°$

$\tan 47° = \dfrac{x}{22}$, $x = 22 \tan 47° \approx 23.6$

$\cos 47° = \dfrac{22}{z}$, $z = \dfrac{22}{\cos 47°} \approx 32.3$

2. $x = \sqrt{24^2 - 10^2} \approx 21.8$

$\cos \alpha = \dfrac{10}{24}$, $\alpha = \arccos \dfrac{10}{24} \approx 65.4°$

$\sin \beta = \dfrac{10}{24}$, $\beta = \arcsin \dfrac{10}{24} \approx 24.6°$

3.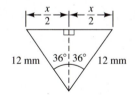

$\sin 36° = \dfrac{x/2}{12}$, so $x/2 = 12 \sin 36°$.

Thus $x = 24 \sin 36° \approx 14.1$ mm.

4.

$\sin 15.5° = \dfrac{1800}{s}$, so $s = \dfrac{1800}{\sin 15.5°} \approx 6735.6$ ft.

$t = \dfrac{s}{v} = \dfrac{6735.6 \text{ ft}}{3 \text{ mi/hr}} \cdot \dfrac{1 \text{ mi}}{5280 \text{ ft}} \approx 0.425$ hr or 25.5 min.

5.

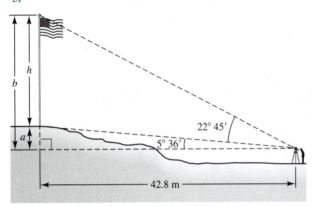

$\tan 5°36' = \dfrac{a}{42.8}$, so $a = 42.8 \tan 5°36' \approx 4.197$ m

$\tan 22°45' = \dfrac{b}{42.8}$, so $b = 42.8 \tan 22°45' \approx 17.948$ m

$h = b - a \approx 13.8$ m

6. Substituting $t = 2$ into the equation $d = -15 \cos 8\pi t$, we find
$$d = -15 \cos 16\pi = -15(1) = -15 \text{ cm}.$$

Exercises

1. $\theta = 52°$, $x \approx 14.2$, $y \approx 11.1$ **3.** $\theta = 28.2°$, $x \approx 7.08$, $y \approx 15.0$ **5.** $\theta = 47°42'$, $x \approx 761$, $y \approx 512$
7. $\alpha \approx 56.3°$, $\beta \approx 33.7°$, $x \approx 21.6$ **9.** $\alpha \approx 38.6°$, $\beta \approx 51.4°$, $x \approx 56.7$ **11.** $\alpha \approx 49.9°$, $\beta \approx 40.1°$, $x \approx 1480$
13. 338 ft **15.** 2.87 km **17.** 147 m
19. 4.31 cm **21.** 145.3° **23.** 2.08 cm
25. $\alpha = 34.3°$, $\beta = 111.4°$ **27.** 199 m **29.** 29.2 s
31. 745 ft **33.** (a) 8 (b) 3 (c) $\frac{1}{12}$ **35.** (a) $\frac{2}{3}$
(b) $\frac{1}{20}$ (c) 10 **37.** (a) $d = 6 \sin \dfrac{2\pi}{3} t$ (b) 3 inches
below the equilibrium position **39.** (a) $d = 20 \cos 5\pi t$
(b) 0 cm (c) 1.1 s

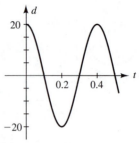

41. $x = 32$ cm, $\theta \approx 73.7°$ **43.** 75°
45. about 13.8 cm **47.** $d \approx 0.227 \sin 3.14t$
49. (a) $x = \cos \theta$, $y = \sin \theta$ (b) $A = \sin \theta \cos \theta + \sin \theta$
(c) $\theta \approx 1.047$ rad $\approx 60°$ (d) about 1.30 sq ft
51. Since $e^{-kt} \to 0$ as $t \to \infty$, the coefficient e^{-kt} causes a decrease in amplitude of successive oscillations.

Section 3.2

Problems

1. The longest side (52) is opposite the largest angle ($\theta = 70°$) and the shortest side ($x = 41.1$) is opposite the smallest angle (48°).

2. $\theta = 180° - (108° + 28°) = 44°$. Hence, by the law of sines,
$$\dfrac{125}{\sin 108°} = \dfrac{x}{\sin 28°} = \dfrac{y}{\sin 44°}.$$
$$\dfrac{125}{\sin 108°} = \dfrac{x}{\sin 28°} \quad \text{and} \quad \dfrac{125}{\sin 108°} = \dfrac{y}{\sin 44°}$$
$$x = \dfrac{125 \sin 28°}{\sin 108°} \approx 61.7 \quad y = \dfrac{125 \sin 44°}{\sin 108°} \approx 91.3$$

3. By the law of sines, we have
$$\dfrac{22}{\sin 38°} = \dfrac{35}{\sin \beta} \quad \text{or} \quad \sin \beta = \dfrac{35 \sin 38°}{22}.$$
Hence $\beta' = \sin^{-1}\left(\dfrac{35 \sin 38°}{22}\right) \approx 78.3676°$ and, therefore,
$\beta \approx 180° - 78.3676° = 101.6324°$. Thus,
$\alpha \approx 180° - (38° + 101.6324°) = 40.3676°$ and again, by the law of sines,
$$\dfrac{22}{\sin 38°} = \dfrac{x}{\sin 40.3676°} \quad \text{or} \quad x = \dfrac{22 \sin 40.3676°}{\sin 38°} \approx 23.1444.$$
Rounding α and β to the nearest tenth of a degree and x to three significant digits, we obtain $\alpha = 40.4°$, $\beta = 101.6°$, and $x = 23.1$.

4. By the law of sines, $\dfrac{CD}{\sin 36°15'} = \dfrac{AC}{\sin 90°}$. However, since $\sin 90° = 1$, we have
$$CD = AC \sin 36°15' = \left(\dfrac{100 \sin 118°30'}{\sin 25°15'}\right) \sin 36°15' \approx 121.8 \text{ ft}.$$

Exercises

1. $\theta = 73°$, $x \approx 35.7$, $y \approx 33.9$ 3. $\theta = 61°$, $x \approx 107$, $y \approx 294$ 5. $\theta = 101°$, $x \approx 9.31$, $y \approx 6.75$
7. $\alpha \approx 22.7°$, $\beta \approx 66.3°$, $x \approx 52.2$ 9. $\alpha \approx 143.7°$, $\beta \approx 17.3°$, $x \approx 1260$ 11. $\alpha \approx 94.4°$, $\beta \approx 37.6°$, $x \approx 47.9$
13. $\theta \approx 37°$, $a \approx 77.2$, $b \approx 91.8$ 15. $\alpha \approx 84.3°$, $\theta \approx 33.7°$, $a \approx 39.4$ 17. two possibilities: $\theta \approx 63.7°$, $\beta \approx 78.3°$, $b \approx 1080$; $\theta \approx 116.3°$, $\beta \approx 25.7°$, $b \approx 479$
19. two possibilities: $\alpha \approx 33.9°$, $\beta \approx 79.1°$, $a \approx 5.91$; $\alpha \approx 12.1°$, $\beta \approx 100.9°$, $a \approx 2.22$ 21. No such triangle exists   23. No such triangle exists   25. 42.7 m
27. 466 m   29. 64.3°   31. (a) 416 m (b) 403 m
33. (a) 201 ft (b) 127 ft   35. 1210 ft
37. (a) 33.7 ft (b) 40.0 ft   39. (a) 184 m (b) 74.6 m   41. $B < A < C$; smallest angle is opposite shortest side and largest angle is opposite longest side.
43. $\sin \theta = \dfrac{a}{c} = \dfrac{\text{opp}}{\text{hyp}}$, right triangle definition of sine
47. appears to be a solution of a triangle
49. does not appear to be a solution of a triangle
51. $v_2 \approx 2.02 \times 10^{10}$ cm/s

Section 3.3

Problems

1. Since the side opposite β is the largest side of the triangle, we cannot tell whether β is acute or obtuse. Thus, we begin by finding the reference angle β' as follows:

$$\beta' = \sin^{-1}\left(\frac{33 \sin 26°}{14.5612}\right) \approx 83.5°.$$

Hence, either $\beta \approx 83.5°$ or $\beta \approx 180° - 83.5° = 96.5°$. If we choose $\beta \approx 83.5°$, then $\alpha \approx 180° - (26° + 83.5°) = 70.5°$. However, this would contradict the law of sines, since

$$\sin 70.5° \neq \frac{28 \sin 26°}{14.5612}.$$

Therefore, we conclude that $\beta \approx 96.5°$.

2. Solving the equation
$16.2^2 = 22.5^2 + 14.9^2 - 2(22.5)(14.9) \cos \alpha$ for $\cos \alpha$, we obtain $\cos \alpha = \dfrac{16.2^2 - 22.5^2 - 14.9^2}{-2(22.5)(14.9)}$ which implies

$$\alpha = \cos^{-1}\left(\frac{16.2^2 - 22.5^2 - 14.9^2}{-2(22.5)(14.9)}\right) \approx 46.0°.$$ Solving the equation $14.9^2 = 16.2^2 + 22.5^2 - 2(16.2)(22.5) \cos \beta$ for $\cos \beta$, we obtain $\cos \beta = \dfrac{14.9^2 - 16.2^2 - 22.5^2}{-2(16.2)(22.5)}$ which implies

$$\beta = \cos^{-1}\left(\frac{14.9^2 - 16.2^2 - 22.5^2}{-2(16.2)(22.5)}\right) \approx 41.4°.$$

3. (a) $A = \dfrac{ab \sin \theta}{2} = \dfrac{(16.2)(22.5) \sin 41.418°}{2} \approx 121$ square units

(b) $A = \dfrac{ab \sin \theta}{2} = \dfrac{(22.5)(14.9) \sin 45.994°}{2} \approx 121$ square units

4. For a triangle with sides $a = 3$, $b = 4$, and $c = 5$, the semiperimeter $s = \tfrac{1}{2}(3 + 4 + 5) = 6$. Thus, by Hero's formula, the area of the triangle is

$$\begin{aligned} A &= \sqrt{s(s-a)(s-b)(s-c)} \\ &= \sqrt{6(6-3)(6-4)(6-5)} \\ &= \sqrt{36} = 6 \text{ square units.} \end{aligned}$$

Since this triangle is a right triangle with base $b = 4$ and height $h = c = 3$, we have

$$A = \frac{bh}{2} = \frac{(4)(3)}{2} = 6 \text{ square units,}$$

which agrees with the answer we obtained by Hero's formula.

5. The information is shown in the sketch.

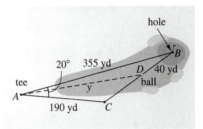

Using the fact that $BC \approx 188.04$ yd (see Example 5), we can apply the law of sines to triangle ABC and determine angle B:

$$\frac{188.04}{\sin 20°} = \frac{190}{\sin B} \quad \text{or} \quad \sin B = \frac{190 \sin 20°}{188.04}.$$

Hence, $B = \sin^{-1}\left(\dfrac{190 \sin 20°}{188.04}\right) \approx 20.218°$. Now by applying the law of cosines to triangle ABD, we can find the distance y from the tee to the ball:

$$\begin{aligned} y^2 &= 40^2 + 355^2 - 2(40)(355) \cos 20.218° \\ y &= \sqrt{40^2 + 355^2 - 2(40)(355) \cos 20.218°} \approx 318 \text{ yds.} \end{aligned}$$

Exercises

1. (a) $x \approx 37.0$, $\alpha \approx 34.5°$, $\beta = 51.5°$ (b) 304 square units   3. (a) $x \approx 198$, $\alpha \approx 133.9°$, $\beta \approx 17.1°$ (b) 8550 square units   5. (a) $x \approx 1090$, $\alpha \approx 129.6°$, $\beta \approx 33.4°$ (b) 860,000 square units   7. (a) $\alpha \approx 86.9°$, $\beta \approx 67.4°$, $\theta \approx 25.7°$ (b) 9000 square units   9. (a) $\alpha \approx 37.6°$, $\beta \approx 96.3°$, $\theta \approx 46.1°$ (b) 221 square units
11. $c \approx 14.5$, $\alpha \approx 61.6°$, $\beta \approx 76.4°$ 13. $a \approx 7.02$, $\beta \approx 25.5°$, $\theta \approx 42.5°$ 15. $b \approx 474$, $\alpha \approx 30.7°$, $\theta \approx 101.3°$

17. $\alpha \approx 107.6°$, $\beta \approx 49.0°$, $\theta \approx 23.4°$ **19.** $\alpha \approx 58.0°$, $\beta \approx 78.1°$, $\theta \approx 43.9°$ **21.** $\alpha \approx 130.8°$, $\beta \approx 27.9°$, $\theta \approx 21.3°$ **23.** no such triangle exists   **25.** 88.0°   **27.** 6.32 ft   **29.** 172 ft   **31.** 4.00 mi   **33.** (a) 9.40 cm   (b) 15.2 cm   **35.** (a) 55.4 m and 139 m   (b) 3430 sq m   **37.** (a) 158 ft   (b) 136 ft   **39.** 117 ft   **41.** 8.10 acres   **43.** No, not if we are given only three angles   **45.** (a) about 133.2°   (b) about 16.5 square units   **47.** $\theta = 120°$ **49.** $b = 6$, $c = 3$ **51.** $x^2 + y^2 = 2a^2 + 2b^2$ **53.** (a) $A \approx 29.30°$, $B \approx 44.55°$, $ACB \approx 106.15°$ (b) $x \approx 3.44$, $y \approx 4.94$, $z \approx 3.02$ (c) $\dfrac{x}{y} \approx 0.698$, $\dfrac{a}{b} \approx 0.698$. It appears that $\dfrac{x}{y} = \dfrac{a}{b}$ (e) $\sqrt{ab - xy} \approx 3.02$. It appears that $\sqrt{ab - xy} = z$

Section 3.4

◆ Problems

1. (a) Standard position is obtained by moving the initial point $(-2, -3)$ and terminal point $(-5, 1)$ two units right and three units up, as shown in the sketch.

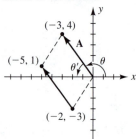

(b) Since the x-component is -3 and the y-component is 4, we write $\mathbf{A} = \langle -3, 4 \rangle$.

(c) $|\mathbf{A}| = \sqrt{(-3)^2 + (4)^2} = \sqrt{25} = 5$

(d) Since θ is in quadrant II and

$$\theta' = \tan^{-1}\left|\dfrac{4}{-3}\right| \approx 53.1°,$$

we have $\theta \approx 180° - 53.1° = 126.9°$.

2. $x = |\mathbf{A}| \cos \theta = 8 \cos 212° \approx -6.78$ and $y = |\mathbf{A}| \sin \theta = 8 \sin 212° \approx -4.24$. Hence, $\mathbf{A} \approx \langle -6.78, -4.24 \rangle$.

3. If $\mathbf{A} = \langle -3, 4 \rangle$, then $-\mathbf{A} = \langle -(-3), -(4) \rangle = \langle 3, -4 \rangle$.

4. $2\mathbf{A} + 3\mathbf{B} = 2\langle -2, 3 \rangle + 3\langle -4, -4 \rangle$
$= \langle -4, 6 \rangle + \langle -12, -12 \rangle$
$= \langle -4 + (-12), 6 + (-12) \rangle$
$= \langle -16, -6 \rangle$

5. $\mathbf{A} - \mathbf{B} = \langle -2, 3 \rangle - \langle -4, -4 \rangle$
$= \langle -2 - (-4), 3 - (-4) \rangle$
$= \langle 2, 7 \rangle$

6. $-\mathbf{A} + 2\mathbf{B} = -(2\mathbf{i} - 3\mathbf{j}) + 2(\mathbf{i} - 6\mathbf{j})$
$= (-2\mathbf{i} + 3\mathbf{j}) + (2\mathbf{i} - 12\mathbf{j})$
$= (-2 + 2)\mathbf{i} + (3 - 12)\mathbf{j}$
$= 0\mathbf{i} - 9\mathbf{j}$ or $\langle 0, -9 \rangle$

7. In order to just lift the sled from the surface, the vertical component of a force $\mathbf{F}$, applied at an angle of 30°, must be 75 lb. Hence, $75 = |\mathbf{F}| \sin 30°$ or $|\mathbf{F}| = \dfrac{75}{\sin 30°} = 150$ lb.

8. The force $\mathbf{F}_3$ that produces equilibrium must have the same magnitude as the resultant force $\mathbf{F}$ but opposite direction. Hence, $|\mathbf{F}_3| \approx 89$ lb with direction angle $\theta \approx 193° - 180° = 13°$.

9. Using the information shown in the sketch, we can find the magnitude of the displacement $\mathbf{D}$ by using the law of cosines.

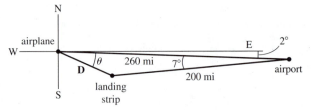

$|\mathbf{D}|^2 = 200^2 + 260^2 - 2(200)(260) \cos 7°$
$|\mathbf{D}| = \sqrt{200^2 + 260^2 - 2(200)(260) \cos 7°} \approx 66.1453$.

Rounding to two significant digits, we obtain $|\mathbf{D}| \approx 66$ miles. Now by the law of sines,

$$\dfrac{200}{\sin \theta} = \dfrac{66.1453}{\sin 7°} \quad \text{or} \quad \sin \theta = \dfrac{200 \sin 7°}{66.1453}$$

which implies

$$\theta = \sin^{-1}\left(\dfrac{200 \sin 7°}{66.1453}\right) \approx 21.6°.$$

Measured from the South axis, we obtain $90° - (21.6° + 2°) = 66.4°$. Thus, the bearing is S-66.4°-E.

◆ Exercises

1.

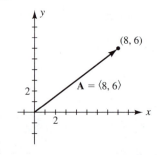

3.

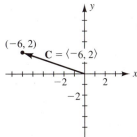

5.

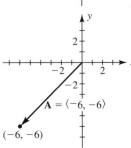

7.

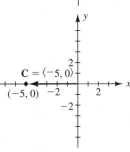

9. $|\mathbf{A}| = 10$, $|\mathbf{C}| = 2\sqrt{10} \approx 6.32$ **11.** For **A**, $\theta = 225°$; for **C**, $\theta = 180°$ **13.** $|\mathbf{A}| = 13$, $\theta \approx 292.6°$
15. $|\mathbf{C}| = \sqrt{117} \approx 10.8$, $\theta \approx 236.3°$
17. $|\mathbf{A}| = \sqrt{5}/4 \approx 0.559$, $\theta \approx 26.6°$
19. $|\mathbf{C}| = 2\sqrt{2} \approx 2.83$, $\theta \approx 110.7°$
21. $x = 310\sqrt{2} \approx 438$, $y = 310\sqrt{2} \approx 438$
23. $x = -42\sqrt{3} \approx -72.7$, $y = -42$ **25.** $x = -1$, $y = -1$
27. $x = 0$, $y = -\frac{9}{2}$ **29.** $\langle 4, 10 \rangle$ **31.** $\langle 4, 1 \rangle$
33. $\langle 1, 11 \rangle$ **35.** $\langle 10, 7 \rangle$ **37.** $\langle 32, 26 \rangle$
39. $\langle 0, -\frac{3}{2} \rangle$ **41.** $\langle 0, 0 \rangle$ **43.** $\langle -26\sqrt{29}, -11\sqrt{29} \rangle$
45.

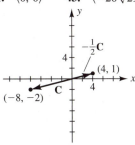

47.

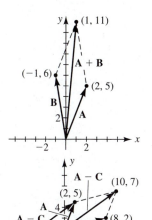

49.

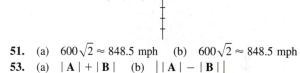

51. (a) $600\sqrt{2} \approx 848.5$ mph   (b) $600\sqrt{2} \approx 848.5$ mph
53. (a) $|\mathbf{A}| + |\mathbf{B}|$ (b) $||\mathbf{A}| - |\mathbf{B}||$
55. $\left\langle \dfrac{x}{\sqrt{x^2+y^2}}, \dfrac{y}{\sqrt{x^2+y^2}} \right\rangle$; a unit vector, since its magnitude is 1   **57.** (a) $c = 3$, $d = -2$ (b) $c = -1$, $d = -4$ **61.** $|\mathbf{A}| \approx 1.46$, $\theta \approx 148.7°$
63. $|\mathbf{C}| \approx 2.01$, $\theta \approx 79.8°$ **65.** $x \approx -3.23$, $y \approx 4.58$
67. $x \approx 76.0$, $y \approx -23.8$ **69.** magnitude: 94.3 lb, direction angle: 58.0°   **71.** magnitude: 46.1 lb, direction angle: 317.7°   **73.** magnitude: 72.3 lb, direction angle: 197.7°   **75.** $\mathbf{F}_3 = \langle -50, -80 \rangle$; $|\mathbf{F}_3| \approx 94.3$ lb, $\theta \approx 238°$
77. (a) magnitude: 5.7 mph, direction angle: 66° from the right-hand bank of the river   (b) about 84 ft   (c) 66° from the left-hand bank of the river   **79.** about 307 mi, S-48°-E
81. about 24.4 mi, S-21°9'-W

Chapter 3 Review Exercises

1. (a) $x \approx 34.2$, $y \approx 19.6$, $\theta = 55°$ (b) about 274 square units   **3.** (a) $x \approx 56.2$, $y \approx 35.0$, $\theta = 103°$ (b) about 958 square units   **5.** (a) $\alpha \approx 139.3°$, $\beta \approx 19.7°$, $x \approx 3000$ (b) about 833,000 square units
7. (a) $\alpha \approx 138.0°$, $\beta \approx 17.0°$, $x \approx 20.2$ (b) about 94.7 square units   **9.** (a) $\alpha \approx 88.2°$, $\beta \approx 60.3°$, $\theta \approx 31.5°$
(b) about 10,400 square units   **11.** $a \approx 38.2$, $\alpha \approx 38.2°$, $\theta \approx 63.8°$ **13.** $\alpha \approx 28.9°$, $\beta \approx 117.0°$, $\theta \approx 34.1°$
15. $b \approx 922$, $\theta \approx 73.5°$, $\beta \approx 68.5°$ or $b \approx 575$, $\theta \approx 106.5°$, $\beta \approx 35.5°$ **17.** No such triangle exists
19. $\alpha \approx 22.3°$, $\beta \approx 109.7°$, $c \approx 5.48$ **21.** No such triangle exists   **23.** $\dfrac{a^2}{4}\sqrt{3}$

25. (a)

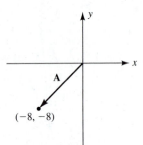

(b) $\langle -8, -8 \rangle$ (c) $8\sqrt{2}$ (d) $225°$

27. (a)

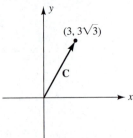

(b) $\langle 3, 3\sqrt{3} \rangle$ (c) 6 (d) $60°$
29. (a) $|\mathbf{A}| = 13, \theta \approx 337.4°$ (b) $|\mathbf{C}| = 4, \theta \approx 48.6°$
31. $\langle 3, 7 \rangle$ **33.** $\langle 0, -3 \rangle$ **35.** $\langle -21, 43 \rangle$
37. $\langle 75, -135 \rangle$
39.

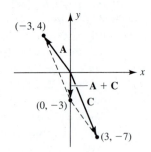

41. 30.0 ft **43.** About 6.2 min
45. (a) $d = \frac{3}{2} \sin \pi t$ (b) $\frac{3}{2\sqrt{2}} \approx 1.06$ inches below the equilibrium position **47.** 101.1 ft
49. (a) about 206 ft (b) about 186 ft
(c) about 276 ft **51.** about 387 m
53. (a) about 37.3 lb (b) about 25.2 lb
55. $\mathbf{F}_4 \approx \langle 161.86, -160.68 \rangle; |\mathbf{F}_4| \approx 228$ lb, $\theta = 315.2°$
57. approximately 171 mi, N-1.3°-E
59. approximately 75 mi, N-49°20'-E

CUMULATIVE REVIEW EXERCISES FOR CHAPTERS 1, 2, AND 3

1. $A = \dfrac{b\sqrt{4a^2 - b^2}}{4}$ **3.** (a) $\dfrac{1 + \sqrt{3}}{2}$ (b) $\dfrac{-3 + \sqrt{2}}{2}$

(c) $\dfrac{5\pi}{12}$ (d) $\dfrac{5}{12}$ **5.** $(-\infty, 1) \cup (1, 6) \cup (6, \infty)$;
$(f \circ f)(2) = -\dfrac{7}{39}$ **7.** $p = 12{,}500i^2$ **9.** (a) about
56.8 (b) $\dfrac{52\pi}{3} \approx 54.5$ **11.** 70 sq units **13.** about
$71\tilde{0}$ ft **15.** (a) $A = 1000 - \tfrac{2}{3}t$ (b) Domain:
[0, 1500]; range: [0, 1000] **17.** (a) about 136 ft and
285 ft (b) approximately 59,700 sq ft **19.** 14
21. (a)

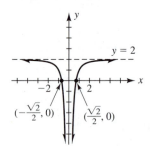

(b)

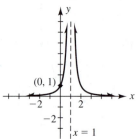

23. $f(0)$ **25.** about 142 ft
27. (a) $|\mathbf{A} - \mathbf{B}| = \sqrt{41} \approx 6.40, \theta \approx 321.3°$
(b) $|\mathbf{B} - \mathbf{A}| = \sqrt{41} \approx 6.40, \theta \approx 141.3°$
29. (a) $x \approx 29.2, y \approx 19.1, \theta = 49°$
(b) $\alpha \approx 55.3°, \beta \approx 86.7°, x \approx 21.0$

CHAPTER 4

Section 4.1

Problems

1. $\dfrac{4}{\csc \theta - 1} + \dfrac{3}{1 - \csc \theta} = \dfrac{4}{\csc \theta - 1} + \dfrac{-3}{\csc \theta - 1}$
$= \dfrac{1}{\csc \theta - 1}$

2. Factoring as a trinomial square, we have
$\csc^2 x - 3 \csc x - 10 = (\csc x - 5)(\csc x + 2)$

3. $\sin^2\theta \cot^2\theta + \sin^2\theta = \sin^2\theta(\cot^2\theta + 1)$
$= \sin^2\theta \csc^2\theta$
$= (\sin\theta \csc\theta)^2$
$= 1^2$
$= 1$

4. $\sin x + \cos x \cot x = \sin x + \cos x \dfrac{\cos x}{\sin x}$
$= \sin x + \dfrac{\cos^2 x}{\sin x}$
$= \dfrac{\sin^2 x + \cos^2 x}{\sin x}$
$= \dfrac{1}{\sin x} = \csc x$

5. $\tan\theta + \cot\theta = \sec\theta \csc\theta$

$\dfrac{\sin\theta}{\cos\theta} + \dfrac{\cos\theta}{\sin\theta}$
$\dfrac{\sin^2\theta + \cos^2\theta}{\cos\theta \sin\theta}$
$\dfrac{1}{\cos\theta \sin\theta}$
$\dfrac{1}{\cos\theta} \cdot \dfrac{1}{\sin\theta}$
$\sec\theta \csc\theta$

6. $\sec^4 x - \tan^4 x = \dfrac{1 + \sin^2 x}{\cos^2 x}$

$(\sec^2 x - \tan^2 x)(\sec^2 x + \tan^2 x)$
$1 \cdot (\sec^2 x + \tan^2 x)$
$\dfrac{1}{\cos^2 x} + \dfrac{\sin^2 x}{\cos^2 x}$
$\dfrac{1 + \sin^2 x}{\cos^2 x}$

7. $\sqrt{4 - x^2} = \sqrt{4 - (2\sin\theta)^2} = \sqrt{4 - 4\sin^2\theta} = \sqrt{4(1 - \sin^2\theta)} = \sqrt{4\cos^2\theta} = 2\cos\theta$ for $0 < \theta < \pi/2$

Exercises

1. $27 \cos^6 y \sin^3 y$ 3. $\dfrac{4\cos x}{\cos 2x}$
5. $9\tan^2\theta + 12\tan\theta + 4$ 7. $24 \tan x \sin x$
9. $\dfrac{1}{\cos x(1 - \cos x)}$ 11. $3\tan\theta(1 - 5\tan^2\theta)$
13. $(\cos\theta + \sin\theta)(\cos\theta - \sin\theta)$
15. $(\sin x - 3)(\sin x + 1)$ 17. $\dfrac{1}{\sin x - 1}$
19. $\sin x - \cos x$ 21. $\cot x$ 23. $\tan x$ 25. 1

27. 0   29. 2   31. $\tan x$ 33. $\cot x$ 35. 1
61. $\cos\theta$ 63. $\csc\theta$ 65. $\csc^3\theta$
67. $\tan\theta - \tfrac{1}{2}\sec\theta$ 69. The power of a sum is not the sum of the powers, that is, $(a + b)^2 \neq a^2 + b^2$.
71. Quadrants I and II   73. Quadrants II and IV
75. $x^2 + y^2 = 25$, a circle with center (0, 0) and radius 5
77. No. For example,
$\sin(30° + 60°) = \sin 90° = 1$, whereas
$\sin 30° + \sin 60° = \dfrac{1}{2} + \dfrac{\sqrt{3}}{2} = \dfrac{1 + \sqrt{3}}{2} \approx 1.37$.
79. Appears to be a trigonometric identity
81. Does not appear to be a trigonometric identity
83. Appears to be a trigonometric identity
85. Appears to be a trigonometric identity

Section 4.2

Problems

Note: In each of the following solutions, *n* represents an integer.

1. $\sqrt{3} \csc x - 2 = 0$
$\csc x = \dfrac{2}{\sqrt{3}}$
$x = \dfrac{\pi}{3}, \dfrac{2\pi}{3}$ in the interval $[0, 2\pi)$

Hence, $\dfrac{\pi}{3} + 2\pi n$ and $\dfrac{2\pi}{3} + 2\pi n$, describe the general form of all solutions.

2. Letting $n = 0$ and $n = 1$ in the general form
$$x = \dfrac{5\pi}{12} + \pi n,$$
we obtain
$$\dfrac{5\pi}{12} \text{ and } \dfrac{17\pi}{12},$$
as solutions in the interval $[0, 2\pi)$.

3. $\sin x \cos x - \cos x = 0$
$\cos x(\sin x - 1) = 0$
$\cos x = 0$ or $\sin x - 1 = 0$
$\sin x = 1$
$x = \dfrac{\pi}{2}, \dfrac{3\pi}{2}$ $x = \dfrac{\pi}{2}$

in the interval $[0, 2\pi)$. Thus, $\dfrac{\pi}{2} + \pi n$ describes all solutions.

4. Squaring both sides of $\sin x - \cos x = 1$, we obtain
$$\sin^2 x - 2\sin x \cos x + \cos^2 x = 1$$
$$1 - 2\sin x \cos x = 1$$
$$2\sin x \cos x = 0$$
$$\sin x = 0 \quad \text{or} \quad \cos x = 0$$
$$x = 0, \pi \qquad x = \frac{\pi}{2}, \frac{3\pi}{2}$$

as possible solutions in the interval $[0, 2\pi)$. However, 0 and $3\pi/2$ are extraneous roots. Hence, $\pi + 2\pi n$ and $\frac{\pi}{2} + 2\pi n$, describe all solutions.

5. For $\tan^2 x - 4 = 0$, we obtain $\tan x = \pm 2$. Now the reference angle x' (in radians) associated with this equation is
$$x' = \tan^{-1}|\pm 2| \approx 1.107$$

Since the tangent function is positive in quadrant I and negative in quadrant II, we place the reference angle in these quadrants as shown in the sketch and obtain
$$x \approx 1.107 + \pi n \quad \text{and} \quad x \approx 2.034 + \pi n$$

as the general form of all solutions. Hence, in the interval $[0, 2\pi)$ the solutions are
$$1.107, \ 2.034, \ 4.249, \ \text{and} \ 5.176.$$

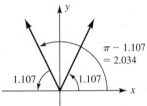

6. Replacing N with $10\frac{1}{2}$, we solve for d as follows:
$$10\tfrac{1}{2} = 12 + 3\sin\left[\frac{2\pi}{365}(d - 80)\right]$$
$$-\frac{3}{2} = 3\sin\left[\frac{2\pi}{365}(d - 80)\right]$$
$$-\frac{1}{2} = \sin\left[\frac{2\pi}{365}(d - 80)\right]$$
$$\frac{2\pi}{365}(d - 80) = \frac{7\pi}{6} \quad \text{or} \quad \frac{2\pi}{365}(d - 80) = \frac{11\pi}{6}$$
$$d = \frac{7\pi}{6} \cdot \frac{365}{2\pi} + 80 \qquad d = \frac{11\pi}{6} \cdot \frac{365}{2\pi} + 80$$
$$\approx 293 \qquad \qquad \approx 415$$

Thus, 10 hours 30 minutes of daylight occurs on day 293 (October 20) and day $415 - 365 = 50$ (February 19).

Exercises

Note: In each of the following answers, n represents an integer.

1. $\frac{\pi}{2} + 2\pi n$ 3. πn 5. $\frac{\pi}{6} + 2\pi n, \frac{11\pi}{6} + 2\pi n$

7. $\frac{\pi}{4} + 2\pi n, \frac{3\pi}{4} + 2\pi n$ 9. $\frac{\pi}{3} + \pi n$ 11. $\frac{\pi}{3} + 2\pi n, \frac{5\pi}{3} + 2\pi n$ 13. $\frac{\pi}{6} + \pi n, \frac{5\pi}{6} + \pi n$ 15. no solution

17. (a) $\frac{\pi}{4} + \pi n$ (b) $\frac{\pi}{4}, \frac{5\pi}{4}$ 19. (a) $3\pi + 4\pi n$ (b) none 21. (a) $\frac{3\pi}{16} + \frac{\pi}{2}n, \frac{5\pi}{16} + \frac{\pi}{2}n$

(b) $\frac{3\pi}{16}, \frac{5\pi}{16}, \frac{11\pi}{16}, \frac{13\pi}{16}, \frac{19\pi}{16}, \frac{21\pi}{16}, \frac{27\pi}{16}, \frac{29\pi}{16}$

23. (a) $\frac{\pi}{6} + \frac{\pi}{2}n, \frac{\pi}{3} + \frac{\pi}{2}n$

(b) $\frac{\pi}{6}, \frac{\pi}{3}, \frac{2\pi}{3}, \frac{5\pi}{6}, \frac{7\pi}{6}, \frac{4\pi}{3}, \frac{5\pi}{3}, \frac{11\pi}{6}$ 25. (a) $\frac{5\pi}{24} + \frac{\pi}{2}n$

(b) $\frac{5\pi}{24}, \frac{17\pi}{24}, \frac{29\pi}{24}, \frac{41\pi}{24}$ 27. (a) $\pm\frac{2\pi}{3} + 2\pi n$

(b) $\frac{2\pi}{3}, \frac{4\pi}{3}$ 29. (a) $\frac{5\pi}{36} + \frac{2\pi}{3}n, \frac{13\pi}{36} + \frac{2\pi}{3}n$

(b) $\frac{5\pi}{36}, \frac{13\pi}{36}, \frac{29\pi}{36}, \frac{37\pi}{36}, \frac{53\pi}{36}, \frac{61\pi}{36}$ 31. (a) $\frac{5\pi}{3} + 2\pi n, \frac{7\pi}{3} + 2\pi n$ (b) $\frac{\pi}{3}, \frac{5\pi}{3}$ 33. $\frac{3\pi}{2} + 2\pi n, \frac{\pi}{6} + 2\pi n, \frac{5\pi}{6} + 2\pi n$ 35. $\frac{\pi}{4} + \frac{\pi}{2}n$ 37. $\frac{\pi}{2} + \pi n, 2\pi n$

39. $\pi n, \frac{\pi}{3} + 2\pi n, \frac{5\pi}{3} + 2\pi n$ 41. $\frac{\pi}{2} + \pi n, \frac{\pi}{4} + \frac{\pi}{2}n$

43. $2\pi n, \frac{2\pi}{3} + 2\pi n, \frac{4\pi}{3} + 2\pi n$ 45. $\frac{\pi}{3} + \pi n, \frac{2\pi}{3} + \pi n$

47. $\frac{\pi}{6} + 2\pi n, \frac{5\pi}{6} + 2\pi n, \frac{3\pi}{2} + 2\pi n$ 49. $\pi + 2\pi n, \frac{2\pi}{3} + 2\pi n, \frac{4\pi}{3} + 2\pi n$ 51. $\frac{\pi}{3} + \pi n, \frac{2\pi}{3} + \pi n$

53. πn 55. $\frac{\pi}{2} + \pi n$ 57. $2\pi n, \frac{3\pi}{2} + 2\pi n$

59. $\pi + 2\pi n$ 61. in seconds: $\frac{\pi}{3}, \frac{2\pi}{3}, \frac{4\pi}{3}, \frac{5\pi}{3}$

63. 1990, 1994 65. $x_0 + \frac{2\pi}{b}n$ 67. Since the graph of $y = \sin nx$ completes n cycles in the interval $[0, 2\pi)$, the equation $\sin nx = k$ for $-1 < k < 1$ must have $2n$ solutions.

69. $\frac{\pi}{2} + \pi n, \frac{\pi}{6} + 2\pi n, \frac{5\pi}{6} + 2\pi n$ **71.** Since $\sin^2 x + \cos^2 x = 1$, $\sin^2 x = 1$ implies $\cos^2 x = 0$. Hence, dividing both sides by $\cos^2 x$ does not generate an equivalent equation.
73. 0.7297, 2.412   **75.** 2.944, 6.086   **77.** 1.231, 1.911, 4.373, 5.052   **79.** 0.3091, 0.7381, 2.403, 2.832, 4.498, 4.927   **81.** 0.2527, 2.889   **83.** 1.249, 2.034, 4.391, 5.176   **85.** 0.6818, 2.460   **87.** 0.4636, 3.605
89. $-1.11, 0.322$ **91.** 2.47, 3.82   **93.** 0.393, 1.96, 3.53, 5.11   **95.** day 127 (May 7), day 215 (August 3)

Section 4.3

Problems

1. $\cos 3x \cos 2x + \sin 3x \sin 2x = \cos(3x - 2x) = \cos x$

2. $\sin\left(x + \frac{\pi}{6}\right) \cos x - \cos\left(x + \frac{\pi}{6}\right) \sin x$
$= \left(\sin x \cos \frac{\pi}{6} + \cos x \sin \frac{\pi}{6}\right) \cos x$
$\quad - \left(\cos x \cos \frac{\pi}{6} - \sin x \sin \frac{\pi}{6}\right) \sin x$
$= \cos^2 x \sin \frac{\pi}{6} + \sin^2 x \sin \frac{\pi}{6}$
$= \sin \frac{\pi}{6} (\cos^2 x + \sin^2 x)$
$= \sin \frac{\pi}{6} = \frac{1}{2}$

3. With a calculator in degree mode, a typical keying sequence is

Rounding the display to four significant digits, we obtain $1.732 \approx \sqrt{3}$.

4. With a calculator in degree mode, we find that $\cos 75° \approx 0.2588$. For $(\sqrt{6} - \sqrt{2})/4$, a typical keying sequence is

Rounding the display to four significant digits, we find that $(\sqrt{6} - \sqrt{2})/4 \approx 0.2588$.

5. From Example 5, $\sin u = \frac{3}{5}$ and $\cos u = -\frac{4}{5}$. Also, if $\tan v = -\frac{1}{4}$ with v in quadrant II, then
$$\sec v = -\sqrt{1 + \tan^2 v} = -\sqrt{1 + \left(-\frac{1}{4}\right)^2} = -\frac{\sqrt{17}}{4}$$

Hence, $\cos v = -\frac{4}{\sqrt{17}}$ and $\sin v = \frac{1}{\sqrt{17}}$. Thus,

$\sin(u - v) = \sin u \cos v - \cos u \sin v$
$= \left(\frac{3}{5}\right)\left(-\frac{4}{\sqrt{17}}\right) - \left(-\frac{4}{5}\right)\left(\frac{1}{\sqrt{17}}\right)$
$= -\frac{8}{5\sqrt{17}}.$

6. $\sin 3x \cot x + \cos 3x = \sin 4x \csc x$

$\begin{array}{c|c}
\sin 3x \cdot \dfrac{\cos x}{\sin x} + \cos 3x & \\
\dfrac{\sin 3x \cos x + \cos 3x \sin x}{\sin x} & \\
& \dfrac{\sin 4x}{\sin x} \\
& \sin 4x \csc x
\end{array}$

7. $\sin 3x \cos x = 1 - \cos 3x \sin x$
$\sin 3x \cos x + \cos 3x \sin x = 1$
$\sin 4x = 1$
$4x = \frac{\pi}{2} + 2\pi n$
$x = \frac{\pi}{8} + \frac{\pi}{2} n$

8. If $i(t) = 1$ amp. then
$$1 = 2 \sin\left(4t + \frac{\pi}{6}\right) \quad \text{or} \quad \frac{1}{2} = \sin\left(4t + \frac{\pi}{6}\right),$$

which implies

$4t + \frac{\pi}{6} = \frac{\pi}{6} + 2\pi n \quad \text{or} \quad 4t + \frac{\pi}{6} = \frac{5\pi}{6} + 2\pi n$
$t = \frac{\pi}{2} n \qquad\qquad\qquad t = \frac{\pi}{6} + \frac{\pi}{2} n$

Thus, the only time in the interval $[0, \pi/2)$ for which $i(t) = 1$ amp is when $t = 0$ or $\pi/6$ s.

Exercises

1. $-\sin x$ **3.** $\sin x$ **5.** $\tan x$ **7.** 0   **9.** 1
11. $\cos 8x$ **13.** $\cos x$ **15.** 1   **17.** $\dfrac{1 + \sqrt{3}}{2\sqrt{2}}$ or
$\dfrac{\sqrt{2} + \sqrt{6}}{4}$ **19.** $\dfrac{1 + \sqrt{3}}{2\sqrt{2}}$ or $\dfrac{\sqrt{2} + \sqrt{6}}{4}$ **21.** $\dfrac{1 + \sqrt{3}}{1 - \sqrt{3}}$
23. $-\frac{5}{3}$ **25.** $\dfrac{4\sqrt{3} - 3}{10}$ **27.** $-\frac{16}{65}$ **29.** $-\dfrac{13}{5\sqrt{10}}$
43. $\frac{\pi}{4} + \frac{\pi}{2} n$ **45.** $\frac{\pi}{4} + 2\pi n, \frac{3\pi}{4} + 2\pi n$
47. $\frac{\pi}{12} + \frac{\pi}{3} n$ **49.** πn **51.** $\frac{\pi}{3} + 2\pi n, \frac{5\pi}{3} + 2\pi n$
53. $\frac{\pi}{4} + 2\pi n, \frac{7\pi}{4} + 2\pi n$

55. (a) $f(x) = 3\sqrt{2} \sin\left(2x + \dfrac{\pi}{4}\right)$ (b) amplitude: $3\sqrt{2}$; period: π; phase shift: $\pi/8$ unit left

57. (a) $f(x) = 2 \sin\left(5x + \dfrac{5\pi}{6}\right)$ (b) amplitude: 2; period: $2\pi/5$; phase shift: $\pi/6$ unit left

59. (a) $\sin\left(\dfrac{\pi}{6} + \dfrac{\pi}{3}\right) = \sin\dfrac{\pi}{2} = 1$, but

$\sin\dfrac{\pi}{6} + \sin\dfrac{\pi}{3} = \dfrac{1}{2} + \dfrac{\sqrt{3}}{2} = \dfrac{1 + \sqrt{3}}{2} \approx 1.366$

(b) Yes. For example, let $u = \pi/6$ and $v = 11\pi/6$, then

$\sin\left(\dfrac{\pi}{6} + \dfrac{11\pi}{6}\right) = \sin 2\pi = 0$, and

$\sin\dfrac{\pi}{6} + \sin\dfrac{11\pi}{6} = \dfrac{1}{2} + \left(-\dfrac{1}{2}\right) = 0.$ **61.** (a) $\dfrac{1}{\sqrt{2}}$

(b) $\dfrac{2 + \sqrt{15}}{4\sqrt{5}}$ or $\dfrac{2\sqrt{5} + 5\sqrt{3}}{20}$

63. $\sin C = a\sqrt{1 - b^2} + b\sqrt{1 - a^2}$ **67.** (a) 0.999
(b) 0.999 **69.** (a) -0.914 (b) -0.914
71. (a) 28.6 (b) 28.6
73. (a) $f(x) = 16.3 \sin(5.2x + 5.60)$
(b) $f(x) = 3.18 \sin(3.7x + 2.01)$ **75.** (a) $a = 13$, $c \approx 1.176$ (b) in seconds: 0.067, 0.309, 0.400, 0.642, 0.733, 0.975

Section 4.4

◆ Problems

1. $\dfrac{2 \tan 3x}{1 - \tan^2 3x} = \tan 2(3x) = \tan 6x$

2. $2 \cos x \csc 2x = \csc x$

$\begin{array}{c|c} 2 \cos x \cdot \dfrac{1}{\sin 2x} & \\ 2 \cos x \cdot \dfrac{1}{2 \sin x \cos x} & \\ \dfrac{1}{\sin x} & \\ \csc x & \end{array}$

3. $\cos x + \sin 2x = 0$
$\cos x + 2 \sin x \cos x = 0$
$\cos x(1 + 2 \sin x) = 0$
$\cos x = 0 \quad \text{or} \quad 1 + 2 \sin x = 0$
$\sin x = -\dfrac{1}{2}$

Hence, the general form of the solutions is

$x = \dfrac{\pi}{2} + \pi n, \quad \dfrac{7\pi}{6} + 2\pi n, \quad \dfrac{11\pi}{6} + 2\pi n.$

4. $\sin 3\left(\dfrac{\pi}{6}\right) = 3 \sin\dfrac{\pi}{6} - 4 \sin^3\dfrac{\pi}{6}$?
$\sin\dfrac{\pi}{2} = 3\left(\dfrac{1}{2}\right) - 4\left(\dfrac{1}{2}\right)^3$?
$1 = \dfrac{3}{2} - \dfrac{1}{2}$?
$1 = 1$ ✓

5. $\cos^4 x = \cos^2 x \cos^2 x$
$= \dfrac{1}{2}(1 + \cos 2x) \cdot \dfrac{1}{2}(1 + \cos 2x)$
$= \dfrac{1}{4}(1 + 2 \cos 2x + \cos^2 2x)$
$= \dfrac{1}{4}\left[1 + 2 \cos 2x + \dfrac{1}{2}(1 + \cos 4x)\right]$
$= \dfrac{1}{4}\left(\dfrac{3}{2} + 2 \cos 2x + \dfrac{1}{2} \cos 4x\right)$
$= \dfrac{1}{8}(3 + 4 \cos 2x + \cos 4x)$

6. With a calculator in radian mode, we find that $\sin(\pi/12) \approx 0.2588$. For $\sqrt{2 - \sqrt{3}}/2$, a typical keying sequence is

Rounding the display to four significant digits, we find that $\sqrt{2 - \sqrt{3}}/2 \approx 0.2588.$

7. If θ is a third-quadrant angle, then

$\pi < \theta < \dfrac{3\pi}{2} \quad \text{or} \quad \dfrac{\pi}{2} < \dfrac{\theta}{2} < \dfrac{3\pi}{4}.$

Hence, $\sin\dfrac{\theta}{2}$ is *positive* and

$\sin\dfrac{\theta}{2} = \sqrt{\dfrac{1 - \cos\theta}{2}} = \sqrt{\dfrac{1 - (-0.28)}{2}}$
$= \sqrt{0.64}$
$= 0.8$

8. $\tan\dfrac{x}{2} + \cot\dfrac{x}{2} = 2 \csc x$

$\begin{array}{c|c} \dfrac{1 - \cos x}{\sin x} + \dfrac{\sin x}{1 - \cos x} & \\ \dfrac{(1 - 2 \cos x + \cos^2 x) + \sin^2 x}{\sin x(1 - \cos x)} & \\ \dfrac{2(1 - \cos x)}{\sin x(1 - \cos x)} & \\ 2 \csc x & \end{array}$

9. If $A = 170$ sq ft, then $170 = 200 \sin\theta$, or $\sin\theta = \dfrac{170}{200}$. Hence, $\theta = \arcsin\dfrac{170}{200} \approx 58.2°.$

◆ Exercises

1. $5 \sin 6x$ **3.** $\cos 4x$ **5.** -1 **7.** $\cot 6x$
9. $\cos x$ **11.** $\cos x$ **13.** 1 **15.** $\sec^2 3x$

29. πn **31.** $\pi n, \dfrac{\pi}{6} + 2\pi n, \dfrac{5\pi}{6} + 2\pi n$

33. $\frac{\pi}{6} + \frac{\pi}{2}n, \frac{\pi}{3} + \frac{\pi}{2}n$ **35.** $\pi n, \frac{2\pi}{3} + 2\pi n, \frac{4\pi}{3} + 2\pi n$

37. $\pi n, \frac{\pi}{3} + \pi n, \frac{2\pi}{3} + \pi n$

39. $\cos 3x = 4\cos^3 x - 3\cos x$
41. $\sin 5x = 5\sin x - 20\sin^3 x + 16\sin^5 x$
43. $\cos^2 3x = \frac{1}{2}(1 + \cos 6x)$
45. $\sin^4 x \cos^2 x = \frac{1}{16}(1 - \cos 2x - \cos 4x + \cos 4x \cos 2x)$
47. $\frac{1}{2}\sqrt{2 + \sqrt{3}}$ **49.** $\frac{1}{2}\sqrt{2 - \sqrt{2}}$ **51.** (a) $\frac{24}{25}$
(b) $\frac{7}{25}$ (c) $\frac{1}{\sqrt{10}}$ (d) $\frac{3}{\sqrt{10}}$ **53.** (a) $\frac{3}{4}$ (b) $-\frac{5}{4}$
(c) $\frac{1}{10}\sqrt{50 - 5\sqrt{10}}$ (d) $-\frac{1}{3}\sqrt{11 + 2\sqrt{10}}$ **61.** $2\pi n$
63. $2\pi n, \frac{2\pi}{3} + 2\pi n$ **65.** $\frac{\pi}{2} + \pi n, \frac{\pi}{4} + \frac{\pi}{2}n$
67. (a) $V = 36\sin\theta$ (b) 36 cu ft (c) $\theta \approx 56.4°$
69. 30° and 60° **71.** (a) amplitude: 3; period: $\frac{\pi}{4}$
(b) amplitude: $\frac{1}{2}$; period: π **73.** (a) $\frac{3}{5}$ (b) $-\frac{7}{8}$
77. $\cos 20° \approx 0.940$, $\sin 20° \approx 0.342$, $\tan 20° \approx 0.364$
79. 0.253, 1.57, 2.89, 4.71 **81.** (a) 0.262, 1.31, 3.40,
4.45 (b) $\frac{\pi}{12}, \frac{5\pi}{12}, \frac{13\pi}{12}, \frac{17\pi}{12}$ **83.** (a) 1.05, 1.57, 2.09,
4.19, 4.71, 5.24 (b) $\frac{\pi}{3}, \frac{2\pi}{3}, \frac{4\pi}{3}, \frac{5\pi}{3}$. Note: The double-angle
formula does not pick up the solutions $\pi/2$ and $3\pi/2$, since
$\tan(\pi/2)$ and $\tan(3\pi/2)$ are undefined. **85.** Appears to be
a trigonomtric identity **87.** Appears to be a trigonometric
identity

Section 4.5

Problems

1. $\sin 3x \cos 2x = \cos 2x \sin 3x$
$= \frac{1}{2}[\sin(2x + 3x) - \sin(2x - 3x)]$
$= \frac{1}{2}[\sin 5x - \sin(-x)]$.
However, since the sine function is an odd function and
$\sin(-x) = -\sin x$, we have
$\cos 2x \sin 3x = \frac{1}{2}(\sin 5x + \sin x) = \frac{1}{2}\sin 5x + \frac{1}{2}\sin x$.

2. $\sin 5x + \sin 7x = 2\sin\frac{5x + 7x}{2} \cos\frac{5x - 7x}{2}$
$= 2\sin 6x \cos(-x)$.
However, since the cosine function is an even function and
$\cos(-x) = \cos x$, we have $\sin 5x + \sin 7x = 2\sin 6x \cos x$.

3. $\cos 75° - \cos 15° = -2\sin\frac{75° + 15°}{2} \sin\frac{75° - 15°}{2}$
$= -2\sin 45° \sin 30°$
$= -2 \cdot \frac{1}{\sqrt{2}} \cdot \frac{1}{2} = -\frac{1}{\sqrt{2}}$

4. $2\cos^2 x \sin x - 2\sin^3 x = \sin 3x - \sin x$

	$2\cos\frac{3x+x}{2}\sin\frac{3x-x}{2}$
	$2\cos 2x \sin x$
	$2(\cos^2 x - \sin^2 x)\sin x$
	$2\cos^2 x \sin x - 2\sin^3 x$

5. To find the six solutions in the interval $[0, 2\pi)$, we let
$n = 0, 1, 2, 3, 4, 5$ in the general form $\frac{\pi n}{3}$, and obtain 0,
$\frac{\pi}{3}, \frac{2\pi}{3}, \pi, \frac{4\pi}{3}, \frac{5\pi}{3}$.

6.
$3\sin 3x + \sin x = 0$
$3(3\sin x - 4\sin^3 x) + \sin x = 0$
$10\sin x - 12\sin^3 x = 0$
$2\sin x(5 - 6\sin^2 x) = 0$
$2\sin x = 0 \quad \text{or} \quad 5 - 6\sin^2 x = 0$
$\sin x = 0 \qquad\qquad \sin x = \pm\sqrt{\frac{5}{6}}$

Hence, $x = \pi n$ and $x \approx 1.15 + \pi n, 1.99 + \pi n$. Thus, in the
interval $[0, 2\pi)$ we have
Relative maxima: $(0, 2), (1.99, 0.544), (4.29, 0.544)$
Relative minima: $(1.15, -0.544), (\pi, -2), (5.13, -0.544)$

Exercises

1. $\cos 2x - \cos 4x$ **3.** $\cos 8x + \cos 4x$
5. $\frac{1}{2}\sin 7x + \frac{1}{2}\sin x$ **7.** $3\sin 3x + 3\sin 7x$
9. $2\cos 3x \cos x$ **11.** $2\sin x \cos 4x$
13. $2\sin x \sin\frac{x}{3}$ **15.** $-\cos\frac{9x}{2}\sin\frac{3x}{2}$ **17.** $\frac{1}{4}$
19. $\frac{\sqrt{3} + 2}{4}$ **21.** $-\frac{\sqrt{6}}{2}$ **23.** $-\frac{1}{\sqrt{2}}$ **37.** $\frac{\pi}{2}n$
39. $\frac{\pi}{4}n$ **41.** $\frac{\pi}{3}n, \frac{\pi}{2} + 2\pi n$ **43.** $\frac{\pi}{3}n$
45. $\frac{3\pi}{4} + \pi n$ **47.** $\frac{\pi}{4} + \frac{\pi}{2}n, \pi n$ **49.** (a) $\frac{\pi}{3} + \frac{2\pi}{3}n$
(b)

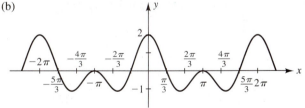

51. (a) $\pi n, \frac{\pi}{6} + \frac{\pi}{3}n$

51. (b)

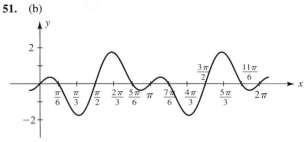

53. Relative maxima: $(0, 2)$, $(\pi, 0)$; relative minima: $\left(\arccos\left(-\frac{1}{4}\right), -\frac{9}{8}\right)$, $\left(2\pi - \arccos\left(-\frac{1}{4}\right), -\frac{9}{8}\right)$ **55.** $\sin 50°$
57. (a) 0; for supplementary angles x and y, $\cos x = -\cos y$.
(b) 0; for supplementary angles x and y, $\sin x = \sin y$.
59. Yes, because the least common multiple for their periods $(2\pi/n$ and $2\pi)$ is 2π.
61.
(a) $2 \cos 56.3° \cos 25.7° = \cos 82.0° + \cos 30.6° \approx 0.9999$
(b) $2 \sin 147°46' \sin 112°52' = \cos 34°54' - \cos 260°38'$
≈ 0.9829
(c) $2 \sin \frac{3\pi}{10} \cos \frac{\pi}{5} = \sin \frac{\pi}{2} + \sin \frac{\pi}{10} \approx 1.309$
(d) $2 \cos 3.02 \sin 2.31 = \sin 5.33 - \sin 0.71 \approx -1.467$
63. (a) π (b) $0.00, 1.05, 2.09, 3.14, 4.19, 5.24$
(c) Relative maxima: $(0.00, 0.00), (1.57, 2.00), (3.14, 0.00)$, $(4.71, 2.00)$; relative minima: $(0.659, -1.12), (2.48, -1.12)$, $(3.80, -1.12), (5.62, -1.12)$ (d) $0, \frac{\pi}{3}, \frac{2\pi}{3}, \pi, \frac{4\pi}{3}, \frac{5\pi}{3}$
65. (a) 2π (b) $0.00, 0.785, 2.36, 3.14, 3.93, 5.50$
(c) Relative maxima: $(0.421, 0.544), (2.72, 0.544)$, $(4.71, 2.00)$; relative minima: $(1.57, -2.00), (3.56, -0.544)$, $(5.86, -0.544)$ (d) $0, \frac{\pi}{4}, \frac{3\pi}{4}, \pi, \frac{5\pi}{4}, \frac{7\pi}{4}$

29. $\frac{1}{16}(1 + \cos 2x - \cos 4x - \cos 4x \cos 2x)$ or, applying a product-to-sum formula, $\frac{1}{32}(2 + \cos 2x - 2 \cos 4x - \cos 6x)$
61. (a) $\frac{2\pi}{3} + 2\pi n, \frac{4\pi}{3} + 2\pi n$ (b) $\frac{2\pi}{3}, \frac{4\pi}{3}$
63. (a) $\frac{\pi}{3} + \pi n, \frac{2\pi}{3} + \pi n$ (b) $\frac{\pi}{3}, \frac{2\pi}{3}, \frac{4\pi}{3}, \frac{5\pi}{3}$
65. (a) $\frac{3\pi}{8} + \frac{\pi}{2} n$ (b) $\frac{3\pi}{8}, \frac{7\pi}{8}, \frac{11\pi}{8}, \frac{15\pi}{8}$
67. (a) $\frac{2\pi}{9} + \frac{2\pi}{3} n$ (b) $\frac{2\pi}{9}, \frac{8\pi}{9}, \frac{14\pi}{9}$
69. (a) $\pi n, \frac{\pi}{6} + 2\pi n, \frac{5\pi}{6} + 2\pi n$ (b) $0, \frac{\pi}{6}, \frac{5\pi}{6}, \pi$
71. (a) $\pi + 2\pi n, \frac{\pi}{3} + 2\pi n, \frac{5\pi}{3} + 2\pi n$ (b) $\frac{\pi}{3}, \pi, \frac{5\pi}{3}$
73. (a) $\frac{3\pi}{2} + 2\pi n$ (b) $\frac{3\pi}{2}$ **75.** (a) $\frac{\pi}{2} + 2\pi n$
(b) $\frac{\pi}{2}$ **77.** (a) $\pi n, \frac{2\pi}{3} + 2\pi n, \frac{4\pi}{3} + 2\pi n$
(b) $0, \pi, \frac{2\pi}{3}, \frac{4\pi}{3}$ **79.** (a) $\frac{\pi}{4} + \pi n$ (b) $\frac{\pi}{4}, \frac{5\pi}{4}$
81. (a) $2\pi n$ (b) 0   **83.** (a) $\frac{\pi}{4} n$
(b) $0, \frac{\pi}{4}, \frac{\pi}{2}, \frac{3\pi}{4}, \pi, \frac{5\pi}{4}, \frac{3\pi}{2}, \frac{7\pi}{4}$
85. (a) $\frac{\pi}{12} + \frac{\pi}{2} n, \frac{5\pi}{12} + \frac{\pi}{2} n$
(b) $\frac{\pi}{12}, \frac{5\pi}{12}, \frac{7\pi}{12}, \frac{11\pi}{12}, \frac{13\pi}{12}, \frac{17\pi}{12}, \frac{19\pi}{12}, \frac{23\pi}{12}$ **87.** 3.553,
5.872   **89.** 1.017, 2.034, 4.249, 5.176   **91.** 0.3218,
0.4636, 3.463, 3.605   **93.** 0.2846, 2.857   **95.** 0.3398,
2.802, 3.481, 5.943   **97.** (a) $f(x) = 2 \sin\left(3x + \frac{\pi}{3}\right)$

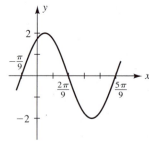

amplitude: 2; period: $\frac{2\pi}{3}$; phase shift: left $\frac{\pi}{9}$

(b) $f(x) = 2\sqrt{2} \sin\left(4\pi x + \frac{7\pi}{4}\right)$

Chapter 4 Review Exercises

1. $\sin 2x$ **3.** 1   **5.** $\cos 9x$ **7.** $\cos x$
9. $\sin 6x$ **11.** $\tan 8x$ **13.** $|\sin 5x|$
15. $\sin 6x + \sin 4x$ **17.** $-2 \sin \frac{5x}{2} \sin \frac{x}{2}$
19. $\frac{\sqrt{2 - \sqrt{2}}}{2}$ **21.** $\frac{2 + \sqrt{3}}{4}$ **23.** (a) $\frac{7}{5\sqrt{2}}$
(b) $\frac{4\sqrt{3} + 3}{10}$ (c) $\frac{24}{25}$ (d) $\frac{3}{\sqrt{10}}$
25. (a) $\frac{12\sqrt{5} + 10}{39}$ (b) $\frac{5\sqrt{5} + 24}{39}$ (c) $\frac{120}{119}$
(d) $\frac{\sqrt{18 + 6\sqrt{5}}}{6}$ **27.** $3 \sec \theta$

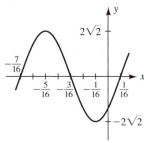

amplitude: $2\sqrt{2}$; period: $\frac{1}{2}$; phase shift: left $\frac{7}{16}$

99. (a) $a = 40$, $c \approx 0.9273$ (b) in seconds: 0.078, 0.223, 0.328, 0.473, 0.578, 0.723, 0.828, 0.973 **101.** 28.2°, 61.8°

CHAPTER 5

Section 5.1

◆ Problems

1. (a) $3\sqrt{-80} - \frac{1}{3}\sqrt{-45} = 3 \cdot 4i\sqrt{5} - \frac{1}{3} \cdot 3i\sqrt{5} = 11i\sqrt{5}$

(b) $\dfrac{9}{\sqrt{-36}} = \dfrac{9}{6i} = \dfrac{3}{2i} = -\dfrac{3}{2}i$

2. (a) $i^{22} = (i^4)^5 i^2 = (1)^5(-1) = -1$

(b) $\dfrac{1}{i^{51}} = \dfrac{1}{(i^4)^{12} i^3} = \dfrac{1}{(1)^{12}(-i)} = \dfrac{1}{-i} = i$

3. $(2 + 3i) - (5 - 4i) + (-3 + i)$
$= (2 - 5 - 3) + (3 + 4 + 1)i = -6 + 8i$

4. $(1 + 3i)(2 + 5i) = 2 + 11i + 15i^2 = -13 + 11i$

5. $\dfrac{1 + \sqrt{-9}}{1 - \sqrt{-9}} = \dfrac{1 + 3i}{1 - 3i} \cdot \dfrac{1 + 3i}{1 + 3i} = \dfrac{1 + 6i + 9i^2}{1 - 9i^2}$

$= \dfrac{-8 + 6i}{10} = -\dfrac{4}{5} + \dfrac{3}{5}i$

◆ Exercises

1. $9i$ **3.** -18 **5.** $3i\sqrt{5}$ **7.** 3 **9.** $-\dfrac{3}{2}i$
11. $-48i$ **13.** -5 **15.** $2i$ **17.** 0 **19.** $108i$
21. $15 + 3i$ **23.** $-2 + 33i$ **25.** $24 + 16i$
27. $30 + 19i$ **29.** $-\dfrac{3}{5} - \dfrac{4}{5}i$ **31.** $-\dfrac{\sqrt{6}}{5} + \dfrac{2}{5}i$
33. $2 - 4i$ **35.** $-99 - 20i$ **37.** $-9 - 46i$
39. $-\dfrac{21}{2} + \dfrac{51}{2}i$ **41.** $\dfrac{7}{25} - \dfrac{1}{25}i$
43. Since $(1 + 0i)(a + bi) = a + bi$, $1 + 0i$ is the multiplicative identity of $a + bi$.
45. (a) 0 (b) 0 **47.** (a) 0 (b) 0

49. Since $\left(\dfrac{\sqrt{2} + \sqrt{2}\,i}{2}\right)^2 = i$, $\dfrac{\sqrt{2} + \sqrt{2}\,i}{2}$ is a square root of i.

51. (a) $4.24i$ (b) $22.5i$ (c) $12.4i$ (d) $0.936i$
53. 1.07 A

Section 5.2

◆ Problems

1.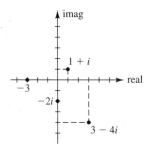

(a) The complex conjugate of $3 + 4i$ is $3 - 4i$.
(b) The complex conjugate of $1 - i$ is $1 + i$.
(c) The complex conjugate of $2i$ is $-2i$.
(d) The complex conjugate of -3 is -3.

2. The modulus r of the complex number $-\sqrt{3} + i$ is
$$r = \sqrt{(-\sqrt{3})^2 + (1)^2} = 2$$

Since $-\sqrt{3} + i$ lies in quadrant II, θ must be a second quadrant angle that satisfies the equation
$$\tan\theta = \dfrac{1}{-\sqrt{3}}$$

Therefore, $\theta = 5\pi/6$ and we conclude that
$$-\sqrt{3} + i = 2\left(\cos\dfrac{5\pi}{6} + i\sin\dfrac{5\pi}{6}\right).$$

3. Using a calculator set in the radian mode, we find $2\sqrt{5}\cos 5.82 \approx 4.00$ and $2\sqrt{5}\sin 5.82 \approx -2.00$. Hence, $2\sqrt{5}(\cos 5.82 + i\sin 5.82) \approx 4.00 - 2.00i$.

4. $\dfrac{6 + 6\sqrt{3}\,i}{-\dfrac{3\sqrt{3}}{2} - \dfrac{3}{2}i} = \dfrac{6 + 6\sqrt{3}\,i}{-\dfrac{3\sqrt{3}}{2} - \dfrac{3}{2}i} \cdot \dfrac{-\dfrac{3\sqrt{3}}{2} + \dfrac{3}{2}i}{-\dfrac{3\sqrt{3}}{2} + \dfrac{3}{2}i}$

$= \dfrac{-9\sqrt{3} + 9i - 27i - 9\sqrt{3}}{\dfrac{27}{4} + \dfrac{9}{4}}$

$= \dfrac{-18\sqrt{3} - 18i}{9} = -2\sqrt{3} - 2i$

5. For $R = 6\,\Omega$, $X_L = 5\,\Omega$, and $X_C = 3\,\Omega$, we have $Z = 6 + (5 - 3)i = 6 + 2i$.

(a) $|Z| = \sqrt{6^2 + 2^2} = \sqrt{40} = 2\sqrt{10} \approx 6.32\,\Omega$.

(b) Since $6 + 2i$ lies in quadrant I, θ must be a first quadrant angle that satisfies the equation $\tan\theta = \frac{2}{6} = \frac{1}{3}$.

Hence, $\theta = \arctan \frac{1}{3} \approx 18.4°$. Therefore, we conclude that the voltage *leads* the current by 18.4°.

Exercises

1. (a)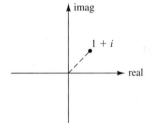

(b) $\sqrt{2}$ (c) $\frac{\pi}{4}$ (d) $\sqrt{2}\left(\cos\frac{\pi}{4} + i\sin\frac{\pi}{4}\right)$

3. (a)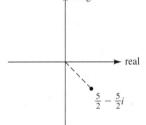

(b) $\frac{5}{2}\sqrt{2}$ (c) $\frac{7\pi}{4}$ (d) $\frac{5}{2}\sqrt{2}\left(\cos\frac{7\pi}{4} + i\sin\frac{7\pi}{4}\right)$

5. (a)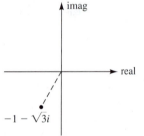

(b) 2 (c) $\frac{4\pi}{3}$ (d) $2\left(\cos\frac{4\pi}{3} + i\sin\frac{4\pi}{3}\right)$

7. (a)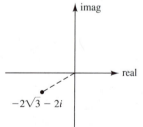

(b) 4 (c) $\frac{7\pi}{6}$ (d) $4\left(\cos\frac{7\pi}{6} + i\sin\frac{7\pi}{6}\right)$

9. (a)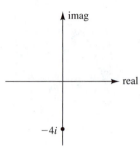

(b) 4 (c) $\frac{3\pi}{2}$ (d) $4\left(\cos\frac{3\pi}{2} + i\sin\frac{3\pi}{2}\right)$

11. (a)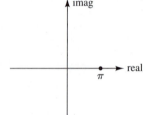

(b) π (c) 0 (d) $\pi(\cos 0 + i \sin 0)$

13. (a)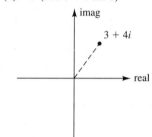

(b) 5 (c) $\arctan\frac{4}{3} \approx 0.927$
(d) $5(\cos 0.927 + i \sin 0.927)$

15. (a)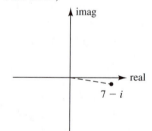

(b) $5\sqrt{2}$ (c) $2\pi - \arctan\frac{1}{7} \approx 6.14$
(d) $5\sqrt{2}(\cos 6.14 + i \sin 6.14)$

17. (a)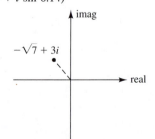

CHAPTER 5

(b) 4   (c) $\pi - \arctan\dfrac{3}{\sqrt{7}} \approx 2.29$

(d) $4(\cos 2.29 + i \sin 2.29)$

19. (a)

$-\tfrac{1}{2} - \tfrac{2}{3}i$

(b) $\tfrac{5}{6}$ (c) $\pi + \arctan\tfrac{4}{3} \approx 4.07$
(d) $\tfrac{5}{6}(\cos 4.07 + i \sin 4.07)$ 21. $6\sqrt{3} + 6i$

23. $-\dfrac{1}{4} + \dfrac{\sqrt{3}}{4}i$ 25. $-1 - i$ 27. $9i$ 29. $-\tfrac{2}{3}$

31. (a) $18\left(\cos\dfrac{3\pi}{2} + i\sin\dfrac{3\pi}{2}\right)$ (b) $-18i$

33. (a) $4\left(\cos\dfrac{5\pi}{4} + i\sin\dfrac{5\pi}{4}\right)$ (b) $-2\sqrt{2} - 2\sqrt{2}\,i$

35. (a) $10\left(\cos\dfrac{5\pi}{3} + i\sin\dfrac{5\pi}{3}\right)$ (b) $5 - 5\sqrt{3}\,i$

37. (a) $4\left(\cos\dfrac{\pi}{2} + i\sin\dfrac{\pi}{2}\right)$ (b) $4i$

39. (a) $6\left(\cos\dfrac{3\pi}{4} + i\sin\dfrac{3\pi}{4}\right)$ (b) $-3\sqrt{2} + 3\sqrt{2}\,i$

41. (a) $\cos\dfrac{3\pi}{2} + i\sin\dfrac{3\pi}{2}$ (b) $-i$ 43. (a) $5\sqrt{2}\,\Omega$

(b) $45°$ 45. (a) $8\,\Omega$ (b) $-30°$

47. $\dfrac{1}{r}[\cos(-\theta) + i\sin(-\theta)]$ or $\dfrac{1}{r}(\cos\theta - i\sin\theta)$

49. (a) $\dfrac{\sqrt{2} - \sqrt{6}}{4}$ (b) $\dfrac{\sqrt{2} + \sqrt{6}}{4}$

51. $8.34(\cos 1.25 + i \sin 1.25)$
53. $1.39(\cos 5.20 + i \sin 5.20)$ 55. $3.00 + 6.00i$
57. $-5.83 + 12.7i$ 59. $-2.77 - 3.01i$
61. $-1.14 + 1.32i$ 63. $-2.86 + 0.192i$
65. magnitude ≈ 43.3 volts, phase angle $\approx 35.2°$

Section 5.3

Problems

1. $(\sqrt{3} + i)^9 = \left[2\left(\cos\dfrac{\pi}{6} + i\sin\dfrac{\pi}{6}\right)\right]^9$

$= 2^9\left(\cos\dfrac{3\pi}{2} + i\sin\dfrac{3\pi}{2}\right)$

$= -512i$

2. $(\sqrt{3} + i)^{-9} = \left[2\left(\cos\dfrac{\pi}{6} + i\sin\dfrac{\pi}{6}\right)\right]^{-9}$

$= 2^{-9}\left[\cos\left(-\dfrac{3\pi}{2}\right) + i\sin\left(-\dfrac{3\pi}{2}\right)\right]$

$= \dfrac{1}{512}i$

3. The four fourth roots of $\sqrt{3} + i = 2\left(\cos\dfrac{\pi}{6} + i\sin\dfrac{\pi}{6}\right)$ are

$w_0 = 2^{1/4}\left[\cos\left(\dfrac{\pi/6 + 2\pi(0)}{4}\right) + i\sin\left(\dfrac{\pi/6 + 2\pi(0)}{4}\right)\right]$

$= 2^{1/4}\left(\cos\dfrac{\pi}{24} + i\sin\dfrac{\pi}{24}\right) \approx 1.179 + 0.155i$

$w_1 = 2^{1/4}\left[\cos\left(\dfrac{\pi/6 + 2\pi(1)}{4}\right) + i\sin\left(\dfrac{\pi/6 + 2\pi(1)}{4}\right)\right]$

$= 2^{1/4}\left(\cos\dfrac{13\pi}{24} + i\sin\dfrac{13\pi}{24}\right) \approx -0.155 + 1.179i$

$w_2 = 2^{1/4}\left[\cos\left(\dfrac{\pi/6 + 2\pi(2)}{4}\right) + i\sin\left(\dfrac{\pi/6 + 2\pi(2)}{4}\right)\right]$

$= 2^{1/4}\left(\cos\dfrac{25\pi}{24} + i\sin\dfrac{25\pi}{24}\right) \approx -1.179 - 0.155i$

$w_3 = 2^{1/4}\left[\cos\left(\dfrac{\pi/6 + 2\pi(3)}{4}\right) + i\sin\left(\dfrac{\pi/6 + 2\pi(3)}{4}\right)\right]$

$= 2^{1/4}\left(\cos\dfrac{37\pi}{24} + i\sin\dfrac{37\pi}{24}\right) \approx 0.155 - 1.179i$

As shown in the sketch, these roots form the vertices of a square in the complex plane.

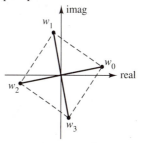

4. The solutions of the equation $x^3 + 1 = 0$, or $x^3 = -1$, are the three cube roots of -1. By the nth root formula, the three cube roots of $-1 + 0i = 1(\cos\pi + i\sin\pi)$ are

$w_0 = 1^{1/3}\left[\cos\left(\dfrac{\pi + 2\pi(0)}{3}\right) + i\sin\left(\dfrac{\pi + 2\pi(0)}{3}\right)\right]$

$= 1^{1/3}\left(\cos\dfrac{\pi}{3} + i\sin\dfrac{\pi}{3}\right) = \dfrac{1}{2} + \dfrac{\sqrt{3}}{2}i$

$w_1 = 1^{1/3}\left[\cos\left(\dfrac{\pi + 2\pi(1)}{3}\right) + i\sin\left(\dfrac{\pi + 2\pi(1)}{3}\right)\right]$

$= 1^{1/3}(\cos\pi + i\sin\pi) = -1$

$w_2 = 1^{1/3}\left[\cos\left(\dfrac{\pi + 2\pi(2)}{3}\right) + i\sin\left(\dfrac{\pi + 2\pi(2)}{3}\right)\right]$

$= 1^{1/3}\left(\cos\dfrac{5\pi}{3} + i\sin\dfrac{5\pi}{3}\right) = \dfrac{1}{2} - \dfrac{\sqrt{3}}{2}i$

Exercises

1. (a) $64\left(\cos\dfrac{\pi}{2}+i\sin\dfrac{\pi}{2}\right)$ (b) $64i$
3. (a) $8\left(\cos\dfrac{5\pi}{4}+i\sin\dfrac{5\pi}{4}\right)$ (b) $-4\sqrt{2}-4\sqrt{2}\,i$
5. (a) $\cos 0+i\sin 0$ (b) 1
7. (a) $\dfrac{27}{8}\left(\cos\dfrac{\pi}{3}+i\sin\dfrac{\pi}{3}\right)$ (b) $\dfrac{27}{16}+\dfrac{27\sqrt{3}}{16}i$
9. (a) $\cos\dfrac{11\pi}{6}+i\sin\dfrac{11\pi}{6}$ (b) $\dfrac{\sqrt{3}}{2}-\dfrac{1}{2}i$
11. (a) $16(\cos 0+i\sin 0)$ (b) 16
13. (a) $32\left(\cos\dfrac{2\pi}{3}+i\sin\dfrac{2\pi}{3}\right)$ (b) $-16+16\sqrt{3}\,i$
15. (a) $128(\cos\pi+i\sin\pi)$ (b) -128
17. (a) $\dfrac{1}{18}\left(\cos\dfrac{\pi}{2}+i\sin\dfrac{\pi}{2}\right)$ (b) $\tfrac{1}{18}i$
19. (a) $\dfrac{1}{64}\left(\cos\dfrac{\pi}{2}+i\sin\dfrac{\pi}{2}\right)$ (b) $\tfrac{1}{64}i$
21. (a) $7\left(\cos\dfrac{5\pi}{6}+i\sin\dfrac{5\pi}{6}\right), 7\left(\cos\dfrac{11\pi}{6}+i\sin\dfrac{11\pi}{6}\right)$
(b) $-\dfrac{7\sqrt{3}}{2}+\dfrac{7}{2}i, \dfrac{7\sqrt{3}}{2}-\dfrac{7}{2}i$
23. (a) $2\left(\cos\dfrac{\pi}{6}+i\sin\dfrac{\pi}{6}\right), 2\left(\cos\dfrac{5\pi}{6}+i\sin\dfrac{5\pi}{6}\right),$
$2\left(\cos\dfrac{3\pi}{2}+i\sin\dfrac{3\pi}{2}\right)$ (b) $\sqrt{3}+i, -\sqrt{3}+i, -2i$
25. (a) $\sqrt{5}\left(\cos\dfrac{\pi}{3}+i\sin\dfrac{\pi}{3}\right), \sqrt{5}\left(\cos\dfrac{5\pi}{6}+i\sin\dfrac{5\pi}{6}\right),$
$\sqrt{5}\left(\cos\dfrac{4\pi}{3}+i\sin\dfrac{4\pi}{3}\right), \sqrt{5}\left(\cos\dfrac{11\pi}{6}+i\sin\dfrac{11\pi}{6}\right)$
(b) $\dfrac{\sqrt{5}}{2}+\dfrac{\sqrt{15}}{2}i, -\dfrac{\sqrt{15}}{2}+\dfrac{\sqrt{5}}{2}i, -\dfrac{\sqrt{5}}{2}-\dfrac{\sqrt{15}}{2}i,$
$\dfrac{\sqrt{15}}{2}-\dfrac{\sqrt{5}}{2}i$ 27. (a) $\cos\dfrac{\pi}{4}+i\sin\dfrac{\pi}{4}, \cos\dfrac{5\pi}{4}+i\sin\dfrac{5\pi}{4}$
(b) $\dfrac{1}{\sqrt{2}}+\dfrac{1}{\sqrt{2}}i, -\dfrac{1}{\sqrt{2}}-\dfrac{1}{\sqrt{2}}i$ 29. (a) $\cos 0+i\sin 0,$
$\cos\dfrac{2\pi}{3}+i\sin\dfrac{2\pi}{3}, \cos\dfrac{4\pi}{3}+i\sin\dfrac{4\pi}{3}$
(b) $1, -\dfrac{1}{2}+\dfrac{\sqrt{3}}{2}i, -\dfrac{1}{2}-\dfrac{\sqrt{3}}{2}i$
31. (a) $2\left(\cos\dfrac{\pi}{4}+i\sin\dfrac{\pi}{4}\right), 2\left(\cos\dfrac{3\pi}{4}+i\sin\dfrac{3\pi}{4}\right),$
$2\left(\cos\dfrac{5\pi}{4}+i\sin\dfrac{5\pi}{4}\right), 2\left(\cos\dfrac{7\pi}{4}+i\sin\dfrac{7\pi}{4}\right)$
(b) $\sqrt{2}+\sqrt{2}\,i, -\sqrt{2}+\sqrt{2}\,i, -\sqrt{2}-\sqrt{2}\,i, \sqrt{2}-\sqrt{2}\,i$

33. (a) $2\sqrt{2}\left(\cos\dfrac{\pi}{3}+i\sin\dfrac{\pi}{3}\right), 2\sqrt{2}\left(\cos\dfrac{4\pi}{3}+i\sin\dfrac{4\pi}{3}\right)$
(b) $\sqrt{2}+\sqrt{6}\,i, -\sqrt{2}-\sqrt{6}\,i$
35. an equilateral triangle

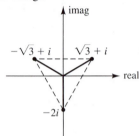

37. a square

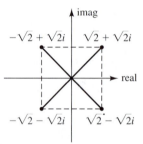

39. $-2, 1+\sqrt{3}\,i, 1-\sqrt{3}\,i$ 41. $3, -3, 3i, -3i$
43. $-i, \dfrac{\sqrt{3}}{2}+\dfrac{1}{2}i, -\dfrac{\sqrt{3}}{2}+\dfrac{1}{2}i$
45. $\dfrac{\sqrt{6}}{2}+\dfrac{\sqrt{2}}{2}i, -\dfrac{\sqrt{6}}{2}-\dfrac{\sqrt{2}}{2}i$ 47. The eight solutions of
$x^8=1$, when plotted in the complex plane, form the vertices of
a regular octagon. 49. $5\left(\cos\dfrac{25\pi}{24}+i\sin\dfrac{25\pi}{24}\right),$
$5\left(\cos\dfrac{41\pi}{24}+i\sin\dfrac{41\pi}{24}\right)$ 53. $2.47-0.329i$
55. $-0.124-0.0176i$ 57. $-11.0+2.00i$
59. $(-5.62\times 10^{-6})+(3.06\times 10^{-6})i$ 61. $1.85+1.85i,$
$-2.53+0.678i, 0.678-2.53i$ 63. $-3.00+2.00i,$
$3.00-2.00i$ 65. $0.966+0.259i, 0.259+0.966i,$
$-0.707+0.707i, -0.966-0.259i, -0.259-0.966i,$
$0.707-0.707i$ 67. $1.62+0.204i, -0.204+1.62i,$
$-1.62-0.204i, 0.204-1.62i$

Chapter 5 Review Exercises

1. $22i\sqrt{5}$ 3. -6 5. -64 7. $-\dfrac{1}{3}i$
9. $12-i$ 11. $47+i$ 13. $-\dfrac{7}{25}+\dfrac{24}{25}i$

15. $-\dfrac{6}{169} + \dfrac{5}{338}i$

17. (a)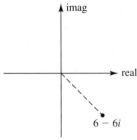

(b) $6\sqrt{2}$ (c) $\dfrac{7\pi}{4}$ (d) $6\sqrt{2}\left(\cos\dfrac{7\pi}{4} + i\sin\dfrac{7\pi}{4}\right)$

19. (a)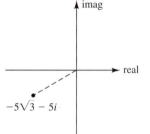

(b) 10 (c) $\dfrac{7\pi}{6}$ (d) $10\left(\cos\dfrac{7\pi}{6} + i\sin\dfrac{7\pi}{6}\right)$

21. (a)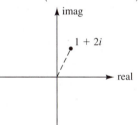

(b) $\sqrt{5}$ (c) 1.107 (d) $\sqrt{5}(\cos 1.107 + i\sin 1.107)$

23. (a)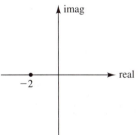

(b) 2 (c) π (d) $2(\cos\pi + i\sin\pi)$

25. (a) $18\left(\cos\dfrac{7\pi}{12} + i\sin\dfrac{7\pi}{12}\right)$ (b) approximately $-4.66 + 17.4i$ **27.** (a) $\cos\dfrac{5\pi}{3} + i\sin\dfrac{5\pi}{3}$

(b) $\dfrac{1}{2} - \dfrac{\sqrt{3}}{2}i$ **29.** (a) $256\left(\cos\dfrac{4\pi}{3} + i\sin\dfrac{4\pi}{3}\right)$

(b) $-128 - 128\sqrt{3}\,i$ **31.** (a) $\dfrac{1}{32}\left(\cos\dfrac{3\pi}{2} + i\sin\dfrac{3\pi}{2}\right)$

(b) $-\dfrac{1}{32}i$ **33.** (a) $\dfrac{1}{\sqrt{2}}\left(\cos\dfrac{7\pi}{4} + i\sin\dfrac{7\pi}{4}\right)$

(b) $\dfrac{1}{2} - \dfrac{1}{2}i$ **35.** (a) $\dfrac{1}{2}\left(\cos\dfrac{\pi}{3} + i\sin\dfrac{\pi}{3}\right)$,

$\dfrac{1}{2}\left(\cos\dfrac{4\pi}{3} + i\sin\dfrac{4\pi}{3}\right)$ (b) $\dfrac{1}{4} + \dfrac{\sqrt{3}}{4}i, -\dfrac{1}{4} - \dfrac{\sqrt{3}}{4}i$

37. (a) $\cos 0 + i\sin 0, \cos\dfrac{\pi}{2} + i\sin\dfrac{\pi}{2}, \cos\pi + i\sin\pi,$

$\cos\dfrac{3\pi}{2} + i\sin\dfrac{3\pi}{2}$ (b) $1, -1, i, -i$

39. (a) $2\left(\cos\dfrac{7\pi}{12} + i\sin\dfrac{7\pi}{12}\right), 2\left(\cos\dfrac{5\pi}{4} + i\sin\dfrac{5\pi}{4}\right),$

$2\left(\cos\dfrac{23\pi}{12} + i\sin\dfrac{23\pi}{12}\right)$ (b) approximately

$-0.518 + 1.93i, -\sqrt{2} - \sqrt{2}\,i$, approximately $1.93 - 0.518i$

41. $1 + \sqrt{3}\,i, -\sqrt{3} + i, -1 - \sqrt{3}\,i$

43. $\sqrt{2}\left(\cos\dfrac{\pi}{4} + i\sin\dfrac{\pi}{4}\right), \sqrt{2}\left(\cos\dfrac{7\pi}{4} + i\sin\dfrac{7\pi}{4}\right)$

45. (a) $a + bi$ (b) $-a - bi$ (c) $a - bi$
(d) $(a^2 - b^2) + 2abi$ **47.** Magnitude: 15Ω;
phase angle: $-36.9°$
49. $I = 8(\cos 36.9° + i\sin 36.9°)$ amps

CHAPTER 6

Section 6.1

Problems

1. The equation of a parabola with vertex $(2, 3)$ and horizontal axis of symmetry must have the form $x = a(y - 3)^2 + 2$. Since the parabola passes through $(4, 1)$, we have $4 = a(1 - 3)^2 + 2$, which implies $a = 1/2$. Hence, the equation of the parabola is $x = \tfrac{1}{2}(y - 3)^2 + 2$.

2. Solving for y then completing the square, we obtain

$$y = \tfrac{1}{2}x^2 - 2x + 5 = \tfrac{1}{2}(x - 2)^2 + 3$$

which is the equation of a parabola with a *vertical* axis of symmetry.

Vertex: (2, 3), *x - intercept*: none, *y-intercept*: 5

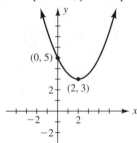

3. Solving the equation $3y^2 - 8x = 0$ for x, we obtain $x = (3/8)y^2$. Hence, $1/(4p) = 3/8$, which implies $p = 2/3$. Thus, the coordinates of the focus are $(2/3, 0)$ and the directrix is the vertical line $x = -2/3$.

4. We can change to general form as follows:

$$x = -\frac{1}{14}(y-1)^2 - \frac{3}{2}$$
$$-14x = (y-1)^2 + 21$$
$$-14x = (y^2 - 2y + 1) + 21$$
$$y^2 + 14x - 2y + 22 = 0$$

5. Replacing p with 9/16, we obtain

$$x = \frac{1}{4p}y^2 = \frac{1}{4(9/16)}y^2 = \frac{4}{9}y^2$$

Hence, the equation of the parabola is $x = (4/9)y^2$.

Exercises

1.

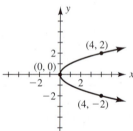

3.

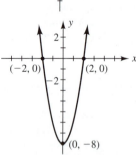

5.

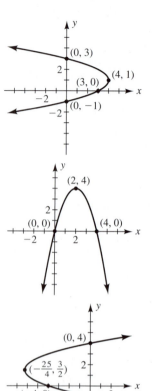

7.

9.

11.

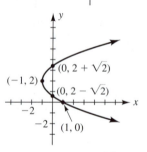

13.

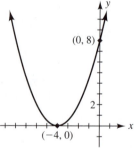

15.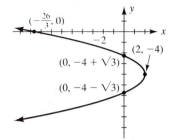

17. Focus: $\left(0, -\dfrac{63}{8}\right)$, directrix: $y = -\dfrac{65}{8}$ **19.** Focus: $\left(-6, \dfrac{3}{2}\right)$, directrix: $x = -\dfrac{13}{2}$ **21.** Focus: $\left(\dfrac{13}{8}, -4\right)$, directrix: $x = \dfrac{19}{8}$ **23.** $x = \dfrac{1}{8}y^2$

25. $y = 2(x + 1)^2 - 1$ **27.** $x = -\dfrac{1}{4}(y - 1)^2 + 1$

29. $y = -\dfrac{1}{20}(x + 3)^2 + 2$ **31.** $x = \dfrac{1}{16}(y + 1)^2 + 3$

33. 30 meters **35.** 9 inches **37.** $\dfrac{a + b}{2}$

39. $c = 0$ **41.** narrower for $0 < p < \dfrac{1}{4}$; wider for $p > \dfrac{1}{4}$ **43.** $\left(0, -\dfrac{E}{4A}\right)$ **45.** (0.744, 7.34)

47. $(-1.78, 0.979)$ **49.** (0.744, 7.34)
51. $(-1.78, 0.979)$ **53.** $y = 0.425(x - 2.67)^2 - 1.98$
55. $y = -0.00969(x - 12.7)^2 + 2.7$ **57.** $x = 0.152y^2$

Section 6.2

1. The y-intercepts are obtained by letting $x = 0$:
$$\dfrac{(0 + 1)^2}{16} + \dfrac{(y - 3)^2}{4} = 1$$
$$\dfrac{(y - 3)^2}{4} = \dfrac{15}{16}$$
$$(y - 3)^2 = \dfrac{15}{4}$$
$$y = \dfrac{6 \pm \sqrt{15}}{2}$$

2. Completing the square, we have
$$3(x - 1)^2 + 2(y + 2)^2 = -4.$$
Since $-4 < 0$, this equation does *not* define an ellipse. Note that no point (x, y) with real coordinates can satisfy this equation.

3. Writing the equation in standard form, we obtain
$$\dfrac{x^2}{1^2} + \dfrac{y^2}{3^2} = 1,$$
which we recognize as an ellipse with center at the origin, vertical major axis of length $2(3) = 6$, and horizontal minor axis of length $2(1) = 2$, as shown in the sketch.

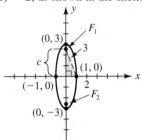

Applying the Pythagorean theorem to the shaded right triangle in the sketch, we find $3^2 = c^2 + 1^2$, which implies $c = 2\sqrt{2}$. Hence, the coordinates of the foci are $F_1(0, 2\sqrt{2})$ and $F_2(0, -2\sqrt{2})$. The eccentricity is $e = \dfrac{2\sqrt{2}}{3} \approx 0.94$.

4. To change to general form, we proceed as follows:
$$\dfrac{(x - 2)^2}{12} + \dfrac{(y - 1)^2}{16} = 1$$
$$4(x - 2)^2 + 3(y - 1)^2 = 48$$
$$4(x^2 - 4x + 4) + 3(y^2 - 2y + 1) = 48$$
$$4x^2 + 3y^2 - 16x - 6y - 29 = 0.$$

 Exercises

1.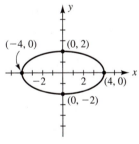

3.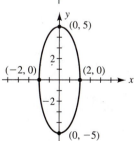

5.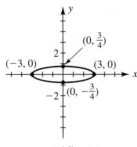

7.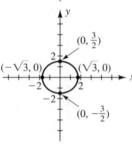

9.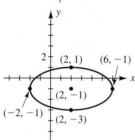

11.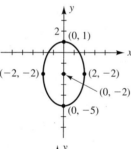

13.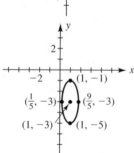

15.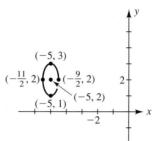

17. foci: $(0, \sqrt{21}), (0, -\sqrt{21})$; eccentricity: $\dfrac{\sqrt{21}}{5} \approx 0.92$

19. foci: $(2 + 2\sqrt{3}, -1), (2 - 2\sqrt{3}, -1)$; eccentricity: $\dfrac{\sqrt{3}}{2} \approx 0.87$ **21.** foci: $\left(1, \dfrac{-15 + 2\sqrt{21}}{5}\right), \left(1, \dfrac{-15 - 2\sqrt{21}}{5}\right)$; eccentricity: $\dfrac{\sqrt{21}}{5} \approx 0.92$

23. $\dfrac{4x^2}{25} + y^2 = 1$

25. $\dfrac{4(x-2)^2}{9} + \dfrac{(y+2)^2}{16} = 1$ **27.** $\dfrac{(x+2)^2}{8} + \dfrac{y^2}{9} = 1$

29. $\dfrac{x^2}{9} + \dfrac{y^2}{5} = 1$ **31.** $\dfrac{(x-\frac{3}{2})^2}{9} + \dfrac{4y^2}{27} = 1$

33. $8\sqrt{2} \approx 11.3$ ft **35.** (a) 40 ft by 20 ft (b) along the major axis and $10\sqrt{3} \approx 17.3$ feet on either side of the center **37.** Tack down the ends of the string with the thumbtacks and insert the pencil point into the loop of the string. Now, keeping the string taught, move the pencil to trace out an ellipse. **39.** $x^2 + y^2 = a^2$

43. Domain: $[-3, 3]$
Range: $[0, 2]$

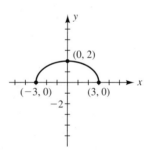

45. $(0, 2.11), (0, -2.11)$ **47.** $(0, 13.2), (0, -4.38)$
49. $(0, 2.11), (0, -2.11)$ **51.** $(0, 13.2), (0, -4.38)$
53. $\dfrac{(x - 3.54)^2}{1.44} + \dfrac{(y + 2.07)^2}{12.96} = 1$ **55.** $\dfrac{x^2}{3856} + \dfrac{y^2}{1121} = 1$
57. approximately 8.1 m

Section 6.3

Problems

1. If the transverse axis is vertical instead of horizontal, then we have

$$2b = 6 \quad \text{or} \quad b = 3,$$

and

$$\pm \frac{3}{a} = \pm \frac{1}{2} \quad \text{or} \quad a = 6.$$

Hence, the equation is $\dfrac{(y-1)^2}{9} - \dfrac{x^2}{36} = 1$.

2. Completing the square, we obtain

$$(x - 3)^2 - 4(y - 1)^2 = 0.$$

Since the right-hand side of this equation is zero, we conclude this equation does *not* define a hyperbola. The graph of this equation is the pair of intersecting lines $y = \frac{1}{2}x - \frac{1}{2}$ and $y = -\frac{1}{2}x + \frac{5}{2}$.

3. Writing the equation in standard form, we obtain

$$\frac{x^2}{1^2} - \frac{y^2}{3^2} = 1,$$

which we recognize as a hyperbola with center at the origin, horizontal transverse axis of length $2(1) = 2$, and vertical conjugate axis of length $2(3) = 6$, as shown in the sketch.

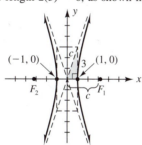

Applying the Pythogorean theorem to the shaded right right triangle in the sketch, we find $c^2 = 3^2 + 1^2$, which implies $c = \sqrt{10}$. Hence, the coordinates of the foci are $F_1(\sqrt{10}, 0)$ and $F_2(-\sqrt{10}, 0)$. The eccentricity is $e = \dfrac{\sqrt{10}}{1} \approx 3.2$.

4. To change to general form, we proceed as follows:

$$\frac{(y-1)^2}{4} - \frac{(x-2)^2}{12} = 1$$
$$3(y-1)^2 - (x-2)^2 = 12$$
$$3(y^2 - 2y + 1) - (x^2 - 4x + 4) = 12$$
$$3y^2 - x^2 + 4x - 6y - 13 = 0.$$

5. Writing the equation $5x^2 + 9y^2 - 180 = 0$ in standard form, we obtain $(x^2/6^2) + [y^2/(2\sqrt{5})^2] = 1$, which we recognize as an ellipse with center $(0, 0)$, horizontal major axis of length $2(6) = 12$, and vertical minor axis of length $2(2\sqrt{5}) = 4\sqrt{5} \approx 8.9$, as shown in the sketch.

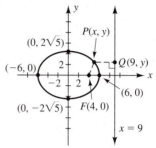

Exercises

1.

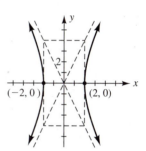

3.

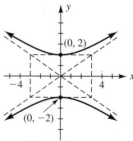

5.

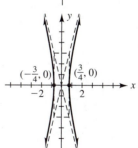

7.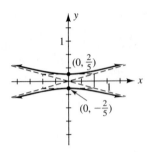

9.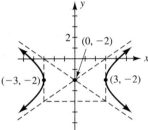

11.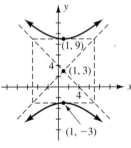

13.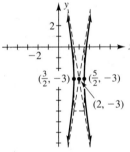

15.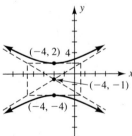

17. foci: $(0, \sqrt{13}), (0, -\sqrt{13})$; eccentricity: $\dfrac{\sqrt{13}}{2} \approx 1.8$

19. foci: $(\sqrt{13}, -2), (-\sqrt{13}, -2)$; eccentricity: $\dfrac{\sqrt{13}}{3} \approx 1.2$

21. foci: $\left(\dfrac{4+\sqrt{37}}{2}, -3\right), \left(\dfrac{4-\sqrt{37}}{2}, -3\right)$; eccentricity: $\sqrt{37} \approx 6.1$ **23.** $\dfrac{y^2}{16} - \dfrac{x^2}{256} = 1$ **25.** $\dfrac{x^2}{4} - \dfrac{y^2}{4} = 1$
27. $\dfrac{y^2}{9} - \dfrac{(x+2)^2}{16} = 1$ **29.** $\dfrac{x^2}{1} - \dfrac{y^2}{3} = 1$
31. $4\left(x-\dfrac{3}{2}\right)^2 - \dfrac{y^2}{2} = 1$ **33.** ellipse **35.** parabola
37. hyperbola **39.** ellipse **41.** $y^2 + 4x - 20 = 0$, parabola **43.** $4x^2 + 3y^2 - 24x - 14y + 31 = 0$, ellipse
45. $x^2 - 3y^2 - 4x + 12y + 40 = 0$, hyperbola
47. 50 cm **49.** (a) The opening of each branch of the hyperbola is very narrow. (b) The opening of each branch of the hyperbola is very wide. **51.** (a) No effect (b) No effect **53.** (a) Degenerate circle: a single point $(3, 4)$ (b) Degenerate ellipse: a single point $(-3, 2)$ (c) Degenerate hyperbola: a pair of intersecting lines, $y = 2x - 6$ and $y = -2x + 2$. (d) Degenerate hyperbola: a pair of intersecting lines, $y = 3x - 1$ and $y = -3x + 11$
55. Domain: $(-\infty, -2] \cup [2, \infty)$
Range: $[0, \infty)$

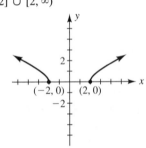

57. Domain: $(-\infty, \infty)$
Range: $(-\infty, -3]$

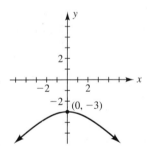

59. $(1.61, 0), (-1.61, 0)$ **61.** $(0, -15.2), (0, -4.34)$
63. $(1.61, 0), (-1.61, 0)$ **65.** $(0, -15.2), (0, -4.34)$
67. $\dfrac{x^2}{(3.12)^2} - \dfrac{y^2}{(1.06)^2} = 1$ **69.** $\dfrac{x^2}{(17.3)^2} - \dfrac{y^2}{(13.4)^2} = 1$
71. (a) $PA - PB = 1430$ ft: P must lie on a branch of a hyperbola since the difference in distances from two fixed points A and B is constant (1430 ft).
(b) $\dfrac{x^2}{511,225} - \dfrac{y^2}{1,738,755} = 1$ (c) $PA = 2340$ ft, $PB = 910$ ft

Section 6.4

Problems

1. $u = 2\sqrt{3} \cos 30° + 4 \sin 30° = 2\sqrt{3}\left(\dfrac{\sqrt{3}}{2}\right) + 4\left(\dfrac{1}{2}\right) = 5$

   $v = -2\sqrt{3} \sin 30° + 4 \cos 30° = -2\sqrt{3}\left(\dfrac{1}{2}\right) + 4\left(\dfrac{\sqrt{3}}{2}\right) = \sqrt{3}$

   Hence, the coordinates of point P in the uv-coordinate system are $(5, \sqrt{3})$.

2. For the equation $3x^2 - \sqrt{3}xy + 2y^2 + 3x - 7 = 0$, we have

   $$\cot 2\theta = \dfrac{A - C}{B} = \dfrac{3 - 2}{-\sqrt{3}} = -\dfrac{1}{\sqrt{3}},$$

   which implies $2\theta = 120°$ or $\theta = 60°$. Thus the rotation angle that eliminates the xy-term is $60°$.

3. For the equation $8y^2 + 6xy - 9 = 0$, we have

   $$\cot 2\theta = \dfrac{A - C}{B} = \dfrac{0 - 8}{6} = -\dfrac{4}{3}.$$

   Therefore, $r = \sqrt{(-4)^2 + (3)^2} = 5$, as shown in the sketch. Hence, $\cos 2\theta = -\tfrac{4}{5}$, which implies $\theta = \tfrac{1}{2} \cos^{-1}\left(-\tfrac{4}{5}\right) \approx 71.6°$.

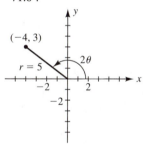

4. For the equation $8y^2 + 6xy - 9 = 0$, the discriminant is

   $$B^2 - 4AC = 6^2 - 4(0)(8) = 36 > 0.$$

   Hence, the graph of this equation (if it exists and is not degenerate) is a hyperbola.

5. For $(2, 2)$, we have $x = 2\cos 30° - 2\sin 30° = \sqrt{3} - 1$ and $y = 2\sin 30° + 2\cos 30° = 1 + \sqrt{3}$.
   For $(2, -2)$, we have $x = 2\cos 30° + 2\sin 30° = \sqrt{3} + 1$ and $y = 2\sin 30° - 2\cos 30° = 1 - \sqrt{3}$.
   Hence, the coordinates of the vertices of the ellipse in the xy-coordinate system are $(\sqrt{3} - 1, 1 + \sqrt{3})$ and $(\sqrt{3} + 1, 1 - \sqrt{3})$.

6. Recall from Problem 3, $\cos 2\theta = -4/5$. Hence,

   $$\sin \theta = \sqrt{\dfrac{1 - \cos 2\theta}{2}} = \sqrt{\dfrac{1 - (-4/5)}{2}} = \dfrac{3}{\sqrt{10}},$$

   and

   $$\cos \theta = \sqrt{\dfrac{1 + \cos 2\theta}{2}} = \sqrt{\dfrac{1 + (-4/5)}{2}} = \dfrac{1}{\sqrt{10}}.$$

   Now, substituting $x = u\cos\theta - v\sin\theta = (u - 3v)/\sqrt{10}$ and $y = u\sin\theta + v\cos\theta = (3u + v)/\sqrt{10}$ into the equation $8y^2 + 6xy - 9 = 0$, gives us $9u^2 - v^2 - 9 = 0$ or $(u^2/1^2) - (v^2/3^2) = 1$, which we recognize as a hyperbola in standard form with center $(0, 0)$, transverse axis of length 2 and conjugate axis of length 6, as shown in the sketch.

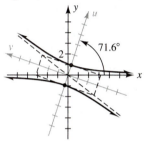

Exercises

1. $P(x, y) = P(2, 2\sqrt{3})$ 3. $P(x, y) = P(-4, -2)$
5. $P(x, y) = P(5, -10)$ 7. $P(u, v) = P(1, 5\sqrt{3})$
9. $P(u, v) = P(-6\sqrt{2}, 2\sqrt{2})$ 11. $P(u, v) = P(13, -9)$

13. (a) hyperbola (b) $45°$ (c) $\dfrac{v^2}{4^2} - \dfrac{u^2}{4^2} = 1$
    (d)

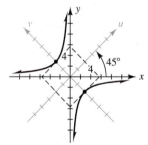

15. (a) ellipse (b) $45°$ (c) $\dfrac{u^2}{2^2} + \dfrac{v^2}{(2\sqrt{3})^2} = 1$
    (d)

17. (a) ellipse (b) 30° (c) $\dfrac{u^2}{2^2} + \dfrac{v^2}{1^2} = 1$
(d)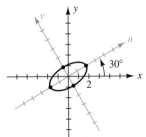

19. (a) parabola (b) 60° (c) $v = -u^2$
(d)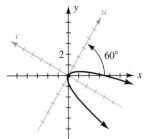

21. (a) hyperbola (b) 30° (c) $\dfrac{u^2}{1^2} - \dfrac{v^2}{1^2} = 1$
(d)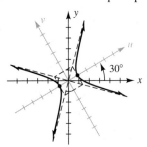

23. (a) ellipse (b) 45° (c) $\dfrac{(u-1)^2}{1^2} + \dfrac{v^2}{2^2} = 1$
(d)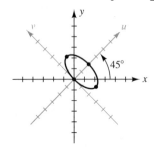

25. (a) parabola (b) 45° (c) $u = (v+1)^2 - 2$
(d)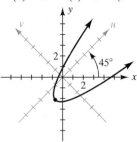

27. (a) hyperbola (b) 45°
(c) $\dfrac{(u-\sqrt{2})^2}{(\sqrt{3})^2} - \dfrac{(v+2\sqrt{2})^2}{3^2} = 1$
(d)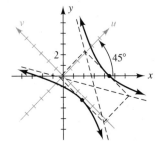

29. (a) hyperbola (b) about 18.4°
(c) $\dfrac{u^2}{(\sqrt{2})^2} - \dfrac{v^2}{(\sqrt{2})^2} = 1$
(d)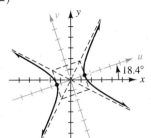

31. (a) hyperbola (b) about 53.1°
(c) $\dfrac{v^2}{(\sqrt{2})^2} - \dfrac{u^2}{(\sqrt{3})^2} = 1$
(d)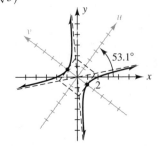

33. (a) parabola (b) about 36.9° (c) $u = \frac{1}{2}v^2$

(d)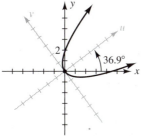

35. (a) ellipse (b) about 36.9° (c) $\frac{u^2}{1^2} + \frac{(v-1)^2}{2^2} = 1$

(d)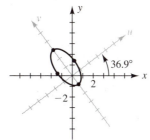

37. (a) $(\sqrt{3}, 1), (-\sqrt{3}, -1)$ (b) $\left(\frac{3}{2}, \frac{\sqrt{3}}{2}\right), \left(-\frac{3}{2}, -\frac{\sqrt{3}}{2}\right)$

39. $(u^2/b^2) + (v^2/a^2) = 1$; The graphs coincide. 41. The graph of $\sqrt{x} + \sqrt{y} = 1$ is the portion of the parabola $x^2 - 2xy + y^2 - 2x - 2y + 1 = 0$ from $x = 0$ to $x = 1$, as shown in the sketch.

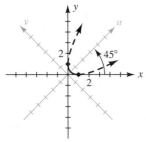

43. (a) a single point $(0, 0)$ (b) a pair of intersecting lines, $y = \frac{2 + \sqrt{3}}{2\sqrt{3} - 1}x$ and $y = \frac{2 - \sqrt{3}}{2\sqrt{3} + 1}x$

(c) a pair of parallel lines, $y = -x + 2\sqrt{2}$ and $y = -x - 2\sqrt{2}$

(d) a pair of parallel lines, $y = 3x \pm \sqrt{10}$

45. hyperbola; approximate coordinates of vertices are $(0.7, -1.3), (-0.8, 1.5)$ 47. ellipse; approximate coordinates of vertices are $(1.7, 3.6), (13.5, 20.4)$

49. parabola; approximate coordinates of vertex are $(2.3, 0.3)$

Section 6.5

Problems

1. The point $P(3, \pi)$ may be represented by $P(3, \pi + 2\pi n)$, for any integer n. Letting $n = 1$, we have $P(3, 3\pi)$ as another name for point P. Also, the point $P(3, \pi)$ may be represented by $P(-3, (\pi + \pi) + 2\pi n)$ for any integer n. Letting $n = -1$, we having $P(-3, 0)$ as another name for point P.

2. From Example 2(b), the point $(-2, 2\sqrt{3})$ has polar coordinates $(4, 2\pi/3)$. The point $(4, 2\pi/3)$ may also be represented by $(-4, (2\pi/3 + \pi) + 2\pi n)$, for any integer n. Since we want $0 \le \theta < 2\pi$, we choose $n = 0$ and obtain $(-4, 5\pi/3)$.

3. (a) The polar equation $r = 5$ represents a circle with center at the pole and radius 5, as shown in the sketch.

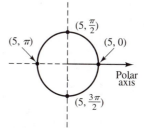

(b) The polar equation $\theta = 2\pi/3$ represents a straight line that passes through the pole and makes an angle of $2\pi/3$ radians with the polar axis, as shown in the sketch.

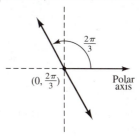

4. (a) Multiplying both sides of the equation $r = 3/\cos \theta$ by $\cos \theta$, gives us $r \cos \theta = 3$. Now, replacing $r \cos \theta$ with x, we obtain the equivalent Cartesian equation $x = 3$.

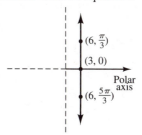

(b) Multiplying both sides of the equation $r = -6\cos\theta$ by r gives us $r^2 = -6r\cos\theta$. Now replacing r^2 with $x^2 + y^2$ and $r\cos\theta$ with x, we obtain the equivalent Cartesian equation $x^2 + 6x + y^2 = 0$ or $(x + 3)^2 + (y - 0)^2 = 3^2$.

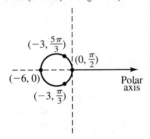

5. Plotting some points, we obtain the three-leafed rose that is shown in the sketch.

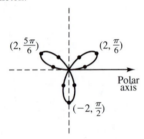

6. We can convert to a Cartesian equation as follows:
$$r = \frac{3}{1 - 2\cos\theta}$$
$$r - 2r\cos\theta = 3$$
$$r - 2x = 3$$
$$r = 2x + 3$$
$$r^2 = (2x + 3)^2$$
$$x^2 + y^2 = 4x^2 + 12x + 9$$
$$3x^2 - y^2 + 12x + 9 = 0$$

Now, completing the square and writing the equation in standard form gives us
$$\frac{(x + 2)^2}{1} - \frac{y^2}{3} = 1,$$
which we recognize as a hyperbola with center $(-2, 0)$ and vertices $(-1, 0)$ and $(-3, 0)$.

7. We can convert to a Cartesian equation as follows:
$$r = \frac{3}{2 + \cos\theta}$$
$$2r + r\cos\theta = 3$$
$$2r + x = 3$$
$$2r = 3 - x$$
$$4r^2 = (3 - x)^2$$
$$4x^2 + 4y^2 = 9 - 6x + x^2$$
$$3x^2 + 4y^2 + 6x - 9 = 0$$

Now, completing the square and writing the equation in standard form gives us
$$\frac{(x + 1)^2}{4} + \frac{y^2}{3} = 1,$$
which we recognize as an ellipse with center $(-1, 0)$ and vertices $(1, 0)$ and $(-3, 0)$.

8. Replacing θ with $\pi/2$, we obtain
$$r = \frac{9.29 \times 10^7}{1 + 0.016\cos(\pi/2)} \approx 9.29 \times 10^7 \text{ miles}$$

Exercises

1.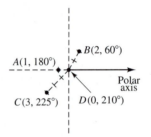

3.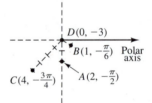

5.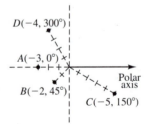

7.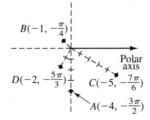

9. $A(3, 180°)$, $A(-3, 360°)$; $B(2, 225°)$, $B(-2, 405°)$; $C(5, 330°)$, $C(-5, 510°)$; $D(4, 120°)$, $D(-4, -60°)$
11. $(-1, 0)$ 13. $(-4, 4)$ 15. $(2, 90°)$
17. $(5\sqrt{2}, 5\pi/4)$ 19. $(-6, 210°)$
21. $x^2 + y^2 = 1$

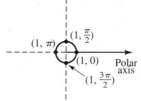

23. $y = \sqrt{3}\, x$

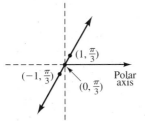

25. $y = -2$

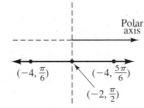

27. $x = 4$

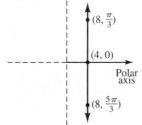

29. $(x - 5)^2 + y^2 = 25$

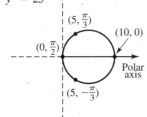

31. $x^2 + \left(y + \dfrac{5}{2}\right)^2 = \dfrac{25}{4}$

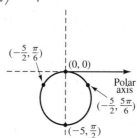

33. $(x - 1)^2 + (y - 2)^2 = 5$

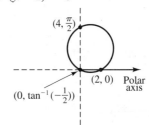

35. $(x - 4)^2 + \left(y + \dfrac{1}{2}\right)^2 = \dfrac{65}{4}$

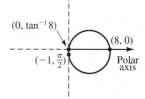

37. $x + y = 2$

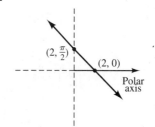

39. $3x + 5y = -15$

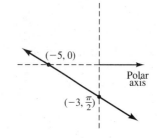

41.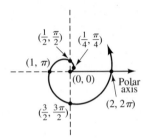

43.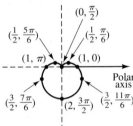

45.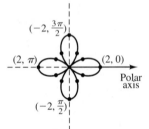

47.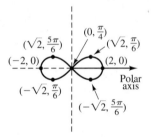

49. (a) parabola
(b)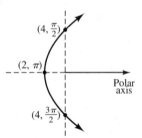
(c) $x = \dfrac{1}{8}y^2 - 2$

51. (a) hyperbola
(b)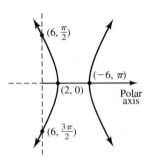
(c) $\dfrac{(x-4)^2}{4} - \dfrac{y^2}{12} = 1$

53. (a) ellipse
(b)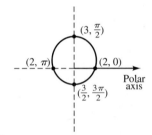
(c) $\dfrac{x^2}{9/2} + \dfrac{(y - \frac{3}{4})^2}{81/16} = 1$

55. (a) parabola
(b)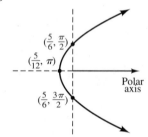
(c) $x = \dfrac{3}{5}y^2 - \dfrac{5}{12}$

57. (a) ellipse
(b)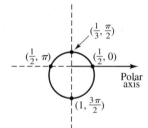
(c) $\dfrac{x^2}{1/3} + \dfrac{(y + \frac{1}{3})^2}{4/9} = 1$

59. (a) hyperbola
(b)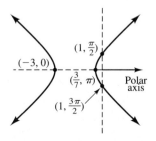

(c) $\dfrac{(x+\frac{12}{7})^2}{81/49} - \dfrac{y^2}{9/7} = 1$ **61.** x-axis: $\theta = 0$, y-axis: $\theta = \pi/2$ **63.** (a) $r = 2\cos\theta + 2\sin\theta$
(b) $r = \dfrac{2}{1-\sin\theta}$ (c) $r = \dfrac{-1}{2-\sin\theta}$
(d) $r = \dfrac{-1}{1-2\sin\theta}$ **67.** (a) $(2\sqrt{2}, \pi/4)$ (b) No.
Although the graphs of both equations pass through the pole, the coordinates at the pole are different. For $r = 4\sin\theta$, the coordinates at the pole are $(0, 0)$, but for $r = 4\cos\theta$, the coordinates at the pole are $(0, \pi/2)$.
69. $r = 2a\cos(\alpha - \theta)$ **71.** $(2.06, 1.33)$
73. $(-0.315, -0.889)$ **75.** $(19.7, 0.890)$
77. $(268, 2.06)$ **79.** (a) $r \approx \dfrac{3.43 \times 10^7}{1 + 0.205\cos\theta}$
(b) 0.205 **81.** (a) $r \approx \dfrac{1.40 \times 10^8}{1 + 0.092\cos\theta}$
(b) 0.092 **83.** perihelion: 0.58 AU, aphelion: 36 AU

Section 6.6

◆ Problems

1. The equation $x = 5 - t$ implies $t = 5 - x$. We now eliminate the parameter t to obtain

$$y = 7 - \frac{2}{3}t = 7 - \frac{2}{3}(5-x) = \frac{11}{3} + \frac{2}{3}x,$$

which we recognize as a line with slope $2/3$. For $t = 0$, we have $(5, 7)$ and for $t = 6$, we have $(-1, 3)$. Hence we conclude that the parametric equations

$$x = 5 - t \text{ and } y = 7 - \frac{2}{3}t, \text{ with } 0 \leq t \leq 6,$$

define the same line segment as shown in Figure 6.67 but with opposite direction.

2. The equation $y = 2t$ implies $t = y/2$. We now eliminate the parameter t to obtain

$$x = t^2 = \left(\frac{y}{2}\right)^2 = \frac{1}{4}y^2,$$

which we recognize as a parabola with horizontal axis of symmetry, vertex $(0, 0)$, and focus $(1, 0)$. The portion of the parabola we traverse and the direction along that portion is shown in the sketch.

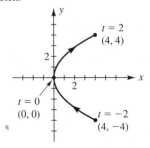

3. The equations $x = 2\sin\pi t$ and $y = 4\cos\pi t$ imply $x/2 = \sin\pi t$ and $y/4 = \cos\pi t$. Now,

$$\left(\frac{x}{2}\right)^2 + \left(\frac{y}{4}\right)^2 = \sin^2\pi t + \cos^2\pi t = 1,$$

or

$$\frac{x^2}{4} + \frac{y^2}{16} = 1,$$

which we recognize as an ellipse with center at the origin, horizontal axis of length 4 and vertical axis of length 8. The direction of travel along the ellipse is shown in the sketch.

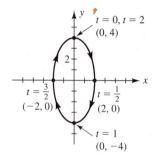

4. For $x = 2\tan\dfrac{\pi}{3}t$ and $y = 4\sec\dfrac{\pi}{3}t$, we have

$$\left(\frac{y}{4}\right)^2 - \left(\frac{x}{2}\right)^2 = \sec^2\frac{\pi}{3}t - \tan^2\frac{\pi}{3}t = 1,$$

or

$$\frac{y^2}{16} - \frac{x^2}{4} = 1,$$

which we recognize as the equation of a hyperbola with center at the origin, vertical transverse axis of length 8 and horizontal conjugate axis of length 4. The portion of the hyperbola we traverse and the direction along that portion is shown in the sketch.

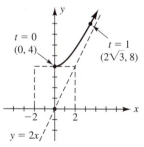

5. The parametric equations that describe the three-leafed rose $r = f(\theta) = 2\sin 3\theta$ are $x = 2\sin 3t \cos t$ and $y = 2\sin 3t \sin t$. Selecting parametric mode and radian mode on a graphing calculator, we enter these equations to obtain the graph shown in Problem 5, Section 6.5.

6. The equation $x = 40t$ implies $t = x/40$. We now eliminate the parameter t to obtain

$$y = 40\sqrt{3}\, t - 16t^2 = 40\sqrt{3}\left(\frac{x}{40}\right) - 16\left(\frac{x}{40}\right)^2$$
$$= \sqrt{3}\, x - \frac{1}{100} x^2$$
$$= -\frac{1}{100}(x - 50\sqrt{3})^2 + 75,$$

which we recognize as a parabola with vertex $(50\sqrt{3}, 75)$. Hence, the maximum height that the ball rises in its parabolic path is 75 ft.

Exercises

1. line: slope 2, passes through (0, 0)

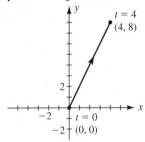

3. line: slope -3, passes through $(-2, 1)$

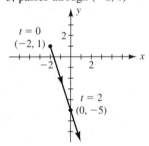

5. parabola: vertex (0, 0), focus (0, -1)

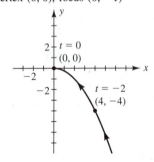

7. parabola: vertex (0, 0), focus (3, 0)

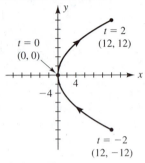

9. circle: center (0, 0), radius 1

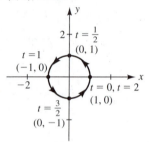

11. circle: center (0, 0), radius 3

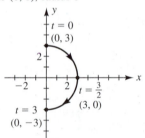

13. ellipse: center (0, 0), horizontal minor axis 2, vertical major axis 4

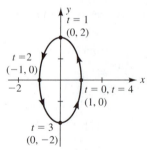

15. ellipse: center $(0, 0)$, horizontal major axis 6, vertical minor axis 4

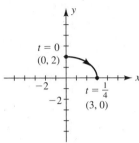

17. hyperbola: center $(0, 0)$, horizontal transverse axis 2, vertical conjugate axis 2

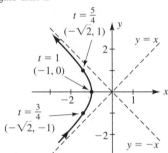

19. hyperbola: center $(0, 0)$, vertical transverse axis 4, horizontal conjugate axis 6

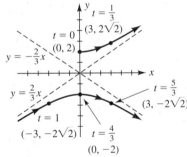

21. $x = -1 + t$ and $y = 4 - 4t$ **23.** $x = 4t$ and $y = 2t^2$
25. $x = 4 \cos t$ and $y = 4 \sin t$ **27.** $x = 3 \cos t$ and $y = 2 \sin t$ **29.** $x = 2 \sec t$ and $y = 2 \tan t$
31. (a) $y = |x| - 1$
(b)

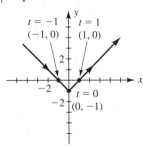

33. (a) $y = \dfrac{1}{x + 2}$
(b)

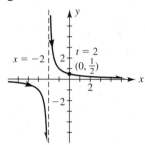

35. (a) $y = -(x - 1)^3$
(b)

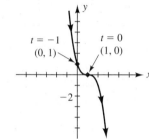

37. (a) $y = \sqrt{x - 3} - 1$
(b)

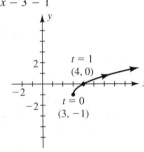

39. (a) $y = 4 - (x + 5)^2$
(b)

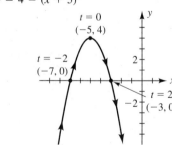

41. (a) $x = 45\sqrt{2}\,t$ and $y = 45\sqrt{2}\,t - 16t^2$ (b) $253\frac{1}{8}$ ft
(c) $y = x - \dfrac{8}{2025}x^2$ (d) $63\frac{9}{32}$ ft **43.** (a) $x = \dfrac{4}{1 + t^2}$
and $y = \dfrac{4t}{1 + t^2}$ (b) $x = \dfrac{4t^2}{1 + t^2}$ and $y = \dfrac{4t}{1 + t^2}$
47. (a) same as C (b) same as C but traced in opposite direction (c) same as C but shifted 2 units right and 1 unit down (d) same as C but reflected about the line $y = x$

49. $x = 2 + 2\cos t$ and $y = 1 + \sin t$

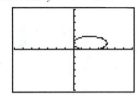

51. $x = -1 + \sec t$ and $y = -2 + \tan t$

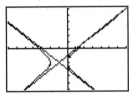

53. $x = \dfrac{t \cos t}{\pi}$ and $y = \dfrac{t \sin t}{\pi}$ **55.** $x = (1 - \sin t) \cos t$ and $y = (1 - \sin t) \sin t$ **57.** $x = 2 \cos 2t \cos t$ and $y = 2 \cos 2t \sin t$ **59.** $x = 2\sqrt{\cos 2t} \cos t$ and $y = 2\sqrt{\cos 2t} \sin t$ **61.** $R \approx 25{,}400$ ft, $h \approx 8740$ ft

Chapter 6 Review Exercises

1.

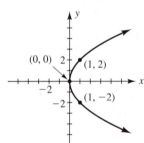

3.

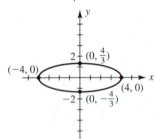

5.

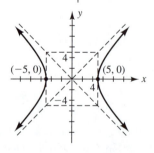

7.

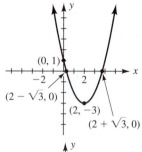

9.

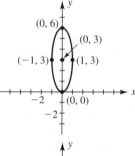

11.

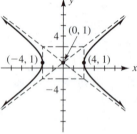

13.

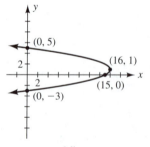

15.

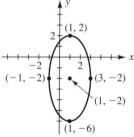

17.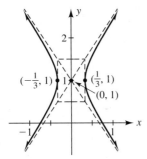

19. A degenerate ellipse: The graph is the single point $(1, -2)$
21. focus: $(1, 0)$, directrix: $x = -1$ **23.** foci: $(\pm 5\sqrt{2}, 0)$, eccentricity: $\sqrt{2} \approx 1.4$ **25.** foci: $(1, -2 \pm 2\sqrt{3})$, eccentricity: $\frac{\sqrt{3}}{2} \approx 0.87$ **27.** $y = x^2 - 4$
29. $x = \frac{2}{9}(y + 3)^2 - 2$ **31.** $y = \frac{1}{2}(x - 1)^2 - \frac{1}{2}$
33. $\frac{(x - 3)^2}{9} + \frac{y^2}{4} = 1$ **35.** $\frac{(x - 2)^2}{16} + \frac{y^2}{12} = 1$
37. $\frac{y^2}{9} - \frac{4x^2}{81} = 1$ **39.** $\frac{x^2}{9} - \frac{(y + 1)^2}{7} = 1$
41. (a) hyperbola (b) $45°$ (c) $\frac{u^2}{8} - \frac{v^2}{8} = 1$
(d)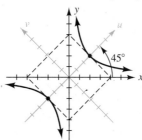

43. (a) ellipse (b) $45°$ (c) $\frac{u^2}{4} + \frac{v^2}{1} = 1$
(d)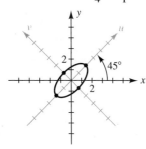

45. (a) parabola (b) $60°$ (c) $u = -v^2$
(d)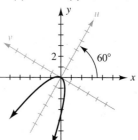

47. (a) ellipse (b) about $63.4°$ (c) $\frac{u^2}{16} + \frac{v^2}{4} = 1$
(d)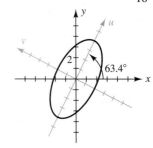

49. (a) $(\sqrt{2}, \sqrt{2}), (-\sqrt{2}, -\sqrt{2})$
(b) $\left(\sqrt{\frac{3}{2}}, \sqrt{\frac{3}{2}}\right), \left(-\sqrt{\frac{3}{2}}, -\sqrt{\frac{3}{2}}\right)$
51. (a) $(-\sqrt{2}, -\sqrt{2})$ (b) $(3, -3\sqrt{3})$
53. $x^2 + y^2 = 4$

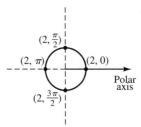

55. $y = 1$

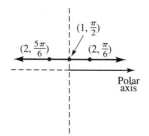

57. $\left(x - \frac{5}{2}\right)^2 + y^2 = \frac{25}{4}$

59. $(x + 4)^2 + (y - 3)^2 = 25$

61. $x = -\frac{1}{4}y^2 + 1$

63. $\frac{x^2}{1/8} + \frac{(y + \frac{1}{8})^2}{9/64} = 1$

65. line: slope -2, passes through $(2, 3)$

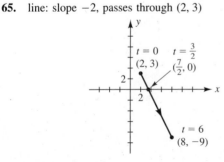

67. parabola: vertex $(0, 0)$, focus $(0, -2)$

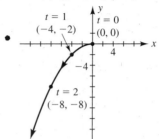

69. circle: center $(0, 0)$, radius 2

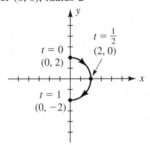

71. ellipse: center $(0, 0)$, vertical major axis 8, horizontal minor axis 6

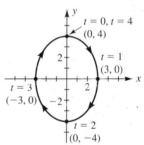

73. hyperbola: center $(0, 0)$, horizontal transverse axis 4, vertical conjugate axis 6

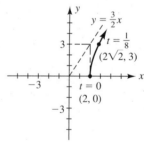

75. $x = 2 + t$, $y = 3 + t$ **77.** $x = 2 \sec t$, $y = 2 \tan t$
79. $x = -2 + 4 \cos t$, $y = 4 + 4 \sin t$
81. $d = 18$ inches   **83.** 24 cm
85. (a) $r = \dfrac{4.81 \times 10^8}{1 + 0.0477 \cos \theta}$ (b) 0.0477
87. (a) perihelion ≈ 0.134 AU, aphelion ≈ 3585 AU
(b) major axis $\approx 3.32 \times 10^{11}$ miles, minor axis $\approx 4.07 \times 10^9$ miles

CHAPTER 7

Section 7.1

Problems

1. $\dfrac{125^{x-2}}{25^{2x-3}} = \dfrac{(5^3)^{x-2}}{(5^2)^{2x-3}} = \dfrac{5^{3x-6}}{5^{4x-6}} = 5^{-x} = \left(\dfrac{1}{5}\right)^x$

2.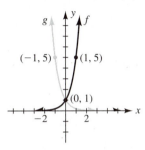

3. The domain and range of $G(x) = 3 - 4^{-x}$ may be determined from its graph (see Figure 7.4). Thus,

   Domain: $(-\infty, \infty)$, Range: $(-\infty, 3)$.

4. The x-intercept is found by solving the equation

   $0 = e^{-x} - 2$ or $e^{-x} = 2$.

   However, at this time, we do not have a procedure for solving an equation in which the unknown appears as an exponent.

5. If interest is compounded monthly, then $n = 12$. Hence,
$A = 1000\left(1 + \dfrac{0.08}{12}\right)^{12 \cdot 5} \approx \$1489.85$.

Exercises

1. 1   3. $\left(\tfrac{1}{36}\right)^x$ 5. 1   7. e^x 9. 9^x
11. $\tfrac{1}{3}$ 13. 5
15.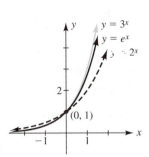

17.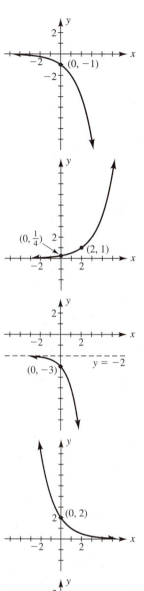

19.

21.

23.

25.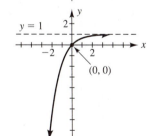

27.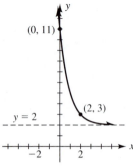

29.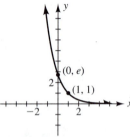

31. $7057.91 **33.** $27,437.17 **35.** $3432.01
37. $124,897.29 **39.** (a) $4^x - 4^{-x}$ (b) 4 **41.** 2
43. 2 **45.** ± 1 **47.** $\frac{1}{4}$

49.
x	0	1	2	3	4	-1	-2	-3	-4
$f(x)$	1	-2	4	-8	16	$-1/2$	$1/4$	$-1/8$	$1/16$

The function f is not defined when $x = 1/n$, where n is an even integer. Thus, we cannot simply connect the points to form the graph.

51. (a) $k = 15$ (b) 6.74 psi **53.** (a) 18 bacteria
(b) 133 bacteria
(c)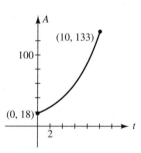

55. (a) $12,000 (b) $4466
(c)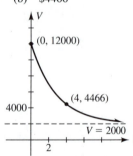

(d) scrap value **57.** (a) 0.211 (b) 0.0393
(c) 23.1 (d) 22.5 **59.** (c) approaches 1

Section 7.2

Problems

1. Since $(\frac{1}{2})^{-3} = 8$, we conclude that $\log_{1/2} 8 = -3$.

2. Using the [LOG] key on a calculator, we find that for log 0.01 the display reads -2.
Using the [LN] key on a calculator, we find that for ln (-1) the display shows *error*.

3. Since $\log_b b^x = x$ for all real x, we have $\log_3 3^{12} = 12$.

4. $(\log_2 2)^{4t-8} = (1)^{4t-8} = 1$.

5. $3 \log_2 (5^2 - 2(5) + 1) = 3 \log_2 16 = 3 \cdot 4 = 12$ and $3 \log_2[(-3)^2 - 2(-3) + 1] = 3 \log_2 16 = 3 \cdot 4 = 12$.

6. $7e^{1-2[(1-\ln 4)/2]} = 7e^{\ln 4} = 7 \cdot 4 = 28$.

7.
$$2 = e^{r(5.5)}$$
$$r(5.5) = \ln 2$$
$$r = \frac{\ln 2}{5.5} \approx .126 \text{ or } 12.6\%$$

8.
$$540 = 18e^{0.1354t}$$
$$30 = e^{0.1354t}$$
$$0.1354t = \ln 30$$
$$t = \frac{\ln 30}{0.1354} \approx 25 \text{ min}$$

9.
$$27 = 30e^{-0.02476t}$$
$$0.9 = e^{-0.02476t}$$
$$-0.02476t = \ln 0.9$$
$$t = \frac{\ln 0.9}{-0.02476} \approx 4.26 \text{ yr}$$

Exercises

1. 2 **3.** 3 **5.** 1 **7.** -4 **9.** -2 **11.** 2
13. 0 **15.** 2 **17.** undefined **19.** $\frac{3}{2}$
21. $-\frac{4}{3}$ **23.** 3 **25.** 5 **27.** 10 **29.** 0
31. $x - 3$ **33.** $x + 2$ **35.** $x^2 + 4$ **37.** $x^2 + 2x$
39. 1 **41.** $3x + 6$ **43.** xe^x **45.** x^2
47. $b = 3$ **49.** $b = \frac{1}{16}$ **51.** $x = \frac{1}{100}$
53. $x = 8$ **55.** $x = 2e$ **57.** $x = \frac{1}{2}, -3$
59. $x = 1 \pm \sqrt{10}$ **61.** $x = \log 5 \approx 0.699$
63. $x = \frac{\ln (2/3)}{3} \approx -0.135$ **65.** $x = \frac{3 - \log 28}{2} \approx$
0.7764 **67.** $x = -\frac{\ln (15/4)}{7} \approx -0.1888$
69. $x = -\log 60 \approx -1.778$ **71.** (a) 5.8 yrs
(b) 25.6 yrs (c) 65.8 yrs **73.** 11%

75. (a) $A(t) = 24\,e^{0.1792t}$ (b) 1996
77. approximately 2140 B.C.   79. (a) 3 (b) 4
81. (a) $(0, e) \cup (e, \infty)$ (b) $(0, 10^{10}]$
83. (a) $i(t) \to 20$ amps (b) approximately 79 ms
85. (a) $T = 70 + 330e^{-0.1805t}$ (b) 262 °F (c) about 7.85 min   87. 3.99   89. 2.24   91. 2.01
93. (a) It appears that $\ln(xy) = \ln x + \ln y$ (b) It appears that $\ln(x/y) = \ln x - \ln y$ (c) It appears that $\ln(x^y) = y \ln x$ 95. (a) pH $\approx$ 6.4 (b) pH $\approx$ 3.2

Section 7.3

Problems

1. For example, $(\log_2 8)^2 \neq 2 \log_2 8$. Note that $(\log_2 8)^2 = 3^2 = 9$, whereas $2 \log_2 8 = 2 \cdot 3 = 6$.

2. For most scientific calculators, the keying sequence is as follows:

   $\boxed{50}\,\boxed{\log}\,\boxed{+}\,\boxed{2}\,\boxed{\times}\,\boxed{4}\,\boxed{\log}\,\boxed{-}\,\boxed{3}\,\boxed{\times}\,\boxed{2}\,\boxed{\log}\,\boxed{=}$.

3. $\log_4 24 = \dfrac{\ln 24}{\ln 4} \approx 2.292$

4. $(\log_2 5)(\log_5 8) = \left(\dfrac{1}{\log_5 2}\right)(\log_5 2^3)$
   $= \left(\dfrac{1}{\log_5 2}\right)[3 \log_5 2] = 3$

5. Letting $I_a =$ the intensity of the 1933 earthquake, and $I_b =$ the intensity of the 1989 earthquake, we have
   $$8.9 = \log I_a - \log I_0$$
   $$7.1 = \log I_b - \log I_0$$
   $$1.8 = \log I_a - \log I_b$$
   Hence, $\dfrac{I_a}{I_b} = 10^{1.8} \approx 63$ times as intense.

Exercises

1. $\log_2(8+8) = \log_2 16 = 4$, whereas $\log_2 8 + \log_2 8 = 3 + 3 = 6$ 3. $\dfrac{\log_2 8}{\log_2 2} = \dfrac{3}{1} = 3$, whereas $\log_2 \dfrac{8}{2} = \log_2 4 = 2$ 5. 10
7. $3 + \log_2 x + \log_2 (x+2)$ 9. $2 + \ln x - \ln 10$
11. $\tfrac{2}{3} \log_3 x - 1$ 13. $2 \log_b x - \tfrac{2}{3} \log_b (x+1)$
15. $-(2 + \tfrac{3}{2} \log_3 x + \tfrac{1}{3} \log_3 y)$ 17. $2x^2 - 2\ln(e^x + 1)$
19. $\ln 24$ 21. 2   23. -2 25. $\log_5 \dfrac{x^2 + 2x - 3}{x^2}$
27. $\ln \dfrac{x-2}{x+2}$ 29. $\ln \dfrac{9\sqrt{x^2-9}}{x^2}$ 31. 3.585

33. 2.757   35. -0.3691 37. -8.697 39. 1
41. 4   43. 2   45. 1   47. about 32 times more intense   49. No. The domain of f is $(-\infty, 0) \cup (0, \infty)$, whereas the domain of g is $(0, \infty)$.   51. (a) $\tfrac{1}{2}$ (b) $\tfrac{1}{3}$ (c) 1 (d) $\tfrac{1}{6}$ (e) $\tfrac{1}{5}$ (f) $-\tfrac{1}{5}$ (g) -1 (h) 1
53. about 1.32×10^{10} times brighter   55. 10^9 times more intense   57. $\log(6.2 \times 10^k) = \log 6.2 + \log 10^k = \log 6.2 + k$ 59. (a) 1 (b) $\ln e = 1$

Section 7.4

Problems

1.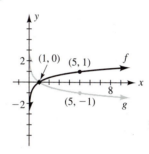

2. The graph of $H(x) = 1 - \log_4 x$ is the same as the graph of $f(x) = \log_4 x$ reflected about the x-axis and then shifted vertically upward 1 unit.

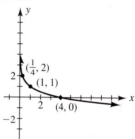

3. $\log_4 16x = \log_4 16 + \log_4 x = 2 + \log_4 x$. Thus, the graph of $F(x) = \log_4 16x$ is the same as the graph of $f(x) = \log_4 x$ shifted upward 2 unit.

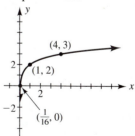

4. $F(x) = \log_4 x^3 = 3\log_4 x$. Hence, the graph of F may be obtained by stretching the graph of $f(x) = \log_4 x$ vertically by a factor of 3.

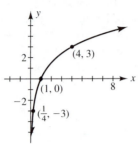

5. $R(10) = 100 - 29 \log [27(10) + 1]$
$= 100 - 29 \log 271 \approx 29.4\%$
$R(20) = 100 - 29 \log [27(20) + 1]$
$= 100 - 29 \log 541 \approx 20.7\%$
$R(30) = 100 - 29 \log [27(30) + 1]$
$= 100 - 29 \log 811 \approx 15.6\%$

Exercises

1.

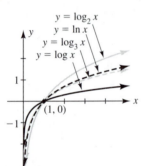

3.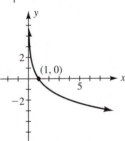

5.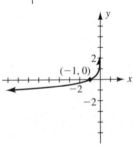

7.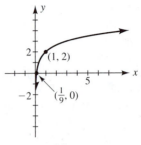

9.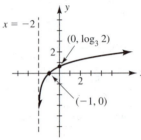

11.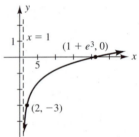

13.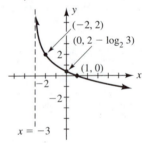

15.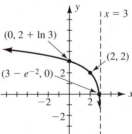

17. f and h **19.** f and g **21.** f and g

23.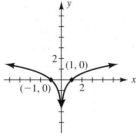

25.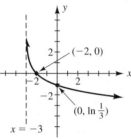

27.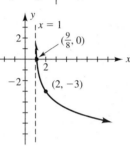

29.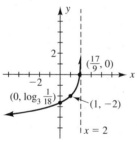

31. (a) 100% (b) 69% (c) about 4 days
(d)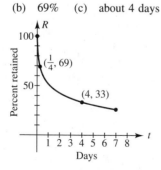

33. (a) $f^{-1}(x) = \log_6 x$
(b)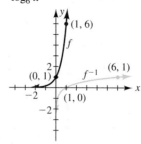

35. (a) $h^{-1}(x) = 2 \ln x$
(b)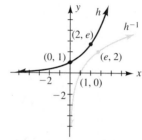

37. (a) $G^{-1}(x) = \frac{1}{2}\log(1 - x)$
(b)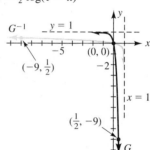

39. (a) $f^{-1}(x) = 8^x$
(b)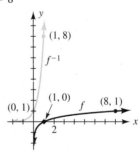

41. (a) $h^{-1}(x) = 2e^{-x}$
(b)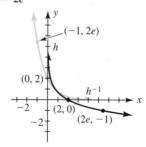

43. (a) $G^{-1}(x) = \frac{1}{3}[10^{(x-1)/2}]$
(b)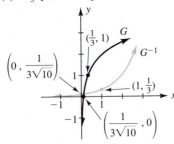

45. (a) 1000 (b) 5615 (c) about $19,000
(d)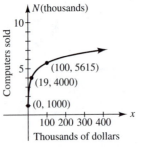

47. (a) $(0, \infty)$ (b) there are no zeros (c) $f(x) \to 0$
(d) $f(x) \to -\infty$
(e)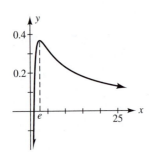

(f) $x = e$

Section 7.5

Problems

1.
$$2 \ln x - \ln 9 = 4$$
$$\ln \frac{x^2}{9} = 4$$
$$e^4 = \frac{x^2}{9}$$
$$x^2 = 9e^4$$
$$x = 3e^2$$

2.
$$x + \ln y = \ln c$$
$$\ln y - \ln c = -x$$
$$\ln \frac{y}{c} = -x$$
$$e^{-x} = \frac{y}{c}$$
$$y = ce^{-x}$$

3.
$$4^{3x+2} = 8^{4x}$$
$$(2^2)^{3x+2} = (2^3)^{4x}$$
$$2^{6x+4} = 2^{12x}$$
$$6x + 4 = 12x$$
$$x = \frac{2}{3}$$

4.
$$5^{x-1} = 325$$
$$(x-1)\log 5 = \log 325$$
$$x - 1 = \frac{\log 325}{\log 5}$$
$$x = 1 + \frac{\log 325}{\log 5} \approx 4.594$$

5.
$$\log \sqrt{x} = \sqrt{\log x}$$
$$\frac{1}{2}\log x = \sqrt{\log x}$$
$$\frac{1}{4}(\log x)^2 = \log x$$
$$(\log x)^2 - 4 \log x = 0$$
$$\log x (\log x - 4) = 0$$
$$\log x = 0 \quad \text{or} \quad \log x - 4 = 0$$
$$x = 1 \qquad\qquad x = 10^4$$

6. On most scientific calculators, the keying sequence is
$\boxed{2}\; \boxed{y^x}\; \boxed{1.620}\; \boxed{+}\; \boxed{3}\; \boxed{y^x}\; \boxed{1.620}$. The display reads $9.002135812 \approx 9$.

Exercises

1. 2 **3.** $3e - 1$ **5.** 4 **7.** 2 **9.** 4 **11.** 1
13. $y = ce^{-2x}$ **15.** $y = \dfrac{10}{(x+a)^2}$ **17.** $y = \dfrac{kx - x}{1 + k}$, where $k = e^c$ **19.** $y = \dfrac{ka}{1 + kb}$, where $k = e^{ac}$ **21.** $\frac{3}{2}$
23. $-\frac{1}{3}$ **25.** $\dfrac{\ln 35}{\ln 7} \approx 1.827$ **27.** $\dfrac{\ln 120}{\ln 9} \approx 2.179$
29. $\dfrac{1}{2 + \log 4} \approx 0.384$ **31.** $\dfrac{\ln 4}{\ln 4 - 1} \approx 3.589$
33. $\dfrac{\ln (3/4)}{\ln 3} \approx -0.2619$ **35.** $\dfrac{\ln 3072}{\ln 16} \approx 2.896$
37. $1, e^2$ **39.** $\dfrac{\ln 4}{2}$ **41.** 0
43. $\ln(10 \pm 3\sqrt{11}) \approx \pm 2.993$ **45.** 3, 9
47. $t = \dfrac{\ln (A/P)}{n \ln (1 + (r/n))}$ **49.** (b) a straight line
51. (a) $P = P_0\, 2^{-(1/3)t}$ (b) 25% **53.** 2.123
55. 1.202 **57.** $-1.637, 1.000$ **59.** (a) 50 ft
(b) $50 \ln \left(\dfrac{3 \pm \sqrt{5}}{2}\right) \approx \pm 48.1$ ft

Chapter 7 Review Exercises

1. 1 **3.** e **5.** $(\frac{1}{2})^x$ **7.** 2 **9.** $\frac{2}{3}$
11. -3 **13.** $\frac{1}{3}$ **15.** $-\frac{2}{3}$ **17.** 4 **19.** 3
21. $2x$ **23.** $x^2 + 1$ **25.** $3 - 3x$ **27.** xe^{2x}
29. 1 **31.** $2 + 2\log_3 x + \log_3(x-1)$
33. $-\log x - \frac{1}{2}\log(2x-3)$
35. $-x^2 + \ln x - \ln(e^x - 1)$ **37.** $\ln 8$
39. $\log_3(x+1)^2$ **41.** 0 **43.** 5 **45.** $-2 \pm e^3$
47. no solution **49.** 3 **51.** $y = kx^2$, where $k = e^c$
53. $\frac{\log 80}{2} \approx 0.9515$ **55.** $-\frac{3}{4}$ **57.** $\frac{\log 4}{\log 36} \approx 0.3869$
59. $\frac{\ln 6}{1 + \ln 9} \approx 0.5604$ **61.** ± 1 **63.** $3, \sqrt{3}$

65.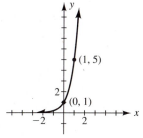

67.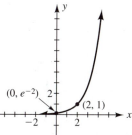

69.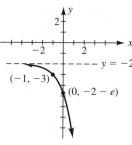

71.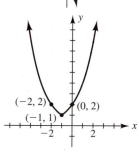

73.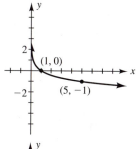

75.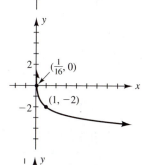

77.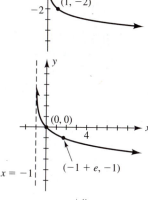

79.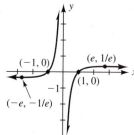

81. 3.465 **83.** -6.644
85. $f^{-1}(x) = \log_8 x$

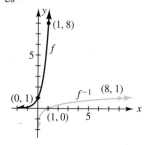

87. $h^{-1}(x) = \dfrac{2 + \ln(x/2)}{3}$

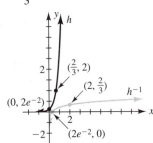

89. $G^{-1}(x) = 2^{-(x+1)}$

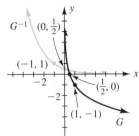

91. $f^{-1}(x) = \sqrt{e^{1-x}}$

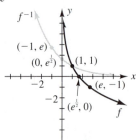

93. (a) $12,058.57 (b) $12,256.79 (c) $12,298.02
95. approximately 7 yrs **97.** (a) $A(t) = 10^3 e^{1.535t}$, t in hours (b) approximately 27 min **99.** about 794 times more intense **101.** (a) $k \approx 0.0405$ (b) approximately 99 °F (c) about 35 min
(d)

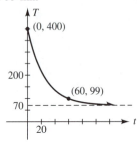

CUMULATIVE REVIEW EXERCISES FOR CHAPTERS 4, 5, 6, AND 7

1. e^{-bla} **3.** (a) line (b) parabola (c) circle (d) hyperbola (e) ellipse (f) hyperbola **5.** (a) 3 (b) e^2, e^{-1} **7.** (a) 1 (b) 0 (c) 1 (d) 1
9.

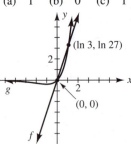

11.

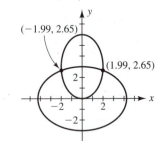

15. (a) $(f \circ g)(x) = -x$, domain: $(-\infty, \infty)$
(b) $(g \circ f)(x) = \dfrac{-x}{x + 1}$, domain: $(-1, \infty)$

17. (a) $\dfrac{\pi}{9}, \dfrac{4\pi}{9}, \dfrac{7\pi}{9}, \dfrac{10\pi}{9}, \dfrac{13\pi}{9}, \dfrac{16\pi}{9}$ (b) $\dfrac{\pi}{6}, \dfrac{5\pi}{6}$
(c) $\dfrac{\pi}{2}, \dfrac{3\pi}{2}$ (d) $\dfrac{\pi}{2}, \dfrac{7\pi}{12}, \dfrac{11\pi}{12}, \dfrac{3\pi}{2}, \dfrac{19\pi}{12}, \dfrac{23\pi}{12}$
(e) $\dfrac{\pi}{6}, \dfrac{\pi}{3}, \dfrac{2\pi}{3}, \dfrac{5\pi}{6}, \dfrac{7\pi}{6}, \dfrac{4\pi}{3}, \dfrac{5\pi}{3}, \dfrac{11\pi}{6}$ (f) $\dfrac{2\pi}{3}, \dfrac{4\pi}{3}$

19. $\sin 3\theta = 3 \sin \theta - 4 \sin^3 \theta$,
$\cos 3\theta = 4 \cos^3 \theta - 3 \cos \theta$ **21.** (a) $y = e^{-2}x^4$
(b) $y = \dfrac{e^x}{x + 2}$ (c) $y = 2$ **25.** (a) $16i$ (b) $-\dfrac{1}{8}$

27. (a) πn (b) $\dfrac{\pi}{6} + 2\pi n, \dfrac{5\pi}{6} + 2\pi n$

29. (a) $y = 3(1 + \ln x)$ (b) $y = \sin e^x$
31. major axis: $8\sqrt{2} \approx 11.3$ inches; minor axis: 8 inches

INDEX

A

Abscissa, 11
Absolute value:
  of a complex number, 299
  of a real number, 4, 5
Absolute value function, 35, 36
Acute angle, 94
Addition:
  of complex numbers, 294
  of ordinates, 280
  of vectors, 206
Additive identity of a complex number, 294
Additive inverse of a complex number, 294
A-frame structure, 271
Algebraic definition:
  of an ellipse, 336
  of a hyperbola, 349
  of a parabola, 323
Algebraic function, 74
Alternating current (AC) circuit, 305, 306
Ambiguous case, 181–184
Amplitude, 117
Analytic geometry, 13, 14
Analytic trigonometry, 228
Angle:
  acute, 94
  central, 80
  complementary, 96
  coterminal, 75, 76
  definition of, 74
  degree measure of, 75–77
  of depression, 170
  of elevation, 170
  initial side of, 74
  measure of, 75
  negative, 75
  obtuse, 179
  positive, 75
  quadrantal, 99–101
  radian measure of, 78–80
  reference, 103–107
  right, 75, 94
  special (30°, 45°, and 60°), 101–103
  in standard position, 74
  straight, 75
  terminal side of, 74
  trigonometric ratio for, 86
  vertex of, 74
Angular speed, 81, 82
Aphelion, 384
Applied functions, 58–61
Approximate number, A1
Arc length formula, 80
Arccosine function (arccos), 148
Arcsine function (arcsin), 146
Arctangent function (arctan), 150
Area:
  of a sector, 84 (Exercise 55)
  of a triangle, 195, 196
Argument of a complex number, 299
Associative property:
  of scalar multiplication, 208
  of vector addition, 208
Asymptote:
  horizontal, 40, A5
  of a hyperbola, 348
  vertical, 40, A4
Autumnal equinox, 247
Axis (axes):
  coordinate, 10
  of an ellipse, 335
  of a hyperbola, 348
  imaginary, 299
  polar, 373
  real, 299
  of symmetry of a parabola, 323
Axis intercept, 37

B

Base:
  of an exponential function, 407, 410
  of a logarithmic function, 417, 418
Bearing, 211
Boyle's law, 66 (Exercise 26)

Brightness scale, 435, 437 (Exercises 53, 54)

C

Calculator usage:
  with inverse trigonometric functions, 147, 149, 151
  with trigonometric equations, 246, 247
  with trigonometric functions, 110, 111
  (see also Graphing calculator usage)
Capacitive reactance, 305
Cardioid, 387 (Exercise 43)
Cartesian coordinates, 10
Cartesian plane, 10
Cassegrain telescope, 358 (Exercise 47)
Catenary, 454 (Exercise 59)
Center:
  of a circle, 17
  of an ellipse, 335
  of a hyperbola, 348
Central angle, 80
Change of base formula, 433, 434
Chord of a circle, 162 (Exercise 93), 250 (Exercise 68)
Circle:
  center of, 17
  chord of, 162 (Exercise 93), 250 (Exercise 68)
  concentric, 22 (Exercise 75)
  as a conic section, 322
  equation of,
    Cartesian, 17–19
    parametric, 391, 392
    polar, 377, 379, 380
  radius of, 17
  sector of, 84 (Exercise 55)
  unit, 108
Classification of the conic sections, 355, 356, 366, 367
Clockwise rotation, 75
Closed interval, 6, 7
Cofunction, 96
Cofunction identities, 282
Comet:
  Halley's, 346 (Exercise 58), 388 (Exercise 83, 84)
  Kohoutek, 404 (Exercise 87)
Common definition of the conic sections, 356, 357
Common logarithm, 418
Commutative property of vector addition, 208
Complementary angles, 96
Completing the square, 18, 326
Complex conjugate, 296, 299, 307 (Exercise 48)
Complex number:
  absolute value of, 299
  argument of, 299
  conjugate of, 296, 299, 307 (Exercise 48)
  definition of, 293
  exponential form of, 459 (Exercise 23)
  modulus of, 299
  nth root of, 312–314
  operations with, 293–297, 302–305
  power of, 308–312
  reciprocal of, 307 (Exercise 47)
  in standard form, 298
  trigonometric form of, 300
Complex number system, 290, 293
Complex plane, 299
Component form of a vector, 202
Components of a vector, 202
Composite function, 47
Composition of functions, 47–49
Composition rules for trigonometric functions, 152
Compound interest, 412–414
Computing the functional value, 26
Concentric circles, 22 (Exercise 75)
Concurrent force, 210
Conditional equation, 238
Conic sections:
  classification of, 355, 356, 366, 367
  common definition of, 356, 357
  introductory comments of, 322
  (see also Circle, Ellipse, Hyperbola, Parabola)
Conjugate axis of a hyperbola, 348
Conjugate of a complex number, 296, 299, 307 (Exercise 48)
Constant function, 35, 36
Constant of variation, 61, 62
Constructions for SSA, 184
Continuous compounding, 413
Converse of the Pythagorean theorem, 14
Conversion factor:
  for degrees to radians, 79
  for radians to degrees, 79
Coordinate plane, 10
Coordinate of a point:
  in the Cartesian plane, 10, 11
  in the polar plane, 373
  on the real line, 3
Coplanar force, 210
Cosecant function (csc):
  of an angle, 85, 86
  graph of, 138–142
  inverse of, 157 (Exercise 57)
  properties of, 139
  as the ratio of the sides in a right triangle, 95
  of a real number, 108
  restricted, 157 (Exercise 57)
Cosine function (cos):
  of an angle, 85, 86
  difference quotient for, 284 (Exercise 60)
  graph of, 117–123
  inverse of, 148
  properties of, 115–117
  as the ratio of the sides in a right triangle, 95
  of a real number, 108
  restricted, 148
Cosines, law of, 189–197
Cotangent function (cot):
  of an angle, 85, 86
  graph of, 135–138
  inverse of, 157 (Exercise 59)
  properties of, 136
  as the ratio of the sides in a right triangle, 95
  of a real number, 108
  restricted, 157 (Exercise 59)
Coterminal angle, 75, 76
Counterclockwise rotation, 75
Cross product property, 80
Cube root function, 36
Cubing function, 36
Cycle, 116

D

Decibel scale, 435, 437 (Exercise 55, 56)
Decreasing function, 44, 45
Degenerate conic sections, 322
Degree (°):
  definition of, 75
  minute (') second ("), 76, 77
  to radian conversion, 79
DeMoivre, Abraham, 308
DeMoivre's theorem, 308–312
Dependent variable, 14, 23
Depression, angle of, 170
Descartes, René, 10, 290
Diagonal of a parallelogram, 206
Difference:
  of complex numbers, 294
  of vectors, 207
Difference formula:
  for cosine, 251–253
  for sine, 253–255
  for tangent, 255–256
Difference quotient:
  for the cosine function, 284 (Exercise 60)
  definition of, 34 (Exercises 81–88)
  for the exponential function, 416 (Exercise 59)
  for the logarithmic function, 458 (Exercise 16)
  for the sine function, 263 (Exercise 66)
Direct variation, 61
Direction angle of a vector, 202
Directly proportional, 61
Directrix:
  of a conic section, 356, 381
  of a parabola, 327
Discriminant, 366
Displacement, 211

Distance formula:
  in the Cartesian plane, 12
  in the polar plane, 388 (Exercise 70)
Distance between points, 5, 12, 388 (Exercise 70)
Distributive property:
  of a scalar over vector addition, 208
  of a vector over scalar addition, 208
Division of complex numbers, 296, 302–305
Domain, 23, 26, 27, 87, 88, 109
Dot product, 214 (Exercises 58–60)
Double-angle formulas, 264–267
Double inequality, 4

E

e, the number, 410
Earth, orbit of, 384–386
Eccentricity:
  of a conic section, 356, 381
  of an ellipse, 340
  of a hyperbola, 353
Elastic curve, 334 (Exercise 58), 403 (Exercise 82)
Element of a set, 2
Elevation, angle of, 170
Ellipse:
  algebraic definition of, 336
  axes of, 335
  center of, 335
  as a conic section, 322
  eccentricity of, 340
  equation of,
    Cartesian, 335–339
    parametric, 391, 392
    polar, 382
  foci of, 339
  geometric definition of, 339, 356
  latus rectum of, 345 (Exercise 41)
  reflection property of, 334, 335
  vertices of, 335
Elliptical orbit (see Orbit)
Equation:
  conditional, 238
  exponential, 446, 448–451
  graph of, 14–16
  logarithmic, 446–448
  parametric, 389–396
  polar, 376–384
  polynomial, 314–316
  quadratic type, 451, 452
  trigonometric, 232, 238–248
Equilateral triangle, 102
Equilibrium, 211
Euler, Leonhard, 290
Euler's formula, 459 (Exercise 24)
Even function, 28, 29
Even-odd identities, 282
Exponent, properties of, 407
Exponential decay function, 424, 426

Exponential equation, 446, 448–451
Exponential form:
  of a complex number, 459 (Exercise 23)
  of an equation, 421–423
Exponential function:
  with base b, 407–410
  with base e, 410, 411
  difference quotient for, 416 (Exercise 59)
Exponential growth function, 424, 425
Extraneous root, 244

F

Factor by grouping terms, 27
Factoring trigonometric expressions, 229, 230
Focus (foci):
  of an ellipse, 339
  of a hyperbola, 352
  of a parabola, 327
Force, 209–211
Force system, 210
Four-leafed rose, 381
Fractions, fundamental property of, 234
Frequency, 127 (Exercise 43), 172
Function:
  absolute value, 35, 36
  algebraic, 74
  applied, 58–61
  arccosine (arccos), 148
  arcsine (arcsin), 146
  arctangent (arctan), 150
  composite, 47
  composition of, 47–49, 152
  constant, 35, 36
  cosecant (csc), 85, 86, 95, 108, 138–142
  cosine (cos), 85, 86, 95, 108, 114–123
  cotangent (cot), 85, 86, 95, 108, 135–138
  cube root, 36
  cubing, 36
  decreasing, 44, 45
  definition of, 23
  domain of, 23, 26, 27
  even, 28, 29
  exponential, 407–411
  exponential growth and decay, 424–426
  graph of, 28–30
  greatest integer, 46 (Exercise 60)
  hyperbolic cosine, 416 (Exercise 50)
  hyperbolic sine, 416 (Exercise 50)
  identity, 35, 36
  increasing, 44, 45
  inverse, 49–55
  inverse cosecant (csc^{-1}), 157 (Exercise 57)
  inverse cosine (cos^{-1}), 148
  inverse cotangent (cot^{-1}), 157 (Exercise 59)
  inverse secant (sec^{-1}), 157 (Exercise 58)
  inverse sine (sin^{-1}), 146
  inverse tangent (tan^{-1}), 150
  logarithmic, 417, 438–442

  normal probability distribution, 417 (Exercise 60)
  odd, 29
  one-to-one, 51, 52
  periodic, 115
  range of, 23, 26
  reciprocal, 36, A4
  restricted cosecant, 157 (Exercise 57)
  restricted cosine, 148
  restricted cotangent, 157 (Exercise 59)
  restricted secant, 157 (Exercise 58)
  restricted sine, 145
  restricted tangent, 150
  retention, 442–444
  secant (sec), 85, 86, 95, 108, 138–142
  sine (sin), 85, 86, 95, 108, 114–123
  square root, 36
  squaring, 35, 36
  step, 46 (Exercise 59)
  tangent (tan), 85, 86, 95, 108, 129–135
  transcendental, 74
  trigonometric, 74, 86, 95, 108
  zeros of, 29
Function machine, 23
Functional notation, $f(x)$, 25, 26
Functional value, 26
Fundamental properties:
  of scalar multiplication, 208
  of vector addition, 208
Fundamental property of fractions, 234
Fundamental trigonometric identities, 91–94, 230

G

Galileo, 396
Gauss, Carl Friedrich, 293
General form of the equation:
  of a circle, 18
  of an ellipse, 337
  of a hyperbola, 350, 351
  of a parabola, 325, 326
General form of the solution of a trigonometric equation, 239
General quadratic equation in two unknowns, 322, 361
Geometric definition:
  of an ellipse, 339, 356
  of a hyperbola, 352, 356
  of a parabola, 327, 356
Graph:
  of an equation, 14–16
  of a function, 28–30
  horizontal shift of, 38–40
  of an interval, 6–8
  of an inverse function, 54, 55
  of a parametric equation, 389–396
  of a polar equation, 376–384
  reflection of, 40–42
  vertical shift of, 36–38
  vertical stretch and shrink of, 42–44

Graphing calculator usage:
  introduction to, 31, 32
  with exponential equations, 452, 453
  with parametric equations, 394–396
  with polar equations, 394–396
  (see also Calculator usage)
Graphing a function, techniques of, 35–44
Greater than ($>$), 3
Greater than or equal to ($\geq$), 3
Greatest integer function, 46 (Exercise 60)
Grouping symbols, 38
Grouping terms, factoring by, 27

H

Half-angle formulas, 268–271
Half-life, 426
Half-open interval, 6, 7
Halley's comet, 346 (Exercise 58), 388 (Exercises 83, 84)
Harmonic motion, 171–173
Hero's (Heron's) formula:
  proof of 200 (Exercise 52)
  statement of, 196
Hooke's law, 66 (Exercise 21)
Horizontal asymptote, 40, A5
Horizontal line, polar equation of, 379
Horizontal line test, 51
Horizontal shift rule, 38–40
Hyperbola:
  algebraic definition of, 349
  asymptotes of, 348
  axes of, 348
  center of, 348
  as a conic section, 322
  eccentricity of, 353
  equation of,
    Cartesian, 347–352
    parametric, 393, 394
    polar, 382
  foci of, 352
  geometric definition of, 352, 356
  reflection property of, 346, 347
  vertices of, 348
Hyperbolic cosine function, 416 (Exercise 50)
Hyperbolic sine function, 416 (Exercise 50)
Hypotenuse, 94

I

i, the imaginary unit, 290
Identity:
  logarithmic, 420, 421
  trigonometric, 232, 282, 283
Identity function, 35, 36
Identity property:
  of scalar multiplication, 208
  of vector addition, 208
Imaginary axis in the complex plane, 299
Imaginary number, 293
Imaginary unit i, 290
Impedance of an AC circuit, 305
Increasing function, 44, 45
Independent variable, 14, 23
Induction, mathematical, 317 (Exercise 51)
Inductive reactance, 305
Inequality symbols, 3
Infinity (∞), 6
Initial point of a vector, 202
Initial side of an angle, 74
Input value of a function, 23
Integer, 2
Intercept, axis, 29, 35
Interest, compound, 412–414
Interval, 6
Interval notation, 6–8, 26
Inverse cosecant function ($\csc^{-1}$), 157 (Exercise 57)
Inverse cosine function ($\cos^{-1}$), 148
Inverse cotangent function ($\cot^{-1}$), 157 (Exercise 59)
Inverse function:
  definition of, 50
  graph of, 54, 55
  method for finding, 52–54
Inverse property of vector addition, 208
Inverse secant function ($\sec^{-1}$), 157 (Exercise 58)
Inverse sine function ($\sin^{-1}$), 146
Inverse tangent function ($\tan^{-1}$), 150
Inverse variation, 62
Inversely proportional, 62
Irrational number, 2
Isosceles right triangle, 101
Isoceles triangle, 13

J

Joint variation, 63

K

Kepler, Johannes, 384
Kepler's first law, 384
Kohoutek, the comet, 404 (Exercise 87)

L

Latus rectum:
  of an ellipse, 345 (Exercise 41)
  of a parabola, 333 (Exercise 42)
Law of cosines:
  application of, 191–194, 196, 197
  derivation of, 189–191
Law of sines:
  ambiguous case for, 181–184
  application of, 179–181, 184, 185
  derivation of, 178, 179
Legs of a right triangle, 94
Lemniscate, 387 (Exercise 47)
Length of a line segment, 12
Less than ($<$), 3
Less than or equal to ($\leq$), 3
Line:
  parametric equation of, 389, 390
  polar equation of, 378–380
Line segment:
  length of, 12
  midpoint of, 12
Linear speed, 81
Lithotripter, 344 (Exercise 36)
Logarithm:
  common, 418
  evaluating, 418, 419, 433
  Napierian, 418
  natural, 418
  properties of, 429–433
Logarithmic equation, 446–448
Logarithmic form of an equation, 421–423
Logarithmic function:
  definition of, 417
  difference quotient for, 458 (Exercise 16)
  graph of, 438–442
  properties of, 429–433
Logarithmic identities, 420, 421
Logarithmic scales, 435
Logistic law, 428 (Exercise 84)

M

Magnitude of a vector, 201
Major axis of an ellipse, 335
Malthus, Thomas, 425
Malthusian model, 425
Mathematical induction, proof by, 317 (Exercise 51)
Measure of an angle, 75
Median of a triangle, 20 (Exercise 29)
Midpoint formula, 12
Minor axis of an ellipse, 335
Minute ('), 76, 77
Modulus of a complex number, 299
Mollweide's formula, 189 (Exercises 47–50)
Motion, projectile, 396, 397
Multiple-angle formulas, 264–272
Multiplication:
  of complex numbers, 295, 302–305
  scalar, 204, 205
Multiplicative identity of a complex number, 297 (Exercise 43)
Multiplicative inverse of a complex number, 297 (Exercise 44), 307 (Exercise 47)

N

Napier, John, 418
Napierian logarithm, 418

Natural logarithm, 418
Natural number, 2
Negative angle, 75
Negative infinity ($-\infty$), 6
Negative of a vector, 206
Newton, Sir Isaac, 384, 428 (Exercises 85, 86)
Newton's law of cooling, 428 (Exercises 85, 86)
Normal probability distribution function, 417 (Exercise 60)
n-sided polygon in the complex plane, 314
nth root of a complex number, 312–314
nth root formula:
    applied to polynomial equations, 314–316
    statement of, 313
Number:
    complex, 293
    e, 410
    i, 290
    imaginary, 293
    integer, 2
    irrational, 2
    natural, 2
    pure imaginary, 290–292
    rational, 2
    real, 2
    whole, 2
Number line, real, 2, 3

O

Oblique triangle, 178
Obtuse angle, 179
Odd function, 29
One-to-one correspondence, 2, 11
One-to-one function, 51, 52
Open interval, 6, 7
Orbit:
    of the earth, 384–386
    of Halley's comet, 346 (Exercise 58), 388 (Exercise 84)
    of Kohoutek, 404 (Exercise 87)
    of the planets, 388 (Exercises 79–82), 404 (Exercises 85, 86)
Ordered pair, 11
Ordinate, 11
Origin, 10
Origin, symmetric with respect to, 29
Output value of a function, 23

P

Parabola:
    algebraic definition of, 323
    axis of symmetry of, 323
    as a conic section, 322
    directrix of, 327
    equation of,
        Cartesian, 323–326
        parametric, 390, 391
        polar, 382
    focus of, 327
    geometric definition of, 327, 356
    latus rectum of, 333 (Exercise 42)
    reflection property of, 330, 331
    vertex of, 323
Parallelogram law for vectors, 206
Parameter, 389
Parametric equation:
    definition of, 389
    of an ellipse (or circle), 391, 392
    of a hyperbola, 393, 394
    of a line, 389, 390
    of a parabola, 390, 391
    for projectile motion, 396, 397
Perihelion, 384
Period, 115
Period formula:
    for cosecant, 140
    for cosine, 119
    for cotangent, 136
    for secant, 140
    for sine, 119
    for tangent, 133
Periodic function, 115
pH of a liquid, 429 (Exercises 95, 96)
Phase shift, 121
Planets, orbit of, 388 (Exercises 79–82), 404 (Exercises 85, 86)
Point-plotting method, 15, 16
Point-slope form, 389
Polar axis, 373
Polar coordinates, 373
Polar equation:
    of a circle, 377, 379, 380
    of a conic section, 381–384
    definition of, 376
    of a line, 378–380
    in parametric form, 395
Pole, 373
Polygon in the complex plane, 314
Polynomial equation, 314–316
Positive angle, 75
Positive integer, 2
Power reduction formulas, 267, 268
Powers:
    of complex numbers, 308–312
    of i, 292, 293
Predator-prey relationship, 123–125
Principal square root of $-a$, 290
Product:
    of complex numbers, 295, 302–305
    dot, 214 (Exercises 58–60)
    of a scalar and a vector, 204, 205
Product-to-sum formulas, 275, 276
Projectile motion, 396, 397
Proof by mathematical induction, 317 (Exercise 51)
Properties:
    of cosine function, 116
    of cosecant function, 139
    of cotangent function, 136
    of logarithms, 429–433
    of real exponents, 407
    of scalar multiplication, 208
    of secant function, 139
    of sine function, 116
    of tangent function, 132
    of vector addition, 208
Pure imaginary number, 290–292
Pythagorean theorem:
    converse of, 14
    statement of (*see inside front cover*)

Q

Quadrant, 10
Quadrantal angle, 99–101
Quadratic equation in two unknowns, 322, 361
Quadratic formula (*see inside front cover*)
Quadratic-type equation, 451, 452
Quotient of complex numbers, 296, 302–305

R

Radian:
    definition of, 78
    to degree conversion, 79
Radius of a circle, 17
Range, 23, 26
Ratio, trigonometric, 86, 95
Rational number, 2
Ray, 74
Real axis in the complex plane, 299
Real exponents, 406, 407
Real number, 2
Real number line, 2, 3
Reciprocal:
    of a complex number, 307 (Exercise 47)
    of i, 291
Reciprocal function, 36, A4
Rectangular coordinates, 10
Reference angle, 103–107
Reflecting telescope, 347
Reflection property:
    of an ellipse, 334, 335
    of a hyperbola, 346, 347
    of a parabola, 330
Reflection rule, x-axis and y-axis, 40–42
Resistance of an AC circuit, 305
Restricted cosecant function, 157 (Exercise 57)
Restricted cosine function, 148
Restricted cotangent function, 157 (Exercise 59)
Restricted secant function, 157 (Exercise 58)

Restricted sine function, 145
Restricted tangent function, 150
Resultant force, 210
Resultant of vectors, 206
Retention curve, 443
Retention function, 442–444
Rhombus, 274 (Exercise 75)
Richter, Charles, 435
Richter scale, 435
Right angle, 75, 94
Right triangle:
　applied problems involving, 168–171
　definition of, 94
　isosceles, 101
　solving, 166–168
　trigonometric ratios for, 94–97, 166
Root:
　of a complex number, 312–314
　extraneous, 244
Rotation angle, 363–366
Rotation formulas, 361–363
Rounding error, 180, A2, A3

S

Scalar, 204
Scalar multiplication, 204, 205
Scientific notation, A2
Secant function (sec):
　of an angle, 85, 86
　graph of, 138–142
　inverse of, 157 (Exercise 58)
　properties of, 139
　as the ratio of the sides in a right triangle, 95
　of a real number, 108
　restricted, 157 (Exercise 58)
Second ("), 76, 77
Sector, 84 (Exercise 55)
Set, 2
Set-builder notation, 6, 7
Shift rule:
　horizontal, 38–40
　vertical, 36–38
Shrink and stretch rule, vertical, 42–44
Significant digits, A1–A3
Similar triangles, 60–205
Simple harmonic motion, 171–173
Sine function (sin):
　of an angle, 85, 86
　difference quotient for, 263 (Exercise 66)
　graph of, 117–123
　inverse of, 146
　properties of, 115–117
　as the ratio of the sides in a right triangle, 95
　of a real number, 108
　restricted, 145
Sines, law of, 178–185

Sketch the graph, 15
Snell's law, 189 (Exercises 51, 52)
Solving a right triangle, 166–168
Special angles (30°, 45°, and 60°), 101–103
Speed:
　angular, 81, 82
　linear, 81
Square, completing the, 18, 326
Square root of $-a$, 290
Square root function, 36
Squaring function, 35, 36
Standard form of the equation:
　of a circle, 17
　of an ellipse, 336
　of a hyperbola, 349
　of a parabola, 323, 324
Standard position:
　of an angle, 74
　of a vector, 202
Step function, 46 (Exercise 59)
Straight angle, 75
Straight line (see Line)
Stretch and shrink rule, vertical, 42–44
Subset, 2
Substitution, 58, 235
Subtraction:
　of complex numbers, 295
　of vectors, 207
Sum:
　of complex numbers, 294
　of vectors, 206
Sum and difference formulas:
　for cosine, 251–253
　for sine, 253–255
　for tangent, 255, 256
Summer solstice, 247
Sum-to-product formulas, 277, 278
Surveying, 184, 185
Symmetry:
　axis of, 323
　with respect to the origin, 29
　with respect to the y-axis, 28

T

Table:
　exponential, A6
　logarithmic, A7–A9
　trigonometric, A10–A16
　of values, 14
Tangent function (tan):
　of an angle, 85, 86
　graph of, 132–134
　inverse of, 150
　properties of, 129–132
　as the ratio of the sides in a right triangle, 95
　of a real number, 108

restricted, 150
Telescope:
　Cassegrain, 358 (Exercise 47)
　reflecting, 347
Terminal point of a vector, 202
Terminal side of an angle, 74
Tests for even and odd functions, 29
Transcendental function, 74
Transit, 170
Transverse axis of a hyperbola, 348
Triangle:
　area formulas for, 195, 196
　equilateral, 102
　isosceles, 13
　isosceles right, 101
　median of, 20 (Exercise 29)
　oblique, 178
　right, 94
　similar, 60, 205
　30°-60°-90°, 102
Trigonometric composition rule, 152
Trigonometric equation:
　conditional, 238
　definition of, 232
　general form of solution of, 239
　techniques of solving, 238–247
Trigonometric form of a complex number, 300
Trigonometric formulas (identities):
　cofunction, 96, 282
　double-angle, 264–267
　even-odd, 282
　fundamental, 91–94
　half-angle, 268–271
　power reduction, 267, 268
　product-to-sum, 275, 276
　sum and difference, 251–259
　summary of, 282, 283
　sum-to-product, 277, 278
　triple-angle, for sine, 267
Trigonometric functions:
　algebraic signs of, 88–91
　of an angle, 85, 86
　domains of, 87, 88, 109
　evaluating, 99–111
　fundamental identities of, 91–94
　introductory comments of, 74
　inverse, 145–155
　as the ratio of the sides in a right triangle, 95
　of a real number, 108
　reciprocals of, 87
Trigonometric identity:
　definition of, 232
　verification of, 232–235
　(see also Trigonometric formulas)
Trigonometric ratios:
　of an angle, 85, 86
　for a right triangle, 95, 166
Trigonometric substitution, 235, 236
Triple-angle formula for sine, 267

U

Unbounded interval, 6, 7
Union of two sets, 7
Unit circle, 108
Unit vector, 209

V

Variable:
   dependent, 14, 23
   independent, 14, 23
Variation:
   direct, 61
   inverse, 62
   joint, 63
Variation constant, 61, 62
Vector:
   applications of, 209–213
   component form of, 202
   components of, 202
   difference of, 207
   direction angle of, 202
   dot product for, 214 (Exercises 58–60)
   equality of, 201
   fundamental properties of, 208
   initial point of, 202
   magnitude of, 201
   negative of, 206
   parallelogram law for, 206
   resultant, 206, 210
   scalar multiplication of, 204, 205
   standard position of, 202
   sum of, 206
   terminal point of, 202
   unit, 209
   zero, 202
Vector diagram, 209
Vector difference, 207
Vector quantities, 201
Vector sum, 206
Verifying a trigonometric identity, 232–235
Vernal equinox, 247
Vertex (vertices):
   of an angle, 74
   of an ellipse, 335
   of a hyperbola, 348
   of a parabola, 323
Vertical asymptote, 40, A4
Vertical line, polar equation of, 380
Vertical line test, 23–25
Vertical shift rule, 36–38
Vertical stretch and shrink rule, 42–44

W

Whole number, 2
Winter solstice, 247
Whispering gallery effect, 334, 335

X

x-axis, 10
x-axis reflection rule, 40–42
x-coordinate, 11
x-intercept, 29, 37

Y

y-axis:
   in the coordinate plane, 10
   symmetric with respect to, 28
y-axis reflection rule, 40–42
y-coordinate, 11
y-intercept, 29, 37

Z

Zero of a function, 29
Zero product property, 243
Zero vector, 202

Credits

pages 1 & 61 Henley and Savage/Tony Stone Worldwide; **73 & 155** E. R. Degginger; **165 & 196** Dave Cannon/Tony Stone Worldwide; **169** Keith Gunnar/Photo Researchers; **170** E. R. Degginger; **227 & 247** Jeremy Walker/Tony Stone Worldwide; **271** E. R. Degginger; **289 & 305** Courtesy of Cherry Semiconducter Corp.; **321 & 384** Courtesy of NASA; **397** Focus On Sports; **405 & 426** Earl Roberge/Science Source/Photo Researchers; **435** Mary Evans Picture Library/Photo Researchers.

TRIGONOMETRIC DEFINITIONS

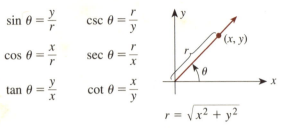

$$\sin\theta = \frac{y}{r} \qquad \csc\theta = \frac{r}{y}$$

$$\cos\theta = \frac{x}{r} \qquad \sec\theta = \frac{r}{x}$$

$$\tan\theta = \frac{y}{x} \qquad \cot\theta = \frac{x}{y}$$

$$r = \sqrt{x^2 + y^2}$$

TRIGONOMETRIC RATIOS FOR RIGHT TRIANGLES

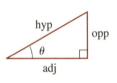

$$\sin\theta = \frac{\text{opp}}{\text{hyp}} \qquad \csc\theta = \frac{\text{hyp}}{\text{opp}}$$

$$\cos\theta = \frac{\text{adj}}{\text{hyp}} \qquad \sec\theta = \frac{\text{hyp}}{\text{adj}}$$

$$\tan\theta = \frac{\text{opp}}{\text{adj}} \qquad \cot\theta = \frac{\text{adj}}{\text{opp}}$$

GRAPHS OF THE TRIGONOMETRIC FUNCTIONS

1. Sine Function

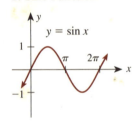

2. Cosine Function

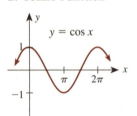

3. Tangent Function

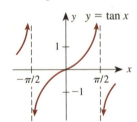

4. Cotangent Function

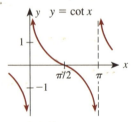

5. Cosecant Function

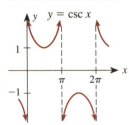

6. Secant Function

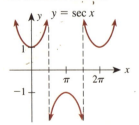

GRAPHS OF THE INVERSE TRIGONOMETRIC FUNCTIONS

1. Inverse Sine Function

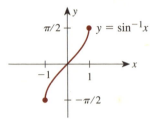

2. Inverse Cosine Function

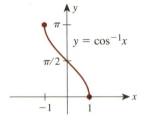

3. Inverse Tangent Function

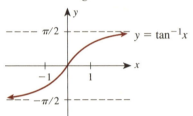

OBLIQUE TRIANGLES

1. The Law of Sines

$$\frac{a}{\sin A} = \frac{b}{\sin B} = \frac{c}{\sin C}$$

2. The Law of Cosines

$$a^2 = b^2 + c^2 - 2bc\cos A$$
$$b^2 = a^2 + c^2 - 2ac\cos B$$
$$c^2 = a^2 + b^2 - 2ab\cos C$$

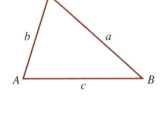

3. Hero's Formula

$$\text{Area} = \sqrt{s(s-a)(s-b)(s-c)},$$

where s is half the perimeter of the triangle.

VECTORS

For the nonzero vector $\mathbf{A} = \langle x, y \rangle$ with magnitude $|\mathbf{A}|$ and direction angle θ,

1. $|\mathbf{A}| = \sqrt{x^2 + y^2}$

2. $\tan\theta = \dfrac{y}{x}$

3. $x = |\mathbf{A}|\cos\theta$

4. $y = |\mathbf{A}|\sin\theta$

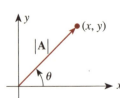